Economic Reform in Sub-Saharan Africa

A World Bank Symposium

Economic Reform in Sub-Saharan Africa

edited by
Ajay Chhibber
Stanley Fischer

The World Bank
Washington, D.C.

First printing December 1991

The complete backlist of publications from the World Bank is shown in the annual *Index of Publications,* which contains an alphabetical title list and indexes of subjects, authors, and countries and regions. The latest edition is available free of charge from the Distribution Unit, Office of the Publisher, at the address in the copyright notice or from Publications, World Bank, 66, avenue d'Iéna, 75116 Paris, France.

Library of Congress Cataloging-in-Publication Data

Economic reform in sub-Saharan Africa / edited by Ajay Chhibber and
Stanley Fischer.
p. cm.
Includes bibliographical references.
ISBN 0-8213-2063-7
1. Africa, Sub-Saharan—Economic policy. 2. Africa, Sub-Saharan—
Economic conditions—1960– . I. Chhibber, Ajay, 1954– .
II. Fischer, Stanley.
HC800.E277 1992
338.967—dc20 92-908
CIP

Preface

The decade of the 1980's witnessed the adoption of programs of economic reform in most countries of Sub-Saharan Africa. Given the enormity of the Sub-Saharan economic crisis, some reforms—particularly macroeconomic reforms designed to reduce large budget and balance of payments deficits—were both unavoidable and well understood; others—which had the purpose of improving efficiency and restoring growth—were crucial but less clearly understood.

While the analysis of economic reform draws on a common set of economic principles, it must be recognized that at the time that African economic reform programs were being implemented, relatively little academic research had been conducted on how to apply these general principles in specific African reform programs. As the reform process has progressed, it has become increasingly clear how vital it is for both African economists and outside researchers interested in African development to pursue further the economic analysis of African reform programs.

Research on the economic problems confronting Africa is important, because the results of that research will help policymakers in each country make better policy choices. But the importance of the research process, in which arguments have to be developed logically and discussed professionally, extends beyond its technocratic impact; the informed public discussion of policy that becomes possible as research results are disseminated improves the public's understanding of the choices confronting society, and thereby improves the policy process.

This volume presents papers on the economics of reform in Sub-Saharan Africa presented at a conference in Nairobi in June 1990. The main aim of the conference was to bring together a group of economists from African and non-African academic and research institutions, as well as international organizations and governments, to discuss economic reform in a professional forum. In addition to the authors, a range of distinguished African academics and policy makers participated actively in the conference.

The conference was organized by the World Bank with substantial assistance from the Africa Economics Research Consortium (AERC) in Nairobi. Financial Assistance was provided by DANIDA-Denmark's Development Agency and the Norwegian Foreign Ministry. The conference could simply not have been put together without the assistance of the AERC, particularly Jeffrey Fine and Benno Ndulu. The support of Peter Eigen, the World Bank's Resident Representative in Nairobi was also critical to the success of the entire effort. At the World Bank, the conference was very ably managed by Carol Sadler, with help from Lesley Davis and Sook Bertelsmeier. The manuscript was edited by Sandra Grain, with desk-topping by Meta de Coquermont and Laura Lee Wilson. The figures and graphs were prepared by Elizabeth Dvorscak and Catherine Cocak. Special thanks are due to Nga Lopez for managing the production of the manuscript. The preparatory work for the conference was managed by a steering committee composed of John Holsen, Johannes Linn and Stephen O'Brien, under the overall guidance of Stanley Fischer and Edward Jaycox.

We hope that this conference, besides stimulating further research on African economic policy issues, will also lead to a biennial conference on African economic issues. The first in this series is scheduled for June 1992 in Lome, Togo.

Contents

1. Introduction — 1
Ajay Chhibber and Stanley Fischer

I. Exchange Rate Policy

2. Exchange Rate Policies and the Social Consequences of Adjustment in Africa — 12
Patrick Guillaumont and Sylviane Guillaumont Jeanneney

3. Membership in the CFA Zone: Odyssean Journey or Trojan Horse? — 25
Shantayanan Devarajan and Jaime de Melo

4. Comments, *Ravi Kanbur and Stephen A. O'Connell* — 34

II. Parallel Markets

5. Exchange Reform, Parallel Markets and Inflation in Africa: The Case of Ghana — 39
Ajay Chhibber and Nemat Shafik

6. The Macroeconomics of the Unofficial Foreign Exchange Market in Tanzania — 50
Daniel Kaufmann and Stephen A. O'Connell

7. Niger and the Naira: Some Monetary Consequences of Cross-Border Trade with Nigeria — 66
Jean-Paul Azam

8. Comments, *Vikram Nehru and Paul Collier* — 76

III. Fiscal Deficit and Expenditure Policy

9. Issues in Public Expenditure Policy in Africa: Evidence from Tanzania's Experience — 80
Laurean Rutayisire

10. Poverty Conscious Restructuring of Public Expenditures 90
Marco Ferroni and Ravi Kanbur

11. Manufacturer's Responses to Infrastructure Deficiencies in Nigeria: Private Alternatives and Policy Options 106
Kyu Sik Lee and Alex Anas

12. Comments, *Nii Kwaku Sowa and Yao Kouadio* 122

IV. Financial Sector Policy

13. The Informal Financial Sector and Markets in Africa: An Empirical Study 125
Ernest Aryeetey and Mukwanason Hyuha

14. Mobilizing Domestic Resources for African Development and Diversification: Structural Impediments to Financial Intermediation 137
Machiko Nissanke

15. Monetary Cooperation in the CFA Zone 148
Patrick Honohan

16. Comments, *F.M. Mwega and Diery Seck* 161

V. Trade Policy

17. Commodity Exports and Real Income in Africa 170
Arvind Panagariya and Maurice Schiff

18. The Response of Firms to Relative Price Changes in Côte d'Ivoire: The Implications for Export Subsidies and Devaluations 182
John L. Newman, Victor Lavy, Raul Salomon and Philippe de Vreyer

19. Do Africans Pay More for Imports? Yes 198
Alexander Yeats

20. Comments, *Ademola Ariyo and C.D. Jebuni* 211

VI. Regional Integration

21. Integration Efforts in Sub-Saharan Africa: Failures, Results and Prospects--A Suggested Strategy for Achieving Efficient Integration 217
Ali Mansoor and Andras Inotai

22. The Record and Prospects of the Preferential Trade Area for Eastern and Southern African States 233
Nguyuru H.I. Lipumba and Louis Kasekende

23. Enhancing Trade Flows Within the ECOWAS Sub-Region: An Appraisal and Some Recommendations 245
Ademola Ariyo and Mufutau I. Raheem

24. Comments, *Atchi Atsain and Sylviane Guillaumont Jeanneney* 259

VII. **Human Capital and Entrepreneurship**

25. Human Resources, Technology and Industrial Development in Sub-Saharan Africa 263
Sanjaya Lall

26. Entrepreneurship and Growth in Sub-Saharan Africa: Evidence and Policy Implications 277
Ademola Oyejide

27. Comments, *S.M. Wangwe and William F. Steel* 285

VIII. **Growth-Oriented Adjustment**

28. Growth and Adjustment in Sub-Saharan Africa 287
Benno J. Ndulu

29. The Liberalization of Price Controls: Theory and An Application to Tanzania 303
Paul Collier and Jan Willem Gunning

30. The Prospects for an Outward Looking Industrialization Strategy Under Adjustment in Sub-Saharan Africa 316
S. Olofin

31. Comments, *Hafez Ghanem and C. Obidegwu* 327

Contributors

Alex Anas
Department of Economics and Engineering
University of Illinois, USA

Ademola Ariyo
University of Ibadan
Ibadan, Nigeria

Ernest Aryeetey
Department of Economics
University of Ghana
Accra, Ghana

Atchi Atsain
Department of Economics
University of Abidjan
Abidjan, Côte d'Ivoire

Jean Paul Azam
Centre d'Etudes et de Recherches sur le Developpement International
Clermont-Ferrand, France

Ajay Chhibber
Development Economics
The World Bank

Paul Collier
Institute for the Study of African Economics
Oxford University, England

Shantayanan Devarajan
Kennedy School of Government
Harvard University

Marco Ferroni
Social Dimensions of Adjustment Unit
The World Bank

Stanley Fischer
Development Economics
The World Bank

Hafez Ghanem
World Bank Office
Abidjan, Côte d'Ivoire

Patrick Guillaumont
Centre d'Etudes et de Recherches sur le Developpement International
Clermont-Ferrand, France

Jan Willem Gunning
Free University
Amsterdam, The Netherlands

Patrick Honohan
Country Economics Department
The World Bank

Mukwanason Hyuha
Department of Economics
University of Dar-es-Salaam
Dar-es-Salaam, Tanzania

Andras Inotai
Country Economics Department
The World Bank

Sylvaine Guillaumont Jeanneney
Centre d'Etudes et de Recherches sur le Developpement International
Clermont-Ferrand, France

Charles D. Jebuni
Department of Economics
University of Ghana, Accra, Ghana

Ravi Kanbur
World Bank Economic Review and Observer
The World Bank

Louis Kasekende
Research Department
Bank of Uganda
Kampala, Uganda

Daniel Kaufmann
South Africa Department
The World Bank

Yao Kouadio
CIRES
Universite Nationale de Cote d'Ivoire
Abidjan, Cote d'Ivoire

Sanjaya Lall
Institute of Economics and Statistics
University of Oxford
Oxford, England

Victor Lavy
Population and Human Resources Department
The World Bank

Kyu Sik Lee
Infrastructure and Urban Development Department
The World Bank

Nguyuru H.I. Lipumba
Department of Economics
University of Dar-es-Salaam
Dar-es-Salaam, Tanzania

Ali Mansoor
Africa Technical Department
The World Bank

Jaime de Melo
Country Economics Department
The World Bank

Francis Mwega
Department of Economics
University of Nairobi
Nairobi, Kenya

John L. Newman
Population and Human Resources Department
The World Bank

Benno J. Ndulu
African Economics Research Consortium
Nairobi, Kenya

Vikram Nekru
Western Africa Department
The World Bank

Machiko Nissanke
International Development Center
University of Oxford
Oxford, England

Chukwuma Obidegwu
Strategic Planning and Review Department
The World Bank

Stephen A. O'Connell
Department of Economics
University of Dar-es-Salaam
Dar-es-Salaam, Tanzania

Samuel Olofin
Department of Economics
University of Ibadan
Ibadan, Nigeria

Ademola Oyejide
Department of Economics
University of Ibadan
Ibadan, Nigeria

Arvind Panagariya
Country Economics Department
The World Bank

Mufutau I. Raheem
Department of Economics
University of Ibadan
Ibadan, Nigeria

Laurean Rutayisire
Department of Economics
University of Dar-es-Salaam
Dar-es-Salaam, Tanzania

Raul Salomon
Population and Human Resources Department
The World Bank

Maurice Schiff
Country Economics Department
The World Bank

Nemat Shafik
International Economics Department
The World Bank

Diery Seck
University of Windsor
Windsor, Ontario, Canada

Nil K. Sowa
Department of Economics
University of Ghana, Accra, Ghana

William F. Steel
Industry and Energy Department
The World Bank

Philippe de Vreyer
Population and Human Resources Department
The World Bank

S.M. Wangwe
Department of Economics
University of Dar-es-Salaam, Tanzania

Alexander Yeats
International Economics Department
The World Bank

1

Introduction

Ajay Chhibber and Stanley Fischer

The Roman Road is the greatest monument ever raised to human liberty by a noble and generous people. It is built broad, straight, and firm. It joins city with city and nation with nation. It is tens of thousands of miles long, and always thronged with grateful travellers. And while the Great Pyramid, a few hundred feet high and wide, awes sight-seers to silence—though it is only the rifled tomb of an ignoble corpse and a monument of oppression and misery ...

Tiberius Claudius, Emperor of the Romans
10 BC - 54 AD

I immediately empowered her to grant a large number of monopolies ... Six months later the removal of competition in the monopoly trades, which included necessaries as well as luxuries, had sent prices up to a most ridiculous height-the merchants were recovering from the consumers what they had paid in bribes to Messalina-and the City became more restless than at any time since the famine winter ... I fixed the prices, for the ensuing twelve months, of the commodities affected ... However, as soon as any complaints reached me that a certain class of goods was not reaching the City in sufficient quantities I added another firm to those already sharing the monopoly.

Tiberius Claudius, Emperor of The Romans
10 BC - 54 AD

The papers in the volume are divided into eight broad and interrelated topics that matter in African economic reform—Exchange Rate Policy, Parallel Markets, Fiscal Deficits and Expenditure Policy, Financial Sector Policy, Trade Policy, Regional Integration, Human Capital and Entrepreneurship, and Growth Oriented Adjustment.

In this introductory paper, we first briefly describe the papers within each section, and then at the end draw lessons from the contributions in the volume.

Exchange Rate Policy

The first section in this volume begins with a discussion of the fundamental question of exchange rate policy of fixed versus floating (or adjusting) exchange rates. The adoption of adjustment programs in a number of African economies, with devaluation as a center-piece, has led to an exceptionally vigorous debate on this subject.

The two papers in the first section deal with the issue of the CFA Franc Zone. In the first, Patrick and Sylviane Guillaumont question the common IMF/-World Bank advice to use nominal devaluation to bring about a realignment of the real exchange rate. The authors do not deny the need to be competitive. Rather they argue that the route of nominal devaluation to achieving a reduction in the overvaluation of the real exchange rate has excessively heavy social costs. These costs arise from the very high inflation which has typically accompanied devaluations in African economies. They argue that the better way to achieve real exchange rate depreciation is through expenditure reduction, which lowers the price of non-tradeables relative to tradeables. They argue that the reduction in expenditure can, through efficient targeting, be controlled in a way that protects the real income of the poor.

While there is no question that internal price adjustment can in principle achieve exactly the same real results as a devaluation, the CFA Franc Zone has not yet achieved the required domestic wage and price reductions. In the second paper on the CFA Zone, Devarajan and de Melo show that the failure to adjust in the 1980's meant low or negative growth rates of both exports and GDP in the CFA zone countries. These countries have benefitted from price stability, but at the cost of severe balance of payments problems (see table 1–1).

The experience of countries which have undertaken exchange rate devaluations has also been mixed. An important factor contributing towards the lack of effective devaluations has been insufficiently strong accompanying fiscal and monetary control. As a result, in many countries an exchange rate-price spiral has been set-off, as is evident in table 1–1. This evidence has allowed policy-makers in CFA Zone countries to argue that in practice devaluations have not generally worked in other African countries, and that fixed exchange rates at least bring the benefits of greater stability not only in prices but also in output.

An important issue which was not addressed in this section—both in the papers and the discussion that followed—is how precisely the fixed exchange rate contributes to low inflation. At least two factors are present. First, on the supply side, the fixed exchange rate keeps the prices of imported inputs constant. Second, the need to maintain convertibility constrains monetary policy, typically keeping monetary growth low. A further factor that is extremely important in considering the CFA experience is to note that most other developing countries that have attempted to operate with fixed rates have not succeeded in maintaining capital account convertibility. Rather, they have typically resorted to capital controls, which in turn led to the development of parallel markets. These are analyzed in the next section.

Parallel Markets

The focus of the three papers on parallel markets is on how to achieve unification of the official and parallel exchange markets. The prolonged overvaluation of the official exchange rate in many countries has led to the widespread emergence of a parallel exchange system. Further, research shows that the parallel exchange rate was the crucial determinant of tradeable goods prices in many African countries.

The first paper by Daniel Kaufmann and Stephen A. O'Connell focuses on three issues related to the parallel exchange rate in Tanzania. First, it examines the transmission of the parallel exchange rate into domestic prices, focusing on the effect of widespread shortages in rural areas in the early 1980s, and the structural change that occurred with the introduction of the own-funds import window in 1984. Second, it specifies and estimates a reduced-form equation for the parallel market premium that allows for both trade and portfolio effects and that takes into account the structural shift that occurred in 1984. The paper shows that as in Ghana (in the next paper by Chhibber and Shafik), both the trade and portfolio effects were important in determining the parallel market premium in Tanzania. However domestic prices did not fully reflect the parallel exchange rate - rationing was effective in some markets, leading to a "money overhang" in 1979-83. The final section of the paper addresses the question of exchange rate unification. It points out that the own-funds scheme, while welfare-improving, is a step away from unification of the official and the parallel rates.

The second paper on parallel markets by Ajay Chhibber and Nemat Shafik focuses on Ghana, which has carried out one of the most thorough structural adjustment programs in Africa. Ghana's increasingly high inflation rate has been attributed to major devaluations of the official exchange rate. Using a macroeconomic model estimated with Ghanaian data, Chhibber and Shafik dispute this conclusion. Their results show that over the period 1965-88 there is no direct relationship between the official exchange rate and inflation; prices had already adjusted to the exchange rate prevailing in parallel markets. Their results also show that official devaluation had a positive effect on Ghana's budget. Revenue improvements came from three channels: the higher grant aid disbursed at a more depreciated exchange rate, a reduction in the subsidies that had accrued to importers through an overvalued exchange rate, and an increase in export taxes as cocoa farmers increasingly marketed their output through official channels. Official devaluations therefore did not produce higher budget deficits, demand pressures did not spill onto the parallel market, and the exchange premium narrowed considerably.

Chhibber and Shafik's results suggest that the key to the success of Ghana's adjustment program was the adequate level of foreign financing, combined with a coherent set of fiscal policies. They argue that although Ghana's inflation had structural causes in the past, the acceleration in recent years is primarily a monetary phenomenon. It also reflects weakness in

Table 1-1. ***Inflation in Sub-Saharan Africa: 1975-89***

Recorded	*1980-89*	*1975-89*	*Highest*
Industrial Countries	5.0	6.4	
Asia	7.9	7.3	
Sub-Saharan Africa	17.2	16.7	
More than 20 percent			
Ghana	43.7	52.2	122.9
Sierra Leone	63.7	43.9	178.7
Somalia*	41.1	32.9	91.2
Sudan*	33.1	25.1	64.7
Tanzania*	30.5	24.1	35.3
Uganda	104.6	—	238.1
Zaire	58.8	62.1	101.0
Zambia*	30.8	24.5	55.6
Between 10-20 percent			
Botswana	10.5	11.0	16.4
Burundi	7.6	10.1	36.6
Gambia	17.9	15.1	56.7
Kenya	10.4	11.3	20.4
Lesotho	13.6	13.9	18.0
Madagascar*	17.2	14.3	31.8
Malawi	16.6	—	33.9
Mauritius	7.6	10.8	42.0
Nigeria	20.5	18.9	40.9
Swaziland*	13.9	13.7	20.8
Zimbabwe	13.5	12.3	23.1
Under 10 percent			
Burkina Faso	4.1	6.4	30.0
Cameroon*	9.4	9.8	17.2
Central African Republic**	3.9	—	14.6
Congo*	8.0	8.5	17.4
Côte d'Ivoire*	5.3	9.5	27.4
Djibouti	4.2	—	18.1
Ethiopia	4.3	8.3	28.5
Gabon*	7.7	9.8	28.4
Liberia*	4.3	6.1	19.5
Niger	2.4	6.7	23.5
Rwanda	4.4	6.8	31.1
Senegal	6.5	6.6	31.7
Togo	4.0	6.3	22.5

Note: *For these countries averages are for 1980-88 and 1975-88. **For these countries averages are for 1981-89.
Source: World Bank data

the financial system that must be tackled to sustain reform.

In the third paper, Jean-Paul Azam shows that in analyzing parallel markets it matters whether the country's currency is convertible or not. Azam shows how Niger can acquire convertible foreign exchange by its parallel trade with Nigeria, despite the latter having an inconvertible currency. This is because the currency of Niger (the CFA franc) is convertible. Although the price of the Naira can be regarded as exogenous for Niger, as it does not adjust to clear the foreign exchange market, it is not constant, and its fluctuations have significant macroeconomic effects. Azam shows that it has pass-through effects on the CPI in Niger, and that Niger can in effect be regarded as a small price taker. The parallel exchange rate for the Naira is shown to have a significant effect on real cash balances in Niger; this is the monetary counterpart of the balance of payments surplus that Niger has had in its cross-border trade with Nigeria.

All three papers on parallel exchange markets show a significant impact of portfolio shifts on the determination of parallel exchange rates"thereby erasing a general perception that parallel markets are driven only by controls on the prices and movement of goods. These results show that the parallel capital market shifts resources across borders in response to changing rates of return. They lead to the important conclusion that unification of exchange markets requires appropriate fiscal and monetary policies, which should take into account the de-facto imperfect control on movements of capital. A further important message, emerging from the papers on Ghana and Tanzania, is that exchange rate unification—official towards parallel—has significant fiscal benefits which can be anti-inflationary. The assumption here is that the Government is a net supplier of foreign exchange to the rest of the economy.

Fiscal Policy and Public Expenditure

One of the most important and controversial questions in designing a reform program is the extent of the needed reduction in the fiscal deficit. Often the disagreement is not over the desirable size of the fiscal deficit, but instead on the speed of adjustment and the composition of budget cuts-over how quickly and in what areas (capital vs. current, social sector vs. military, primary education vs. tertiary) expenditures should be cut. Of course, the current state of expenditure incidence methodology does not allow very precise answers to these questions. However, it is more often than not possible to suggest the needed direction of change. Policy distortions are usually so egregious that the direction of reform and its rough orders of magnitude are easily identifiable. The three papers included in this section demonstrate this on three different fiscal and expenditure questions.

The first, by Laurean Rutayisire, focusses on the macroeconomic and debt implications of the fiscal deficit in Tanzania, and analyzes the sustainable level of the fiscal deficit. Rutayisire first examines the factors underlying the increase in government spending. He then discusses these factors in the context of various theories of the motivation behind government spending, and concludes that unless the government's income increases, Tanzania's present level of government expenditure is unsustainable.

The second paper, by Marco Ferroni and Ravi Kanbur, starts with a brief look at patterns of public expenditure in Africa during the adjustment decade, especially in the social sectors. The authors ask the pertinent question—how should governments prioritize the use of scarce resources for poverty reduction? Their answer is that the decision should be made on the basis of both the direct effects of basic needs expenditures, and the indirect effects of other income-enhancing policies and expenditures. It will also depend on the weights attached to various components of the standard of living, and information on the fraction of social expenditures reaching the poor. Using household survey data for Cote d'lvoire, Kanbur and Ferroni show how the government can set levels, based on poverty considerations, for the consumption of rice, housing, and education.

The third paper in this section, by Kyu Sik Lee and Alex Anas, examines the implications of public sector inefficiencies for the response of the private sector in Nigeria. It asks questions such as: How do firms respond to the constraints caused by deficient infrastructure? What alternatives do firms have, and what do they cost? Is the private provision of services a viable alternative to their public provision? The paper also examines supply side questions, such as the causes of the failure to deliver adequate services. To what extent are such failures caused by a lack of capacity, or by poor operations and maintenance? How do inappropriate pricing and user charges contribute to the problem? What are the options, in terms of investment, technology, institutions, regulations, and financing, for remedying these failures?

Based on empirical observations, Lee and Anas suggest policy options for improving the provision of infrastructure services in Nigeria. These include regulatory changes to encourage the fuller use of existing capacity (for example, allowing the sale of excess private electrical power); participation of the private sector in the supply of infrastructure-related services; and pricing policies that take account of the presence of congestion, system failures, and variations (by firm size and location) in the private provision of services.

Financial Sector

Structural adjustment programs adopted in many Sub-Saharan African countries have placed high priority on the liberalization of financial markets. Given the shallowness of the financial markets in many of these economies, policies relying heavily on the market mechanism alone can, according to Machiko Nissanke, generate unexpected side effects. For example, interest rates are likely to be excessively high if determined by market forces in very thin financial markets. Painstaking efforts at institution building are needed to achieve market depth, so that market forces can operate beneficially. The simple switch of policy from financial repression to financial liberalization can by itself do little to revitalize such economies.

Nissanke shows that because of the absence of effective mechanisms for domestic resource mobilization in these countries, not only has the system been segmented into the informal and formal sectors but no interactions have taken place among formal institutions. The economies are faced with little or no capacity to generate long-term credit or loan provisions for self-sustained diversification efforts. In spite of the high liquidity in the overall banking system, the potential productive sectors in the economy are starved of adequate institutional credit.

In the second paper in this section, Ernest Aryeetey and Mukwanason Hyuha argue that analysis of the impact of monetary and other economic policies in many African economies usually underestimates the role of the informal financial sector, and sometimes leads to the pretence among policy makers that this activity is inconsequential. Even where it is acknowledged that a substantial informal financial sector exists, policy makers usually do not take into account its linkages with the formal financial sector. They argue that while the sector thrives and affects growth processes in several ways, it is usually unaccounted for in both economic analysis and economic statistics.

Current interest in the role the informal financial sector can play in the economic transformation of many African countries derives from the dearth of financial resources. Recent studies carried out in Ghana and Tanzania and reported by Ernest Aryeetey and Mukwanason Hyuha (as well as other studies) have helped throw more light on the internal workings of the informal markets, as well as on the relationships between them and the formal financial markets. The authors argue on the basis of the evidence that the informal financial sector in the region is relatively large and often plays a crucial role in the savings-investment process. Thus, for optimal financial resource allocation, informal financial markets need to be given due consideration in policy design.

The shallowness of financial markets in many countries in SSA is obviously a function of macroeconomic and financial sector policies. However, it is also to a considerable extent due to the small size of these economies. One way to increase the size of the financial market is to join a monetary union. In considering the merits of joining a monetary union, small countries naturally value the credible commitment to exchange rate and price stability that membership represents—and that would be hard to sustain by unilaterally pegging their own currency. Membership offers other potential advantages. Within a monetary union, capital might flow more freely to where it is most needed. If the distribution of union benefits is reasonable, this could benefit all members—even those who because of low capital productivity became net lenders within the union.

Moreover, the operation of monetary policy and the prudential supervision of the banking system might be more effective if the resources of several small countries were pooled in a strong and independent Central Bank. In the third paper in this section, Patrick Honohan asks whether these advantages have been realized in the CFA zone? Honohan argues that the experience has not been encouraging. Despite the fixed exchange rate and an elaborate set of rules for avoiding over-expansion of credit, the CFA zone has almost foundered in widespread bank insolvency. The zone's institutional set-up seems equitable, but in practice the burden of paying for losses appears to have fallen disproportionately on the poorer countries—whereas most of the non-performing credits have been made in some of the zone's more prosperous countries.

Trade Policy

It has been argued frequently that while an individual primary exporter may benefit from increasing its exports, an expansion of primary product exports by several developing countries is likely to lead to a decline in their terms of trade, export revenues and real incomes. The paper by Arvind Panagariya and Maurice Schiff presents a systematic analysis of this issue with respect to a commodity—cocoa—in which many African countries have a large share of world exports. It examines how real incomes and export revenues compare under the existing and alternative (Nash) taxes. It also examines the impact of export expansion resulting from increased efficiency on real income, export revenues and tax revenues under alternative tax regimes, and compares the effects of export expansion by African Countries with that by non-African countries.

The paper shows that there are no easy answers to the choices facing primary commodity producers. An economist can devise a set of welfare maximizing Nash taxes, but the poor record of adherence to international arrangements on production quotas, export quotas or prices means these taxes are unlikely to be instituted or maintained. The entry of Asian cocoa producers, notably Malaysia, makes coordination more complicated. The entry of Asian producers also shows that coordination is unlikely to work in the long run, since there are always potential new entrants. Thus withholding production, or using export taxes to control exports, is likely to be self-defeating, since new entrants will take over market share. The best long-run option may be to pass international price changes on to the farmers and assist them in diversification.

The response of firms in Africa to changes in relative prices and their ability to expand exports as this activity becomes more profitable has been central in the debate over the use of export subsidies. This issue was very relevant in the 1980s for many of the member countries of the West African Monetary Union, who kept a fixed exchange rate with the French Franc and decided not to devalue their currencies. Searching for alternative means to alter the relative price of exportables, and with the recommendation of the World Bank, some of these countries opted to mimic a devaluation by increasing both import tariffs and export subsidies. Opponents of the policy claim that is doomed to fail because supply response in Africa is very sluggish, export elasticities are low, and because the heavy flow of resources through the government budget suggests that the policy is time inconsistent, undermining its credibility and discouraging firms from responding to it.

The second paper on trade policy, by Victor Lavy, John Newman and Phillipe De Vreyer, addresses these issues and provides relevant empirical evidence from Côte d'lvoire, the first country to adopt the tariff cum subsidy program in 1986. Their empirical work is based on data from the six years that preceded the implementation of the program (1980-1985). The paper focuses on modeling the response of firms to exogenous changes in export prices. The new elements introduced are the endogeneity of the domestic price, and the explicit modeling and integration of domestic demand so that it is estimated jointly with the producer behavioral equations. The estimated model is then used to simulate the effects of the subsidy program, compare it to a simulated devaluation, and finally, in light of the two simulations, to evaluate the 1986-1989 export subsidy program of the Côte d'lvoire.

Lavy, Newman and De Vreyer show that the combination of an export subsidy with an import tariff, which mimics a devaluation, would counter some of the short-run negative effects of an overvalued exchange rate. However, they point out that the scheme introduced in Cote d'lvoire in 1986 did not result in a uniform increase in export prices and has also involved heavy fiscal costs since then. Their results, including the heavy fiscal costs, suggest that this complicated administrative attempt to mimic a devaluation will not work in the long-run. It should be pointed out that East Asian countries have used export subsidies and devalued exchange rates as complementary rather than substitute measures: they typically kept undervalued exchange rates even when they were using export subsidies.

Numerous studies have examined the influence of market structure on performance in domestic markets of industrial countries. These investigations show that prices and profits are higher, and resources less efficiently allocated, in markets where aggressive competition is absent. Using techniques similar to those in the earlier studies, Alexander J. Yeats examines the relative prices paid for iron and steel products by selected African and other developing and developed countries. The findings parallel those of the industrial country market studies. Typically, prices are higher in international markets that are more concentrated (less competitive), or that rely on fewer trade contacts.

Yeats' analysis shows very high excess price margins on iron and steel imports from France to 20 former French colonies in Africa. Over the longer term (1962-1987), the African countries paid an average premium of 20 to 30 percent over other importers. The losses from these excess prices came to about $2 billion by 1987. This study also finds that the overpricing extends to other (non-French) African countries. Former colonies of Belgium, Portugal and the United Kingdom still pay premiums of 20 to 30 percent on imports from the former colonial power.

The three papers all point to substantial potential benefits from freer trade and greater competition in Sub-Saharan Africa. On the export side the attempt to manage exports of primary commodities through commodity boards has led to a very substantial loss of market shares. On the import side too, the evidence indicates that managed trading relationships, typically between parastatals and former colonial countries, have cost African countries dearly. A more open trading structure would bring substantial benefits.

Regional Integration

The sixth section of the book deals with the important and much discussed issue of regional integration. In the first paper Ali Mansoor and Andras Inotai argue that attempts at economic integration among developing countries (EIDC) have generally failed. They attribute these failures to (1) the costs of not allowing free trade for products produced in more than one member country; (2) the emphasis on regional import substitution behind high barriers; and (3) the attempts at industrial planning to ensure an equitable distribution of benefits among participating countries.

The costs of these inefficiencies could not be sustained after commodity prices collapsed and debt burdens grew in the early 1980s. Since then most EIDC arrangements have either stagnated or suffered setbacks. However, most developing countries maintain their membership in at least one EIDC initiative, and interest in the subject has been renewed by greater integration among industrialized countries, such as the US-Canada Free Trade Agreement and the EC 1992 single market agreement.

Recent regional integration initiatives among developing countries reflect the new consensus on the importance of an *outward orientation*. They also reflect a more pragmatic approach where flexibility of trading arrangements is more important than consensus among a large (often disparate) membership. This reorientation of EIDC may be particularly important for SSA where economic fragmentation limits many of the opportunities for horizonal and vertical integration that are available in Latin American and Asian countries. Further, EIDC offers the means for the entrepreneurship talent that has developed in the more outward oriented economics to be made available in other member countries.

In Sub-Saharan Africa, the aim of the Lagos Plan of Action of 1980 for an eventual common market, which was restated at the recent summit of the Organization of African Unity, sets out an ambitious objective which will yield economic benefits provided the common market has an outward orientation with substantially lower trade barriers than those in place today at the national level. A common market could yield efficiency gains as a result of greater internal (regional) competition as well as through greater factor mobility. Inotai and Mansoor argue that to ensure that MFN trade liberalization proceeds in parallel with intra-regional liberalization of labor and capital flows, a common external tariff should be avoided until tariffs are generally low. The first step should be to eliminate non-tariff barriers on the basis of mutual concessions among participating countries, with a phased unilateral extension on an MFN basis to avoid maintaining preferences that result in inefficient trade and investment flows.

The second paper, by Nguyuru Lipumba and Louis Kasekende, analyses the rationale and achievements of the Preferential Trade Agreement of Eastern and Southern Africa (PTA) since it entered its operational phase in July 1984. The major focus of PTA activities during this period was the preferential reduction of tariffs, buyer-seller meetings to encourage intra-PTA trade, with the aim of increasing intra-PTA trade and economizing on the use of foreign exchange. PTA has also emphasized the development of transport and communication links among member states.

The record does not show an increase in the ratio of intra-PTA trade to total trade. Lipumba and Kasekende argue that tariffs are not the effective barriers to trade. Rather, non-tariff barriers, particularly foreign exchange allocation schemes and import licensing, are the main hindrance. The authors argue further that as member states liberalize their overall trade and payment regimes, it will be important to reduce tariff preferences in the promotion of intra-PTA trade. A planning approach to a common market for the PTA region is not compatible with a market oriented economy, in the sense that the multinational allocation of industries in unlikely to be feasible in a diverse region. Closer economic integration and expanded trade will require factor mobility and currency convertibility among member states, as well as improvement in transport and communication links and increased availability of trade and market information.

On the other side of the continent, the formation and eventual commencement of operations of ECOWAS in 1975 was seen as one of the surest means of accelerating development in West Africa. The paper by Ademola Ariyo and Mufutau I. Raheem looks at several aspects of ECOWAS. First, the consolidation period for the operation of ECOWAS was expected to last 15 years. Hence, this year marks an important milestone for evaluating its performance. Second, most of the studies so far reported were undertaken before the introduction of stabilization programs by some members of the ECOWAS sub-region. Hence, the possible incremental impact of this development as reflected in the structure and volume of intra-ECOWAS trade is yet to be documented. Third, little attention had been paid to the trade matrix amongst members of the community, a detailed analysis of which should enhance our understanding of current and prospective trade flows within the community. Finally, attention had earlier been focussed on distribution, and less on production, of tradeable goods within the region.

Ariyo and Raheem's findings suggest that there has been a small increase in the volume of trade within the region, perhaps arising from the benefits of stabilization programs which encourage intra-regional economic cooperation. The paper reveals that the structure of trade flows has broken the language and geo-political barriers especially along the Anglophone-Francophone dimension. The paper's findings also suggest that trade within the region is limited, mainly because so much of the required imports are not produced within the region. This finding formed the basis for Adiyo and Raheem's suggestion that the success of ECOWAS requires agreed production arrangements to increase the availability of some goods hitherto not available within the sub-region. This view is quite the opposite of that reached by Kasekende and Lipumba.

Despite political efforts towards the goal of regional integration, the evidence in the three papers on the topic in this volume is that official policies have, if anything, thwarted regional integration. The good news is that despite these contra-integration policies, unofficial trade has expanded, bringing about much greater de-facto integration than is suggested by official statistics. This unofficial trade has contributed to reducing the economic costs of poor trade and pricing policies.

In future work, the issue of regional integration must also be viewed from the angle of its implications for the degree of openness of the overall trade regime"the central question of trade diversion versus trade creation. Will greater regional integration (a la the EEC) come at the expense of less openness towards the rest of the world? Is that becoming necessary in a world of managed trade-blocs? These are issues on which further research by economists interested in Africa is clearly needed.

Human Capital, Entrepreneurship and Industrialization

The seventh section of this volume deals with the interrelated questions of human capital, industrialization, and entrepreneurship. In the first paper, Sanjaya Lall examines the problems of industrialization in Africa, and asks why industrial development remains at fairly low levels of technological sophistication, integration with the domestic economy, and competitiveness in world markets. Lall traces the poor performance of African industry to several causes. External shocks and poor macroeconomic management in many countries have reduced the supply of foreign exchange as well as domestic demand for industry, causing severe dislocations. Highly interventionist, inward-looking policies have created and sustained considerable inefficiencies, while holding back the emergence of industrial exporters.

Exogenous shocks and interventionist strategies do not, however, fully explain the problems of African industrialization. The analysis of industrial development requires a broader approach that takes into account not just macro-economic conditions and the incentive framework, but also the ability to respond to incentives. This ability is determined by a mixture of entrepreneurship, industrial skills, technological effort and institutional development.

Lall suggests that a very large part of the explanation of the African industrial experience lies in the small base of industrial capabilities and institutional support with which most countries started, and the inadequate growth over time in these relative to the industrial facilities set up"in other words, that industrial development was initially too ambitious. Data on education levels and technological effort are used to illustrate the gap between African countries and the newly-industrializing countries of East Asia and elsewhere. The policy conclusions drawn differ significantly from those commonly advanced, which aim to provide "quick-fix" solutions to African industrialization. Changes in price and trade policies are not enough to foster industrialization—although they are necessary to avoid mis-allocation of resources and investment in wrong industries. Lall argues instead for longer-term educational and institutional changes to enhance the quality of entrepreneurship and the ability to respond to price signals.

The second paper in this section, by T. Ademola Oyejide, shifts focus from the question of human capital development to the somewhat less tangible but equally relevant issue of entrepreneurship. Against the background of poor economic performance during the 1970's and 1980's, many structural adjustment programmes are placing increasing emphasis on private initiative and market forces. Advocates of the new policies assume that this new orientation will improve the flexibility with which African economies can respond to rapid and unexpected changes in the international economic environment. The new orientation also implies an expanded role for the private sector in the development process.

However, there is increasing concern that the private sector's supply response appears to be inadequate. This concern suggests the need for studies aimed at acquiring a deeper understanding of the determinants of African entrepreneurship and its capability for response, as well as analysis of the factors which constrain the development of entrepreneurship and the private sector in Africa.

Oyejide's paper makes a contribution to this broad research agenda. Its primary purpose is to survey the available literature with the aim of identifying major elements of the dynamics of private enterprise development in Africa. The main focus of the paper is an analysis of the characteristics of African entrepreneurship, followed by an examination of its pattern of upgrading, graduation and growth, leading to a discussion of the constraints impeding enterprise development and growth. Oyejide shows that very few firms in the informal and micro-enterprise sector have grown to become modern firms of any size. He also shows that special targeted assistance programs for these enterprises have not worked. The results of Oyejide's paper suggest that the emphasis in assistance programs and interventions should not be on firm size, but on the constraints a particular industry faces. The technology in each industry thrown-up an optimal firm size which could be a number of small firms or one large firm which competes internationally. Size of firm is not the relevant issue.

Growth and Adjustment

The last section of the book contains three papers, which examine the impact on growth of the adjustment strategy that has been followed in Sub-Saharan Africa. Benno J. Ndulu focuses on key trade-offs between growth and macroeconomic adjustment in the medium term consistent with longer term sustainable growth. The special characteristics of the growth process in Sub-Saharan Africa are highlighted. Key among these is the fact that reasonable levels of investment during the last two decades were not accompanied by commensurate growth rates. Ndulu distinguishes capacity growth from actual growth, and examines the factors influencing the wedge between them. Ndulu argues that this wedge can be explained to a large extent by import compression. Two factors are responsible for the import compression: (a) the poor export performance; and (b) increased debt burden. The paper discusses how foreign aid can be used to relieve import compression.

Ndulu presents a simple growth model incorporating the above special features. The model highlights the trade-offs between capacity expansion and capacity utilization in an import-compressed economy. It explains the coexistence of high investment and low actual growth and discusses requirements for breaking out of this scenario and implied policy choices. By modelling the financing of the fiscal gap, it also highlights the trade-off between growth and price stability in situations of heavy reliance on the inflation tax to finance growth. The role of foreign resources in filling financing gaps is also discussed and fungibility in the use of such resources to support measures for improved efficiency in the future is emphasized.

The second paper on adjustment, by David Bevan, Paul Collier, and Jan Willem Gunning emphasizes the need to re-examine the recommendations of adjustment programmes in the agricultural sector. Generalized price controls on consumer goods, when enforced, give rise to shortages. Such policies have been fairly common in Africa. This paper develops a theory which argues that shortages of consumer goods have serious consequences for the rural economy. First, the unavailability of consumer goods changes the marketed supply response to price changes (a point that is well-known in the context of the former socialist economies): higher crop prices can actually reduce crop sales. Second, depending upon how the shortages are distributed, they may induce a build-up in money balances which give rise to an overhang problem in the transition back to market clearing. Third, the rural economy is liable to implode into subsistence. Farmers reduce incomes because they are unable to make purchases, but in aggregate these income reductions lower the supply of consumer goods in subsequent periods.

The first part of the Bevan, Collier and Gunning paper sets out the theory, using a variant of a diagram due to Malinvaud. The second part applies the theory to the economic reforms which have taken place in Tanzania since 1984. This paper provides a convincing—and surprising—rationale for the important role for financing of consumer goods (as opposed to capital and intermediate goods) imports as part of the financing package that underpins adjustment programs in Sub-Saharan Africa.

In the last paper of the volume, Samuel Olofin examines the prospects for an outward looking industrialization strategy in Sub-Saharan Africa. This is the Export Oriented Industrialization strategy (EOI), or what is sometimes referred to as the G-4 model, following the remarkable success of its foremost practitioners, the so-called East Asian 'gang of four': Hong Kong, South Korea, Singapore and Taiwan. Most countries in Sub-Saharan Africa have begun to adopt a semblance of the EOI strategy as part of World Bank/IMF assisted Structural Adjustment Programs (SAPs). Very few studies have been undertaken to analyze the implications of SAP for industrialization in Sub-Saharan Africa.

Given the successes of the Asian NICs, and the failures of the inward looking industrialization strategy (ISI) which many African countries have pursued, it is natural to ask whether African countries should pursue an Asian style EOI industrialization strategy. Olofin addresses the extent to which the EOI or G-4 model may be expected to work in individual African countries. The paper points out that serious supply constraints exists in most African economies, which will not allow them to easily duplicate the East Asian experience. It is therefore not enough to simply change relative prices and expect immediate and sizeable export-led growth. Rather, active measures will have to be taken to develop supply response in Africa.

Concluding Remarks

What lessons can be drawn from this particular collection of papers and the comments that follow? No doubt each reader will draw her or his own conclusions, but we believe the balance of the evidence presented here warrants the following conclusions:

First, a realistic exchange rate policy is crucial for growth. The exchange rate can sometimes be used temporarily as a nominal anchor to fight inflation. But fundamentally inflation control requires control over fiscal deficits and money growth. Using the exchange rate as a nominal anchor frequently ends in overvaluation of the currency, which guarantees low growth. Experience shows that most East Asian countries have kept their exchange rates extremely competitive, in order to boost exports and growth. African countries would need to match these policies to remain competitive.

Second, returns to investment are typically very low in Africa. This is due partly to poor policies but, more so than elsewhere, it appears to be due to inefficient management of public capital. One solution is to privatize industries and reduce the share of public investment in total investment. The other is to encourage more joint ventures with foreign companies, in both the public and private sectors, accompanied by strong efforts at capacity building to improve overall managerial capabilities.

Third, trade diversification—into processed and industrial goods—is obviously necessary to get away from the common problem faced by many African countries of declining terms of trade for their major exports. However, the way to diversify is not by taxing primary commodity producers as in the past. This dries up investment in primary commodity exports without encouraging it in other sectors and results in African countries losing market shares to producers in Asia and Latin America. Rather the emphasis should be on encouraging exports through maintenance of appropriate exchange rates, and deregulation of the domestic economy, as well as focussed efforts at export promotion.

Fourth, the role of the informal financial system in resource mobilization and allocation needs to be recognized. Past financial sector policies have tended to concentrate on formal financial systems, often at the expense of well-functioning informal financial markets.

Fifth, formal efforts at regional integration in Africa have so far failed. However, regional trade (mostly illegal) is much larger than is recorded in official statistics. This suggests that current trade and investment policies hinder rather than foster regional integration. African governments would help regional integration by encouraging more free trade (even unilaterally) and by focussing on improving the infrastructure for moving commodities across the various sub-regions such as Western Africa, among the SAADC countries and in East Africa.

Sixth, incentive policies are one important element of the package needed to restore growth, but they are not enough. They need to be accompanied by substantial investments in infrastructure - roads, ports, telecommunication and power in order to reduce the cost of doing business. These complementarities between incentive policies and the provision of infrastructure are well known and understood. An interesting twist to the supply response story in the African context is the need for provision of consumer goods (requiring external financing) and thereby reducing overall shortages to elicit the necessary supply response from the farm sector.

Seventh and last is the issue of human capital. The level of human capital development has been very low—largely because African countries were subject to colonialism which did not encourage education. The consequences of this are widespread—they affect entrepreneurship, public sector management, and the adoption of new technology, as well as what we now call governance. In short it affects the ability to grow and participate effectively in the global economy. Its rectification will require substantial investments over several decades. This is not an area where quick fixes are possible.

It is our hope as editors that this collection of papers and the conference at which they were presented will help encourage further research on African economic issues. Each of the papers raises many further questions. In addition, a number of issues important for African economic development have not been touched on in this volume—among them, labor markets and employment, public enterprise reform, and the broader issue of governance.

These critical issues, together with further work on the topics that are covered in the present volume, deserve further discussion, which can be taken up in the next conference and the volume that will accompany it. This volume also does not include a detailed examination of African agricultural issues on which there is considerable literature as well as the question of African debt which is covered in a recent World Bank publication (Husain and Underwood, 1991).

The challenge for African economic policy, and for researchers on African economies, is the restoration of growth in the 1990's. More and better research can play a crucial role in assisting policymakers and the public to understand more clearly what works and what does not.

References

Husain, I and J. Underwood (1991): African External Finance in the 1990's, World Bank, Washington D.C.

Mellor, J.W., C.L. Delgado, and M.J. Blackie (1987): Accelerating Food Production in Sub-Saharan Africa, John Hopkins Press, Baltimore.

2

Exchange Rate Policies and the Social Consequences of Adjustment in Africa

Patrick Guillaumont and Sylviane Guillaumont Jeanneney

During the 1980s the majority of African countries implemented structural adjustment programs designed to ensure lasting balance of payments equilibrium and the resumption of growth. The adjustment policies were based on a wide variety of economic policy instruments in the context of diverse exchange rate systems. The social consequences of structural adjustment policies are significantly influenced by the choices made in setting the exchange rate and regulating exchange transactions. We try to examine here the social consequences linked to the exchange rate policy implemented during the adjustment.

The paper is divided into four main parts. The aims and social implications of structural adjustment are summarized. The diversity of African exchange rate systems is reduced to three main types and an attempt is made to show their implications for real exchange rate decline and inflation. The social consequences of the different modalities of real exchange rate decline are drawn and the social consequences of exchange rate policy are analyzed through its effects on productivity in structural adjustment.

The Aims and Social Implications of Structural Adjustment

In the case of a major disequilibrium in the balance of payments current account, it is generally necessary to bring about an initial reduction in overall demand. This stabilization reduces economic activity and inevitably has a high social cost. To reduce the current deficit without sacrificing economic growth—to achieve structural adjustment—the structure of production must change in favor of tradable goods. This can be achieved in two ways and these can occur simultaneously.

The first is through a reduction in the real exchange rate, which is the ratio of the price of nontradable goods to that of tradable. The real exchange rate changes in line with the ratio between the price indices in the country and abroad multiplied by the index of the nominal effective exchange rate. In this paper, the International Monetary Fund principles of calculations of the effective exchange rate are used to express the cost of local currency in foreign exchange.

Another way in which structural adjustment can be achieved is to increase productivity. When productivity improves in the production of tradable goods, their profitability is increased directly. When it improves in the nontradable goods sector, the profitability of the production of tradable goods is increased if the prices of nontradable goods decline. This is likely if the price elasticity of the demand for nontradable goods is low.

Structural adjustment by definition has a socially beneficial aspect, since it is intended to restore the economic growth compromised by the external deficit. However the two ways to achieve structural adjustment outlined above have different social implications. Reducing the real exchange rate involves strong control over demand and induces a transitory reduction in national income and the standard of living. Those whose incomes are linked to price changes in tradables, such as small farmers and entrepreneurs, should benefit at the expense of those whose incomes depend on the price of nontradables, such as local businesses and transport companies, and those with fixed incomes (wage earners) irrespective of their sector of activity. Increasing productivity enables the incomes of small farmers and companies in the tradable goods sector to increase without necessarily reducing the incomes of wage earners and the earnings of companies in the home goods sector. Increasing productivity requires more time and sometimes leads to a reduction in employment, but it involves a smaller reduction in the remuneration of labor compared to reducing the real exchange rate.

Three Exchange Rate Policies Aimed at Adjustment

Various exchange rate policies were implemented in Africa during the 1980s.[1] These different situations can, for the simplicity of exposition, be classified under three major exchange rate policies: devaluation in the presence of a significant parallel foreign exchange market, the standard case of devaluation, and adjustment without devaluation. This classification of major exchange rate policies is a simplification compared with the diversity of actual exchange rate systems and policies. Table 2–1 shows the different currencies or units to which African currencies are pegged and those which have flexibility. In the following, each of the above three cases is illustrated by the experiences of African countries.

Devaluation in the Presence of a Significant Parallel Foreign Exchange Market

The majority of African countries have practiced strict exchange control, even applied to current transactions. This has resulted in a parallel foreign exchange market which is more or less tolerated, if not encouraged, by the national authorities. The parallel market covers an appreciable proportion of the foreign exchange transactions. Access to official foreign exchange is limited and depreciation of the official exchange rate appears as a policy aimed at unifying the two exchange markets. Devaluation in the presence of a significant parallel foreign exchange market is the experience in Ghana, Guinea, Nigeria, Tanzania, and Uganda. In some cases the parallel market is authorized; in others it is simply tolerated.

The situation in which the parallel exchange market is more or less tolerated can be illustrated by Nigeria. All along both sides of its far-flung borders, there are concentrations of population belonging to the same ethnic groups which have traditionally maintained intense commercial relations. Due to limited availability of foreign exchange at the official rate, a parallel market developed between the national currency (the naira) and the currencies of the neighboring countries (West and Central African CFA francs, which are convertible). The smooth functioning of this market is apparent from the fact that, over the entire length of Nigeria's long frontiers, the rate for the naira in CFA francs tends to be constant.

Nigeria implemented its adjustment policy in two phases. The first phase was begun in 1983, without assistance from international institutions. It consisted of a drastic contraction in public expenditure and tightening of the quantitative restrictions on imports, without any change in the official exchange rate. Inflation remained high and the overvaluation of the naira became much more marked. The exchange rate on the parallel market, on the other hand, showed a pronounced depreciation which more than offset the difference due to inflation with the neighboring countries.

During the second phase, beginning in 1986, Nigeria was assisted by international institutions. The new policy was characterized by a measure of liberalization of foreign trade and prices, recourse to external capital, and flexibility in the official exchange rate. Since September 1986 a system of foreign exchange auctions has been operating, which at first coexisted with an administered exchange rate. Since 1987 the auction-set rate has been applied to all official transactions. The parallel market for the naira has not disappeared, but the difference between the official rate, set by the auctions, and the parallel market rate is now much smaller than it was in the previous phase.

The depreciation of the official rate was anticipated on the parallel market. For an appreciable part of the economy the actual reduction in the exchange rate was more progressive than might have appeared from the movement of the official rate. As a consequence the rate of inflation was not so high (see table 2–2). The ratio of the variation in the real effective exchange rate to that of the nominal effective exchange rate, a ratio that Edwards calls "effectiveness of devaluation"—what can be called here the "apparent" effectiveness—was 93 percent. Thus the core of the reforms in Nigeria consisted not only of greater flexibility in the official exchange rate, but also in a trend toward unification of the market exchange rate.

Standard Case of Devaluation

This case is the one most frequently referred to in adjustment theory. In this case the country conducts a policy of flexibility in its official nominal exchange

Table 2–1. *Exchange rate arrangements of African countries, 1988*

	Exchange rate pegged to					
	a single currency			*a basket of currencies*		
Country	*US$*	*FF*	*Rand*	*SDR*	*Other*	*Flexible*
Benin		x				
Botswana					x	
Burkina Faso		x				
Burundi				x		
Cameroon		x				
Cape Verde					x	
Central African Republic		x				
Chad		x				
Comoros		x				
Congo		x				
Côte d'Ivoire		x				
Djibouti	x					
Ethiopia	x					
Equatorial Guinea		x				
Gabon		x				
Gambia						x
Ghana						x
Guinea						x
Guinea-Bissau						x
Kenya				x		
Lesotho			x			
Liberia	x					
Madagascar						x
Malawi					x	
Mali		x				
Mauritius					x	
Mauritania					x	
Mozambique	x					
Niger		x				
Nigeria						x
Rwanda				x		
Sao Tomé and Principe					x	
Senegal		x				
Seychelles				x		
Sierra Leone	x					
Somalia	x					
Sudan	x					
Swaziland			x			
Tanzania					x	
Togo		x				
Uganda	x					
Zaire						x
Zambia	x					
Zimbabwe					x	

Source: IMF, Annual Report, 1988

Table 2-2. *Variation in the nominal and real effective exchange rates and inflation in Sub-Saharan African countries that have implemented structural adjustment policies*

Country	*Period*	*Real effective exchange rate at start of adjustment period (index, 1970=100)*	*Variation in nominal effective exchange rate during adjustment period*	*Variation in real effective exchange rate during adjustment period*	*Ration of variation in real effective exchange rate to variation in nominal effective exchange rate*	*Inflation rate during adjustment period (percentage)*	
						Total	*Average Annual*
Burundi	1983-89	142	-52	-44	0.85	35	6.2
Central African Republic	1980-87	110	-10	14	—	77	10.3
Congo	1981-87	96	-00.8	-003	0.38	2	2.2
Côte d'Ivoire	1980-87	125	-12	-17	1.42	48	4.4
Gabon	1981-87	122	00.1	-2.7	—	-0.9	-0.9
Gambia	1983-87	99	-69	-27	0.39	180	29.3
Ghana	1982-87	1,011	-99	-92	0.93	498	43
Guinea	1985-88	—	-96	n.a.	n.a.	—	—
Guinea-Bissau	1983-87	74	-93	-88	0.71	581	61
Kenya	1980-88	91	-58	-33	0.59	121	10.4
Liberia	1980-87	83	9.2	0.5	0.05	28	3.5
	1984-87	113	-25	-26	1.04	8	2.6
Madagascar	1981-87	120	-74	-58	0.78	151	17
Malawi	1979-87	86	-40	-14	0.35	214	15.4
	1985-87	101	-37	-26	0.70	43	19.4
Mali	1982-87	86	-5.5	-15	2.73	23	4.3
Mauritius	1978-88	94	-40	-26	0.65	180	10.8
Niger	1981-87	122	-5.5	-29	5.27	5.5	0.9
Nigeria	1984-88	318	-88	-82	0.93	41	14.1
	1985-88	284	-87	-80	0.93	38	17.1
Senegal	1975-87	127	-10	-14	1.4	148	8.6
	1979-87	110	-8	0	0	94	7.8
Sierra Leone	1982-87	104	-97	-34	0.35	2,395	90.3
	1984-87	158	-95	-56	0.59	791	107.3
Somalia	1981-87	247	-94	-67	0.71	671	40.5
Tanzania	1982-87	172	-88	-63	0.72	297	31.7
	1985-87	205	-81	-69	0.85	115	31.2
Togo	1977-87	114	-16	-25	1.56	74	5.7
Zaire	1978-87	248	-99.8	-77	0.77	4,594	53.4
Zambia	1982-87	104	-89	-67	0.75	328	33.7
Zimbabwe	1982-87	89	-49	-25	0.52	106	15.6

Notes: The countries included in the table are those which had one or several agreements with the IMF between 1980 and 1987. The nominal effective exchange rates are a geometric mean of the official bilateral exchange rate indexes (annual average of the cost of the national currency expressed in foreign exchange) weighted by the official structure of imports (excluding oil) over the period from 1980 to 1986. Only convertible currencies are included. The real effective exchange rates are equal to the product of the nominal effective exchange rates and the ratios of the annual consumer price indexes in the country considered and elsewhere.(geometric mean of the consumer price indexes caluculated with the same weighting as the nominal effective exchange rates). In case of Mali the GDP deflator was used because of the lack of a consumer price indez for the whole period of adjustment. The adjustment periods have been defined by the commencement of the period of nominal depreciation of the currency for the countries that have devalued, and by the start of the demand restriction policy for the Franc Area countries and Liberia (which have not devalued). As the adjustment policy does not generally start at the beginning of the calendar year, and since the exchange rates are annual indexes, the variation percentages calculated provide orders of magnitude for the movements and not a percise meaurement. The ratio of variation in real effective exchange rate to variation in nominal effective exchange rate corresponds to what Edwards (1988) calls the effectiveness of the devaluation.

rate, and this rate is actually applied in the bulk of all foreign transactions. The parallel market is relatively negligible, either because the official rate makes it possible for the supply of and demand for foreign exchange to be kept in balance, or because clandestine operations are hindered by geographic factors or vigorous repression. This is not the most common experience in Africa; however, it can be illustrated by the exchange rate policies followed in Kenya, Malawi, Mauritius, Zimbabwe, and Madagascar.

Since 1982 Madagascar has conducted a policy of dramatic depreciation of the nominal exchange rate. Under the liberalized import regime, when importers applied for foreign exchange they paid a set fee regardless of the degree to which their requests were met. As this fee system was not applied to priority imports or to those financed by international assistance, it led to a sort of dual official exchange rate. It was replaced by the liberalized import system, which is applied to most imports and does not include a fee. All requests for foreign exchange are met at the official rate, which is adapted according to the Central Bank's exchange reserves position.

The real exchange rate did not begin to decrease until 1984. Edwards' effectiveness of devaluation ratio was 78 percent. The decline in the real exchange rate and a large inflow of international aid enabled the progressive removal of foreign exchange restrictions. All applications for goods imports are now fully met. The strong depreciation of the nominal effective exchange rate, which resulted in the decline in the real exchange rate, simultaneously induced a rapid increase in domestic prices.

In other countries, such as Kenya, Mauritius, and Zimbabwe, the official exchange rate was the actual exchange rate for the main part of transactions. These countries devalued an initial real exchange rate which had not increased, compared to that of 1970 (see table 2–2). They did so because of a need for structural adjustment due to a terms of trade decline and to an increase in debt service. They devalued in several steps and succeeded in lowering the real exchange rate. However devaluation caused a rather high inflation in which the price level more than doubled in the three countries during the adjustment period. This inflation resulted both from the rise of tradable goods induced by devaluation and from the monetary and fiscal policy which accompanied the devaluation.

Adjustment Without Devaluation

During the 1980s other African countries, whose currencies are convertible, have implemented adjustment policies without devaluation. The official fixed rate is used for all transactions. Countries in this category are Côte d'Ivoire, Mali, Niger, Senegal, Togo, the Central African Republic, Congo, and Gabon. These countries are members of the CFA Franc Zone.[2]

The fourteen African countries in the Franc Zone have no national currencies. Except for Comoros, they are members of two monetary unions. This means that devaluation would require a joint decision by all the members of each union. Each currency of the two distinct monetary unions is called CFA franc and is pegged to the French franc. The member countries have made monetary agreements with France enabling them to maintain the convertibility of their currency.

The counterpart of this guarantee of convertibility given to the CFA franc consists of the commitment of member countries to conform to certain rules of monetary policy, including a limitation on advances by central banks to national treasuries. Franc Zone countries are subject to certain institutional constraints leading to monetary stability.

Despite the absence of devaluation of CFA francs, the nominal effective exchange rate of Franc Zone countries decreased between 1980 and 1985 as a result of changes in the value of the French franc in terms of other currencies (see table 2–2). However, during the 1980s the CFA franc appreciated strongly on the parallel market in relation to the naira and the cedi. When changes in the value of the naira and the cedi are taken into account, the result is a smaller nominal depreciation of the CFA franc (Guillaumont and Guillaumont 1989).

During the adjustment period monetary and budgetary policies were restrictive. The increase in domestic prices was kept below that in the rest of the world, and prices decreased in some years (see table 2–2).[3] In Côte d'Ivoire, Niger, Senegal, and Togo the real exchange rate was reduced by more than the nominal exchange rate. The ratio of effectiveness of the depreciation of exchange was greater than one.

The Social Implications of Real Exchange Rate Decline With and Without Devaluation

Devaluation is an apparently simple way of reducing the real exchange rate, which by definition derives from the nominal exchange rate and from the ratio of domestic to foreign prices. However, devaluations must be accompanied by monetary and budgetary policy designed to maintain overall demand and to avoid an increase in nominal remunerations. Without devaluation, a decline in the real exchange rate can be brought about only by keeping inflation lower at home than abroad through control of demand. As world inflation is generally low, adjustment without devaluation is only possible in countries which did not experience a too high inflation beforehand and in which the real exchange rate appreciated only moderately.

This section looks at the social consequences of real exchange rate decline when it is obtained with and without devaluation. Transitional income, relative price, and inflation effects of real exchange rate decline are assessed.

The Transitional Income Effect

Demand contraction is intended to induce real currency depreciation. Without devaluation, demand must contract in order to slow inflation down to a point where it is below world inflation. With devaluation, and if there are not inflationary anticipations, the contraction of demand is designed only to curb the inflationary effects of the improvement in the trade balance resulting from the devaluation.

A given decline in the real exchange rate would be less costly with devaluation than without. However, devaluation reduces the real exchange rate only if the increase in tradable goods prices is not offset by the increase in domestic prices and wages. Otherwise, a restrictive policy would be necessary to reduce the real exchange rate, as in the case where there is no devaluation (Guillaumont Jeanneney 1988). When a country is obliged to devalue, it is generally still in the grip of inflation. If this inflation is not curbed by restrictive monetary and budgetary policy, it becomes permanent and contributes to a new real appreciation of the currency. It then becomes more likely that economic agents will anticipate the effect of these devaluations on prices, making a restrictive policy all the more necessary.

A simple test of the restrictiveness of the policy in African countries adjusting with and without devaluation is given in table 2–3. GDP growth rates during 1980 to 1987 are compared for Franc Zone countries and other African countries. Within each category, data is given for adjusting countries. In order to control for other factors of growth (country size, climatic vulnerability, terms of trade, initial debt, etc.) a cross-sectional model of the exogenous (non-policy) determinants of growth has been estimated. Each country residual has been interpreted as an indicator of its growth performance relative to the norm.

There are two main results. First, although the observed growth was higher in CFA Franc Zone countries than in other African countries, it was similar in adjusting countries in both groups. Second, the growth performance in the CFA Franc Zone is higher than in other African countries, for the whole sample and for the adjusting countries. These comparisons do not seem to support the idea that the adjustment without devaluation in Franc Zone countries has implied a more restrictive policy than in the adjustment with devaluation in other African countries.

The Relative Price Effects

Structural adjustment policy is based on a decrease in the real exchange rate, which alters the relative price for home goods in relation to tradable goods. This often causes a shift in the ratio of urban to rural income, to the advantage of the latter. Since rural income is typically lower, this effect could reduce urban bias. However this depends on the manner in which prices and incomes are determined in rural and urban areas. Here we look at how exchange rate policy affects rural and urban incomes.

Exchange rate policy and rural incomes. The adjustment policy can have a preponderant influence on farmer income through agricultural prices. However, many of the rural poor do not produce tradable goods because they do not have enough land. They cannot benefit from higher prices and are gradually becoming the laborers of the landowning farmers. Moreover, there is a division of labor between the genders, with men responsible for tradable goods, and women concentrating on home goods (Azam, Chambas, Guillaumont, and Guillaumont 1989, p. 19-20).

The nominal level of agricultural prices, p, is expressed in the following (Guillaumont 1986, Bonjean and Marodon 1988):

$$p = \frac{p'_w}{r} \cdot (1 - a) - m \qquad (2\text{–}1)$$

in which p'_w is the international price expressed in foreign exchange, r is the exchange rate, a is the public levy ratio, and m is the unit marketing and processing costs. The real price of the good sold, p_r, or its unit purchasing power, depends on the nominal price and on the domestic prices of the goods consumed by the farmer. The price index for domestic goods consumed by the farmer is $\hat{p}_c$. Then,

$$p_r = \frac{p}{\hat{p}_c} = \frac{p'_w}{\hat{p}_c \cdot r}(1 - a) - \frac{m}{\hat{p}_c} \qquad (2\text{–}2)$$

Let us call $\hat{p}'_{wc}$ the foreign consumer price index and r the product of the initial exchange rate r_0 and the exchange rate index ρ. Then $r = r_0 \cdot \rho$ and

$$(2\text{–}3) \qquad p_r = \frac{p'_w \cdot \hat{p}'_{wc}}{\hat{p}'_{wc}} \cdot \hat{p}_c \cdot r\} (1 - a) - \frac{m}{\hat{p}_c} .$$

The real exchange rate, *RER*, is by definition equal to $\hat{p}_c/\hat{p}_{wc}$, therefore

$$(2\text{–}4) \qquad p_r = \frac{p'_w}{\hat{p}'_{wc}} \cdot \frac{1}{r_0 \cdot RER} (1 - a) - \frac{m}{\hat{p}_c} .$$

Thus, the real producer price depends by definition on: the real international price of the product, $p'_w/\hat{p}'_{wc}$ which could be called the international terms of trade for the product; the real exchange rate, *RER*; the public levy ratio, *a*; and the real marketing and processing costs, $m/\hat{p}_c$.

A first requirement for a devaluation to increase the real producer price is that the rise in domestic prices does not offset that in the international price expressed in local currency, provided that there is a decrease in the real exchange rate. A second requirement is that the increase in export prices be passed on to the producer and not seized by the state, a stabilization board, or a marketing agent (the increase is not offset by an increase in *a* or m/p_c). When there is no devaluation, the increase in the producer price involves a decline in either the public levy ratio or the unit marketing and processing costs. The impact of an increase in the nominal producer price is stronger, since there is no inflation induced by devaluation.

In rural areas the social consequences of adjustment policies are not limited to the direct effects of prices on farmer income. Major indirect consequences result from the manner in which the farmers react to the new prices. Higher prices and income normally tend to encourage farmers to increase production; they often also provide the means to do so, as they facilitate access to inputs and equipment, particularly through greater borrowing power. This favorable impact cannot occur if the infrastructure is

Table 2–3. ***Observed GDP growth rate and growth performance, 1980-87***

	Type of avergage	Growth performance		
		Observed growth	*Case 1*	*Case 2*
Africa South of Sahara	a	1.8	0.0	-0.1
(25)	b	1.5	-0.4	-0.4
	c	1.7	0.4	0.1
Franc Zone African countries	a	2.4	1.0	0.7
(9)	b	1.9	1.0	0.5
	c	2.7	1.3	1.0
Of which with adjustment program	a	1.3	0.3	-0.1
(6)	b	1.3	0.6	0.2
	c	1.8	0.3	0.2
Other African countries	a	1.5	-0.6	-0.5
(16)	b	1.4	-0.8	-0.8
	c	1.6	-0.9	0.0
Of which with adjustment program	a	1.2	-0.7	-0.7
(12)	b	1.4	-0.8	-0.8
	c	1.5	-0.9	-0.5

Note:
a) simple average
b) weighted average
c) median
Growth performance is defined as the difference between observed growth and growth estimated as a function of "environmental" factors on a sample of 58 developing countries. For growth performance case (1) these factors are the following: population size in 1980; climatic vulnerability as measured by the instability in agricultural value added during 1970 to 1987 (weighted by the ratio of agricultural value added to GDP); variation in the terms of trade (during 1980 to 1987, relative to 1975 to 1979), weighted by the exports to GDP ratio; and debt structure (private debt to GDP in 1980). Growth performance case (2) includes all of the above factors and a dummy variable for major political disturbances. The six Franc Zone countries with adjustment programs are Central African Republic, Côte d'Ivoire, Mali, Niger, Senegal, and Togo. The twelve other African countries with adjustment programs are Ghana, Kenya, Madagascar, Malawi, Mauritania, Uganda, Sierra Leone, Sudan, Tanzania, Zaire, Zambia, and Zimbabwe.
Source: Gross data from World Bank tables 1988-89.

run down or the marketing and credit channels are not organized, especially if the farmers cannot sell their output and/or cannot buy anything in exchange for their increased monetary income. Such a shortage of goods due to lack of foreign exchange does not occur in the adjusting countries benefitting from the convertibility of their currency, as in the CFA Franc Zone.

In terms of country experiences, a special case occurs in the case of devaluation in the presence of a significant parallel foreign exchange market, when the farmers can sell their harvests in neighboring countries outside the official marketing channels. They are then paid in the currency of those countries, which they exchange on the parallel market at a rate which, especially if that currency is convertible, is much higher than the official rate. Under such circumstances a devaluation of the official exchange rate does not have a major direct effect on their income, but can nevertheless encourage the farmers —if the depreciation of the currency is reflected in the official agricultural prices—to sell their products once again through the official channels, which gives them an advantage since clandestine marketing inevitably involves risks and costs. A return to the official channels did in fact occur in Guinea and Ghana (for Ghana, see Heller et al 1988).

The pattern of change in the real producer prices in the countries of the Franc Zone, where nominal prices are usually administered, is an interesting test of the possible impact of an adjustment policy without devaluation. In most cases during the 1980s, the official producer prices for export crops were first increased and then maintained, whereas the trend in international prices was becoming unfavorable. As a disinflation policy was simultaneously being implemented, real producer prices changed relatively little; real per capita income in the country frequently declined (see table 2-4). This was the trend until 1985 because of the fall in the French franc exchange rate against the dollar. It continued throughout the period because of a decline in the government levy ratio at

Table 2-4. *Movement of real official producer prices in the African Franc Zone countries that have implemented adjustment prices*

	1980 1981	*1981 1982*	*1982 1983*	*1983 1984*	*1984 1985*	*1985 1986*	*1986 1987*	*1987 1988*	*1988 1989*
Central African Republic									
Coffee	100	96	114	78	79	73	77	—	—
Cotton	100	88	92	92	95	97	—	82	—
Côte d'Ivoire									
Cocoa	100	92	86	94	97	101	95	90	83
Coffee	100	92	86	94	98	101	85	90	83
Cotton	100	87	86	101	111	110	103	98	90
Palm Nuts	100	138	128	121	116	145	136	129	119
Niger									
Groundnuts (shelled	100	92	97	100	101	124	135	149	—
Cotton	100	105	157	145	134	145	150	137	—
Cowpeas	100	118	138	149	153	167	129	138	—
Millet	100	114	130	120	138	112	—	—	—
Paddy	100	104	113	116	107	115	91	100	—
Senegal									
Groundnuts (unshelled)	100	132	112	100	90	102	96	100	80
Cotton	100	107	94	64	75	95	90	93	95
Millet	100	118	101	99	98	100	95	98	—
Paddy	100	117	100	104	104	118	111	116	118
Togo									
Cocoa	100	85	80	86	97	109	114	144	—
Coffee	100	90	89	100	113	133	140	140	—
Cotton	100	90	82	87	107	127	122	122	—

Note: The nominal prices have been deflated by the African consumer price index (annual index relating to the start of the season).

Sources: BCEAO, Bulletin d'information et statistiques, and Ministry of Coopération and Development - Caisse Centrale de Coopération Economique, Bulletin de conjoncture, various numbers.

the same time that the countries were attempting to cut their budget deficits. Nevertheless, in some countries, after international prices for some commodities (coffee, cocoa and groundnuts) fell again, maintaining producer prices has had to be dramatically reconsidered, owing to the government's financial straits (for example in Côte d'Ivoire).

Exchange Rate Policy and Urban Incomes

Urban remuneration is especially affected by the adjustment policy when it decreases the real exchange rate. Urban production of nontradable goods and services (financial services, transportation, trade, public works and construction, etc.) is relatively high. Moreover, in general a larger share of remunerations in urban than in the rural areas consists of fixed income (business and government salaries, rents, etc.). In the event of devaluation, their real value is reduced by the increase in prices. That effect is attenuated when there is a parallel foreign exchange market where the currency is depreciated before the official devaluation.

Without devaluation, a decrease in income results from the restrictive policy aimed at blocking or even diminishing the nominal level of remunerations. This is achieved through monetary policy for business salaries, budget policy for government salaries, and regulation of minimum salaries. The price of tradable goods increases in tandem with world inflation and sometimes even more rapidly when those prices—after having been controlled—are deregulated. Thus, a marked drop in real wages was seen in the African countries that implemented adjustment policies (see table 2–5, for a more detailed analysis of the situation in Côte d'Ivoire, Levy and Newman 1989).

Another important determinant of the urban standard of living during adjustment is the level of employment. It is often reduced by the stabilization policy that is generally a prerequisite of structural adjustment. The decrease in demand primarily affect urban activities because of the high proportion of home goods production. Structural adjustment involves a shift of jobs from nontradable goods production towards production of tradable goods.

Analysis of the urban social consequences of exchange rate policy must take into account the dualism within the urban manufacturing and tertiary sector. In addition to the modern sector there is an informal sector, which remains largely outside official regulation and taxation. The modern sector has in general been more affected by adjustment policies than has the informal sector. The latter is better able to adapt because of its greater flexibility in determining the real remuneration of workers and the number of people it employs. The informal sector participates in an informal structural adjustment. One of the most visible aspects of the urban consequences of adjustment policies is precisely growth in the informal sector. This phenomenon can be heightened when the real exchange rate is decreased without devaluation, as in the Franc Zone countries (for the case of Niger, see CERDI 1988a). In those countries, a drop in nominal wages (or unit remunerations) is more easily achieved in the informal sector than in the modern sector, under the pressure of foreign competition or the effect of a decline in domestic demand (as far as home goods are concerned).

Table 2–5. ***Declining real wages in adjusting countries***

Country	*Adjustment period*	*Change in minimum wage during adjustment (percentage)*
Côte d'Ivoire	1980-1987	-25.6
Niger	1981-1987	-5.5
Senegal	1979-1987	-11.6
Togo	1977-1987	-27.0

Source: BCEAO, Notes d'information et statistiques (1985-88).

The Inflation Effects

Adjustment policies in Africa have led to inflation rates that differ widely from one country to another, depending on how devaluation is used. There is variation ranging from nearly stable prices to rates of inflation in the vicinity of 100 percent a year (table 2–2).

One effect of inflation is to increase the burden imposed by progressive income taxes, and to reduce that imposed by lump taxation. This is true insofar as tax brackets and lumps are not indexed. Wage earners in developing countries often pay a graduated tax, even at comparatively low income levels, while some rich merchants are lump taxed. In this case, the inflation that accompanies an adjustment policy involving devaluation might well add to the distortions between wage earners in the modern sector, whose tax burden would increase, and individuals active in the informal sector, who would find their tax obligations reduced.

Another general consequence of inflation is its wealth redistribution effect. The increase in prices drastically reduces the value of monetary and quasi-monetary balances, while real estate assets maintain their value. The tax charge payable on cash balances works to the advantage of those who benefit from money creation (the state and private borrowers), which is why it is described as an inflation tax. It is a type of taxation that appears to weigh proportionally more on poor groups, since a higher percentage of their wealth, if not most of it, is in monetary form. It is also possible that the inflation tax hits small farmers harder than urban workers because of the usually seasonal nature of inflows of farm income and the resulting need to hold cash balances. However, this situation may be mitigated when farmers have the opportunity to set up savings in the form of goods (grain, livestock, etc.). Urban groups would be less affected by inflation than are their rural counterparts because they are more likely to invest their money profitably and have more monetary debt. Certain people only in urban centers are in a position to profit from devaluations by investing in foreign exchange. Inflation may cause urban real estate prices to rise, which would result in reduced urban property taxes during the lag in reassessing the tax base.

The Social Consequences of the Exchange Rate Policy Through its Effects on Productivity Growth

After stabilization and after getting the prices right, productivity increase can be seen as the next stage of adjustment. When there is a large nominal depreciation of the currency, adjustment is sought predominantly through a decrease in the real exchange rate. In countries where price distortions require only a moderate decline in the real exchange rate, adjustment programs rely more on productivity increases.

In the following, the relation between exchange rate policy and productivity is explored. We discuss the negative effects of reduced real wages, fixed parity as an incentive to improve efficiency, exchange rate policy and private investment, and exchange rate policy and public social expenditures.

Real Exchange Rate and Labor Productivity

Reduction of the real exchange rate has been viewed in the foregoing as an objective. But in many African countries the large reduction in real wages has impacted negatively on the motivation of wage earners. These effects have not been offset by improved social protection. In low-income African countries, wage earners' ability to work has been adversely affected by poor nutrition and health.[4] There are similar negative productivity effects in the nontradable goods production sector.

As shown a quarter of century ago by Balassa (1964) and more recently by Kravis et al (1975), countries with higher relative income per capita have a higher real exchange rate. In the long run a relatively rapid rate of growth involves an increase of the real exchange rate. In this perspective, the strong decline of the real exchange rate in some African countries, such as Ghana and Madagascar, although considered as a success of adjustment policies, is a symptom of the failure of long run development.

It appears that the relation between the real exchange rate and average labor productivity runs in two directions. In the short and medium term, the real exchange rate decline may induce a decline in labor productivity. In the long term, a decline in the relative level of labor productivity implies a decline in the real exchange rate.

Fixed Parity as an Incentive to Improve Efficiency

From a social standpoint, decreasing the real exchange rate is reflected by a decrease in the remuneration for labor and a change in the employment structure. In the case of devaluation, companies can easily protect their profits by adjusting their prices; they therefore have less incentive to enhance their management efficiency. Conversely, the constraint of stable prices is an incentive to reduce costs.

The Franc Zone countries lack the means of raising prices in the agroindustrial export subsectors. They have made significant productivity gains which have contributed to adjustment. Although international prices for cotton (in Chad, Central African Republic, Mali, and Togo) have been low, producer prices have not been reduced much. This has been possible due to improvements in the productivity of extension, processing, transportation and marketing.

Increasing productivity does not entail a decrease in remunerations, but generally an increase in the intensity of labor and a modification in the employment structure. Efforts to increase productivity may prompt some businesses to make layoffs that are not immediately offset by new jobs stemming from the increase in production. The change in the employment structure may lead—at least temporarily—to an increase in urban unemployment. Therefore productivity improvement has at least a temporary social cost, which is less drastic than that resulting from a decrease in the real exchange rate.

The social cost of increased productivity is directly felt by the businesses whose productivity increases, which explains their resistance to it. An open question is whether in Africa reducing employment is socially more or less costly and politically more or less feasible than lowering wages.

Moreover, when adjustment is based more on an increase in productivity than a decrease in the real exchange rate, the informal sector has a more important role. This is because of its greater flexibility in labor remuneration and the number of people employed. This would explain why during the adjustment period Franc Zone countries seem to have experienced an especially rapid expansion of their informal sector.

Exchange Rate Policy and Private Investment

The social consequences of structural adjustment depend on the country's ability to sustain growth. Lasting growth requires a sufficient level of investment especially in the modern sector. Therefore as far as the exchange rate policy influences investment, it has a long term social impact.

The influence of exchange rate policy on investment depends on the relationship between inflation and savings. It depends also on the effects of the real exchange rate level and instability in the inducement to invest. Investment may be increased by getting the prices right and by an appropriate real exchange rate. However investment is likely to decrease due to exchange rate instability, such as that resulting from a floating exchange rate or discontinuous devaluations. Moreover, strict exchange control makes for low investment and inefficiency because of its administrative complications and the risks of breakdowns in supplies associated with it.

Exchange Rate Policy and Public Social Expenditures

The exchange rate policy exerts a range of effects on the real level of social spending per capita. The level of public social expenditures has a direct social impact. It also effects workers' productivity, which is sensitive to the workers' health and education status.[5]

The factors which in the course of adjustment act on the per capita level of social expenditure, expressed in constant prices, can be examined by means of the following equation:

$$\frac{D_s}{P \cdot p_s} = \frac{D_s}{D} \cdot \frac{D}{Y} \cdot \frac{Y}{P \cdot p} \cdot \frac{p}{p_s} \qquad (2\text{–}5)$$

where D_s is government social expenditure, D is total government expenditure, P is population, Y is GDP, p is an index of the general price level, and p_s is an index of the price of social goods. This equation shows that the level of social expenditure depends on four factors: GDP per capita, the ratio of government expenditure to GDP, the structure of government expenditure, and relative prices (see CERDI 1988b). All four factors may be affected by exchange rate policies.

The ratio of government expenditure to GDP is determined by the rate of taxation and by the sustainable budget deficit. A drop in the exchange rate, whether sought through devaluation or not, implies strict control of the budget deficit. This is why the level of government expenditure is limited by the level of government revenue.

Exchange rate policy can exert a significant impact on the tax ratio. In the African countries, taxation on tradable goods, and particularly on external trade, appears to be heavier than taxation on home goods and incomes. A drop in the real exchange rate has a favorable effect on the tax/GDP ratio.

This pattern can be seen in the African countries that have devalued. When the drop in the real exchange rate has been sharp, public revenue generated by foreign trade has increased more quickly than the general level of prices. Increases in public revenue drawn from trade have tended to be greater in countries with a stabilization board system and where the official producer prices have not reflected the increases in the local currency export prices. The expected favorable effect of devaluation on agricultural income and output has then been moderated and sometimes even reversed. Hence, there is a trade-off between generating more revenue in order to finance government expenditure and increasing primary agricultural income. Even if devaluation allows for greater fiscal pressure, this effect can be compensated by an output decline caused by progressive taxation and variable income imposition.

A change in the rate of government expenditure D/Y has repercussions on the level of social expenditure. This depends on the structure of government expenditure D_s/D and the relative price of social goods p_s/p, which are both influenced by the exchange rate policy.

Because of debt service obligations expressed in foreign currencies, exchange rate policy acts directly on the structure of government expenditure. If there is a devaluation, the relative share of public expenditure earmarked for debt service increases automatically. The proportion available for other spending, particularly social expenditures, shrinks. In the course of the adjustment the authorities often have to cut back more on capital than on recurrent expenditures and more on operating expenses than on labor costs.

In countries facing foreign exchange shortages, reducing expenditures is also a matter of import content. Under a given nominal government (non-debt) expenditure structure, exchange rate policy may have an effect on the real level of social spending. A decline in the real exchange rate, by altering relative prices, changes the distribution of government expenditure. The volume of social expenditure is affected according to its content of tradable goods. Education expenditure has a relatively low tradable goods content, unlike health expenditure, which includes a quite high proportion of tradable goods and services (drugs, equipment). For this reason, and also because it involves a smaller labor component than education expenditure, real health expenditure is likely to be particularly affected during adjustment. This seems to have happened in various African countries including those which have not devalued, as far as their real exchange rate has declined.

Concluding Remarks

This study has aimed to compare the social consequences of different exchange rate policies based on the adjustment experience of African countries.

The African exchange rate policies have been classified in three categories. Some countries have dramatically devalued or adopted floating rates after experiencing a very strong overvaluation of the official rate associated with a large parallel market for foreign exchange. Their policy was targeted to unify the foreign exchange market, since depreciation of the currency had already been achieved for a large part of the economy. In those countries, devaluation was unavoidable.

Some other countries have experienced moderate or no real appreciation of their currency. They chose to devalue their currency in order to facilitate their balance of payments adjustment. Other countries, because they belong to monetary unions, adjusted without devaluation. Countries in these two categories have achieved a decline in their real exchange rate, which has been on average a little more pronounced in the devaluing countries. But in these countries inflation was much higher than in the non-devaluing adjusting countries.

Real exchange rate decline, which is the main way to achieve structural adjustment, has important social consequences, partly depending on the nominal exchange rate policy. The transitional income effect of the control of demand needed to decrease the real exchange rate is probably weaker when devaluation is used and there is some money illusion. However, this advantage of devaluation and the extent of money illusion are often overstated. The relative price effect of a given decline of the real exchange rate does not seem to differ greatly whether the country adjusts with or without devaluation. Real exchange rate decline has the favorable social impact of reducing urban bias. A main social difference between the two ways of decreasing the real exchange rate results from the effects of inflation.

Productivity improvement is the other major way to achieve structural adjustment. Adjustment without devaluation leads to greater emphasis on this alternative. Decreasing the real exchange rate involves decreasing labor remuneration, which may have a negative impact on labor productivity. Maintaining a fixed parity is an inducement to efficiency. Hence, once major distortions are eliminated, there may be a trade-off between the real exchange rate decline and productivity increase.

The effects of the exchange rate policy on investment in material and human capital are complex. However, inflation and instability of relative prices are often associated with successive devaluations and floating exchange rates. These may have a negative impact on investment. Monetary illusion may help to correct distorted relative prices; however, it neither increases productivity nor favors the poor.

In the short run productivity increase involves greater labor intensity and flexibility. In the long run it is socially beneficial. In the African countries where the real exchange rate has reached a sustainable level, productivity increase constitutes the priority for structural adjustment.

Notes

1. A presentation on the different exchange policies of sixteen Sub-Saharan African countries that have devalued will be found in Ministère de la Coopération et du Développement – Caisse Centrale de Coopération Economique (1988, vol. II and III), and in Jacquemont and Assidon (1988).

2. Other countries, however, have conducted their economic policy without devaluing and without a lasting recourse to the international lending institutions for adjustment aid (Eithiopia, Rwanda, and some other Franc Area countries).

3. The adjustment experiences of Congo and Gabon (see table 2-2) are too recent to be illustative.

4. The negative effect of low wages on productivity has been especially noted in the public sector (Lindauer et al 1988 and Klitgaard 1989).

5. Despite the dificulty of accurately distinguishing public spending of a social nature from other categories of government expenditure, the convention is to regard education and health costs and welfare payments as the most characteristic items of social expenditure. A good many infrastructure facilities (rural roads, urban developments, etc.) obviously have a direct influence on the standard of living of poor groups.

6. In cases where exchange control is stringent and a parrallel exchange market exists, a considerable proportion of trade eludes official channels and taxation. The major effect of a new exchange policy (devaluation and lighter controls) which helps to bring some part of import and export traffic back into official channels is to increase government revenue.

References

Azam, J. P., G. Chambas, P. Guillaumont, and S. Guillaumont. 1989. *Impact of Macroeconomic Policies on the Rural Poor*, UNDP, New York.

Balassa, B. 1964. "The Purchasing Power Parity Doctrine: A Reappraisal," *Journal of Political Economy*, Dec. p. 584-596.

BCEAO. Various years. Notes d'information et statistiques.

Bonjean, C. and R. Marodon. 1988. *Contraintes et efficacite de la politique des prix agricoles. Examples de la Côte d'Ivoire, du Kenya, de Madagascar, du Niger, du Rwanda et du Sénégal*, CERDI, University of Clermont I.

CERDI. 1988a. *La politique d'ajustement au Niger* (1982-87), report produced at the request of the Government of the Republic of Niger, under the direction of Sylviane Guillaumont (with J. P. Azam, T. Montalieu, J. Mathonnat, G. Chambas, A. M. Geourjon, P. Plane, C. Bonjean, C. Dejou), November, 650 p.

CERDI. 1988b. *Les conséquences sociales des politiques d'ajustement. Elements d'analyse, portant plus particulièrement sur le cas de la Côte d'Ivoire et du Sénégal.* Study made for the Ministry of Research and Technology, under the direction of P. Guillaumont, with F. Andre, M. Bergougnoux, C. Bonjean, G. Chambas, M. Demeocq, C. Fournioux, A. M. Geourjon, R. Marodon, T. Montalieu, P. Plane, December, 520 p.

Guillaumont, P. 1986. Facteurs déterminant le prix réel payé au producteur, CERDI, June.

Guillaumont, P. and S. Guillaumont. 1989. *La politique d'ajustement au Niger,*" to be published.

Guillaumont Jeanneney, S. 1988. "Devaluer en Afrique?," Observations et diagnostics économiques, Revue de l'OFCE, No. 25, October.

Heller, P. S. et al. 1988. *The Implications of Fund-Supported Adjustment Programs for Poverty: Experiences in Selected Countries*, IMF Occasional Paper No. 58, Washington, D.C.

IMF. 1988. *Annual Report.*

IMF. Various years. *International Financial Statistics.*

IMF. Various years. *Direction of Trade Statistics.*

Jaquemot, P. and E. Assidon, in collaboration with A. H. Akanni. 1988. *Politiques de change et adjustement en Afrique. L'expérience des 16 pays d'Afrique subsaharienne et de L'Océan Indien*, Ministry of Cooperation and Development, Etudes et documents, distr. La Documentation Française.

Klitgaard, R. 1989. "Incentive Myopia," *World Development*, Vol. 17, No. 14, April, p. 447-460.

Kravis, I. B. and alii. 1975. *A System of International Comparisons of Gross Product and Purchasing Power*, The Johns Hopkins University Press, 1975.

Levy, V. and J. L. Newman. 1989. "Wage Rigidity: Micro and Macro Evidence on Labor Market Adjustment in the Modern Sector," *The World Bank Economic Review*, Vol. 3, No. 1, January, p. 97-128.

Lindauer, D. L., O. A. Meesook, P. Snebsaeng. 1988. "Government Wage Policy in Africa: Some Findings and Policy Implications " *The World Bank Research Observer*, Vol. 3, No. 1, January, p. 1-26.

Ministry of Cooperation and Development, Caisse Centrale de Cooperation Economique. 1988. *Politique de change et adjustement en Afrique. Une étude sur 16 pays d'Afrique sub-saharienne hors zone monétaire* (Coordinator: P. Jacquemot) (3 vol.).

World Bank. 1989. *World Bank Tables 1988-89.*

3

Membership in the CFA Zone: Odyssean Journey or Trojan Horse?

Shantayanan Devarajan and Jaime de Melo

For most of the thirteen African members of the CFA Franc Zone, the 1980s have been a decade of slow or negative growth in per capita GDP, worsening balance of payments, debt crises, financial crises, declining competitiveness and—most distressing of all—an apparent lack of adjustment to the changed external environment they inherited from the 1970s. Of the few recently-documented "success stories" of adjustment in Africa, none is a member of the CFA Zone (World Bank 1989).

This disappointing performance is curious in light of the cautious optimism about CFA Zone membership voiced earlier in the decade by, among others, Guillaumont and Guillaumont (1984), and Devarajan and de Melo (1987a). Their optimism stemm-ed from the notion that participation in the CFA Zone would foster growth and reduce the need for adjustment. Guaranteed convertibility of the CFA franc and the fixed exchange rate with the French franc would lead to a stable investment climate for domestic and foreign investors, thereby stimulating economic growth. As for adjustment, the rules of the CFA Zone led to monetary and fiscal discipline. By avoiding some of the excesses of their African neighbors, CFA Zone members' need to adjust would be less—even though they lacked an important instrument of adjustment, namely, currency devaluation. Furthermore, as Devarajan and de Melo (1987b) pointed out, CFA countries had enough instruments with which to depreciate the real exchange rate which was, after all, the relevant signal for structural adjustment.

The purpose of this paper is to reassess the benefits and costs of the CFA Zone in light of the poor performance of its members in the 1980s. We ask whether on average CFA countries fared worse than a group of comparator countries. Since there is no single clear-cut group of comparators, we look at three: other Sub-Saharan African countries, other low- and middle-income countries and other primary and fuel exporters. Recognizing that a comparison of averages neglects differences within a group of countries, we perform some statistical estimations. Assuming that year-to-year GDP growth rates for all countries are drawn from a random distribution, we investigate whether there is evidence that the distribution of CFA countries' growth rates is significantly different from that of the comparators. We take a closer look at the adjustment experience of CFA countries vis-a-vis their comparators. Did CFA countries adjust less, controlling for the size of the external shock, than other countries? Did they adjust differently? One argument is that, given that they cannot devalue their nominal exchange rate, CFA countries cannot levy an inflation tax to finance a fiscal deficit. We ask whether this led to lower inflation and higher current account deficits. Another argument is that the fixed exchange rate makes expenditure-switching more difficult, so that CFA countries rely more on expenditure-reduction as a means of adjustment. We test this hypothesis.

Growth and Adjustment in the CFA Zone: An Overview

In this section we compare the average performance of CFA members with that of three groups of

Table 3–1. ***A comparison between the 1970s and 1980s***

Country group	*1973-81*	*1982-89*
Average Annual Real GDP Growth Rate *(percentage)*		
CFA (11)	3.7	2.6
Other:		
SSA (20)	2.7	2.0
Low-income (41)	4.4	2.9
Primary (52	4.6	3.9
Real Total Investment/Real GDP *(percentage)*		
CFA	24.3	18.9
Other:		
SSA	20.3	17.8
Low-income	21.6	19.8
Primary	22.5	19.4
Debt/GDP (debt service/exports in parenthesis) *(percentage)*		
CFA	30.6 (7.7)	62.5 (19.2)
Other:		
SSA	28.6 (9.7)	70.5 (20.9)
Low-income	26.0 (13.0)	58.4 (22.3)
Primary	24.9 (15.1)	56.4 (25.2)
Average annual inflation rate *(percentage)*		
CFA	12.0	4.3
Other:		
SSA	24.3	29.7
Low-income	18.4	33.3
Primary	24.4	44.9
Real exchange rate *(index, 1980=100)*		
CFA	107.0	108.0
Other:		
SSA	115.0	147.0
Low-income	103.0	121.0
Primary	103.0	119.0
Average annual export growth rate *(percentage)*		
CFA	6.8	1.5
Other:		
SSA	1.9	2.6
Low-income	4.9	5.0
Primary	4.8	7.6

Note: Unweighted averages. Number of countries in parenthesis.

Source: World Bank data

comparator countries. Rather than undertake a detailed analysis, we look for broad patterns that will suggest the statistical evaluations of subsequent sections. This approach is based on simple, unweighted averages of countries' performance indicators over different periods. The approach does not recognize that during the last two decades countries have been subjected to external and internal shocks that have varied in timing and magnitude across countries. Our method of aggregation by country groupings implies that the shocks were uniform within each group. Later we will allow for some diversity in comparisons based on an error components model.

We compare CFA members with three groups of comparators. The most important comparator group is other Sub-Saharan African countries. These countries have most in common with CFA Zone members in terms of economic structure, history and culture. Furthermore, being their neighbors, they provide CFA Zone members with the best perspective on life outside the CFA Zone. We also compare CFA countries with other low- and middle-income countries and other primary and fuel exporters. Except for Gabon, all CFA countries had a per capita income in 1980 below $1200. We use this figure as the cut-off for low- and middle-income countries. Higher income countries, such as those in Latin America and East Asia, tend to have very different economic structures and human capital endowments; hence their exclusion from the comparator set. In the same vein, there is some evidence that countries producing and exporting primary and fuel products have different adjustment histories from those which emphasize manufactured goods (Faini and de Melo forthcoming). Since every CFA country is either a primary or fuel exporter, we also compare them with other primary and fuel exporters.

Table 3–1 displays the averages for seven indicators for CFA countries and their comparators during 1973 to 1989. We look at two periods, corresponding (roughly) to the pre- and post-adjustment periods for most countries. The results broadly confirm our earlier results (Devarajan and de Melo 1987a) for the sample period from 1960 to 1982. There we showed that CFA countries' average GDP growth rate was slightly higher than that of other Sub-Saharan African countries, but lower than that of other developing countries. Furthermore, we found that the relative performance of CFA countries improved after 1973. We attributed the differences to the benefits from the stability of a fixed exchange rate regime, especially in the turbulent post-1973 era of floating exchange rates.

The pattern we discerned using data up to 1982 appears to have persisted through the late 1980s. Table 3–1 indicates that CFA countries enjoyed an average annual real GDP growth rate of over half a percentage point higher than their African neighbors. Their performance relative to other low- and middle-income and primary and fuel exporting countries continued to be inferior. While the basic trends established up to 1982 appear to be sustained, the gap between CFA countries and other African countries seems to be narrowing. The difference in GDP growth rates was a full percentage point in the 1970s.

In the 1970s CFA Zone members had a higher investment ratio than any of the comparators (and four percentage points higher than other African countries). The gap in investment-to-GDP ratios between CFA countries and their African counterparts narrowed during the 1980s. The decline in investment and the narrowing of the gap between CFA countries and others is a two-edged sword. Lower investment ratios could be associated with an increase in the efficiency of investment, as various white elephants are abandoned. However, if there is no improvement in the marginal efficiency of investment, the shortfall would signal a further slowdown in GDP growth in the future. As we will see, the latter possibility appears more likely when we look at the period from 1982 to 1989 more closely.

The debt-to-GDP and inflation indicators highlight particular aspects of the CFA Zone rather sharply. Having a fixed and rigid nominal exchange rate with the French Franc—changes in parity require the unanimous consent of CFA Zone members—effectively limits the seignorage tax that CFA countries can levy. Consequently, the CFA countries have experienced dramatically lower inflation rates than their African counterparts and other low-and middle-income and primary and fuel exporting countries. While average inflation rates increased in other parts of the developing world, as a consequence of the deterioration in the external environment which began around 1982, they decreased in the CFA Zone where the average rate in the 1980s was under 5 percent.

The CFA countries have had, until recently, an unlimited line of credit from the French treasury, known as the *compte d'operations*. This enabled them to have an average debt-to-GDP ratio in the 1970s that was higher than that of all of the comparator groups. As it did for other low- and middle-income countries, this ratio doubled for the CFA Zone in the 1980s. However, the debt-to-GDP ratio increased less rapidly than in the Sub-Saharan African group. Thus, while the fixed exchange rate may have exerted some monetary discipline in the CFA Zone, it did not decrease CFA countries' reliance on external finance.

The debt service-to-exports and real exchange rate indicators should be examined together with the debt-to-GDP ratio. A real exchange rate depreciation should cause a decrease in the debt service ratio if there is an export supply response -- although it will also cause an increase in the debt-to-GDP ratio (see, for example, Rodrik 1989). While CFA countries' average debt-to-GDP ratio over the 1982 to 1989 period was lower than that of their African counterparts, the average debt service ratios of the two groups were almost the same. In other words, the debt servicing needs and, implicitly, the creditworthiness of these two groups of countries are roughly comparable. Yet, while the real exchange rate of the other African countries depreciated almost 30 percent from the 1970s to the 1980s, that of the CFA Zone stayed the same. This may provide a clue to the debt service puzzle mentioned earlier. By depreciating their real exchange rates, the other African countries were able to increase exports so that their average debt service-to-exports ratio was comparable to that of CFA countries, however their average debt-to-GDP ratio was much higher than that of the CFA countries. By not depreciating their real exchange rate, the CFA countries have probably not generated a comparable export supply response. Their debt service ratio is the same as that of other African countries, although their debt-to-GDP ratio is lower.

Some of the speculation in the above paragraph is vindicated by a comparison of export growth in CFA countries vis-a-vis their comparators. CFA Zone members experienced faster growth in their exports than all of their comparators during 1973 to 1981; during 1982 to 1989, their average export growth rate was slower than those of the comparators, perhaps because of the lack of real exchange rate depreciation.

This comparison between the 1970s and 1980s masks the evolution of adjustment throughout the 1980s. When we look at the two subperiods, 1982 to 1985 and 1986 to 1989, the deterioration in the CFA Zone's position becomes clearer. The results appear in table 3–2. Several striking patterns emerge. First, while the CFA countries' average GDP growth was higher than that of other African countries in the first subperiod, it was actually lower in the second. When the growth rate in the rest of Africa accelerated after 1985, that in the CFA Zone declined. Second, the investment rate in the CFA Zone fell sharply during the 1980s, to the point where it was (marginally) lower than that of all of the comparators (in the 1973 to 1981 period it was the highest). Third, while the real exchange rate in other African countries depreciated sharply in the second subperiod, it appreciated in the CFA Zone, partially as a reflection of the depreciation of the dollar vis-a-vis the French franc.

One interpretation of this change in relative positions is that, by undertaking adjustment programs that emphasized a real exchange rate depreciation, the other African countries were able to benefit from the improvement in world commodity prices and demand after 1985. In particular, they were able to enjoy export growth which translated into faster GDP growth. Another interpretation would emphasize that the competitiveness of CFA countries was undermined by the continued depreciation of their neighbors' currencies. There is also some evidence that after 1982 the line of credit from the *compte d'operations* was no longer completely open. CFA Zone members had to cut back their current account deficits. Given the lack of real exchange depreciation, this must have come through expenditure-reduction. In particular, they reduced investment sharply. This led to a reduction in GDP growth rates which may continue into the future.

Table 3–2. *A closer look at the 1980s*

Country group	*1982-85*	*1986-89*
Average Annual Real GDP Growth Rate *(percentage)*		
CFA (11)	3.5	1.8
Other:		
SSA (20)	1.0	3.0
Low-income (41)	2.4	3.4
Primary (52)	4.8	2.9
Real Total Investment/Real GDP *(percentage)*		
CFA	21.3	16.6
Other:		
SSA	18.4	17.1
Low-income	20.7	18.8
Primary	20.6	18.2
Debt/GDP (debt service/exports in parenthesis) *(percentage)*		
CFA	58.0 (16.1)	67.1 (21.5)
Other:		
SSA	57.1 (17.1)	83.5 (24.9)
Low-income	49.3 (19.5)	67.6 (25.0)
Primary	47.1 (22.4)	65.9 (27.9)
Average annual inflation rate *(percentage)*		
CFA	8.6	1.0
Other:		
SSA	26.2	35.7
Low-income	19.5	50.4
Primary	28.9	64.6
Real exchange rate *(index, 1980=100)*		
CFA	115.0	100.0
Other:		
SSA	124.0	177.0
Low-income	109.0	136.0
Primary	106.0	136.0
Average annual export growth rate *(percentage)*		
CFA	3.0	0.1
Other:		
SSA	0.1	5.0
Low-income	1.2	8.8
Primary	7.6	7.7

Note: Unweighted averages. Number of countries in parenthesis.

Source: World Bank data

Are CFA Zone Growth Rates Different from Those of the Comparators?

As mentioned earlier, a comparison of averages assumes implicitly that all countries within a group are uniform. We now relax that assumption. Specifically, we assume that the GDP growth rate for each country in each year is drawn from a random distribution. We then ask whether the distribution from which CFA countries' growth rates are drawn is significantly different from that of the comparators. This is a strong assumption since it assumes away the role of a host of other factors which influence growth.

A common method for answering the question asked by the title to this section is by pooling the cross-section and time-series data and using least squares regression with dummy variables. To control for country-specific differences, this method requires a dummy variable for each country, which severely restricts the number of degrees of freedom. Instead, we use a modified approach, known as the "error-components framework" which assumes that the intercept term in the regression of the logarithm of GDP on the time trend (and a dummy variable) is also a random variable. This random variable is assumed to pick up the influence of omitted variables in determining growth. The error-components method requires the use of a generalized least squares estimator to get efficient estimates (see Fuller and Battese 1974), but results in greater degrees of freedom. As this was the model estimated in our previous work on the CFA Zone (Devarajan and de Melo 1987a), it has the advantage of providing a basis for comparison.

Table 3–3 presents the estimated coefficients for and ' in the following model:

$$Y_{it} = a_i D_{it} + a'_i D'_{it} + b D_{it} T + b' D_{it} T + u_{it}$$

where Y_{it} is the logarithm of GDP of country i in year t, D_{it} is the dummy variable for CFA members, D'_{it} is the dummy variable for comparator group members, T is the time trend, and u_{it} is the composed error term. The estimates of b and b' represent the growth rates for CFA and comparator countries, respectively.

The regression results confirm the pattern suggested by the comparisons of averages above. While they enjoyed significantly higher growth rates than their African counterparts in the 1970s, CFA countries fell behind in the 1980s. The estimated growth rate for CFA countries is lower than that for other Sub-Saharan African countries for the 1982 to 1989 period as a whole. The comparison of averages above showed this to be true only for the latter half of this period, i.e., for 1986 to 1989. Furthermore, the gap between CFA countries and other low- and middle-income countries and other primary and fuel exporters appears to have widened.

Table 3–3 shows that the growth rates of CFA countries and their comparators were all significantly different from zero, and that the growth rate of CFA countries fell more in the second period than did those of the others. It does not, however, indicate whether the growth rate of CFA countries was significantly different from those of its comparators. To determine this, we need to test whether the differences in the coefficients are significantly different from zero. Table 3–4 reports the results of such a test, where "significantly different from zero" is defined at the 95 percent confidence level.

The results corroborate the fact that the CFA Zone's performance has declined in relative terms during the 1980s. The CFA members' growth rate was significantly higher than that of Sub-Saharan Africa during the 1970s. During the 1980s it was lower, but not significantly so. If we think of economic growth as a running race, CFA countries in the 1970s were clearly ahead of the pack of all African countries. By the 1980s they were indistinguishable from the rest of the pack. Similarly, while CFA members' growth rate was not significantly different from that of other low- and middle-income countries during 1973 to 1981, it became significantly lower during 1982 to 1989. That is, in the race with other low- and middle-income countries, CFA Zone members fell behind the pack in the 1980s.

Table 3–3. *Estimated growth rates from error-components model* *(standard errors in parenthesis)*

Country group	*1973-81*	*1982-89*
CFA	3.9	2.1
	(0.33)	(0.43)
Other:		
SSA	2.5	2.3
	(0.25)	(0.33)
Low-income	4.2	3.0
	(0.19)	(0.23)
Primary	4.5	2.8
	(0.17)	(0.20)

Note: All the coefficients are significantly different from zero at the 95% confidence level.

A Control Group Approach to Adjustment

The statistical approach in the previous section does not control for factors that are likely to affect performance. In particular, developing countries were differentially affected by external shocks during the 1970s and 1980s. Could the differences in growth rates between the CFA Zone and its comparators be attributable to these left-out factors? In this section we attempt a partial answer to that question by relying on a statistical approach often used in the evaluation of adjustment programs, namely, the modified control-group approach (see, for example, Faini et al forthcoming). This method amounts to looking for a fixed effect (in this case, belonging to the CFA Zone) in explaining performance after controlling for changes in the external environment facing each country. As before, our comparisons are for the two periods, 1973 to 1981 and 1982 to 1989. We now specify that the change in performance between the two periods—where performance is measured by the average value of selected indicators during each period—is a function of autonomous policy changes

Table 3–4. *Growth comparisons*

	1973-81	*1982-89*
SSA	+	NS-
Low-income	NS-	–
Primary	–	NS-

Note: + (-) indicates that growth in the CFA Zone was significantly higher (lower) than in the comparator group; NS+ (NS-) indicates the growth rate in the CFA Zone was higher (lower) but not significantly so.

after controlling for changes in the external environment between the two periods. The set of performance indicators *j* for country *i* is denoted by y_{ij}. We postulate that changes in the value of each performance indicator depend on the vector of autonomous policy changes, x_i; on changes in the external environment, SH_i; and on membership in the CFA Zone:

$$y_{ij} = a_{oj} + x_i' a_i + SH_i a_{2j} + a_{3j} d + e_{ij}$$

where *a* prime denotes *a* transpose and *d* is a dummy variable that takes the value 1 for countries that belong to the CFA Zone and 0 otherwise.

Since autonomous policy changes are difficult to observe, we postulate that they are a function of the difference between the realized value of a performance indicator and some target value. Hence, we use lagged values of the performance indicators as proxies for the changes in policy (see Faini et al forthcoming, for details). The statistical results reported below show the coefficients for a model in which each observation is an average over the first or second period.

The shock variable is a weighted average of changes in the world real interest rate *(R)* and the export price index *(PX)* and import price index *(PM)* for each country. The weights are the ratios to GDP (Y) of debt *(D)*, exports *(X)* and imports *(M)*, respectively. By "changes" we mean the difference in the average value of these variables between 1973 to 1981 and 1982 to 1989. In symbols,

$$SH = (R_2 - R_1)(D/Y)_1 - (PX_2 - PX_1)(X/Y)_1 + (PM_2 - PM_1)(M/Y)_1$$

where the subscripts refer to averages over the first and second periods, and country subscripts have been omitted.

The estimate of the shock variable turns out to be lower for CFA countries than for any of the comparator groups. Indeed, the shock faced by other Sub-Saharan African countries was over twice that facing CFA countries. Hence, some of the lack of adjustment in the CFA Zone may be explained by a reduced need for adjustment. The question is whether, given the reduced need for adjustment, the observed adjustment was still too little.

The results are presented in table 3–5. The CFA dummy variable has a negative coefficient for all comparisons. The two coefficients that are consistently significant for the CFA Zone are those in the inflation and current account equations. The interpretation is that CFA countries, after controlling fordifferences in the shocks they faced between the two periods, had lower inflation and less current account improvement in the 1980s relative to the 1970s. The coefficients on the dummy variables for the GDP growth and investment equations are also lower, but not significantly so. The overall impression from this set of regression coefficients is that, while average GDP growth and investment-to-GDP ratios may have been lower in the 1980s in the CFA Zone, the more significant differences appear in the average inflation rates and current account improvement. As mentioned earlier, the rules of CFA Zone membership prevent CFA governments from levying an inflation tax, so that the fixed effect on inflation for CFA countries would be lower. The current account coefficient is much more troubling. It implies that in all of the comparator groups, CFA countries are conspicuous by their inability to reduce their current account deficits, even taking into account the fact that the shocks they face (i.e., the need to adjust) may have been different.

This observation is consistent with our earlier result about the lack of real exchange rate depreciation in the CFA Zone. While the rest of Sub-Saharan Africa depreciated its real exchange rate by an average of 28 percent during the 1980s, the CFA Zone's real exchange rate appreciated. Such a result would not have been a problem if the CFA countries did not need to depreciate their real exchange rate. However, the coefficients in table 3–5 show that they needed to do so. Controlling for differences in the shocks they faced, the amount of current account improvement in the CFA Zone was systematically lower than that in the comparator groups.

The coefficients of the lagged variables are usually significant with the expected sign. The higher the value of the own-lagged variable, the lesser the change in performance between the two periods. For example, other things equal, the higher is average growth in the period 1973 to 1981, the less is the increase in average growth during 1982 to 1989 compared with 1973 to 1981. While the external shock variable has the expected sign for the inflation and current account equations, it enters positively with the change in growth and investment, which is unexpected. Although, in most instances, these coefficients are not statistically significant, the low explanatory power of the proxy measure for external shocks suggests some measurement inaccuracies.

Another potential problem with the results in table 3–5 is that we may have misspecified the model. In particular, by controlling for autonomous policy changes separately from CFA Zone membership, we may be neglecting those aspects of Zone membership that are associated with policy. An alternative formu-

Table 3-5. *Performance in the CFA Zone*

Dependent variable	*Y-1*	*I/Y-1*	*INF-1*	*CA-1*	*CFA*	*SH*
Comparison with other Sub-Saharan countries						
Y	-.889*	.094	.021	.057	-.009	.049
	(-4.73)*	(1.47)	(.80)	(.54)	(-.82)	(1.01)
I/Y	1.032*	-.656*	.010	.061	-.022	.235
	(2.37)	(-3.72)	(.09)	(.27)	(-.89)	(1.70)
INF	-2.151*	-.012	.212	-1.397*	-.138*	.135
	(-2.47)	(-.05)	(.67)	(-4.24)	(-3.01)	(.63)
CA	.282*	.052	.021	-.661*	-.032*	-.106*
	(1.73)	(.85)	(.95)	(-5.84)	(-3.24)	(-2.18)
Comparison with other low-income countries						
Y	-.433*	.035	.039	-.093	-.021	.014
	(-2.63)	(.58)	(1.27)	(-.95)	(-1.58)	(.37)
I/Y	.457*	-.36*	-.004	.185	-.01	-.078
	(2.07)	(-3.74)	(-.06)	(1.54)	(-.77)	(-1.28)
INF	-1.70*	-.276*	.163	-.827*	-.134*	.081
	(-4.4)	(1.75)	(.65)	(-4.64)	(-3.85)	(1.17)
CA	.111	-.025	.015	-.765*	-.024*	-0.56
	(.76)	(-.48)	(.62)	(-6.87)	(-2.48)	(-1.40)
Comparison with primary exporting countries						
Y	-.63*	.069	.025	.01	-.015	.02
	(-5.1)	(1.60)	(1.58)	(.16)	(-1.30)	(.66)
I/Y	.55*	-.391*	-.027*	-.091	-.005	.116
	(2.77)	(-5.41)	(-1.85)	(-.63)	(-.36)	(2.00)
INF	-.75*	-.003	-.123	-.537*	-1.06*	.046
	(-1.72)	(-.002)	(-.59)	(-2.74)	(-2.58)	(.59)
CA	-.033	.032	-.001	-.781*	-.026*	-.068
	(-.23)	(.58)	(-.06)	(-7.01)	(-2.40)	(1.51)

Notes: The constant term is omitted from the results. Asterisks denote those coefficients which are significant at least at the 90 percent level. Y=GDP growth; I/Y=investment/GDP; CA=current account/GDP; INF=inflation rate. The subscript (-1) denotes lagged values. Results are corrected for heteroskedasticity by weighing each observation by the inverse of its estimated standard error. Extreme influential observations are excluded.

lation would be to leave out the independent variables for autonomous policy changes or their proxy, the lagged dependent variables. Unfortunately, this specification yielded significant coefficients only for the CFA variable in the inflation equation. The effect of CFA membership on current account improvement was not statistically significant. Thus, isolating autonomous policy changes sharpens our estimates of the fixed effects due to CFA Zone membership, possibly because there was no systematic relationship in the use of these policies among CFA Zone members.

Despite these qualifications, there is some evidence that adjustment in the CFA Zone has been insufficient in the 1980s, arguably the era when almost everybody else in Africa was undertaking major adjustment programs. We have suggested one reason for the CFA Zone's lack of adjustment, namely, the inability of CFA Zone members to effect a nominal devaluation. But the nominal exchange rate is but one instrument of adjustment. In principle, CFA Zone members have enough instruments with which to adjust their economies (Devarajan and de Melo 1987b). In practice, they have been reluctant to use these other instruments or, when they have chosen to use them, the results have been disppointing. The advantage of a nominal devaluation is that it permits expenditure-switching to accompany the necessary expenditure-reduction of an adjustment program. In this case, the amount of expenditure reduction required would be less (Corden 1988). It follows, therefore, that CFA members would have had to rely more on expenditure-reduction as opposed to expenditure-switching in reducing their current account deficits. We now test this hypothesis.

Using the data from the two periods, 1973 to 1981 and 1982 to 1989, we estimate an equation which links the target of adjustment—the resource balance expressed as a ratio to GDP—with the investment rate and the real exchange rate. The investment rate is used here as an instrument for expenditure changing policies. Of course, there are other such instru-

Table 3–6. *Changes in resource balance, investment and the real exchange rate betweem 1978/79 and 1987/88*

	Average predicted change in resource balance/GDP	*Investment component*	*Real exchange rate component*
CFA	.046-	.071	-.033
Other low-income	.050-	.053	.232

Note: Column 1 is equal to the sum of colums 2 and 3 plus the intercept term.

ments, including government consumption expenditure. However, it is easier to cut investment first, especially in a slow-growing economy. Therefore, it is worth examining the extent to which improvements in the resource balance reflected declines in investment.

The regression linking the resource balance with investment and the real exchange rate also contains a dummy variable for each country, to capture country-specific effects, and a time trend. The results of the regression yield a negative—and highly significant—coefficient on the investment variable for both CFA and non-CFA countries. By contrast, the coefficient on the real exchange rate variable is small and indistinguishable from zero. This is consistent with other studies that have attempted to link the real exchange rate to the resource balance or the trade balance. For example, Pritchett (1990) found only a weak relationship between the merchandise trade balance and the real exchange rate, even when controlling for terms of trade movements. His reasoning, which may also apply here, was that while exports may respond to changes in the real exchange rate, imports were determined by foreign exchange availability, i.e., exports, and hence may move perversely.

Undaunted by the insignificance of the real exchange rate variable, we use the estimated regression coefficients to compute the predicted resource balance in 1978-79 and 1987-88. For both CFA and non-CFA countries, the resource balance improved between the first and second two-year periods. This is a reflection of the cutback in foreign lending in the 1980s and the simultaneous need for these countries to make increasingly higher debt service payments. We can then ask how much of the improvement in the resource balance between 1978-79 and 1987-88 was due to investment reduction, and how much to real exchange rate depreciation. Were the relative proportions different between CFA and non-CFA countries? The results are reported in table 3–6.

First, although the improvement in the resource balance-to-GDP ratio was roughly comparable for the two groups of countries, the investment components were quite different. CFA countries relied more heavily on cutting investment than did other low-income countries. Second, the real exchange rate component has the wrong sign for CFA countries. Instead of contributing to the reduction in the current account deficit, the real exchange rate may have worked against it in the CFA Zone, though the lack of statistical significance of the coefficient on the real exchange rate variable calls for caution in interpreting this result. Finally, and most importantly, the relative contributions of investment reduction and real exchange rate depreciation are very different between the two groups of countries. For other low-income countries, it is about one-quarter whereas for CFA members it is over two (and with the wrong sign).

Conclusions

The purpose of this paper has been to assess whether the particular institutional arrangement of the CFA Zone has aided or hurt its members. It has been argued that the convertible currency with a fixed exchange rate results in monetary and fiscal discipline which, in turn, benefits CFA Zone members. Just as Ulysses tied himself to the mast, CFA governments abdicated the right to levy an inflation tax so that they would never be tempted to do so. The evidence of the relative performance of CFA countries' economies vis-a-vis their comparators shows that this argument was persuasive until the early 1980s.

After 1981, changes in the world environment and persistent current account deficits meant that CFA countries needed to adjust their economies along with most other developing countries. Our statistical results show that they did not adjust as much as they needed. Furthermore, their growth performance was disappointing. Under every estimate, CFA Zone members' GDP growth rates fell behind those of their counterparts, including the other African states. Finally, the burden of adjustment appears to have fallen disproportionately on expenditure reduction in general, and investment reduction in particular—an ominous sign for future growth.

Of course a change in external circumstances does not necessarily mean that the original commitment to a fixed exchange rate was unwise. It is possible that

CFA Zone members took all these contingencies into account in making the original decision to join the CFA Zone and hence forego the opportunity to devalue their nominal exchange rate in the future. In this case, there are no policy implications from the recent deterioration in the CFA Zone's economic performance. The members may have drawn a bad hand, but not one which renders their original decision sub-optimal.

An alternative interpretation is that the current circumstances facing the CFA Zone lie outside the set of events which were considered when the original decision to join the CFA Zone was made. In particular, the adverse terms of trade shocks of the 1970s and 1980s, and the attendant need to shift resources from nontradables to tradables, may not have been expected in the 1960s. Such an argument is compelling because the exchange rate is both an instrument for transforming resources from nontradables to tradables as well as an inflation creating (or controlling) tool. It could be that the founders of the CFA Zone calculated the inflation controlling benefits of a fixed exchange rate without anticipating the costs in terms of the countries' inability to adjust to unfavorable external circumstances. Thus, the very institutional arrangement which enabled these countries to enjoy faster and more stable growth in the 1970s is preventing them from adjusting to the external and internal shocks of the 1980s. In short, what began as an Odyssean journey may have turned into a Trojan horse.

References

Corden, W. M., (1988) "Macroeconomic Adjustment in Developing Countries," Research Department, International Monetary Fund.

Devarajan, S. and J. de Melo, (1987a) "Evaluating Participation in African Monetary Unions: A Statistical Analysis of the CFA Zones," *World Development*, Vol. 15, No. 4, pp. 483-96.

Devarajan, S. and J. de Melo, (1987b) "Adjustment with a Fixed Exchange Rate: Cameroon, Cote d'Ivoire and Senegal," *World Bank Economic Review*, Vol. 2, No. 2.

Faini, R., J. de Melo, A. Senhadji-Semlali and J. Stanton, (forthcoming) "Growth-Oriented Adjustment Programs: A Statistical Analysis", *World Development*

Faini, R. and J. de Melo, (forthcoming) "Adjustment, Investment and the Real Exchange Rate in Developing Countries," *Economic Policy*.

Fuller, W. and G. Battese, (1974) "Estimation of Linear Models with Crossed Error Structure," *Journal of Econometrics*, Vol. 2, pp. 67-78.

Guillaumont, P. and S. (1984) *Zone Franc et Developpement Africain*, Paris: Economica.

Guillaumont, P., S. Guillaumont and P. Plane, (1988) "Participating in African Monetary Unions: An Alternative Evaluation," *World Development*, Vol. 16, No. 5, pp. 569-76.

Guillaumont, P. and S. Guillaumont, (1988b) eds., *Strategies de Developpement Comparees: Zone Franc et Hors Zone Franc*, Paris: Economica.

Honohan, P. (1990) "Monetary Cooperation in the CFA Zone," Policy, Research and External Affairs Working Paper No. WPS 389, World Bank.

O'Connell, S. (1989) "Uniform Trade Taxes, Devaluation and the Real Exchange Rate: A Theoretical Analysis," Policy Planning and Research Working Paper No. WPS 185, World Bank.

Pritchett, L., (1990) "The Merchandise Trade Balance and the Real Exchange Rate in LDC's," Country Economics Department, World Bank.

Rodrik, D., (1989) "The Welfare Economics of Debt Service," John F. Kennedy School of Government, Harvard University.

World Bank, (1989) *Sub-Saharan Africa: From Crisis to Sustainable Growth*, Washington, D.C.

4

Comments on Exchange Rate Policy

Ravi Kanbur and Stephen O'Connell

Ravi Kanbur

First I would like to bring up two criticisms of recent treatments of the exchange rate and real income distribution and poverty. One of the criticisms has been that the notion of thinking of households as having been involved in only tradable activities or only nontradable activities is clearly empirically not true. Through diversification each household can get its income from both tradable and nontradable activities. Taking this to the extreme, suppose there is a nationally representative household, which is in fact exactly representative of the national composition of tradable and nontradable activities. Then, for a given total expenditure, altering the composition of national income should not affect the standard of living of this typical representative household. So to the extent that households are diversified, to the extent that the household's portfolio of sources of earnings is exactly the same as the nation's portfolio of sources of earnings, expenditure-switching should have no effect at all on poverty. So I think this is one thing that would go against the simple way of looking at it, which is that all households get their incomes either from tradables or from nontradables.

Another criticism which actually goes in the other direction is related to the very important area of intrahousehold inequality. The analyses that have been done so far are focused entirely on looking at the distribution of income at the household level. So we take the sources of income of the household as a whole and we take that as being somewhat equally distributed within the household. But in fact, one of the points that is coming out, not from household surveys so much but from anthropological investigations in the African context, is the gender division of labor within the household. In particular, females seem to have more control over the food crop plot and males seem to have more control over the cash crop plot. In other words, there is a division within the household between the control and use of income from tradable goods versus nontradable goods. If in this case food is a nontradable good, like root crops for example, then we are back to the old story in which some individuals derive their incomes primarily from tradable goods and other individuals derive their incomes primarily from nontradable goods. As the share of tradable goods income is increased relative to nontradable goods income, in fact intrahousehold inequality will be increased. I think this is a feature that has not been taken into account in the literature so far. This has not been directly analyzed because the data is not available; however, there is plenty of indirect evidence, firstly that these types of social structures exist, and secondly that the actual division of income between different members of a household affects the consumption patterns of that household. Paul Collier and a number of other people have looked at indirect evidence which supports this.

In terms of the social consequences of adjustment following through from the impact of real exchange rate changes, I essentially see three stages to the argument. In the first stage, we assumed that there was a complete division between tradables and nontradables. On the basis of that assumption, most of us came to the conclusion that real exchange rate changes would not be detrimental to poverty. In the second stage, people said that households tend to have diversified sources of income. The third stage is that, yes, it is true that households may be diversified, but within the household there are individuals who focus on one rather than the other source of income and there is real conflict between these individuals in the household so actually the impact of structural adjustment on inequality is understated.

The Guillaumonts' analysis leads to these issues, which are not covered in the paper.

The Guillaumont and Guillaumont Jeanneney Paper

The first part of the Guillaumonts' paper discusses the relationship between the nominal exchange rate and the real exchange rate. The problem here is that two relative prices are affected by the nominal exchange rate at the same time. One is the relative price between tradables and nontradables, the real exchange rate, and the other is the relative price between present goods and future goods, the inflation rate. The nominal exchange rate affects simultaneously the real exchange rate and the inflation rate. All of their discussion is based around this, as is a lot of the Homeric discussion (in the Devarajan and de Melo paper).

The Devarajan and de Melo Paper

Essentially the point is that you have an instrument which you can either use or abdicate the use of completely. If you retain use of it, you may be tempted to use it to levy the inflation tax. The point made by Devarajan and de Melo is that, by abdicating the use of the nominal exchange rate, you are abdicating the ability to levy an inflation tax. But, if circumstances are such that you need to use that instrument to affect the relative price between tradables and nontradables, and if you have already abdicated its use beca use you do not want to use it to change the relative price between present goods and future goods, then of course you cannot use it to change the first relative price either. There is a certain difficulty in the way the paper presents this, which is the notion of Ulysses tying himself to the mast. Circumstances change so it is now worthwhile to untie Ulysses from the mast, but as soon as you untie him, instead of rescuing the damsel in distress, he is liable to go off toward the siren.

The basic problem is that in abdicating the use of this instrument to be able to change one relative price, you abdicate its use to change the other relative price. In conditions or circumstances where you need to change the other relative price, if you use it, you remove the precommitment not to use that instrument to change the second relative price.

Suppose now that nominal devaluation is indeed tried, and that commitment has been lost. Then what problem is being solved by a rational government or what is the problem that Devarajan and de Melo see rational governments solving in this context? One line of argument seems to be that, in the early part of the period, there is a commitment to a fixed exchange rate and it seems to work. There are high growth rates, etc. Now the external circumstances change, so we should abandon that commitment and use this instrument to affect these relative price changes. But of course once you abandon it, you will have lost the earlier benefits. Devarajan and de Melo may even argue that the fixed exchange rate policy commitment should never have been made in the first place, but I think that is an argument that has to be made quite separately.

This discussion can be cast in terms of time consistency. The difficulty is that once the commitment to the fixed exchange rate is abandoned, it is lost virtually forever. What are the costs of that from the view of commitment?

Stephen A. O'Connell

Though differing in motivation and scope, these two stimulating papers share a concern for the relationship between policy regime and macroeconomic performance in Sub-Saharan Africa. I will restrict my comments to this area of common interest, leaving the second half of the Guillaumonts' paper, on the social consequences of alternative adjustment modes, to my fellow discussant.

The Guillaumont and Guillaumont Jeanneney Paper

This paper begins with a useful overview of exchange rate arrangements in Sub-Saharan Africa. The authors then go on to make four main points related to the implications of the exchange rate system for structural adjustment. First, exchange rate policy helps determine how much inflation will accompany real exchange rate adjustment. Second, in many cases a devaluation will not reduce the amount of expenditure reduction required for external (i.e., current account) adjustment. Third, external adjustment does not require real exchange rate depreciation, since it can take place through productivity improvements without any change in the real exchange rate. Fourth, low inflation programs are more favorable to investment, and therefore more likely to produce productivity-led external adjustment.

However broad the authors' intent, it is difficult not to read this line of argument as an advertisement for maintenance of the current parity in the CFA Zone. This is a position that the authors have vigorously defended in other work and on other grounds. While the presence of this sub-theme makes for lively reading (all the more so since the two sets of authors appear to differ on both the facts and their interpretation), it undercuts the paper somewhat as a survey of the implications of exchange rate policy in Sub-Saharan Africa. Thus, for example, the authors spend very little time discussing the role of exchange rate policy in stabilization, an issue of extreme relevance outside of the Zone; and they lump the Zone together with other fixed exchange rate regimes in Sub-Saharan Africa, without discussing the mechanisms that have allowed the Zone to accumulate such a remarkable history of nominal exchange rate stability.

In discussing real exchange rate depreciation and inflation, the authors point to a correlation between devaluation and inflation across African countries. But this observation does not establish a causal role for nominal exchange rate policy. What role, if any, does the exchange rate regime play in determining inflation and the path of the real exchange rate? What we know from outside of the CFA Zone is that fixed exchange rates often fail to deliver stability of prices except over very short time horizons. While this is most dramatic in the devaluation-cum-stabilization cycles of Latin America, the African record is also full of examples of fixed parities that were eventually changed or abandoned altogether under the pressure of domestic inflation. These cases suggest that at least over the medium run (e.g., the time horizon of the typical adjustment program), inflation and the rate of official depreciation are the joint result of other policies, such as those determining fiscal deficits, external borrowing, and the rate of domestic credit creation. From a CFA Zone perspective, it may appear that these other policies are subordinated to the commitment to maintain the parity. But this is evidently not a generalizable feature of the Zone. One is therefore left wondering what feature, or combination of features, of the Zone accounts for the ability of these countries to maintain policies consistent with a fixed parity for so long.[1]

The question of whether a nominal devaluation allows external adjustment to take place at lower cost in terms of expenditure reduction is clearly essential to any assessment of external adjustment strategies in a managed exchange rate system. The authors claim that nominal devaluations are likely to have minimal transitional benefits, basing their claim on the example of Madagascar and an appeal to lack of money illusion on the part of wage earners. Both arguments deserve further scrutiny. In the Madagascar case, we need the counterfactual: what contraction would have been required if the exchange rate had not been used? Moreover, since the transitional benefits depend on the credibility of the new nominal exchange rate, what lessons does the Madagascar case have in the very different institutional context of the CFA Zone? With regard to the presence or absence of money illusion, this is an empirical question. Elsewhere in the paper, the authors stress the effects of nominal effective movements in the French franc on the real exchange rate in CFA countries; it is difficult to square this particular nonneutrality (which seems to suggest the presence of some wage-price stickiness in the CFA countries) with the claim that a policy-induced nominal devaluation would have no transitional benefits.

The authors make an important contribution by emphasizing the potential role of productivity increases in generating external adjustment without excessive cost in terms of growth. One has only to contrast the expansionary adjustment pattern of some of the Asian debtors with the contractionary adjustment pattern of their Latin American and African counterparts to appreciate the significance of this point. I was less impressed, however, by the argument that adjustment programs that avoid nominal devaluation are more conducive to productivity increase than others. While there is ample theoretical and empirical work relating productivity growth to trade policy and even fiscal policy (e.g., government expenditure patterns), there is relatively little evidence linking productivity growth to the exchange rate regime, much less the level of the nominal exchange rate. And given the authors' statement that adjustment with a fixed parity is out of the question for high inflation countries, it appears that we are again working within the narrow confines of the CFA Zone debate, with few lessons for the rest of Sub-Saharan Africa.

The relationship between exchange rate policy and productivity growth is well worth further investigation on both the analytical and empirical ends. As the authors suggest in their discussion, it is probably uncertainty regarding real exchange rates and inflation, rather than overvaluation or inflation per se, that discourages investment and productivity growth. Recent empirical work provides some support for this notion.[2]

Overall, (the first half of) this paper brings a fresh perspective to the debate on exchange rate policy in Sub-Saharan Africa. Future contributions will undoubtedly develop some of the analytical arguments further and strengthen their empirical content.

The Devarajan and de Melo Paper

The CFA countries have long been noted for their choice of monetary stability and low inflation (e.g., Mundell 1972). Has this stability delivered benefits in terms of increased growth? Both sets of authors represented here have recently given a (qualified) yes. Devarajan and de Melo (1987) found that the CFA countries had grown more rapidly than the rest of Sub-Saharan Africa (though not quite as rapidly as developing countries outside of Africa) over the period from 1960 to 1982. Similar methodology, including more extensive attempts to control for structural differ- ences and external shocks, was applied by Guillaumont, Guillaumont and Plane (1988), with similar conclusions.

The present paper by Devarajan and de Melo shakes up the debate by suggesting that Zone membership may not be such a bargain after all. In the first half of the paper, the authors extend their earlier work comparing macroeconomic performance in the Zone to performance in three alternative comparator groups. When data from 1983 to 1989 are included, the relative growth performance of the Zone deteriorates noticeably, particularly after 1986. Moreover, relative performance on the external front, as measured by real exchange rate depreciation and export growth, deteriorates even more markedly. While some portion of the export recovery observed outside of the Zone may simply be a shift of existing exports onto official markets in response to declining direct and indirect taxation of official exports, it is revealing that the average export growth rate in the Zone declines dramatically in the 1980s, while that of other developing country groups remains roughly the same or improves. The authors point out that while inflation was extremely low in the Zone in the 1980s, real exchange rates were essentially unchanged over the period. Outside of the Zone, in contrast, real exchange rates depreciated substantially. In short, the Zone appears to have lost much of its magic in the 1980s.

Performance comparisons are subject to the critique that Zone membership is only one of many factors affecting performance. The use of comparator groups (low-income countries, primary exporters, and Sub-Saharan Africa) addresses this and allows the authors to control to some degree for level of development, economic structure, and external environment. Nonetheless, it is possible that the results are driven by chance correlations between Zone membership and other unmeasured factors like the terms of trade. Guillaumont, Guillaumont and Plane (1988) approached this problem by estimating a cross-sectional model of the determinants of growth, and then interpreting each country's residual as an indicator of its performance relative to the norm. In the second half of their paper, Devarajan and de Melo take an alternative route (the control group approach), regressing performance indicators on a composite economic shocks variable, a set of macroeconomic policy indicators, and a CFA-specific dummy variable.

I found the control group approach somewhat less appealing than the rest of the paper, for two reasons. First, in order to econometrically identify the unspecified CFA Zone effect, the approach assumes that the effects of Zone membership do not operate through differing impacts of autonomous shocks or different channels of influence of autonomous policy. This is a troubling starting point, since we know that policy regimes differ and we suspect that economic responses differ due to the different policy framework.

Second, instead of measuring autonomous policy changes directly, the authors adopt a target-adjustment model in which the lagged performance indicator serves as a proxy for policy changes. However, since the target itself is unobservable, the fixed effect ends up picking up (among other things) differences in targets between CFA and non-CFA countries. Thus the finding that the current account deficit fixed effect is higher in the CFA Zone than outside may not be an indication of relatively poor performance in the Zone; it may simply be evidence of a less binding external constraint. In particular, the arrangements of the Zone, including the special economic and political ties to France, may give these countries better access to international capital flows (including prospective debt forgiveness) than otherwise, leading them to choose an optimal mix of public finance that is weighted more heavily towards external finance. This would be consistent with a more appreciated real exchange rate and a lower fixed effect in the inflation equation.

Given the difficulties associated with the control group approach, a more satisfactory approach might be to run a fixed-effects model controlling only for external shocks. The authors do something similar "they re-run the control group regressions with the policy variables omitted" and find that only the inflation effect remains as a clear distinction between CFA and non-CFA countries. While the authors characterize this finding as unfortunate, it does support their basic point that the CFA sparkle has disappeared in the 1980s. Further work attempting to control for external factors, including perhaps an update of the paper by Guillaumont, Guillaumont and Plane (1988), would seem highly justified.

It is clear that the arrangements of the CFA Zone came under unprecedented pressure in the 1980s, and that many of these pressures are continuing into the 1990s. The authors trace the relative deterioration of performance to an excessive reliance on expenditure reduction. This focuses attention on the question of the parity: can the CFA countries devalue without endangering the very existence of the Zone? Given the severity of external shocks in the 1980s (including the first Zone-wide decline in export prices and the substantial appreciation of the French franc), it seems at least conceivable that an adjustment tied to these unusual circumstances might be viewed as consistent with a continued commitment to a fixed parity. More broadly, the issue is whether there are policy rules for the nominal exchange rate that would provide more flexibility but still be consistent with other important features of the Zone, such as absence of severe capital controls and continuation of the convertibility guarantee from France. The paper by Devarajan and de Melo provides essential background for any further analysis of these issues.

Notes

1. There are numerous possibilities, including (1) the availability of short-term external finance through the operations account, which shelters these countries from external liquidity crises; (2) the convertibility guarantee, which imposes a responsibility on France to monitor macroeconomic policy in the Zone; (3) the institutional rule requiring unanimity in deciding on any parity change; (4) the institutional rule limiting government borrowing from the central bank; and (5) the high degree of diversification of commodity exports, implying relatively low terms of trade volatility for the Zone as a whole.

2. For example, Caballero and Corbo (1989) show that exports are negatively related to volatility of the real exchange rate in a group of Latin American countries.

References

Caballero, R. J. and Corbo (1989), How Does Uncertainty About the Real Exchange Rate Affect Exports?, The World Bank, PPR Working Papers Series No. 221, June.

Devarajan and de Melo (1987), Evaluating Participation in African Monetary Unions: A Statistical Analysis of the CFA Zones, *World Development* Vol. 15, No. 4: 483-496.

Guillaumont, Patrick, Sylviane Guillaumont, and Patrick Plane (1988), Participating in African Monetary Unions: An Alternative Evaluation, *World Development* Vol. 16, No. 5: 569-576.

Mundell, Robert A. (1972), African Trade, Politics and Money, in R. Tremblay, ed., *Africa and Monetary Integration* (Montreal, Les Editions HRW).

5

The Inflationary Consequences of Devaluation with Parallel Markets: the Case of Ghana

Ajay Chhibber and Nemat Shafik

A common concern in macroeconomic adjustment programs is the potential inflationary effects of a combination of devaluation, trade liberalization, subsidy reduction and price decontrol. This issue has become critically important in the African context where inflation has been accelerating in a number of countries, particularly in those which are not members of the CFA Franc zone. The high inflation in the non-CFA zone countries is alleged to result from the devaluations many of these countries have undergone as part of the reform programs. The link between devaluation and inflation has been analyzed in considerable detail, both theoretically and empirically. The empirical evidence suggests that devaluation is inflationary, but the degree of inflation depends on accompanying fiscal and monetary policies. The analysis of the inflationary effects of devaluation is complicated by the presence of parallel markets.

This paper presents a theoretical framework to address the question of exchange rate devaluation in the presence of parallel markets in the second section. It then uses Ghanaian data to empirically estimate a model in the third section which is used to analyze policy trade-offs for Ghana in the fourth section. The last section draws together the lessons from the Ghanaian experience.

Ghana has carried out one of the most thorough adjustment programs in Africa. Its inflation rate, which is depicted in figure 5–1, is high and has been rising in recent years, although it is still lower than during the worst crisis years. These high rates of inflation have coincided with, and have often been blamed on, the large official devaluations of over 40 percent per annum implemented between 1983 and 1988.

This paper disputes the conclusion that the recent surge in inflation in Ghana is due to the large official devaluations. The conclusion is based on a model which makes it possible to quantify the effects of exchange rate changes in the presence of parallel markets. The results show that there is no direct relationship between the official exchange rate and inflation since prices had already adjusted to the parallel exchange rate. In fact, official devaluation in Ghana had a positive effect on the government budget. Because of this improvement in the fiscal

Figure 5-1. Mark-Up Model of Inflation in Ghana

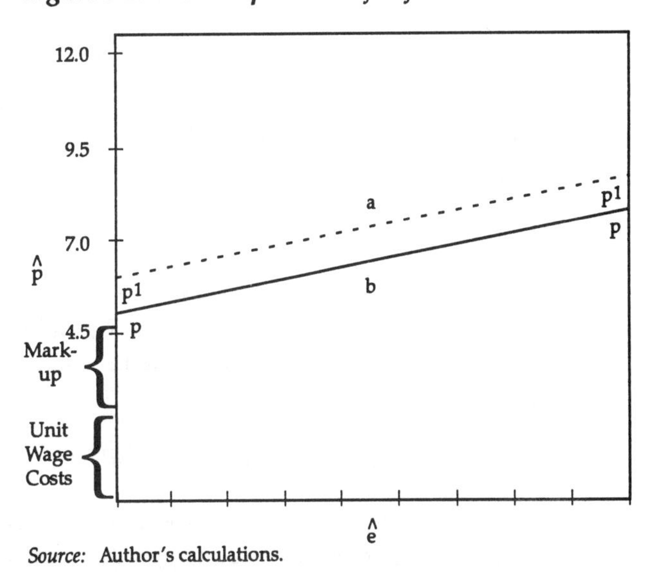

Source: Author's calculations.

deficit, the official devaluation had favorable effects on the rate of monetization, on inflation and on the rate of depreciation of the parallel exchange rate.

The Model

The model simultaneously determines the rate of inflation and the parallel market exchange rate premium and links them to the monetary, fiscal and real sides of the economy through several channels. A segmented goods market is hypothesized. There is an official market at which goods are available at a subsidised price (P_3). There are traded goods ($P1$) and nontraded goods ($P2$) in the parallel market and the uncontrolled legal market. Traded goods prices are equal to foreign goods prices (P_f) converted at the parallel exchange rate, plus a mark-up (s) for the unit costs of smuggling.

$$P_1 = (1 + s)(e_p \cdot P_f).$$

We assume for expository convenience that s = 0

$$\hat{P}_1 = \hat{e}_p + \hat{P}_f$$

A third category of final goods transacted in the parallel market is nontraded goods whose production requires labor and imported inputs. Foreign exchange for the purchase of imported inputs comes from official sources with exchange rate eo as well as from the parallel market exchange rate e_p. The following variable mark-up model is used to explain nontraded goods prices (P_2):

$$P_2 = (1 + u)[e_p \cdot P_f + (1\text{-}k)\, e_0 P_f\, a^1 (wp)\, a^2.]$$

where u denotes the mark-up, wp denotes unit labor costs, and k is the parallel market's share of foreign exchange used for imported inputs. The size of the mark-up is a function of excess demand (ED) in the economy. So that the rate of change in P_2 is;

$$\hat{P}_2 = b_0\, ED + b_1 k\, [\hat{ep} + \hat{P}_f] + b_1 (1\text{-}k)(\hat{e}_o + \hat{P}_f) + b_2 \hat{wp}$$

The average rate of inflation is the weighted average of the changes in prices of traded (P_1) and nontraded (P_2) goods and goods transacted at controlled prices (P_3):

$$\hat{P} = C_1\hat{P}_1 + C_2\hat{P}_2 + (1 - C_1 - C_2)\hat{P}_3 P$$

We assume that there is a surplus in the labor market such that an increase in aggregate demand decreases the size of the labor surplus, but has no effect on unit labor costs. In the asset markets, there is a domestic asset, cedis, and a foreign asset, foreign exchange (F). Overall demand is a function of total financial assets which are defined as:

$$A = M + e_p F.$$

Where M represents the money supply, e_p is the parallel exchange rate, and F is the foreign asset. The demand for foreign exchange is composed of transactions demand and portfolio demand. The transactions demand for foreign exchange is a negative function of the exchange premium e_p/e_o, a positive function of the difference between prices in the open market and controlled prices and a positive function of total financial assets, A. The portfolio demand for foreign exchange is a negative function of the difference between domestic and foreign interest rates and a positive function of total financial assets, A.

$$e_p.\ F_d = h(e_p/e_o, P/P_3, A) + j\,[(i - i^* - {}_pe), A]$$

where $h_1 > 0, h_2 > 0, h_3 > 0; j_1 < 0, j_2 > 0.$

The supply of foreign exchange is a function of the exchange premium, the real exchange rate (r), and the ratio of open market prices to the controlled official price (P/P_3):

$$e_p F_s = n(e_p/e_o, r, P/P_3)$$

where $n_1 > 0, n_2 < 0, n_3 > 0.$

A rise in P increases the transactions demand for foreign exchange as the difference between the open and controlled price grows. A rise in P also increases expectations that the parallel exchange rate will depreciate. This leads to an increase in the portfolio demand for foreign exchange, given domestic and foreign interest rates. The effect on the supply of foreign exchange is also positive, but the net effect is an increase in the demand for foreign exchange. A rise in e_p is required to offset this net demand increase and restore equilibrium. The direct price effect of e_p on F_d is negative. But there is a positive wealth effect. If the substitution effect is greater than the wealth effect, the overall demand effect is negative and the ee curve unambiguously slopes upwards (figure 5–2). An increase in e_p also lowers the expectation of future devaluation and thereby lowers the portfolio demand for foreign exchange.

Figure 5–2 describes the dynamic equilibium in e_p and P. In quadrant I we have an excess supply of goods but an excess demand for foreign exchange so that equilibrium is restored with a decline in P and an increase in e_p. In quadrant II we have an excess

Table 5-1. ***The complete model***

A. Inflation

1.1. $\hat{P} = x_1 \hat{P}_1 + x_2 \hat{P}_2 + (1 - x_1 - x_2) \hat{P}_3$

1.2. $\hat{P}_1 = \hat{P}_f + \hat{e}$

1.3. $\hat{P}_2 = \hat{U} + [a_1 \hat{w}_p + a_2 \hat{mc}]$

1.4 $\hat{mc} = \hat{P}_f + \hat{e}$.

1.5. $\hat{U} = f(ED); f_1 > 0$

1.6. $ED = \log M/p - \log Md/p$

1.7. $\log Md/p = d(\log y, i, \hat{p}_e); d_1 > 0, d_2 < 0, d_3 < 0$

1.8. $\hat{e} = k\hat{e}_0 + (1-k)\hat{e}_p$

B. Parallel exchange rate

1.9. $F = F^t + F^P$

1.10. $e_pF^t = h(e_p/e_0, P/P_3, A); h1 < 0, h2 > 0, h3 > 0$

1.11. $e_pF^P = j[i - i^* - e^e_p] . A; j_1 > 0$

1.12. $e_p.Fs = n(e_p/e_0, r, P/P_3); n_1 > 0, n_2 > 0, n_3 > 0.$

C. Money and fiscal aspects

1.13. $M_2 = \hat{m} + \hat{R}$

1.14. $R = COG + NFA + OA$

1.15. $COG = COG^{-1} + GBB$

1.16. $GBB = GDEF - GFB - GDB$

1.17. $GDEF = GEXP - GREV$

1.18. $GREV = CUDT + EXPT + GOREV + AID$

1.19. $CUDT = IMPQ.Pf.e_0.tm$

1.20. $EXPT = EXPQ.Px.e0.tx$

1.21. $AID = \$AID.e_0$

1.22. $GEXP = WBILL + SUBS + INTD + INTF + GOVI + GOEXP$

1.23. $WBILL = Wg.Lg$

1.24. $INTF = FDEBT.if.e_o$

1.25. $FDEBT = FDEBT_{-1} + GFB/e_o$

1.26. $INTD = DDEBT.i$

1.27. $DDEBT = DDEBT_{-1} + GDB$

1.28. $PCRED = PCRED_{-1} + PCRED$

1.29. $\Delta PCRED = \Delta TCRED - \Delta GCRED$

D. Real side

1.30. $gy = a_0 + a_1 gk + a_2 gPoP$

1.31. $K = K_p + K_g$

1.32. $K_p = K_{p-1}(1-d) + Ip$

1.33. $Kg = Kg_{-1}(1-d) + Ig$

1.34. $Ip = f(PCRED)$

where
AID=Foreign aid in cedis
\$AID=Foreign aid in US dollars
COG=Govt. borrowing from the Central Bank
CUDT=Customs duty earnings
DDEBT=Domestic debt
e_0=Official exchange rate (cedis/US dollars)
e_p=Parallel exchange rate (cedis/US dollars)
e=Weighted average exchange rate
ED=Excess demand
EXPT=Export tax revenue
EXPQ=Export volume
F=Foreign exchange in parallel market
F^T=Transactions demand for foreign exchange in parallel market
F^P=Portfolio demand for foreign exchange in parallel market
Fs=Supply of foreign exchange
PDEBT=Foreign debt
GBB=Government borrowing from the Central Bank
GDEF=Fiscal deficit
GFB=Government foreign borrowing
GDB=Government domestic borrowing
GEXP=Government expenditure
GREV=Government revenue
GOEXP=Other government expenditure
GOVI=Government investment (nominal)
GDEXP=Other government expenditure
gy=Growth rate of real output
gk=Growth rate of capital stock
gPoP=Growth rate of population
IMPQ=Import volume
INTF=Interest on foreign debt
INTD=Interest on domestic debt
i=Domestic interest rate on six-month deposits
if=Foreign interest rate
Ip=Real private investment
Ig=Real public investment
K=Capital stock
Kp=Private capital stock
Kg=Government capital stock
Lg=Government labor growth
M=Money supply (M2); includes 30-day deposits
Md=Money demand
mc=Costs of intermediate imports
P=CPI
P_1=CPI for traded goods
P_2=CPI for nontraded goods
P_3=CPI for controlled goods
PCRED=Credit to private sector (real)
Pf=Import price index
Px=Export price index
r=Real official exchange rate
TCRED=Total credit (real)
tm=Customs duty rate
tx=Export tax rate
U=Mark-up
wp=Unit labor costs
wg=Government wage level
WBILL=Government wage bill

Figure 5-2. Comparative Statics with Devaluation of the Official Exchange Rate

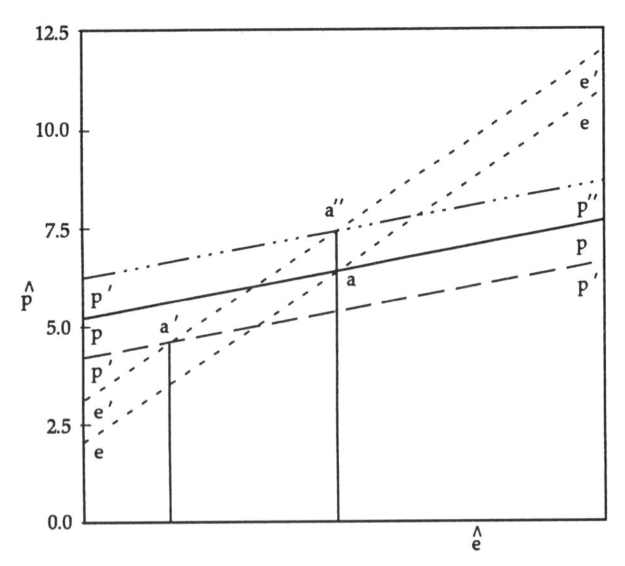

Source: Author's calculations.

supply of goods and foreign exchange and both *P* and e_p decline. The direction of movements to equilibrium in quadrants III and IV is self-explanatory.

What happens if the official exchange rate is devalued? There are two effects that could potentially go in opposing directions. First, with an official devaluation there is a direct cost-push effect which shifts *PP* upwards to P^1P^1.

In addition, there are budgetary effects which affect private asset holdings. If the budget deficit improves with an official devaluation and the rate of monetary creation declines, then the *ee* curve shifts inwards to *e'e'* due to a declining wealth effect. The P^1P^1 curve also shifts inwards as the lower excess demand reduces the average level of mark-up in the economy. How far P^1P^1 shifts depends on the size of the wealth effect on demand and on the mark-up in comparison with the higher cost-push effect of a devaluation (figure 5–2). As long as the cost push effect of the official devaluation is not larger than P^1P^1, then e_p, the rate of parallel market depreciation, unambiguously declines, but the net effect on is an empirical question that depends on the relative magnitudes of different variables. This empirical question will be explored in the simulation model that follows. On the other hand if the official devaluation leads to a larger budget deficit both *ee* and *PP* shift outwards and e_p and rise unambiguously.

The Empirical Estimation of the Model

The detailed empirical model with all the equations and identities is presented in table 5–1. The model is divided into four blocks: inflation, the parallel exchange rate, monetary and fiscal aspects, and the real side. Most of the relationships in table 5–1 are self-explanatory, but those that are not are discussed below.

Block A: Inflation

The approach to modelling excess demand in the goods market was to use Walras' law to assume that excess demand in the goods market is equivalent to excess supply in the money market. This implies that the substitution between money and goods is far more important than that between money and other financial assets. This is a plausible assumption in a developing economy such as Ghana with a relatively shallow financial system characterized by administered interest rates and a limited number of financial assets.

The excess supply of money is specified as the log difference of real money supply to real money demand (equation 1–6). A standard money demand function is hypothesized where Y is real income, *i* is

Table 5–2. ***Inflation equation with wage variables***

Equation	*Constant*	$e_o + P_f$	*Log (M/P-1)*	*Log y*	*i*	*W*	W_g	*p-1*	CHi^2	*DW*
2.1	18.8822	0.0887	0.3939	-1.9721	0.0372	0.3447		0.4551	10.09	1.83
	(1.64)	(0.28)	(2.34)	(1.76)	(1.92)	(0.64)		(1.85)		
2.2	13.6146	0.4812	0.4420	-1.5027	0.0315		-0.1185	0.4404	12.45	1.90
	(1.47)	(2.04)	(3.90)	(1.68)	(1.00)		(0.26)	(1.65)		

All equations were estimated by two stage least squares using PCGIVE. The instruments used were lagged values of inflation, output, money supply, interest rates, import prices at the official and parallel exchange rates, the parallel market premium, the wage gap, and real public sector wages.

e_o: Change in official exchange rate (cedis/US dollars)
P_f: Change in foreign price index in US dollars
M: Money supply (M2)
y: Real GDP
i: Six-month deposit rate
W: Real wage minus productivity
W_g: Real public sector wages
p-1: Lagged inflation

the rate of interest on deposits, and e is the expected rate of inflation.

Changes in wages can have inflationary consequences when they exceed the growth in labor productivity, i.e. a "wage gap" exists. In order to ass- ess this channel for the acceleration of inflation, an index of nominal wages was constructed.

Regressions of real wage growth on a constant, a time trend, and a lagged dependent variable always resulted in insignificant coefficients, implying that it was not plausible to assume that productivity followed a time trend in the Ghanaian case. Instead, it was necessary to analyze labor productivity more directly in order to evaluate the inflationary consequences of the wage gap. The proxy used for labor productivity was the growth rate of real output per capita (*gypc*). The following equation was estimated:

$$\underset{}{WGPG} = \underset{(0.84)}{-0.05}\ \text{CONSTANT} + \underset{(1.83)}{1.81}\ (gypc).$$

The residual that resulted from this regression was defined as the "wage gap." The coefficient on labor productivity is not significantly different from unity.

Table 5–2 contains the results of econometric estimates of the reduced form inflation equation for Ghana. All of the regressions are two stage least squares estimates. The model performs well with all of the variables significant and appropriately signed with the exception of unit wage costs. This is not surprising in Ghana given that a very small share of the labor force is in the formal sector and unionized and that there has been tremendous downward pressure on wages in recent years.

An alternative argument might be that it is public sector wages that rise at a faster rate than productivity and therefore fuel inflation. Given the difficulties in measuring public sector productivity, it has been assumed to be constant so only the growth rate in public sector wages has been included in the inflation equation. The results reported in equation 2.2 of table 5–2 also confirm that public sector wages did not contribute to the inflationary process in Ghana. This is consistent with a declining trend in real public sector wages over much of the high inflation period.

Equation 3.1 in table 5–3 drops the wage variables as there is no evidence that wages were an important factor in overall price determination in Ghana. The resulting equations are very robust with significant coefficients and diagnostic statistics. Expected inflation was specified using both perfect foresight and adaptive lagged inflation. The empirical results indicate that lagged inflation worked better.

Inflation and Exchange Rates

The cost-push literature posits that the relationship between the exchange rate and inflation operates through the inelastic demand for imports and inelastic supply of exports that characterize many developing economies. Because of such rigidities, a devaluation implies cost increases for importers without a concommitant rise in the income of exporters, at least in the short run.

Econometric evidence on the relationship between inflation and exchange rates is presented in table 5–3. Equation 3.1 in table 5–3 shows the significance of the import price index at the official exchange rate (e_o+P_f). Equation 3.2 includes only the parallel market version of import prices (e_p + P_f) along with the basic inflation equation. Again, the coefficient is highly significant. These results would seem to imply that since both the official and parallel market rates matter, the overall rate of inflation is a weighted average of changes in both exchange rates since the market is segmented and clearing at two different prices.

In order to test this hypothesis, a grid search was conducted using varying values of k to determine

Table 5–3. ***Inflation equations, 1965-1988***

Equation	*Constant*	$e_o + P_f$	$e_p + P_f$	*Log (M/P-1)*	*Log y*	*i*	*p-1*	*CHI²*	*DW*
3.1	14.3193	0.2909		0.4036	-1.5513	0.0390	0.3209	22.62	1.80
	(2.07)	(2.03)		(3.80)	(2.30)	(2.97)	(2.17)		
3.2	8.8233		0.1649	0.3566	-1.0049	0.0258	0.5872	69.60	1.39
	(1.97)		(4.52)	(6.90)	(2.32)	(3.25)	(5.22)		
3.3	9.1805	0.0691	0.1474	0.3832	-1.0533	0.0281	0.5596	42.74	1.35
	(1.80)	(0.39)	(1.60)	(5.31)	(2.12)	(2.86)	(3.32)		

All equations were estimated by two stage least squares using PCGIVE. The instruments used were lagged values of inflation, output, money supply, interest rates, import prices at the official and parallel exchange rates, the parallel market premium, the wage gap, and real public sector wages.

e_o: Change in official exchange rate (cedis/US dollars)
e_p: Change in parallel market exchange rate (cedis/US dollars)
P_f: Change in foreign price index in US dollars
M: Money supply (M2)
y: Real GDP
i: Six-month deposit rate
p-1: Lagged inflation

whether the overall fit of the inflation equation improved with varying weights on a composite foreign price variable. The result was that the fit deteriorated as the relative weight of the official exchange rate increased. This was true even when the sample period for the grid search was restricted to the pre-1983 period. This is an indication that consumer prices reflected the parallel market price of foreign exchange even during the period prior to the introduction of the auction. This is also confirmed by equation 3-3 in which both the official and parallel market exchange rates are included in the inflation equation and the parallel rate is far more significant. The inflation equation was also estimated recursively and the results of the one-step Chow tests indicated that the parameters were stable at the 1 percent confidence level over the entire sample period. To summarize, the results on the determinants of inflation indicate that the relevant cost-push variable is the parallel market exchange rate.

Block B. Exchange Rate Premium

A dynamic variant of the theoretical model in Section 2 is estimated to identify the factors which determine the differential between the official exchange rate and the parallel market exchange rate. The demand for foreign exchange in the parallel market is divided into transactions demand and portfolio demand (equation 1.9). The transactions demand depends on the exchange premium and the degree of price controls (equation 1.10). In determining their asset portfolio demand, agents choose between holding Ghanaian cedis and foreign exchange. The relative shares of cedis and foreign exchange in the public's asset holdings will depend on the relative returns between cedis and foreign exchange. This will depend on domestic and foreign interest rates and the rate of expected depreciation in the parallel market (equation 1.11).

The supply of foreign exchange to the parallel market will depend on the exchange premium, the real official exchange rate, and the difference between prices in the open market and controlled prices. The resulting empirical model which determines the parallel market exchange rate is based on the supply and demand factors for foreign exchange in the parallel market. This model posits that the parallel market premium depends on the real effective official exchange rate, the depreciation-adjusted interest differential and the stock of real money balances in the economy.

This model was estimated for Ghana using the cedi/dollar exchange rate in the parallel market on the left hand side and the real official exchange rate, the depreciation-adjusted interest rate differential between Ghana and the United Kingdom, and the real stock of cedi assets on the right hand side. The resulting estimate is reported as equation 4.1 in table 5-4. The model shows that the coefficient of the relative yield variable is significant but that of real money balances is insignificant. The significance of the relative yield variable implies that as long as real interest rates in Ghana diverge widely from world rates, there will always be a parallel market premium. Exchange rate unification therefore implies eventual opening up of the capital account so that there is legal arbitrage between Ghanaian cedis and foreign assets. This result highlights the important linkages between exchange rate management and domestic interest rates. Although a number of other factors, such as confidence and credibility, offset movements in the premium, the interest rate constitutes an accessible policy instrument for influencing the exchange rate premium.

The real exchange rate and the real stock of domestic assets are insignificant in equation 4.1. The equation performs much better when a dummy variable is included for 1978, the year in which the cedi's link

Table 5-4. *Exchange rate premium equations, 1965-1988*

Equation	*Constant*	*r*	*Ry*	*M2P*	*Dum 78*	*PVAR*	*CHI²*	*DW*
4.1	-0.5157	-0.0004	0.9992	0.0040			2.87	2.38
	(0.74)	(0.17)	(2.56)	(0.63)				
4.2	-0.0303	-0.0012	1.1486		-1.8476		8.54	2.33
	(0.20)	(0.85)	(4.00)		(4.01)			
4.3	-0.4874	-0.0038	1.3173		-2.4904	1.8525	13.96	1.73
	(2.81)	(2.87)	(5.67)		(6.21)	(3.56)		

All equations were estimated by two stage least squares using PCGIVE. The instruments used were lagged values of the exchange rate differential, the real exchange rate, relative yields, price variability, inflation, and foreign exchange as a proportion of imports.

r: Real exchange rate index (cedis/US dollars)
Ry: Relative yield between cedis and UK Stg.
DUM 78: Dummy variable for 1978
PVAR: Index of monthly variation in prices
M2P: M2/CPI

to the US dollar was severed and the official exchange rate was devalued. The resulting estimates are reported as equation 4.2 in table 5–4.

If there is imperfect substitutability between Ghanaian cedis and foreign exchange, the exchange rate premium is also affected by uncertainty about future exchange rates, inflation and government policies. While it is not possible to quantify much of this uncertainty adequately, the effects of price instability have been explored empirically. A variable representing price volatility over time was constructed (PVAR) using monthly consumer price indices. In periods of rapid price changes, there is greater pressure on the authorities to devalue which results in a reduction of the exchange rate differential. The inclusion of PVAR also results in the real exchange rate becoming significant. Because of the major changes in the foreign exchange regime, equation 4.4 was also estimated recursively to analyze the stability of the coefficients. As with the inflation equation, the one-step Chow tests indicated parameter stability at the 1 percent confidence level over the sample period. These test results imply that, despite the policy reforms that have occurred, some of the underlying economic relationships have been captured in the model.

In the case of Ghana, the differential between the official and parallel market exchange rates was so large that the share of transactions that occurred through official channels decreased steadily over time. The spread between the official and parallel market exchange rates is depicted in nominal terms in figure 5–3. Because of this leakage to smuggling and parallel market activities, the efficacy of the overvalued official exchange rate as an export tax diminished. The Ghanaian government's strategy was one of a gradual reduction of the fiscal deficit combined with a relaxation of exchange rationing and large, discrete devaluations. The reduction in the fiscal deficit was necessary to reduce demand in the economy and thereby reduce the rate of inflation and the rate of depreciation of the parallel exchange rate.

Block C: Monetary-Fiscal effects

Monetary growth is modelled conventionally as a function of the money multiplier and the growth of

Figure 5-3. Parallel and Official Nominal Exchange Rates, 1965-87

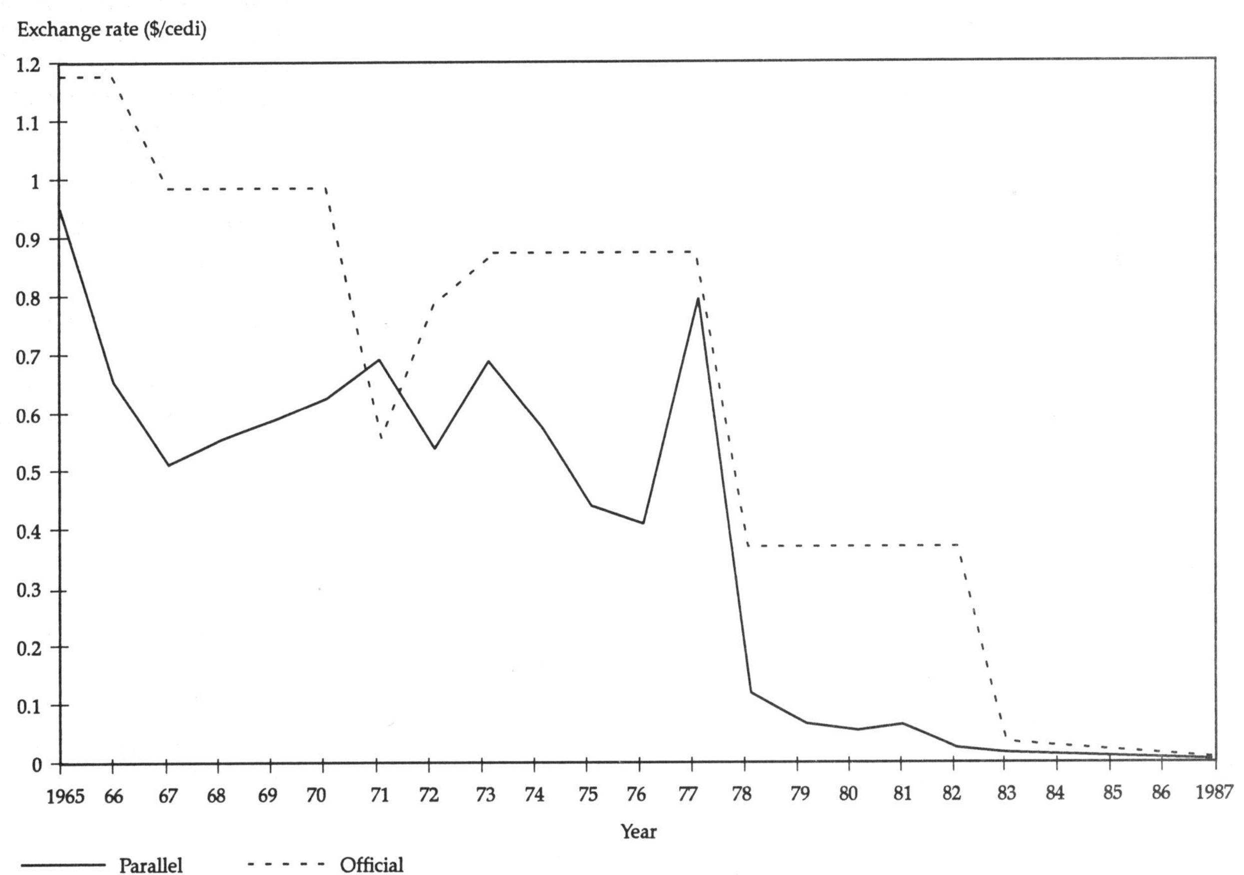

Source: World Bank data.

reserve money, which in turn is affected by the size of the fiscal deficit and the manner in which it is financed. The fiscal block is included in the model to incorporate the effect of price changes on various elements of public expenditure and revenue. Fiscal revenues are a function of the exchange rate, taxes on foreign trade, and output. Government expenditures are disaggregated between wages, subsidies, interest payments, public investment, and other government expenditure. These components of government expenidture depend, in turn, on variables such as the exchange rate and output. The size of the fiscal deficit is thus the endogenous outcome of government policy. The identities in the model are self-explanatory. For the behavioral equations the following empirical results were observed.

Other government revenues, which are essentially non-trade tax receipts, were modelled as a function of nominal GDP:

$$(5\text{–}1)\quad \log(GOREV) = -6.1962 + 1.2793 \log(GDP)$$
$$(6.41\ )\quad (16.15)$$

DW = 1.74; *Chi2* = 5273.6; *TSLS*
Instruments: lags of government expenditure and of GDP.

Government investment is treated as an exogenous policy variable. Other government expenditure is modelled as a function of nominal output:

$$(5\text{–}2)\quad \log(GOEXP) = -2.5374 + 0.9516 \log(GDP)$$
$$(8.2\ 0)\quad (34.03)$$

DW = 2.15; *Chi2* = 11955.0; *TSLS*
Instruments: lags of government revenue and of GDP.

Government expenditure on subsidies and transfers reflects the difference between controlled and market prices as well as operating losses of the enterprises and social security payments. Econo-metrically, transfers and subsidies as a share of GDP *(SUB/GDP)* have been modelled as a function of agricultural output *(GDPAG)* and the import price index relative to the GDP deflator *(PMI/YDEF)*:

$$(5\text{–}3)\quad \log(SUB/GDP) = -3.0443 - 0.2452$$
$$(3.72)\quad (3.13)$$
$$\log(GDPAG) + 0.9258 \log(PMI/YDEF)(-1)$$
$$(3.96)$$

DW = 1.76; *Chi2* = 792.14; *TSLS*
Instruments: lags of GDP in agriculture, subsidies as a share of GDP, and import prices as a ratio of the GDP deflator.

The negative relationship between agricultural output and subsidies shows that the government attempted to cushion the effects of food price inflation due to droughts. This cushion constitutes a greater burden for the government subsidy bill when the relative price of imports is greater than domestic prices.

Block D: The Real Side: Investment and Output

In order to explore the consequences of price dynamics for the real side of the economy, a simple model of output and investment determination was estimated for Ghana. In addition, an export function is defined to analyze the interaction between exchange rates, the balance of payments and inflation.

Exports are hypothesized to be a function of the real exchange rate, agricultural output, and lagged exports:

$$(5\text{–}4)\quad \log(EXPT) = -3.4206 - 0.2560 \log(r)(-1)$$
$$(0.94)\quad (1.67)$$
$$+ 1.0286 \log(GDP(-1)\,AG)$$
$$(1.35)$$
$$+ 0.7978 \log(EXPT)(-1)$$
$$(3.38)$$

DW = 2.02; *R-2* = 0.58, *OLSQ*

The equation indicates a relatively low short run elasticity of exports of about 0.26 with respect to the real exchange rate. This is not surprising given the long gestation periods associated with some of Ghana's major exports—cocoa and forestry products. Note also that the long-run elasticity of real exports to the real exchange rate in Ghana is about 1. This is a very plausible number and close to that obtained in numerous studies on export elasticities.

Output is hypothesized to be determined by a Cobb-Douglas production function. For empirical estimation, it was assumed that the participation rate was constant over the period so population growth could serve as a proxy for the growth of the labor force. The empirical estimate of this production function using TSLS was:

$$(5\text{–}5)\quad \log(Y/L) = 1.7052 + 0.4481 \log(K/L)$$
$$(6.44)\quad (7.47)$$

DW = 0.68; *Chi2(B = 0)* = 27313.41; *TSLS*
Instruments: lags of the capital-labor ratio, output-labor ratio, and deviation of output from trend.

The capital-labor ratio is significant and appropriately signed. The coefficient implies that the share of capital in output is approximately 45 percent.

A number of different models of investment determination were tested on the data for Ghana. Because of the numerous shocks to the economy, the absence of any real growth during much of the period, and the existence of considerable uncertainty, the private investment rate was fairly volatile. In addition, the data for private investment, which is derived as the difference between aggregate investment, which is estimated largely on the basis of imports, and government investment, may not be very reliable. In particular, the private investment series displays a sharp drop in 1976 that is difficult to explain. The following simple equation provides the best explanation of the determinants of private investment over the period:

$$(5\text{–}6)\ PRIVID = \underset{(3.50)}{9.7888} + \underset{(3.08)}{0.5785}\,(PCRED) - \underset{(4.93)}{21.2118}\,(DUM76)$$

$DW = 1.46$; $CHi2 = 97.38$; $TSLS$
Instruments: lags of Private Investment and of Private Credit.

This implies that investment is a function of the real quantity of credit available to the private sector and a dummy variable to capture the unusual collapse of investment in 1976. The irrelevance of the interest rate and government capital formation to private investment determination was verified econometrically. The quantity of credit reflects the rationing that occurs under administered interest rates. The elasticity of private investment to real private credit is about 0.58. This implies that a ten percent reduction in real credit to the private sector will result in a 5.8 percent decline in real private investment. This is a very sizable impact and has obvious implications for the design of a program oriented towards the recovery of private investment. However, the availability of credit is only a partial explanation since banks held excess liquidity during the period of credit rationing. The recovery of private investment in Ghana will depend in part on alleviating the financing constraint, as well as on improving the allocation of credit and addressing an array of issues related to private sector confidence.

Did the Official Devaluations Accelerate Inflation?

The econometric results presented in the previous section showed that a 1 percent devaluation leads to about a 0.14 to 0.16 percent immediate increase in inflation. The long run effects are even larger because of the important role of lagged expectations. To recapitulate the lagged inflation variable was highly significant with a coefficient of about 0.5. This implies that if the short run effect of exchange rate devaluation is about 0.15 the long run effect would be twice that or about 0.30.

Figure 5-4a. CPI with Slower Official Devaluation 1981-88 (1980 = 100)

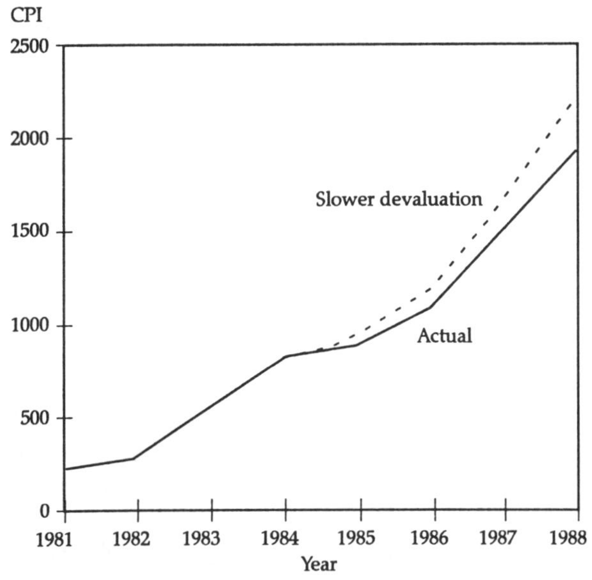

Source: Author's calculations.

Figure 5-4b. Parallel Market Exchange Rate with Slower Devaluation, 1982-88

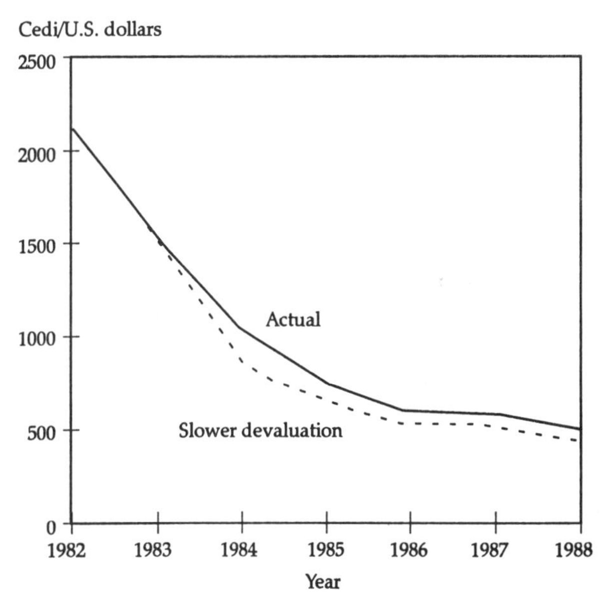

Source: Author's calculations.

These are the effects on inflation of movements in the free market exchange rate. The Ghanaian story is somewhat more complicated because the official rate has been catching up to the parallel market rate and this has had budgetary implications. In Ghana, the devaluation of the official rate led to budgetary improvements in recent years since the government is a net supplier of foreign exchange to the economy. The government's supply of foreign exchange increased as a result of the devalution since a larger proportion of transactions occurred through official channels. In addition, the devaluations reduced the large subsidy to those importers with access to foreign exchange at the official exchange rate.

In order to quantify these effects, the model in the third section was simulated with a slower official exchange rate depreciation of 25 percent in the period from 1984 to 1988. The results of this simulation show that the parallel exchange rate would have depreciated faster (figure 5–4), and the exchange premium would have increased. An additional cost of such a policy would be lower growth as government borrowing to finance the higher fiscal deficit arising from slower exchange rate adjustment would squeeze out private investment.

One would expect lower inflation in the simulation as the parallel exchange rate depreciates more slowly. In fact, because of lower growth, inflation turns out to be higher under this formulation, averaging 29.3 percent per annum as against 27.8 percent per annum in the base case during 1983 to 1988. The net outcome is lower growth, a larger public debt, higher inflation and a more overvalued official exchange rate. This result is somewhat counter-intuitive and is due to three factors: (a) the official devaluations did not have a cost-push effect as prices had already adjusted to the parallel exchange rate; (b) the devaluation had a positive budgetary effect and therefore did not fuel inflation—as a result, a substantial real devaluation took place which narrowed the premium; and (c) relative yields on cedis versus foreign exchange improved because the large official devaluation reduced expectations of future devaluation. In summary, this simulation has shown that the official devaluations could not explain the high and persistent inflation in Ghana since the reforms and particularly since 1986 when devaluations appear to have contributed to a reduction in inflation

Conclusions

This paper has presented a theoretical framework for analyzing the economic consequences of exchange rate reform in the presence of parallel markets. The results address, among other things, a central criticism of adjustment programs in Africa, i.e., that excess-ive reliance on the exchange rate as a tool to bring about relative price adjustments is excessively inflationary. In the case of Ghana, which has carried out one of the most thorough adjustment programs in Africa including very large adjustments in the official exchange rate, high inflation was and continues to be a problem.

However, the recent inflationary surge in Ghana is not due to the exchange reforms. The empirical results for Ghana show that the official devaluations had no direct effect on consumer price inflation. If anything, the official devaluations had a positive budgetary effect which was anti-inflationary. Statistical tests show that both before and after 1983 the relevant exchange rate for pricing decisions was the parallel market price of foreign exchange. In fact, a slower devaluation would have led to higher depreciation of the parallel exchange rate and higher inflation according to our model simulations. The model also explains the evolution of the parallel market premium as a function of the real exchange rate calculated at official prices, interest rate differentials and uncertainty. Devaluation of the official exchange rate leads to a narrowing of the differential. Similarly, reduction in the relative yield through either upward movement of domestic interest rates or through a reduction in the expectation of a devaluation also leads to a narrowing of the exchange rate premium.

The improvement in the budgetary position of the government in response to official devaluation was a crucial element of the ability of the authorities to sustain the exchange unification. An important element of the credibility of the program was the heavy inflow of concessional assistance. This external support strengthened the political hand of those advocating economic reform. This has significant lessons for exchange rate reforms in Africa which can be destabilising where there is not adequate support for the adjustment program. It is useful to contrast the Ghanaian experience with the Zambian reforms where an important factor responsible for the unravelling of the introduction of the exchange rate auction was the underfunding of the program.

References

Azam, J and T. Besley (1989), "General Equlibrium with Parallel Markets for Goods and Foreign Exchange: Theory and Application to Ghana, *World Development*, Vol. 17, No. 12, pp. 1921-1930.

Balassa. B (1985), *Change and Challenge in the World Economy*, MacMillan, London.

Blejer, M. and M. Khan (1984), "Government Policy and Private Investment in Developing Countries," *IMF Staff Papers*, volume 31.

Bruno, M. (1978), Exchange Rates, Import Costs and Wage-Price Dynamics," *Journal of Political Economy*, volume 86, pp. 379-404.

Bruno, M. and J. Sachs (1985), *Economics of Worldwide Stagflation*, Cambridge, Massachusetts: Harvard University Press.

Chhibber A. and S. van Wijnbergen (1988), "Public Poliicy and Private Investment in Turkey", World Bank PPR Working Paper, Washington, D.C.

Chhibber, A., J. Cottani, R. (1989), Firuzabadi and M. Walton, "Inflation, Price Controls and Fiscal Adjustment in Zimbabwe," PPR Working Paper, The World Bank, Washington, D.C.

Chhibber, A., and N. Shafik (1990), "Exchange Reform, Parallel markets and Inflation in Africa: The Case of Ghana", PRE Working Paper number 427, The World Bank, Washington, D.C.

Corbo, V. (1985), "International Prices, Wages and Inflation in an Open Economy: A Chilean Model," *Review of Economics and Statistics*, 68.

Dornbusch, R. (1986), "Special Exchange Rates for Capital Account Transactions," *World Bank Economic Review*, volume 1, number 1.

Dornbusch, R. (1988), "Exchange Rate and Inflation", The M.I.T. Press, Cambridge, Mass. and London, England.

Edwards, S. *Real Echange Rates, Devaluation and Adjustment: Exchange Rate Policy in Developing Countries*, The M.I.T. Press, Cambridge, Mass and London, England.

Fardi, M. (1991), "Zambia's Adjustment Program" in Thomas, Chhibber Dailami and de Melo, ed., *Structural Adjustment and The World Bank*, forthcoming, Oxford University Press.

Fischer, S. (1981), "Relative Shocks, Relative Price Variability, and Inflation," *Brookings Papers on Economic Activity*,.

Gordon, R. (1975), "Alternative Responses of Policy to External Supply Shocks," *Brookings Papers on Economic Activity*, volume 1.

Greene, J. (1989), "Inflation in African Countries: General Issues and Effect on the Financial Sector," IMF Working Paper.

Herbst, J. (1990), "Exchange Rate Reform in Ghana: Strategy and Tactics," mimeograph, Princeton University.

Kalecki, M. (1971), *Selected Essays on the Dynamics of the Capitalist Economy: 1933-1970*, Cambridge University Press.

Khan, M. (1980), "Monetary Shocks and the Dynamics of Inflation," *IMF Staff Papers*, volume 27, number 2.

Khan, M. and J. Lizondo (1987), "Devaluation, Fiscal Deficit, and the Real Exchange Rate," *The World Bank Economic Review*, volume 1, number 2.

Killick, T. (1972), "Price Controls, Inflation and Income Distribution: The Ghanaian Experience," Harvard University Center for International Affairs, Economic Development Report no. 223.

Killick, T. (1978), Development Economics in Action: A Study of Economic Policies in Ghana, New York: St. Martin's Press.

Leechor, C. (1991), "Ghana's Adjustment Program" in Thomas et.al. *Strucutral Adjustment and the World Bank*, Oxford University Press.

Pinto, B. (1989), "Black Market Premia, Exchange Rate Unification and Inflation," *The World Bank Economic Review*, Volume 3, Number 3.

Rocha, R. (1989), "The Black Market Premium in Algeria," mimeograph, The World Bank, Washington, D.C.

Shafik, N. (1990), "Modeling Investment Behavior in Developing Countries: An Application to Egypt," PRE Working Paper number 452, The World Bank, Washington, D.C.

Sundararajan, V. and S. Thakur (1980), "Public Investment, Crowding Out, and Growth: A Dynamic Model Applied to India and Korea," *IMF Staff Papers*, volume 27.

6

The Macroeconomics of the Unofficial Foreign Exchange Market in Tanzania

Daniel Kaufmann and Stephen A. O'Connell

An active parallel foreign exchange market has existed in Tanzania since the early 1970s. The characteristics of the market have varied over time in response to economic shocks and the evolving policy regime, with saving and portfolio decisions featuring importantly in some periods and illegal trade transactions in others. At present, the market is very extensive, with both trade and financial transactions playing important roles.

This paper analyzes the macroeconomics of the parallel foreign exchange market in Tanzania. It focusses on the following questions: What factors led to the emergence of an unofficial market? What factors determine the premium between the unofficial and official exchange rates? What are the linkages between the unofficial foreign exchange market and the rest of the economy? The answers to these questions have important implications for macroeconomic management in Tanzania and provide essential background for consideration of such current policy issues as unification of the foreign exchange markets.

While the exact size of the unofficial foreign exchange market is (by its very nature[1]) difficult to judge, a recent policy measure legalizing one key dimension of the market provides an unambiguous indication of current orders of magnitude. Starting in mid-1984, individuals with access to unofficial foreign exchange were allowed to obtain import licenses without accounting for the source of their funds.[2] Based on official figures, the own-funds window accounted for roughly 40 percent of import licenses issued in 1988. Unofficial estimates suggest that the share of own-funded imports in total imports is even larger perhaps significantly exceeding one half.[3] Given these orders of magnitude, it is clearly critical in the current policy context to have an understanding of the parallel foreign exchange market in Tanzania.

The second section begins with a brief summary of major policy developments since Independence. We then provide a detailed overview of macroeconomic developments in Tanzania since 1967, concentrating on the external sector and the evolution of the premium on foreign exchange in the parallel market. In the third section, we specify and estimate a simple empirical model of the parallel premium using annual data from 1966 to 1988. The fourth section summarizes the key conclusions and indicates our agenda for future work.

Economic Structure and Policy: An Overview

Figure 6–1 shows quarterly movements in the parallel premium on the U.S. dollar in Tanzania since 1970 (table 6–1 gives annual observations back to 1966). Data for the unofficial rate are from the *World Currency Yearbook*, supplemented after 1984 by a small survey carried out in Dar es Salaam by Maliyamkono and Bagachwa (1990).[4] The figure also shows the official exchange rate against the dollar; periods of discrete devaluation against the relevant

Figure 6-1. Tanzania: Parallel Premium vs. Official Exchange Rate, 1970:3 - 1989:4

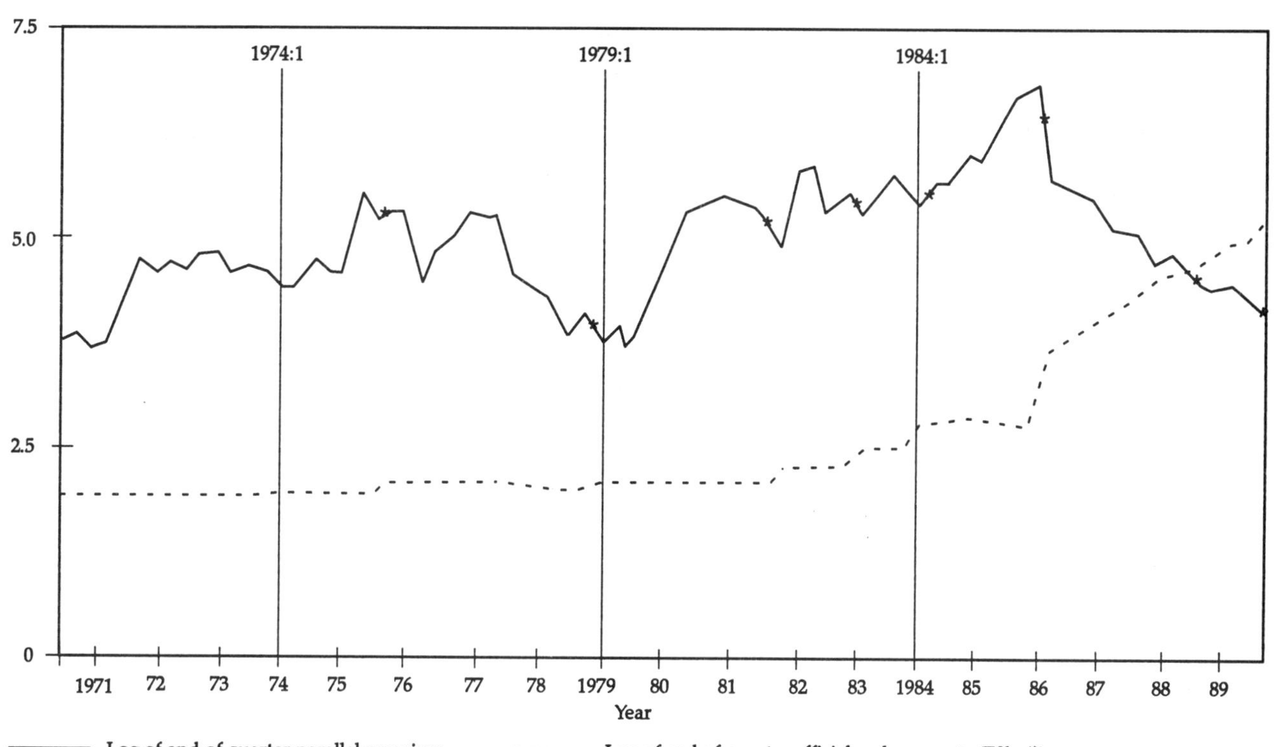

Note: Stars (*) denote discrete devaluations of the official rate.
Source: World Bank data.

currency basket are identified with asterisks. Table 6–2 provides a list of currency realignments over the sample.

The parallel premium shows substantial variations over time in both trend and level. Between July 1970 and March 1986, the premium increased at an average rate of nearly 1 percentage point per month; from April 1986 to the end of the sample (the period of the Economic Recovery Program), it declined at a rate of over 4 percentage points per month.[5] Fluctuations around trend, which are often large and persistent, occur throughout the sample in response to changes in the macroeconomic and regulatory environment.

1967 To 1973

After the Arusha Declaration of 1967, the Government rapidly consolidated its control over all major aspects of the economy. In the external sector, the eight major private import-export firms were nationalized and replaced by the State Trading Corporation; the activities of trade finance were taken over by the National Bank of Commerce after nationalization of the banking sector.

Despite the abrupt and major institutional changes, real GDP per capita grew at an average annual rate of 5.2 in the period from 1967 to 1973. With the implementation of the Second Five-Year Plan (1969 to 1974), the gross investment rate rose to above 20 percent of GDP (from an average of 14.3 percent between 1964 and 1968), and Tanzania achieved significant improvements in the social sectors, particularly in education and health. Structural weaknesses were already beginning to emerge, however, which would become more pronounced and affect future economic performance. Agricultural exports began to stagnate in volume terms in the late 1960s. Domestic savings performance reached a peak of 18.1 percent of GDP in 1970 but fell to 15 percent by 1973, and then dropped as low as 8 to 9 percent of GDP in the crisis years 1974-75. The widening gap between investment and domestic savings was reflected in the external accounts: between 1970 and

Table 6–1. ***Exchange rates and the parallel premium, 1966-1989***

	Official exchange rate		*Unofficial exchange rate*		*Parallel premium*	
	TShs/$				*(percentage)*	
Period	*(avg.)*	*(eop)*	*(avg.)*	*(eop)*	*(avg.)*	*(eop)*
1966	7.14	7.14	8.6	8.6	20.4	20.4
1967	7.14	7.14	8.7	8.8	21.8	23.2
1968	7.14	7.14	8.5	8.3	19.0	16.2
1969	7.14	7.14	8.7	9.1	21.8	27.4
1970	7.14	7.14	10.1	10.5	40.8	46.3
1971	7.14	7.14	11.6	15.0	62.2	110.0
1972	7.14	7.14	15.2	15.4	113.0	115.6
1973	7.02	6.90	14.5	13.5	106.9	94.9
1974	7.14	7.14	13.5	13.0	88.5	96.0
1975	7.41	8.26	20.6	25.0	176.5	202.5
1976	8.38	8.32	21.9	20.4	161.8	145.1
1977	8.27	7.96	21.5	15.1	159.1	89.1
1978	7.69	7.41	13.1	11.8	69.8	58.5
1979	8.25	8.22	12.0	13.5	45.2	64.2
1980	8.19	8.18	21.0	26.5	156.6	223.9
1981	8.29	8.32	27.6	24.4	232.7	192.6
1982	9.33	9.57	32.6	29.2	247.5	204.7
1983	11.26	12.46	39.6	50.0	252.8	301.4
1984	15.51	18.11	55.9	70.0	259.5	286.6
1985	17.35	16.50	100.8	150.0	487.6	809.1
1986	34.26	51.72	165.0	180.0	478.0	248.0
1987	65.62	83.72	180.0	190.0	178.3	127.0
1988	100.39	125.00	210.0	230.0	110.2	84.0
1989	144.47	190.00	254.2	300.0	76.7	57.9

Source: Official exchange rates: IMF, International Financial Statistics, Unofficial exchange rates: *World Currency Yearbook* (formerly *Pick's Currency Yearbook*) for monthly rates from July, 1970 to January,1984; unofficial survey b Maliyamkono and Bagachwa at University of Dar es Salaam for monthly rates from February 1984 to November 1989; our estimate for December, 1989. Unofficial rates before July, 1970 are based on occasional observations reported in *Pick's Currency Yearbook.*

1973, the trade deficit was already 6 or 7 percent of GDP, as compared with balanced trade in the mid and late 1960s.

Under the Currency Board system, balance of payments problems had been virtually nonexistent in East Africa. The currency issue of the East African Currency Board was backed virtually 100 percent by sterling, so that the currency stock moved one-for-one with the sterling reserves of the Board (see Newlyn 1967).[6] Proponents of an independent central bank viewed the introduction of the Bank of Tanzania in 1966 as an opportunity to move to a less conservative monetary policy more in tune with the country's ambitious development program.

Tanzania's balance of payments performed favorably in the first three years of operation of the Bank of Tanzania, with gross reserves rising steadily from 1966 to 1969. Serious pressures first began to emerge in the early 1970s, in response to the rise in internal demand, stagnating exports, and capital flight. Tanzania's first (minor) balance of payments crisis occurred in 1970-71, when international reserves fell by 25 percent between the end of 1969 and the end of 1971. Relative to the rising import bill, the decline in reserves was more drastic, with import coverage dropping from 4 months to less than 2 months over the period. The Government weathered the crisis by tightening import controls, extending exchange controls to Kenya and Uganda, and mobilizing inflows of concessionary financing (Green, Rwegasira and van Arkadie 1980).

The use of direct controls for balance of payments adjustment was consistent with the ongoing transition to socialism and the Government's already-established aversion to exchange rate devaluation;[7] it was institutionalized with the introduction of foreign exchange budgeting in 1970/71. A domestic credit planning apparatus was introduced in the

Table 6–2. *Major party changes since 1970*

Period	Change
July 1973	revaluation 3.4%
January 1974	devaluation 3.4%
October 1975	switch to SDR and devaluation 16%
January 1979	switch to undisclosed basket, and devaluation 10%
March 1982	devaluation 10%
June 1983	devaluation 20%
June 1984	devaluation 36%
June 1986	devaluation 55% and initiation of weekly crawl
November 1988	devaluation 23%
December 1989	devaluation 29%

same year, with the intention of implementing the Government's development priorities and influencing the overall growth of credit.

With the advent of the semi-annual foreign exchange plan, trade policy (as represented by the set of import quotas implicit in foreign exchange allocations) became an endogenous function of foreign exchange revenues. With the exception of a brief period immediately following the Arusha Declaration, the parallel premium in Tanzania was below 30 percent until the 1970-71 mini-crisis. From the experience of other countries, a premium of this magnitude is consistent with the operation of binding capital controls in a stable and otherwise relatively undistorted macroeconomic environment. With the emergence of the crisis, however, the premium moved above 50 percent. By the end of 1971, following the extension of exchange controls to Kenya and Uganda, it exceeded 100 percent. While these short-run movements can plausibly be attributed to intensified desires for capital flight, the premium did not return to its previous low level after the bulk of the nationalizations had been accomplished. This corroborates our observation (see also Green, Rwegasira and van Arkadie (1980)) that although the 1970-71 mini-crisis was successfully contained by short-run measures, the crisis itself was an indication of the emergence of more fundamental imbalances.

1974 To 1978

The 1974 to 1978 subperiod began with the country's first serious balance of payments crisis, brought on by two years of drought and the first oil shock. The crisis exposed some of the longer-term weaknesses in economic performance. Large current account deficits (averaging 14 percent of GDP) were financed by increased aid and capital inflows in 1974 and 1975, as the government froze wages and restricted imports other than oil and food in an attempt to manage the short-term situation. International reserves fell by nearly 70 percent in 1974, and for most of 1975 covered only about three weeks of imports.

The Government's management of the 1974-75 crisis represented a conscious decision not to sacrifice the development program in the face of adverse circumstances (Green, Rwegasira and van Arkadie (1980). In practice, this meant increased reliance on aid and capital inflows. Pressures on the external accounts were eased dramatically in 1976-77 with the recovery of domestic food production and the arrival of the coffee boom. By the end of 1977, reserves were at the unprecedented level of nearly 5 months of imports. Fiscal pressures were eased as well, since the Government chose to tax away most of the windfall in export proceeds.

The weak underlying external situation emerged dramatically in 1978, however, when the government loosened import constraints in response to the boom-related inflow of foreign reserves (as it had done in 1973).[8] As imports expanded dramatically, the coffee boom collapsed; the current account deficit rose to above 15 percent of GDP in 1978 and gross reserves fell by nearly $200 million over the course of the year. External arrears appeared for the first time in 1978.

The parallel premium fluctuated dramatically over the 1974 to 1978 period, rising to above 250 percent by the end of 1975, and nearly as high again in the first half of 1977, and then falling sharply starting in the third quarter of 1977. These movements reflect a number of macroeconomic influences, including increased savings incentives associated with the temporary coffee boom revenues (see Bevan, Collier and Horsnell 1989) and the foreign aid inflows and import liberalization in 1978. The 1977 breakup of the East African Community was a further influence; in that year, the Tanzanian Government closed its border with Kenya, which probably raised the cost of illegal trade between the two countries substantially.[9]

1979 To 1984

This was a period of fiscal and external crisis and cumulative economic collapse. Soon after the ill-fated import liberalization of 1978, the economy was hit by the second OPEC oil price increase and the onset of war with Uganda. In contrast to the balance of payments crises of 1970-71 and 1974-75, when Tanzania managed to maintain consumption per capita through increased inflows of external capital, adjustment to the economic shocks of 1978-79 eventually required a substantial cutback in both consumption and aggregate investment.

The macroeconomic collapse that unfolded in the first half of the 1980s has been analyzed carefully by Bevan et al (1990), Ndulu (1987), and others. Starting in 1979, the government tightened import controls severely, while at the same time raising producer prices for export crops in the hope of improving export incentives. The import compression fell particularly strongly on intermediates and consumption goods, in line with government and donor priorities which remained strongly geared towards the Basic Industrial Strategy objective of increasing manufacturing capacity (and over 1978 to 1981, defense). The compression of intermediate imports drove down capacity utilization in the manufacturing sector; together with the direct compression of consumer imports, this produced a severe shock to the supply of consumer goods.

In an exchange-control regime without domestic price controls or government control of internal trade, a reduction in the supply of consumer goods would be equilibrated by a rise in the domestic relative price of imports and import substitutes (and probably a rise in the real consumption rate of interest, provided the shock were viewed as temporary). In Tanzania, where price controls were pervasive [10] and a "confinement" policy restricted most domestic and foreign trade operations to selected parastatal agencies, the consequence of import compression was the emergence of widespread shortages of consumer goods, particularly in rural areas. Moreover, since the government resisted devaluing the exchange rate, the rise in producer prices meant substantial losses by the exporting parastatals and a corresponding increase in the public sector borrowing requirement and in inflationary pressure.[11] Under the combined pressure of shortages and falling real producer prices, peasants retreated into subsistence production (Bevan et al 1987, 1990) and, to a limited degree, increased smuggling of the export crop.

The collapse of recorded exports in the early 1980s was dramatic: exports declined by roughly 10 percent between 1979/80 and 1981/82, and then by a further 20 percent in 1983. Against this background of macroeconomic collapse, the parallel premium increased dramatically throughout the third subperiod, with only minor and short-lived interruptions in response to devaluations of the official rate. Given the key role of shortages of consumer goods in exacerbating the collapse, it is important to ask why shortages were not averted by inflows of illegal imports financed by unofficial foreign exchange and sold at market-clearing prices. Surveys conducted by Bevan et al (1990) document that especially in rural areas, goods could not in fact be obtained even through illegal channels. One reason is that the elaborate system of controls on distribution meant that the costs of avoiding detection were extremely high. A second factor is that the activity of smuggling is transport-intensive and therefore subject to the serious deterioration of infrastructure that occurred beginning in the late 1970s. Illegal activity was further discouraged by the 1983 "economic saboteurs" campaign during which a large number of businessmen were jailed (Maliyamkono and Bagachwa 1990).

A concomitant of the shortages that emerged starting in 1979 was that the monetary expansion of that year did not immediately push prices up; instead, the velocity of money fell by 40 percent in 1979, and then stayed at the lower level until the introduction of the own-funds scheme and domestic price decontrol in 1984. Although the unofficial exchange rate did not immediately reflect the expansion of real money balances (one would expect a depreciation in the presence of portfolio substitution between domestic money and unofficial foreign exchange holdings), it did begin to rise dramatically by the end of 1979.

During the early 1980s a number of attempts at policy reform (e.g., small devaluations) failed to address the key problems and did not always obtain the necessary political support.[12] The 1984/85 budget represented a turning point, and provided the first indication of a major shift towards pragmatism in the Government's economic management. The exchange rate was devalued by one-third, parastatal subsidies were cut, an import liberalization program was initiated through introduction of the own-funds import scheme, an export retention scheme was introduced allowing exporters to retain a portion of their proceeds to purchase imports, and restrictions on the movement of grain were eased. Simultaneously, cooperatives (which had been abolished in 1976) were reestablished, and took over many of the functions of the parastatal crop authorities.

1985 To Present

This was a period of regime change and gradual recovery. The period from 1985 to the present is one of gradual economic recovery coinciding with a sutained liberalization of economic policy. In mid-1986, the Government produced a medium-term "Economic Recovery Program." The Economic Recovery Program aimed at achieving a positive growth rate in per capita income, reducing the rate of inflation, and restoring a sustainable balance of payments position. Its main thrust was to reduce distortions and encourage more efficient resource allocation while exercising fiscal and monetary restraint. In the public sector, rehabilitation of the transport infrastructure and support for agricultural production were identified as the most urgent priorities. The measures initiated at the time of the 1986/87 budget and continued thereafter include: (1) significant adjustments in the official exchange rate; (2) increases in interest rates, resulting in positive real interest rates by 1988; (3) increases in producer prices for export crops; and (4) a significant reduction in the number of price-controlled items.

Both GDP per capita and trade volumes began to rebound in 1986 after reaching their lowest points in 1985. Real GDP growth averaged 4 percent from 1986 to 1989, with even higher growth evident in the extensive informal sector. The most visible source of growth has been the agricultural sector, where overall production increased between 4 and 5 percent in both 1987 and 1988, reflecting continued increases in production of foodgrains and some traditional export crops.

The cornerstone of the Economic Recovery Program has been the adjustment of the Tanzanian currency. Although the devaluation of mid-1984 was substantial, it did not represent a fundamental change in the Government's approach to exchange rate management. The parallel premium continued its twenty-year rising trend, reaching over 700 percent in early 1986, perhaps in anticipation of the major devaluation and policy shift that accompanied the 1985/86 budget and agreement with the IMF and World Bank. Since March, 1986, the premium has gradually fallen, reaching roughly 50 percent in the first half of 1990, a level not experienced since the early 1970s and briefly following the coffee boom. While a mild premium (e.g., below 30 percent) can be expected to persist reflecting capital controls, convertibility of the exchange rate for current account transactions now appears to be a realistic option.

The Parallel Premium: Some Empirical Results

The chronology presented above suggested that a variety of forces were at work in determining the parallel premium in Tanzania. In this section, we present some empirical results and suggest directions for further work on the determination of the parallel premium.

One of the key questions emerging from the second section is the relative importance of trade and portfolio factors in the determination of the parallel premium, both in the short run and over time. In table 6-1, we address this issue using static and dynamic versions of the Dornbusch, et al (1983) model for the parallel premium. The model builds on two basic relationships. The first is a flow equation in which the change in private holdings of unofficial foreign exchange, dF (the unofficial trade balance), is a function of the incentives for illegal trade, including the parallel premium z, the official real exchange rate *RER*, and other variables w:

$$(6\text{–}1) \qquad dF_t = f(\underset{+}{z_t}, \underset{+}{RER_t}, w_t)$$

We provide a detailed rationale for an equation like (6–1) below.

The second equation is a portfolio equilibrium condition in which the allocation of financial wealth between domestic assets M and unofficial foreign exchange is a function of the uncovered interest parity differential. Letting the notation ${}_ty_{t+1}$ denote the expected value of y_{t+1} conditional on information available at time t, the portfolio balance condition is

$$(6\text{–}2) \quad M_t = g(i^*_t + {}_tdlnU_{t+1} - i_t, x_t)(M_t + u_tF_t), \; g'_t < 0$$

where i^* and i are the foreign and domestic nominal interest rate, respectively, and x is a vector of other variables affecting portfolio behavior. The uncovered interest parity differential measures the difference in expected yields between dollar and TSh-denominated assets (not including the expected penalty, if any, associated with holding illegal foreign assets); a rise in this differential lowers the desired share of domestic bank deposits and other TSh-denominated assets in the overall portfolio. Other influences on relative yields, or on the relative liquidity or risk of domestic and foreign assets, are captured by x. An increase in penalties for violations of exchange or capital controls, for example, would simultaneously reduce the expected yield and decrease the liquidity associated with illegal foreign exchange holdings; at the same time, it might well increase the riskiness of dollar assets. The overall effect would be to lower g for any value of the interest parity differential.

Using the identity ${}_t1nz_{t+1} = {}_t1nU_{t+1} - {}_t1nE_{t+1}$ where U and E are the unofficial and official exchange rates, respectively, equation (6–2) yields the following dynamic equation for the parallel premium:

$$(6\text{–}3) \quad {}_tdlnz_{t+1} = h(\underset{-}{M_t / E_t}, \underset{+}{F_t}, \underset{+}{z_t}, x_t) - (i^*_t + dln_tE_{t+1} - i_t)$$

Equations (6–1) and (6–3) form a second-order dynamic system in which the parallel premium and the private stock of foreign exchange evolve together in response to current and anticipated movements in the real exchange rate, the domestic asset stock (measured in foreign exchange), the interest parity differential, and the other flow and stock determinants, w and x.[13]

It is apparent from equation (6–1) that for fixed values of the right-hand side variables, the model has a steady state in which the parallel premium is a function only of the flow determinants RER and w (simply set $dF_t = 0$). On the other hand, the portfolio determinants, by equation (6–3), clearly affect the parallel premium in the short run. We therefore estimate the following dynamic specification that allows for separate short and long run effects of both groups of determinants:[14]

$$(6\text{–}4)\quad lnz_t = a_0 + a_1 lnz_{t-1} + a_2\, d\,(M/E)_t + a_3 dIPD_t + a_4 dRER_t + a_5 (M/E)_{t-1} + a_6 IPDEV_{t-1} + a_7 RER_{t-1}$$

Table 6–1 gives the results of OLS and instrumental variables estimation of equation (6–4). The data are given in table 6–2.[15] We use the *ex post* interest parity deviation as a proxy for the expected deviation, and apply instrumental variables to handle the implied measurement error.[16] For the value of domestic assets, we use $M2$ (= Currency + Demand Deposits + Time Deposits). The real exchange rate is a trade-weighted index of bilateral real exchange rates with the eight major trading partners.

Given the short sample and the uncertain quality of the data, the basic results (columns 1 and 3) are quite satisfactory. They give strong support to the conclusion that both trade and portfolio factors are at work in determining the premium on unofficial foreign exchange in Tanzania. The short-run effects of the various determinants, given by a_2 - a_4, are all of the expected signs: a rise in the interest parity deviation or an increase in the real value of domestic financial assets leads to portfolio substitution towards unofficial foreign exchange, raising the premium; a real appreciation shifts incentives away from export smuggling and towards import smuggling, raising the premium. An appreciation of the real exchange rate raises the parallel premium in both the short run and the long run, as predicted by the model; moreover, since a_4 and $a_7 > 0$, an unanticipated shock to the real exchange rate produces an "overshooting" of the parallel premium in the short run.

The results also support the conclusion that nominal devaluations are capable of lowering the parallel premium to the extent that they lower the share of domestic financial assets in private portfolios or depreciate the real exchange rate. In both cases, the results clearly indicate the need for complementary macroeconomic policies, since the effect of a nominal devaluation can be nullified by increases in nominal money or domestic prices.

With respect to the long-run effects of the portfolio factors, the results are mixed. We cannot reject the null hypotheses that a_4 and a_5 are simultaneously zero, using standard F-tests. In this sense, the resu- lts support the prediction that portfolio factors influence the parallel premium in the short run only, and that the premium is determined by flow factors alone in the long run.[17] Taken separately, however, it appears that while changes in real money balances have no effect in the long run (i.e., a_5 is negligible), changes in the interest parity deviation do have a cumulative effect over a two-year horizon. And when we drop the long-run portfolio effects from the regression (columns 2 and 4), the overall performance deteriorates noticeably. Caution is clearly appropriate in interpreting the results regarding dynamics.

While the hypothesis of no serial correlation of the residuals cannot be rejected at the marginal 5 percent significance level based on the Box-Pierce statistic, both the Durbin-Watson (which is biased towards 2 given the lagged dependent variable) and the Box-Pierce statistic suggest that further work on the dynamic specification and/or estimation with a serial correlation correction may be in order.

Extending the Analysis

The discussion above emphasized that in an exchange control regime, the parallel premium is jointly determined with the domestic price of imports, since the marginal supply of imported goods enters the country through smuggling and underinvoicing. A rise in the official allocation of foreign exchange, for example, will increase the total supply of imports, thus reducing the domestic price of imports relative to the world prices and depreciating the real exchange rate; simultaneously, it will reduce the value of illegal imports, reducing the flow demand for foreign exchange in the parallel market and lowering the parallel premium.

To incorporate these considerations, we first specify flow equilibrium in the parallel foreign exchange market more carefully and then take care of endogeneity of the real exchange rate. The flow demand for foreign exchange comes from two sources: (1) directly smuggled imports, M^s, or imports through the own-funds window, M^{own}; and (2) imports brought in through the official window, but underinvoiced to avoid payment of tariff. In the underinvoicing case, we denote the officially reported value of imports and the amount by which these imports are underinvoiced by V^{off} and V^{u}, respectively. In a strict foreign exchange budgeting regime, V^{off} is not a choice variable, since it equals the official allocation of foreign exchange for imports. Note V^{u} need not be positive; imports may be overinvoiced as a way of obtaining official foreign exchange for sale on the parallel market.

To derive the flow demand for foreign exchange, consider first the case where there is no own-funds scheme, so that the marginal supply of imports into the economy is through smuggling and underinvoicing. Suppose (1) that the marginal cost of smuggling a unit of imports rises with the amount smuggled, and is denominated in the good being smuggled (cf. Bhagwati and Hansen (1974)); and (2) that the cost of underinvoicing is denominated in foreign exchange and is an increasing function of the distance of $v = V^{u}/V^{off}$ from zero (i.e., an increase in fraud in either direction raises marginal costs; cf. Macedo (1987)). Then letting w^{s} and w^{u} be variables like the government enforcement effort that increase the marginal costs of smuggling and underinvoicing, respectively, the flow demand for foreign exchange for illegal imports takes the form.[18]

$$(6\text{–}5) \qquad FD = P^{*}_{m} M^{s} + V^{u} = p^{*}M^{s}(\underset{+}{q},\underset{-}{z},\underset{+}{t_m},\underset{-}{w^{m^{s}}}) + v^{u}(\underset{+}{q},\underset{-}{z},\underset{+}{t_m},\underset{-?}{w^{u}_{m}})V^{off}$$

$$= FD(\underset{+}{q},\underset{-}{z},\underset{+}{t_m},\underset{+?}{V^{off}},\underset{+}{P^{*}_{m}},\underset{-}{w^{s}_{m}},\underset{-?}{w^{u}_{m}}),$$

where $q = (P_m - EP^{*}_{m})/EP^{*}_{m}$ is the premium of the domestic price of imports over the world price at the official exchange rate, t_m is the import tariff rate, and $z = (U-E)/E$ is the parallel premium.

For smugglers or for underinvoicers with fixed individual foreign exchange allocations, the optimal illegal behavior embodied in equation (6–5) is straightforward: smuggle and/or underinvoice those goods whose price on the domestic market is high enough to offset the parallel premium (i.e., those goods for which $q > z$), and overinvoice the rest (i.e. those for which $q < z$).[19]

Notice that the tariff rate does not affect the demand for foreign exchange of these agents, since it does not affect their marginal incentives for illegal activity once q and V^{off} are given. This is obvious for smugglers, who avoid tariffs altogether; for underinvoicers who receive a fixed allocation of foreign exchange, the value of reported imports (and therefore the base of the import tariff) is fixed in advance, so that changes in tariffs affect overall profits but not the optimal degree of underinvoicing.

We include the tariff rate as an argument in the underinvoicing function to capture the fact that while the overall allocation of official foreign exhange may be determined in advance, individual traders may view themselves as having some influence over their own allocation of foreign exchange. For these traders, cost-minimizing behavior requires trading off the marginal benefit of an additional dollar of underinvoicing, which is t_m, against the marginal cost, which is z plus the marginal increase in expected penalties. Underinvoicing will occur on those goods for which t_m exceeds z; if t_m is less than z, the good will be overinvoiced. Similar behavior is exhibited by individuals who illegally import either a given quantity of goods (e.g., a single car or machine tool) or goods without well-developed markets within the country, such as specialized spare parts. Cost minimization yields an amount of smuggling and underinvoicing that is an increasing function of the tariff rate and a decreasing function of the parallel premium (and zero for $t_m = z$).

Finally, notice that the effect on *FD* of an increase in officially recorded imports is uncertain, since it depends on the sign of v, i.e., on whether imports through the official window are being underinvoiced or overinvoiced on average. If the average domestic price premium q is below the parallel premium, however, the overall incentive will be to *overinvoice*, so that v is negative; in this case a rise in V^{off} will decrease the flow demand for foreign exchange, and a rise in $w_m u$ will increase it.

The flow supply of foreign exchange, *FS*, comes from direct export smuggling in amount X^{s} and from underinvoicing of exports officially reported to the authorities, X_u. The amount of underinvoicing of exports is given by $X^{u} = x^{u}(t_x, x)[X(RER_x, ODA) - X^{s}]$, where $X(RER_x, ODA)$ is total export supply as a function of the real official exchange rate for exports and the level of official development assistance. We use the latter as a proxy for the demand for exports of housing services and tourism, two major channels of illegal exports in Tanzania.[20] Both X^{s} and x^{u} depend positively on the parallel premium and on the gap between the domestic price of exports and the world price at the official exchange rate, t_x. Again letting w^{s} and w^{u} be variables like government enforcement efforts that increase the marginal cost of smuggling and underinvoicing, respectively, we have:

$$(6\text{–}6)\quad FS = P^*{}_x\{x^s(t_x,a,w^s_x) + x^u(t_x,z,w^u_s)\{X(t,z,w^u)$$

+ + -   + + -

$$[X(RER_x,ODA) - X^s(t_x,z,w^s{}_x)]\}$$

-   +

$$= FS(t_x,z,RER_x,ODA,P^*_x,w^s_x,w^u_x)$$

+ + -   + + - -

Notice that there are two sources of increases in t_x, given the world price of exports: (1) decreases in producer prices paid by marketing boards without a change in the exchange rate (e.g., through higher overhead margins being charged by marketing parastatals, or through a policy decision to tax exports more heavily); (2) devaluations in the official exchange rate that are not passed on to producers. Notice also that in Tanzania, the relevant "world price" for key export commodities may in fact be the producer price being paid in neighboring countries (e.g., the producer price for arabica coffee in Kenya).

Taking equations (6–5) and (6–6) together gives us the following version of equation (6–1) (we consolidate the w's into a single vector for notational convenience):

$$(6\text{–}7)\quad z = z(q,t_m,t_x,RER_x,V^{off},ODA,P^*_m,P^*_x,w,FS\text{–}FD)$$

+ + - + -? - + -   +

The real exchange rate for exports, RER_x, is the ratio of the consumer price index to the price received by exporters (we use PP_x, the producer price for agricultural exports):[21]

$$(6\text{–}8)\quad RER_x = \frac{P^{\alpha}_m P^{1-\alpha}_n}{PP_x} = \left[\frac{1+q\alpha}{1-t_x}\right]\alpha\,(P_n/EP^*_x)^{1-\alpha}$$

$$(P^*_m/P^*_x)^{\alpha} = RER_x(q,t_x,P_n/EP^*_x,P^*_m/P^*_x),$$

+ + +   +

Equation (6–7) can therefore be written

$$(6\text{–}9)\quad z = z(q,t_m,t_x,P^*_n/EP^*_x,P^*_m,P_x,ODA,V^{off},w,S\text{–}FD)$$

+ + ? + + - - -? +

The effect of a rise in t_x is uncertain a priori, since there are two opposing effects: the real exchange rate for exports rises, reducing aggregate export supply, while the share of exports that is diverted onto unofficial channels also rises. The net effect on illegal exports, and thus on z, is an empirical question.

Equation (6–9) contains four potentially endogenous variables: q, P_n/EP^*_x, V^{off}, *and* $dF = FS\text{-}FD$. Consider the relative prices first. Both q and P_n/EP^*_x are relative prices of goods that are nontraded at the margin; these prices are determined by overall absorption, A, and by supply conditions. In the case of nontradeds, an upward-sloping supply curve comes from standard general equilibrium considerations (resources must be attracted from other sectors, including the export sector, to increase the supply of nontradeds); for the case of imports under exchange control, the marginal supply of imports comes from smuggling and underinvoicing, so that q is a function of t_m and z as well as A.[22] Substituting for q and P_n in equation (6–9), we have

$$(6\text{–}10)\quad \{z = z_m(t_m,t_x,M^{off},A,P^*{}_m,P^*{}_x,ODA,w,FS\text{–}FD)$$

+ ? - -? + + - - +

Finally, we need models for A, V^{off} and $FS\text{-}FD$. In the Tanzanian context of foreign exchange rationing, V^{off} is a policy variable. We can therefore either take it as exogenous or specify a "reaction function" in which the amount of foreign exchange allocated for imports depends on other variables. Based on actual Tanzanian experience as outlined in the second section, one possibility is to have V^{off} endogenously respond to the reserves position in the previous period, and to current official exports and aid receipts:

$$(6\text{–}11)\quad v^{off}_t = f(V^{off}_{t-1},(R_{t-1}-R_{t-2}),P^*_x(x-x^s),ODA$$

+ + + +

where R_{t-1} is reserves at the end of period $t\text{-}1$.[23] In this case, equation (6–11) becomes

$$(6\text{–}12)\quad z = z(t_m,t_x,V_{t-1},R_{t-1}-R_{t-2},A,P_m,P_x,ODA,w,FS\text{–}FD)$$

+ ? -? -? + + - - +

Equation (6–12), together with (6–13), provides a rich structure for incorporating flow determinants of the parallel premium and analyzing the linkages of the parallel foreign exchange market with the rest of the economy. Intertemporal considerations, for example, enter through the determination of A: a commodity boom that raises desired absorption will tend to raise the parallel premium by driving up the domestic relative price of imports and raising the profitability of import smuggling; on the other hand, it will (i) increase the foreign exchange value of any given volume of smuggled exports, and (ii) rapidly feed into higher allocations of foreign exchange through the official window, with the opposite effect on the premium. The net effect on the parallel premium over time is an empirical question, depending largely on the degree to which the commodity boom is perceived as temporary. In the case of the 1976-77 coffee boom, which was clearly the effect of a temporary supply shock (a frost affecting the Brazilian crop), we would expect a very mild effect on aggregate expenditure, and therefore, given the time path of the domestic money stock and official exchange rate, a tendency for the parallel premium to be driven down by the valuation effect on smuggling volumes and the endogenous trade liberalization.

Aid Inflows. With respect to an increase in Official Development Assistance (ODA), holding desired absorption constant, the effect should be to lower the premium, both by increasing the flow supply of foreign exchange onto the parallel market and by producing an endogenous trade liberalization. On the other hand, increased aid should raise disposable income and therefore desired absorption, again depending in part on how permanent the increase is expected to be; this would tend to raise the domestic relative price of imports and the parallel premium. As in the case of a commodity boom, the net effect is an empirical question.

A major weakness of equation (6–12), in combination with (6–3), is that it leaves out the government budget constraint. It therefore misses the endogenous determination of the domestic money stock and the official exchange rate. This is an important potential direction for extensions. In the commodity boom case, for example, the commodity revenues would be received by the private sector primarily in domestic currency, since the bulk of the export crop is marketed through official channels. In the absence of sterilization, portfolio balance considerations would then tend to push the unofficial exchange rate up in the short run, counteracting the effect of the (actual and anticipated) trade liberalization.

Own funds. Extending the derivation of an equation like (6–12) to the case of an own-funds scheme is relatively straightforward. Since the costs of direct smuggling are positive, imports that were previously

Table 6–3. ***OLS and instrumental variables estimation results for equation (6–4)***

	Dependent variable: Parallel premium (PPREM)[1]			
	OLS		*IV*[2]	
	1967-1988		*1968-1988*	
	1	2	3	4
CONSTANT	191.38	-259.94	-195.95	-285.31
	(-2.27)	(-2.69)	(-1.39)	(-2.47)
PPRT$_{T-1}$EM	0.38	0.45	0.71	0.40
	(1.49)	(2.03)	(1.41)	(1.60)
d(M2/$_t$E)	0.19	0.24	0.22	0.27
	(2.58)	(2.82)	(1.55)	(2.32)
d(IPD$_t$)	2.45	1.20	1.91	0.89
	(2.16)	(1.09)	(0.76)	(0.58)
d(RE$_t$R)	3.32	0.80	6.13	0.48
	(2.24)	(0.69)	(2.03)	(0.37)
(M2/E$_{t-1}$)	-0.05	—	-0.12	—
	(-1.40)	—	(-1.60)	—
IPD$_{t-1}$	3.63	—	4.97	—
	(3.03)	—	(2.74)	—
RE$_{t-1}$R	2.46	3.08	2.56	3.17
	(2.78)	(3.09)	(1.84)	(4.26)
RBA^{3}R	0.86	0.79	0.80	0.78
(Q11)[3]	-14.73	16.76	8.67	14.78
	(0.20)[4]	(0.12)[4]	(0.56)[4]	(0.14)[4]

Note: (t-statistics are in parenthesis) a.The data are in table 6–2. b. Instruments for d(REER)$_t$ and d(IPD)t are M2$_{t-2}$, REER$_{t-2}$ and IPD$_{t-2}$ (along with the other right-hand side variables, which are assumed to be predetermined; note that in the case of PPREM$_{t-1}$, this is only valid if the disturbances are serially uncorrelated). c. Q is the Box-Pierce statistic for testing general serial correlation. For columns 3 and 4, the statistic reported is Q(10).

own-funds scheme.[24] Ignoring the cost of operating in the (still illegal) foreign exchange market, and assuming that expected penalties for underinvoicing are an increasing function of the ratio of underinvoicing to reported imports, there will be a perfectly elastic supply of imports through the own-funds window at the price $q - t^e = z$, where t^e is the effective tariff paid on own-funds imports.[25] The amount of own-funds imports will then be determined residually, as the difference between total import demand M, and the amount of imports brought in through the official window: $M^{own} = M - (V^{off}+V^u)/P^*_m$. We therefore have

$$(6\text{–}13)\ \{FD = P^{*m}M^{own} + V^u = P^*_m M(A, z, t_e, y) - V^{off},$$

$$= FD(\underset{+}{A}, \underset{-}{z}, \underset{-}{t_m}, \underset{-}{w^u_m}, (V^{off}/P^*_m), \underset{+}{y}),$$

where A is aggregate expenditure, y is a vector of other variables influencing the aggregate demand for imports (e.g., government absorption of imports), and V^u (recall) is underinvoicing through the official window. The statutory tariff rate t_m and the enforcement variable v_{mu} enter as determinants *of* t^e, and z enters through the arbitrage relationship $q - t^e = z$. The resulting reduced form can be written exactly as in (6–12), although the parameters will differ reflecting the change in the supply function for imports financed by unofficial foreign exchange.

One conclusion that emerges unambiguously in the current model is that introduction of an own-funds scheme will raise the parallel premium consistent with any given value of the private current account surplus *FD-FS*, and therefore that it must raise the steady state parallel premium, ceteris paribus (since in the steady state *FD-FS* is fixed at zero). The reasoning is straightforward: by reducing the costs of import smuggling, an own-funds scheme shifts out the supply of imports and drives down the gap between the domestic price of imports and their international price at the official exchange rate (i.e., q falls). Total imports are therefore higher under the

Table 6–4. *Data for parallel premium regressions*

Period	*PPREM*	*M2E$*	*IPDOP*	*REER*	*TAXCINV*	*TOT*	*AID*
1966	20.40	190.54	n.a.	97.61	49	96.8	81.49
1967	23.20	215.56	1.96	99.14	50	91.9	70.29
1968	16.20	253.93	2.86	105.99	47	93.0	68.71
1969	27.40	277.39	6.26	104.91	71	112.3	117.32
1970	46.30	310.74	5.02	101.22	53	106.5	105.52
1971	110.00	364.41	3.08	97.09	54	98.8	136.10
1972	115.60	432.56	0.17	91.83	75	95.3	191.48
1973	94.93	529.42	7.13	88.03	39	118.8	170.00
1974	96.00	624.69	11.21	92.94	35	107.9	148.87
1975	202.53	671.95	16.93	99.60	35	92.8	332.93
1976	145.07	834.52	0.25	93.02	27	126.3	331.65
1977	89.07	1048.58	-5.44	92.94	23	140.3	393.36
1978	58.46	1267.25	11.72	93.11	40	115.1	404.55
1979	64.21	1679.35	6.26	87.03	34	114.9	535.91
1980	223.89	2141.34	10.37	100.00	52	100.0	546.61
1981	192.59	2486.09	25.55	131.93	52	85.2	496.25
1982	204.70	2584.86	29.04	158.73	57	88.4	405.17
1983	301.40	2338.33	43.44	171.82	52	91.1	318.69
1984	286.63	1668.01	16.39	177.94	53	96.4	310.80
1985	809.13	2387.37	103.98	206.61	50	90.5	424.92
1986	248.04	972.73	94.50	113.12	43	103.0	447.41
1987	126.95	793.62	42.36	174.52	50	89.6	704.92
1988	84.00	702.24	33.79	49.83	46	94.3	818.79
1989	57.89	n.a.	n.a.	n.a.	n.a.	n.a.	n.a.

Note: n.a. not available. PPREM=100*(U-E)/E is the end-of-the-year parallel premium in percentage points, with the unofficial and official exchange rates U and E taken from table 2–1. M2E$ is the end-of-the-year M2 in TShs (source: IMF, IFS), deflated by the official exchange rate. IDPOP=100*[(1 + i)(E_{t+1}/E_t)-(1 + i)] is the uncovered interest parity differential, with i given by the London Eurodollar deposit rate (source: IFS) and i by the Saving deposit rate in Tanzania (source: Bank of Tanzania). REER is the ratio of the Tanzanian CPI to a trade-weighted average of WPIs of 8 major developed country partners (source: World Bank).

own-funds scheme than previously, for a given supply of official foreign exchange. Smuggled exports must therefore rise to finance the higher imports if the private current account surplus is to remain unchanged. The only way this can occur is for the parallel premium to rise.[26]

Implementing the Extended Model

Table 6–3 shows the results of estimation of a version of the extended model. The parallel premium is regressed on the portfolio determinants (imposing the condition that these operate only in the short run) and on a subset of the flow determinants in equation (6–12). The data are in table 6–4. TAXCINV corresponds to the inverse of t_x in (6–12): it is the ratio of the domestic producer price of coffee to the fob export price (converted to TShs).[27] As discussed above, the effect of export taxes on the premium is theoretically ambiguous; a positive net effect would indicate (i) that coffee growers are actively adjusting the share of the crop sold through official channels in response to taxes and the parallel premium, and (ii) that these adjustments are large enough to have macroeconomic implications.

TOT is the terms of trade; it enters (1) through the endogenous trade liberalization that follows an improvement in the balance of payments; (2) through a direct valuation effect on the illegal trade deficit; (3) through resource movements in favor of exports and away from imports; and (4) through effects on aggregate demand, depending on the savings response. The first three of these would be expected to lower the parallel premium; the third would raise it, to an extent depending on the savings response. Overall, we expect a net negative effect.

AID is net official resource transfers in dollars; a rise in AID should lower the premium both through direct increases in illegal export flows (expatriate housing, etc.) and through the endogenous trade liberalization effect; it should raise the premium to the degree that it raises aggregate demand. Again, we expect a negative effect on balance, although the aggregate demand effect might be rather strong given that changes in aid have a strong permanent component.

Own funds is a dummy variable for the period from 1984 to the end of the sample, during which the own-funds scheme was in operation. By the arguments given above, we expect it to have a positive effect on the premium, given the values of the other variables. Interpretation of the estimated coefficient will be complicated, however, by the fact that the 1984/85 budget, in which the own-funds policy was introduced, was an integrated policy package that included a devaluation of the official rate, decontrol of some prices, and other policy actions. To the degree that the package was perceived as a signal of genuine policy reform, and thus of a prospective increase in the return on domestic real assets, it would have tended to lower, rather than raise, the parallel premium.

Finally, D83 is a dummy variable for the 1983 crackdown on "economic saboteurs"; its effect on the premium is theoretically uncertain. From the illegal trade side, while a crackdown unambiguously reduces the volume of illegal trade, it may either raise or lower the premium, since it simultaneously affects both the supply and the demand for illegal foreign exchange. On the portfolio balance side, a crackdown impairs the liquidity of foreign exchange assets and reduces their expected yield; both effects would tend to lower the parallel premium.

Most of the variables have the expected signs. Both portfolio determinants enter significantly, with magnitudes generally close to those found in table 6–1. Of the flow determinants, only the TOT comes in strongly, with the lagged TOT exerting a strong negative effect on the premium (as observed, for example, during and after the coffee boom). Lagged aid inflows also lower the premium, although the effect is not estimated precisely. The effect of lagged TOT and lagged aid is consistent with a substantial endogenous trade liberalization in response to balance of payments improvements; this corroborates evidence from import equations in Ndulu and Lipumba (1988).

The coefficient on the coffee tax variable is consistently negative but insignificant, implying that any smuggling response is more than offset by an aggregate coffee supply response in the opposite direction. While this finding does not rule out a macroeconomic role for coffee smuggling in determining the parallel premium (cf. Donnelly and Mshomba (1989), who argue that up to 25 percent of the arabica coffee crop has been smuggled to Kenya in some years), it suggests that the elasticity of smuggling supply is low in the coffee sector, at least over the horizon of a year.

The own-funds scheme appears to have raised the premium, ceteris paribus, as predicted by the model. The magnitude of the increase, between 150 and 240 percentage points, is impressive, and suggests that the low elasticity of smuggling response indicated in the coffee case may be a more general phenomenon. More obviously, the results indicate that the lowering of the parallel premium since 1986 has been a function of other developments in policy and external conditions, such as (i) cumulative depreciations of the official exchange rate that reduced the real stock of domestic money, and (ii) large inflows of foreign aid, and not of the own-funds scheme itself.

Implementation of the extended model using annual data is clearly problematic given the short sample. Nonetheless, while the results in table 6–3 should be interpreted with caution, they suggest that there may be significant payoff to further work using equation (6–12).

Conclusions

This paper has taken a macroeconomic perspective on the parallel foreign exchange market in Tanzania. We discussed the emergence and behavior of a premium on foreign exchange over the period since 1966, and presented some preliminary empirical evidence on the determination of the premium. In this conclusion, we briefly discuss selected empirical and policy issues and the agenda for future work.

The results of the third section show that both portfolio and trade factors are at work in determining the parallel premium in Tanzania. They therefore suggest that rises in the premium in the early 1980s may have been in part due to the substantial monetary expansions that occurred starting in 1979. In the context of the model, however, the resulting rise in the premium should have raised the market-clearing domestic prices of imports and import-substitutes. But in this case, real money balances would not have risen as dramatically as they did in 1979, with velocity falling by 40 percent (for both M1 and M2) and remaining at the lower level until 1984 (at which time it equally abruptly returned to trend). Surveys conducted by Bevan, et al (1990) and Ndulu and Hyuha (1989) suggest that while official price indices did not fully reflect transactions prices during the third subperiod, the gaps are not large enough to explain the fall in measured velocity. One of the important issues for further investigation, therefore, is the impact of failure of market clearing on the behavior of the parallel premium and its interaction with other macroeconomic variables.

A second empirical issue has to do with handling the change in regime that occurred between 1984 and 1986. For familiar reasons, a major change in policy regime should be expected to change the parameters of behavioral equations and reduced forms. The short sample, particularly after the regime change, has prevented us from investigating this issue carefully; one possible avenue for future work is to implement simple versions of the model using quarterly data (which are available for prices, interest rates, money stocks and exchange rates).

On the policy side, the key issue regarding the unofficial foreign exchange market is unification. The resource allocation gains of a single an market-determined exchange rate for commercial transactions are well known, and do not require elaboration here. Holding aside the question of capital account liberalization, what are the prospects for adoption of an essentially market-determined official exchange rate for commercial transactions?

First, there is a fiscal impact of unification. Pinto (1987) has emphasized that if the government is a net buyer of foreign exchange from the private sector, unification will worsen the real deficit and raise the inflation rate required to finance the fiscal deficit. Preliminary calculations we have made suggest that Tanzania is on the favorable side of this calculation. Given the large inflows of aid being channeled by the Government, increases in the official exchange rate provide a fiscal bonus, lowering the fiscal deficit in domestic currency terms. Moreover, devaluation will improve the trade tax base, both by moving activity from unofficial channels to official channels (a real devaluation is required here), and by the valuation effect of higher official exchange rate (since import tariffs are levied on the domestic value at the official exchange rate). It therefore appears likely that official exchange rate adjustments in the course of unification will not contribute to inflation via the fiscal channel.

Second, while unification through adoption of an across-the-board floating exchange rate has been tried in other African countries, capital controls are likely to be in place in Tanzania for the foreseeable future. This means that a parallel rate will continue to exist, and that some amount of trade activity will take place at this rate. "Unification" therefore means the removal of rationing of import licenses at the official rate, and the provision of a competitive official outlet for export proceeds, rather than the reduction of the parallel premium to zero. Moreover, unification in this sense is a complex process, involving change in policy institutions and gradual adjustment on the part of market participants. In this context, the parallel rate is likely to play an important allocative and signalling function for a long time to come. Characterizing the nature of this role is an important part of our agenda.

As a final observation, one of the most interesting aspects of the Tanzanian experience, and a potential lesson for other countries similarly situated, is the role of the own-funds scheme. The de facto dismantling of the QR-dominated trade regime through the own-funds policy resulted in a significant inflow of consumer and intermediate imports, providing incentives to farmers to increase production, and channeling scarce spare parts and transport equipment to industry and agriculture. The resulting supply response was significant. This, coupled with the price alignment to reflect scarcity values in virtually all commodities (brought about by the de facto trade liberalization itself and the price decontrol), implied that the official devaluations initiated in 1986 did not result in an acceleration of inflation, even though there was a marked increase in money supply growth during the 1987 to 1989 period. The success of the own-funds scheme suggests that carefully identified policies linking the parallel market for foreign exchange with the official economy can provide a significant supply response and price alignment immediately preceding the adoption of politically controversial structural adjustment efforts.

Notes

1. The parallel foreign exchange market in Tanzania is an illegal or "black" market. We use the terms "unofficial" and "parallel" interchangeably in this paper, although the latter term is broader and in some countries refers to a legal, officially recognized market.

2. Similar schemes have been operated in Ghana and The Sudan in recent years.

3. Ndulu and Hyuha (1989) give three reasons why the share of licenses may underestimate the true share: (1) own-funds consignments under TShs 10,000 (approximately USD 50) do not require licenses, (2) the utilization rate of own-funds licenses is considerably higher than that of licenses accompanied by official foreign exchange; and (3) the incentive to underinvoice own-funds imports to avoid customs duties is much stronger than for officially financed imports, since the cost of foreign exchange on the underinvoiced portion is identical to that on the declared portion.

4. The *World Currency Yearbook* publishes monthly data from July 1970 (with annual observations back to 1966). The series used in this paper is the *World Currency Yearbook* series up to January 1984, and the Maliyamkono and Bagachwa (1990) series thereafter. While the two series move closely together for most of the period since 1984, an exception occurs in 1985, when the *World Currency Yearbook* shows a stabilization of the parallel rate in contrast to the trend depreciation reported by Maliyamkono and Babachwa. We use the latter series based on our own observations during that period, and consultation with other observers, who unanimously regarded the *World Currency Yearbook* data for 1985 as anomalous.

5. The trends are calculated by regressing the monthly parallel premium on a constant and a trend over the period in question. The coefficients on the trend term are 0.0091 for July, 1970 to March, 1986 and -0.041 for April, 1986 to December, 1989.

6. Starting in 1955, the Board was allowed to issue currency against government securities. By the end of 1965, however, 82.5 percent of the currency issue was still backed by sterling.

7. The Tanzanian government decided not to follow the devaluation of sterling in 1967. Arguments given then about the inefficacy of devaluation in the Tanzanian context became an established part of the policy canon until the mid-1980s.

8. A second policy development in 1978 with possible implications for the balance of payments was the amendment of the Bank of Tanzania Act, which abolished the previous limit of 25 percent of recurrent revenue on government borrowing from the Central Bank (Ndulu and Hyuha 1989). It is not clear, however, that this represented a genuine loosening of fiscal constraints, since parastatals had always been able to borrow from the National Bank of Commerce (itself a parastatal), and the National Bank of Commerce from the Bank of Tanzania.

9. The border was re-opened in 1984. The effect of the border closure on the parallel premium is ambiguous, since it should raise the cost of both import smuggling (thus reducing the demand for illegal foreign exchange) and export smuggling (thus reducing the supply of foreign exchange). The theoretical model in the third section captures the effects of macroeconomic and regulatory changes on the parallel premium.

10. The National Price Commission existed from 1973 to 1984. By the mid 1970s, it controlled some 2000 prices of imported and domestically produced goods. Prices were decontrolled starting in 1984; only 12 commodities remained under price control by July, 1988.

11. The fiscal developments largely accounted for the government's failure to meed performance criteria associated with IMF borrowings in the 1979-80. The absence of IMF support meant an additional shock to external funds.

12. Ndulu (1987) gives a chronicle of official responses to the external crisis starting in 1979; there were three unsuccessful short-term government/IMF ventures in 1979-80, and then the unsuccessful government three-year Structural Adjustment Program starting in June 1982.

13. The system is saddle-path stable; in each period, the parallel premium jumps to clear the asset market, and the resulting incentives for illegal trade determine the flow addition to private foreign exchange holdings. Dornbusch et al (1986) and Rocha (1990) give a diagrammatic analysis.

14. See also Rocha (1990) for an application to Algeria, and Fishelson (1988) for an application of the model to 19 developing countries. Fishelson uses the uncovered differential at the unofficial rate, on the argument that movements in the unofficial rate provide a good proxy for expected movements in the official rate. Neither of these

papers makes a distinction between the dynamic effects of the portfolio determinants and those of the flow determinants.

15. The dependent variable actually used in the regressions is the parallel premium, defined as 100*(U-E)/E, where U and E are the unofficial and official exchange rates, respectively, rather than $\ln z = \ln(U/E)$ as defined in the text.

16. Assuming market participants have rational expectations, the forecast error will be uncorrelated with variables observed at time t or earlier. We therefore can use lagged values of the right-hand side variables as instruments for the interest parity deviation. We also instrument for the change in the real exchange rate, on the grounds that the current real exchange rate is jointly determined with the parallel premium. We treat the nominal money stock and the official exchange rate as predetermined.

17. This interpretation of equation (4) reflects a backward-looking interpretation of the dynamic adjustment towards the steady state. In reality, the system formed by (1) and (3) has both a stable and an unstable root, so that the rational expectations solution for z takes the form ${}_t dz_{t+1} = a[z_t - zbar_t]$, $0 < a < 1$, where $zbar_t$ is a function of current and expected future levels of the flow determinants and changes in the stock determinants of the parallel premium. Since future values of these variables must be predicted based on current information, however, the final solution for z would include a distributed lag on current and past values of the determinants, as in equation (4).

18. The exact form of the smuggling function depends on the costs associated with the illegal activity. While our formulation is fairly general, we have made one key simplification in assuming that the costs of smuggling are ultimately denominated in foreign exchange: this implies that only the parallel premium, and not the level of the official exchange rate, will directly affect smuggling incentives.

19. Smuggling profits are equal to $R^s = P_m M^s - U[M^s + C(M^s)]$, where $C(M^s)$ is the smuggling cost function. The first-order condition is $C'(M^s) = (q-z)/(1+z)$, yielding a smuggling function $M^s(q,z)$ with the properties given. Under(or over)invoicing profits are given by $R^u = P_m[(V^{off}+V^u)/P_m^*] - E(1+t)V^{off} - U(V^u+c(v)V^{off}) = \{(q-t) - (q-z)v - (1+z)c(v)\}EV^{off}$ (we have assumed that the invoicing cost function is homogeneous of degree one in V^u and V^{off}). Given V^{off}, the first-order condition for v is similar to the smuggling case: $c'(v) = (q-z)/(1+z)$. This yields the invoicing function $V^u = V^u(q,z)V^{off}$ given in equation (5).

20. We use the real official exchange rate as the relative price influencing export supply decisions under the assumption that peasants continue to voluntarily sell some portion of their export crop to the marketing authorities rather than selling it on the parallel market. In this case, arbitrage between the two markets implies that the domestic price of exportables will equal the producer price. Notice that the same is not true on the import side; there, the relevant marginal price for imports is the unofficial exchange rate (plus the costs of smuggling), because exchange controls act like an import quota and imply that the marginal source of imports is illegal activity.

21. The real exchange rate for exports is usually defined as the domestic price of nontradables relative to exports. We use the CPI relative to the export price to capture the resource pull away from exportables production more comprehensively.

22. The domestic price premium q falls towards t_m, the tariff rate, as the official foreign exchange allocation rises. For sufficiently high tariffs and liberal foreign exchange allocations, the implicit quota on imports becomes nonbinding, and imports become a legally traded good on the margin. We then have $q = t_m$, and aggregate absorption is no longer a determinant of q.

23. One could also make M^{off} an increasing function of q and z, on the argument (suggested earlier) that official foreign exchange allocations may respond to rent-seeking behavior. This does not affect the signs of the derivatives in (10) provided that the effect of q on foreign exchange allocations is not large enough to offset the positive effect of q on z through the incentive to smuggle.

24. Provided, that is, that the authorities credibly commit not to scrutinize the source of funds. This does appear to have been the case in Tanzania. Informal sources suggest that surveillance on the own-funds window has been extremely loose in general.

25. Note that imports brought in through the own-funds window are fully financed at the parallel exchange rate, regardless of the degree of underinvoicing chosen by the importer. The incentive to underinvoice is therefore stronger for own-funds importers than for importers using the official window, since in the latter case an increase in underinvoicing raises the trader's costs by increasing the share of the import that must be financed at the parallel rate. The effective tariff in the own-funds case will therefore be below that on officially financed imports, ceteris paribus. In the Tanzanian case, this difference is exacerbated by the general laxity of surveillance of goods coming in through the own-funds window.

26. Notice the importance of our assumption that the costs of import smuggling are independent of the costs of export smuggling. If import and export smuggling were joint activities, then an own-funds scheme might not drive out all import smuggling. The assumption that smuggling costs are private costs is also important; if the cost of smuggling were a loss of some portion of the good being smuggled, then the own-funds scheme would represent an improvement in the economy's marginal terms of trade and therefore in the productivity of a unit of illegal exports in generating imports, In this case, while an own-funds

scheme would raise total imports, it is an empirical question whether this would require an increase in the flow of illegal exports, and thus in the premium (we are grateful to K. Krumm for pointing this out).

27. We use an average ratio for arabica and robusta, calculated by taking the ratio of payments to producers for the two types of coffee to the total fob export value of the two types. For data availability reasons, we use the advance price for coffee.

References

Azam, Jean-Paul, and Tim Besley (1988) "General Equilibrium With Parallel Markets for Goods and Foreign Exchange: Theory and Application to Ghana", Princeton University Research Program in Development Studies, Discussion Paper #143, December.

Bagachwa, M.S.D., N.E. Luvanga, and G.D. Mjema (1989), "Tanzania: A Study on Non-Traditional Exports (Preliminary Draft)," mimeo, University of Dar es Salaam, September.

Bevan, David, Arne Birgsten, Paul Collier and Jan Willem Gunning (1987), "East African Lessons on Economic Liberalization" Thames Essays no. 48, Trade Policy Research Centre, London.

Bevan, D., P. Collier and Jan Gunning (1990) Peasants and Governments (Oxford, Oxford University Press).

Bevan, D., P. Collier and P. Horsnell (1988), "Supply Response Under Goods Market Rationing in Tanzanian Peasant Agriculture," paper presented at the OECD.

Bhagwati, Jagdish N. and Bent Hansen (1973), "A TheoreticaB Analysis of Smuggling," Quarterly Journal of Economics, May, reprinted in Bhagwati, ed., Illegal Transactions in International Trade (Amsterdam, North Holland, 1974).

Biersteker, Thomas J. (1986), "Self-Reliance in Theory and Practice in Tanzanian Trade Relations," in John Ravenhill, ed., Africa in Economic Crisis (New York, Columbia University).

Cliffe, Lionel and John S. Saul (1972), Socialism in Tanzania: An Interdisciplinary Reader, Vol. 1: Politics (Nairobi, East African Publishing House).

Cliffe, Lionel and John S. Saul (1973), Socialism in Tanzania: An Interdisciplinary Reader, Vol. 2: Policies (Nairobi, East African Publishing House).

Collier, Paul (1988), "Aid and Economic Performance in Tanzania," mimeo.

Collier, Paul and Jan Willem Gunning (1989), "The Tanzania Recovery, 1983-1989", mimeo, November.

Domowitz, Ian and Ibrahim Elbadawi (1987), "An Error-Correction Approach to Money Demand: The Case of the Sudan", Journal of Development Economics.

Donnelly, Lawrence P. and Richard Mshomba (1987), "The Smuggling of Coffee: The Case of Kenya and Tanzania", mimeo, presented at November 1987 annual meeting of the African Studies Association.

Edwards, Sebastian (1990), Real Exchange Rates, Devaluation and Adjustment: Exchange Rate Policy in Developing Countries (Cambridge, Massachusetts, MIT Press).

Fishelson, Gideon (1988), "The Black Market for Foreign Exchange: An International Comparison", Economics Letters 27: 67-71.

Green, Reginald H., Rwegasira and Brian van Arkadie (1980): Economic Shocks and National Policy Making (The Hague, Institute of Social Studies).

Greenwood, Jeremy and Kent P. Kimbrough (1987), "Foreign Exchange Controls in a Black Market Economy", Journal of Development Economics 26: 129-143.

Krumm, Kathie (1987), "Medium-Term Framework for Analysis of the Real Exchange Rate: Application to the Philippines and Tanzania", CPD Discussion Paper No. 1987-9, The World Bank, May.

Macedo, Jorge Braga de (1987): "Currency Inconvertibility, Trade Taxes and Smuggling", Journal of Development Economics 27: 109-125.

Maliyamkono, T. and M.S.D. Bagachwa (1990): The Second Economy in Tanzania (Oxford, Oxford University Press).

Ndulu, Benno J. and Nguyuru H.I. Lipumba (n.d., approx. 1988): "International Tradeand Economic Development in Tanzania", mimeo, Department of Economics, University of Dar es Salaam.

Ndulu, Benno and Mukwanason Hyuha (1989): "Inflation and Economic Recovery in Tanzania: Some Empirical Evidence", Uchumi 2, No. 1, University of Dar es Salaam.

Newlyn, W.T. (1967): Money in an African Context (London, Oxford University Press).

O'Connell, Stephen A. (1990): "Short and Long-run Effects of an Own-Funds Scheme", mimeo, Department of Economics, University of Dar es Salaam, March.

Pinto, Brian (1988): "Black Markets for Foreign Exchange, Real Exchange Rates and Inflation: Overnight vs. Gradual Reform in Sub-Saharan Africa", mimeo, World Bank.

Rocha, Roberto De Rezende (1990): "The Black Market Premium in Algeria", mimeo, World Bank.

Tanzanian Economic Trends, Vol. 1, No. 4, January, 1989 (Economic Research Bureau, University of Dar es Salaam).

7

Niger and the Naira: Some Monetary Consequences of Cross-Border Trade With Nigeria

Jean-Paul Azam

Nigeria is by far the largest economy of West and Central Africa; its macroeconomic behavior affects all the countries in the region. Most of its neighbors belong to the CFA Franc Zone and have a convertible currency, both internally and externally. External convertibility is guaranteed by France, which provides automatic balance of payments support to the two monetary unions, the Union Monétaire Ouest Africaine (UMOA) in West Africa and the Banque des Etats de l'Afrique Centrale (BEAC) in Central Africa. On the contrary, the naira, which is Nigeria's currency, is inconvertible.

This fact, together with various other trade distortions, explains why a parallel market for the naira has developed. Arnoult (1983), Azam (1990a), and Grégoire (1986) describe some of the distortions regarding trade between Niger and Nigeria. These include different tax rates and the outright ban on imports of some goods from some countries. As most trade distortions are impossible to enforce tightly in Africa, there are various illegal and semi-legal transactions involving goods and foreign exchange.

Convertibility makes a difference in how the economy works when there are active parallel markets. When a currency is convertible, its parallel market has to clear at each point in time through price flexibility, i.e. by exchange rate adjustment. Azam and Besley (1989) analyze the parallel market for the cedi, the Ghanaian currency. All the shocks affecting the parallel foreign exchange market are reflected in the exchange rate and hence in the prices of the goods which are traded at this exchange rate. In many African countries where official imports are restricted, most economic agents are affected at the margin by these prices, whereas official prices play only an inframarginal role (Azam and Besley 1989).

By contrast, when the currency is convertible, the parallel market for foreign exchange needs not clear, even if the official exchange rate is flexible, as any surplus or deficit can be exchanged on the official market. Here we neglect the "laundering" costs incurred when the money has been acquired by selling illegal goods. Hence, convertibility, by unifying the foreign exchange market, dampens exchange rate fluctuations by pooling the shocks affecting each segment of the market when the exchange rate is flexible.

In the case of the members of the CFA Franc Zone, the exchange rate is not flexible, and each country can run an external imbalance, at least temporarily. As a result of internal convertibility, any imbalance in the parallel balance of payments will eventually show up in the official balance of payments, as agents buy (sell) the net flow of foreign exchange which they need (have) from (to) the formal banking system.

The contrast between convertible and inconvertible exchange regimes has interesting implications for the analysis of trade between a small CFA Franc Zone member like Niger and its large neighbor, Nigeria, whose currency is inconvertible. Niger's economy has probably run a sizeable surplus in its parallel balance of payments with Nigeria since 1983. We try in the following to offer some explanatory factors for this performance.

We analyze the currency flow below to provide for Niger an indicator of the balance of payments on the parallel market. We show in the second section how Niger's economy can acquire convertible foreign exchange by trading with Nigeria, despite the latter having an inconvertible currency. Although the price of the naira can be regarded as exogenous for Niger, as it does not adjust to clear the foreign exchange market, it is not constant, and its fluctuations have significant macroeconomic effects. The third section focuses on its impact on the consumer price index, and shows that Niger can be regarded as a small price taker. The fourth section shows how the price of the naira affects the demand for cash balances in Niger. It is suggested that the increase in cash balances which occurred during a large part of the 1980s, following the sustained depreciation of the naira on the parallel market, played a favorable role in the improvement of Niger's external balance. The fifth section provides additional evidence in favor of an asset approach to at least partly explaining Niger's parallel balance of payments surplus. The last section provides a conclusion.

Earning Convertible Foreign Exchange on the Naira Market

As the naira is inconvertible, its parallel market has to clear by price adjustment; no central bank intervenes to peg the exchange rate. Hence, from the viewpoint of Nigeria, this market is in equilibrium.

All of Nigeria's neighbors belong to the CFA Franc Zone. The CFA franc is pegged to the French franc; it is thus linked, via the latter, to all the currencies of the European Monetary System and hence to all the main currencies of the world. Therefore the CFA franc price of the naira is determined by Nigeria and by the world economy. No other African economy on its own has a significant influence on this market.

This is illustrated by the following equation, which Azam and Vernhes (1990) have estimated on quarterly data for the period from 1982.1 to 1987.2, using an extension of the model analyzed by Azam and Besley (1989):

$$\text{(7–1)} \quad \log e = \underset{(21.4)}{-41.0} + \underset{(4.93)}{0.97} \log M + \underset{(1.64)}{1.23} \log q + \underset{(5.78)}{5.79} \log p^W - \underset{(1.38)}{0.06} \log Q - \underset{(4.04)}{0.34} \log q^W$$

$$N = 22, R^2 = 0.99, DW = 1.70$$

The number of observations is noted N and R^2 is the usual coefficient of determination. DW is the classic Durbin and Watson test of residuals autocorrelation and the numbers in parentheses below the estimated coefficients are the usual t-ratios. The dependent variable loge is the log of the parallel exchange rate of the naira in Zinder, Niger, expressed in terms of nairas per CFA franc. M is the quantity of local currency held outside the banking sector in Nigeria, as a proxy for cash balances; q is the Nigerian producer price of cocoa; p^W is the consumer price index for industrialized countries; Q is the official imports into Nigeria, in real terms, as a proxy for import quotas; and q^w is the world price of cocoa on the London market. The value of the R^2 shows that nothing much is left to Niger for influencing this exchange rate.

The countries which participate in the naira market can run a surplus or a deficit in their parallel balance of payments with Nigeria, provided the rest of the market participants have a net balance of the same magnitude with the opposite sign. Figure 7–1 illustrates this reasoning. On the left hand panel, the demand (N^{Nd}) for and supply (N^{Ns}) of nairas in Niger are pictured. The former is the sum of imports from Nigeria, investments by residents from Niger in Nigeria, and increases in naira balances held by Nigerien citizens. Its negative slope reflects the fact that an increase in the price of the naira reduces the competitiveness of Nigerian products and reduces the expected capital gains on assets denominated in naira, for a given expected path of the exchange rate in the future. The supply of nairas in Niger is the sum of exports to Nigeria, investment in Niger by Nigerians, and reductions in naira balances held. The positive slope is due to the increased profitability of export to Nigeria when the price of the naira increases and, for given expectations, the reduced attractiveness of naira-denominated assets.

The right-hand panel of figure 7–1 represents the corresponding naira demand and supply curves for the rest of the world, denoted by N^{Rd} and N^{Rs}, respectively. Their slopes are determined by the same arguments as those presented above. Units have been chosen in such a way that the exchange rate between Niger and the rest of the market is equal to one. If the naira market is unified by the arbitrage operations of exchange brokers, the exchange rate clears the market in such a way that any imbalance of the Niger segment of the market is offset by a net imbalance of the same size and the reverse sign in the rest of the market. Evidence reported in Azam (1990a) supports this assumption of arbitrage. Comparing the CFA Franc price of the naira on the parallel market in

Lome, Togo, and Zinder, Niger, using 86 monthly data points between 1980.08 and 1987.12 shows a striking arbitrage performance. Excluding a few outliers during the six months following the closure of the border by the Nigerian government in April 1984, the mean premium on the Lome market is not significantly different from zero. Its value is -0.012 percent, and it has a standard deviation of 2.72 percent. Only seven data points, besides the six post-April 1984 outliers, show a premium larger than plus or minus 5 percent, with a maximum of 7.3 percent in November 1984.

In the case pictured in figure 7–1, there is an excess supply of nairas in Niger, which is offset by excess demand in the rest of the market. If arbitrage did not function properly, the price of the naira would go down in Niger and up in the rest of the market. Assuming perfect competition in the exchange brokers business, any premium appearing in the rest of the world segment of the market triggers a flow of nairas away from the Niger market, until excess demand is cut down to zero everywhere in the market. If we neglect transactions costs and the profits of the brokers, which vanish in a free entry equilibrium, this flow of nairas must be compensated by a flow of currency from the rest of the market in the reverse direction. The market for the naira clears globally. It serves as an indirect channel whereby Niger is connected through arbitrage with the other economies participating in this market. When Niger's parallel balance of payments shows a surplus, Niger can acquire foreign exchange from the international naira market.

In the case of Niger in the 1980s, one can find an indicator of such an inflow of currency via the naira market by looking at the series of banknotes from the BEAC monetary union purchased by the Banque Centrale des Etats de l'Afrique de l'Ouest (BCEAO) in Niger. These CFA francs from Central Africa can be exchanged one for one against the CFA francs from West Africa, without any transactions costs. However, they are not perfect substitutes for the BCEAO banknotes because they are not legal tender. They are not generally accepted in the shops in Niger.

The above analysis of the naira market may be simplified by assuming that the rest of the market comprises only Cameroon. Cameroon has an active border trade with Nigeria and, through it, the other member states of the BEAC zone (see Azam 1990a). The case of figure 7–1, in which Niger runs a surplus and the rest of the market a deficit, would give rise to a flow of nairas from Niger to Cameroon; BEAC CFA francs would flow in the reverse direction. It is reasonable to assume that these flows of currency

Figure 7-1. The Naira Market in Niger and in the Rest of the World

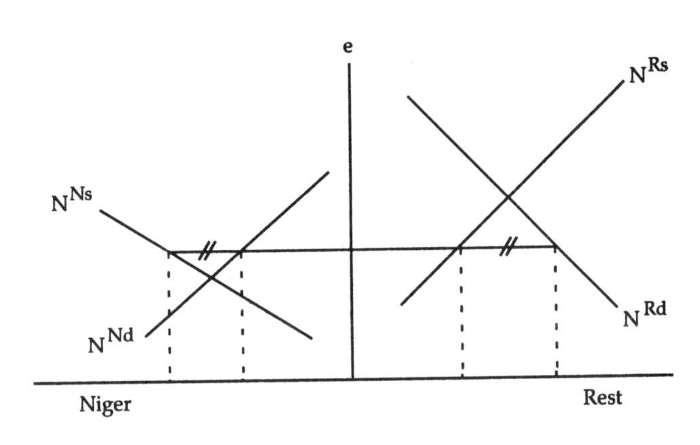

Figure 7-2. Flow of BEAC Banknotes into Niger, 1980-89

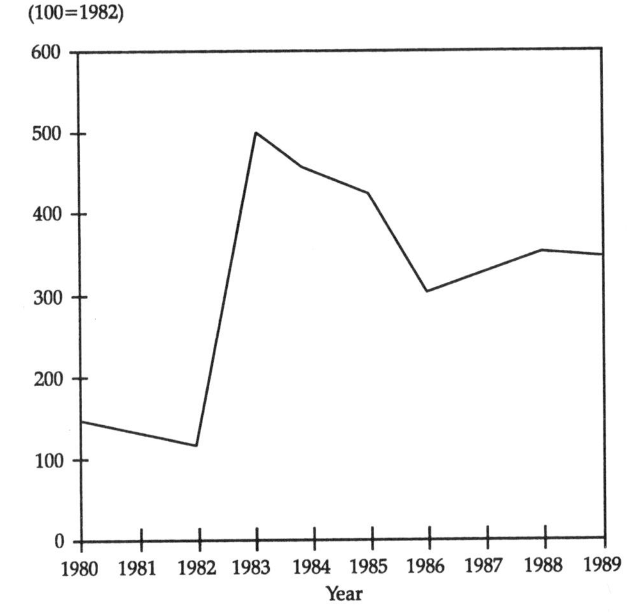

Source: BCEAO.

mainly involve banknotes, as checking accounts are not widely used in this milieu, especially when dealing with illegal transactions.

The resulting net inflow of CFA francs from the BEAC into Niger would eventually be used to acquire BCEAO banknotes from the Central Bank and would then be recorded, thus contributing to the official balance of payments (see Azam 1990b). These banknotes would be acquired by the BCEAO in its three agencies in Niamey, the capital city, and in Maradi and Zinder, the two main trading towns in the Hausa country north of the Nigerian border. On average the latter two towns account for more than

85 percent of this flow. This strengthens the interpretation of this series as an indicator of the balance of cross-border trade between Niger and Nigeria, as Maradi and Zinder are known as the focal points of this trade (Arnoult 1983, Gregoire 1986). Nevertheless, this indicator is somewhat crude and probably involves quite large observation errors (see Azam 1990a for further discussion).

The inflow of BEAC banknotes into Niger during the 1980s is pictured in figure 7–2. There is a downward drift over most of the period and a sharp rise in 1983. The jump in 1983 corresponds probably to a switch from a deficit position in the parallel balance of payments to a surplus. Additional evidence presented in Azam (1990a) supports this explanation. The low volume inflow of BEAC banknotes before 1983 reflects probably a normal level of trade with Chad or Cameroon, while its fivefold increase in 1983 is large enough to indicate a drastic change in the flow of payments.

It is reasonable to assume that the adjustment program launched in 1982 by the Niger government, with the support of the International Monetary Fund and later the World Bank, played a large part in this external improvement (see Guillaumont et al 1988). Real public expenditures were cut by more than 18 percent in 1982 and 1983. Moreover, there was a drought during 1983 to 1985, when peasants and herdsmen were selling their cattle. But it is not obvious why such an improved position carried on until the end of the decade, while public expenditure started to grow again. This is explained in part below, by analysis based on assets holding behavior. But first the relationship between Niger and Nigeria is analyzed in terms of the extent of cross-border market integration. This analysis sheds some light on the way the price level is determined in Niger.

Cross-Border Market Integration

The border between Niger and Nigeria does not divide the Hausa country into two separate markets; there is an integrated market across this border. The price level in Niger is determined by the border price and the tariff rate, and does not depend on monetary policy or aggregate demand management in general. This is consistent with the theory of smuggling costs recently developed by Pitt (1981) and Devarajan, Jones, and Roemer (1989). It is assumed that private firms are engaged both in official trade and smuggling. To keep things simple, let us concentrate on one side of the trading business only, namely imports. We assume that smuggling entails costs described by the function:

$$C(F,G) \quad (7\text{–}2)$$

where F denotes the amount of official transactions and G denotes the amount of smuggling, and with derivatives $C_F < 0$ and $C_G > 0$. The negative impact of official transactions on the costs of smuggling is an assumption due to Pitt (1981) and discussed by Devarajan, Jones and Roemer (1989). It captures the "cover effect" whereby official transactions can cloak illegal ones.

The local price, p, and the border price, p^*, of the imported good are expressed in local currency, and t denotes the rate of ad valorem tax on official transactions. Then the profit function of the smuggling firm is defined as:

$$R(p,p^*,t) = max\,((p-p^*)G + (p-p^*(1+t))F - C(F,G)). \quad (7\text{–}3)$$

This definition entails the following arbitrage equations:

$$0< p - p^* = CG = p^*t + CF < p^*t. \quad (7\text{–}4)$$

This entails that official transactions are made at a loss, if C_F is strictly negative, with the local price below the tax inclusive border price. The rationale is that official transactions have a positive effect on the profitability of illegal transactions. The illegal profit pays for the loss incurred by selling below the tax inclusive cost of the legally imported good.

Assuming that free entry prevails in the smuggling sector, then

$$R(p,p^*,t) = 0. \quad (7\text{–}5)$$

Hence, in equilibrium, we obtain a kind of purchasing power parity relation by inverting equation (7–5). This implies that the domestic price level p is determined by the border price and the tariff rate, and does not depend on domestic monetary policy, etc. By totally differentiating profit function (7–5), the following type of purchasing power parity relationship is obtained:

$$d\,\log p = (1 + at)\,(p^*/p)\,d\,\log p^* + a(p^*/p)\,dt, \quad \text{where } 0< a = F/(G+F) \quad (7\text{–}6)$$

This type of argument, which we have restricted to the case of illegal imports for the sake of simplicity, could be reproduced for illegal exports, mutatis mutandis. Moreover, one could assume the presence of economies of scale in smuggling, so that illegal exports would entail a reduced marginal cost for illegal imports *(see Azam and Besley 1989)*. Hence, provided the smuggling sector is reasonably competitive, with approximately free entry, the prices of traded goods are determined by the border prices, irrespective of monetary and budgetary policy, even if official imports are restricted. Such an assumption seems reasonable for the case of Niger, where nontraded goods do not seem to play a significant part in the African Consumer Price Index (CPI). (The European Consumer Price Index is not considered here.)

The following econometric exercise shows this result quite clearly, using monthly data from 1982.91 to 1989.10:

$$\text{(7–7)}\quad \log p = \underset{(2.29)}{0.67} + \underset{(5.95)}{0.47}\log p(-1) + \underset{(4.95)}{0.28}\log p(-12) + \underset{(5.99)}{0.08}\log ep^* - \underset{(2.40)}{0.25}AR(12)$$

$N = 94$, $R^{2a} = 0.75$, $DW = 1.92$, $F = 70.8$

Here p is the African CPI in Niamey (Niger), e is the parallel market exchange rate in Zinder and p^* is the rural CPI in Nigeria. The one month and twelve months lagged values of p are denoted by p(-1) and p(-12), respectively. $AR(12)$ denotes residuals autoregression of the 12th order, estimated by the Cochrane-Orcutt method. The dynamic specification of equation (7–7) was selected by a simplification procedure going from the general to the simple (McAleer, Pagan, and Volker 1985). The numbers in parentheses below the estimated coefficients are the standard t-ratios. N is the number of data points used, R^{2a} is the usual coefficient of determination adjusted for degrees of freedom, DW is the classic Durbin-Watson test of residuals autocorrelation, and F is the usual test of the estimated equation against the assumption of no linear relationship at all.

The fit is reasonably good for monthly price data. The price index of Nigerian goods, ep^*, which is computed using the parallel market exchange rate, is highly significant. Adding other lagged values of logp and other AR terms does not reduce its significance. But the purchasing power parity-type hypothesis implies not only that prices in Nigeria affect the CPI in Niger; it also implies that the CPI in Niger is not affected by the macroeconomic policies pursued in Niger.

To test this second part of the assumption, we have added to equation (7–7) an indicator of monetary policy, namely "currency held outside the banking sector" (circulation fiduciaire). This indicator is preferable to a broader measure of the quantity of money for two reasons. First, one can argue that the representative agents of the sectors involved in cross-border trade do not make much use of bank accounts. And second, during the period under study, the rules concerning the bank accounts of the mining firms were changed, with a significant impact on the quantity of money series.

We included in the equation several current and lagged values of the monetary policy indicator, separately and jointly. The most nearly significant impact appears in the following equation:

$$\text{(7–8)}\quad \log p = \underset{(1.77)}{1.04} + \underset{(5.57)}{0.46}\log p(-1) + \underset{(4.14)}{0.26}\log p(-12) + \underset{(4.36)}{0.07}\log ep^* - \underset{(0.74)}{0.03}\log M(-12) - \underset{(2.14)}{0.24}AR(1)$$

$N = 94$, $R^{2a} = 0.75$, $DW = 1.90$, $F = 56.52$

Here $M(-12)$ denotes the nominal quantity of currency, lagged one year. It is not significant and has the wrong sign. Moreover, inclusion of this additional variable in the equation has not modified noticeably the estimated coefficient of $\log ep^*$. This confirms the intuitive result that e and p^* are exogenous with respect to the macroeconomic policies of Niger. It also confirms the result reported in Azam (1990a), which is based on a shorter sample with quarterly data. Niger is too small to have a significant impact on the market for the naira, which is presumably dominated by the relationship between Nigeria and the entire CFA Franc Zone. In other words, there is probably a causal relation from ep^* to p. Figure 7–3 illustrates the price deflation which occurred in Niger over the second half of the 1980s as a consequence of the naira having depreciated on the parallel market faster than the CPI increased in Nigeria.

Therefore, it seems acceptable to regard Niger as a small open economy which is price taker in all its markets. Nontraded goods do not seem to play a significant role in this economy. This entails as a corollary that the surplus of the parallel balance of payments which we have presented in the previous section cannot be explained by applying the "elasticities approach". Competitiveness is not an explanatory factor if prices are determined abroad. Moreover, figure 7–3 illustrates that the CPI in Niger has increased relative to the price index of Nigerian goods. Therefore Niger has experienced, if anything, a loss of competitiveness rather than an improvement. The divergence between the two series in figure 7–3 can be explained by the fact that Niger's CPI is based on prices and consumption patterns in the capital, whereas the rural CPI is used for Nigeria.

Figure 7-3. Price Indexes in Niger and Nigeria

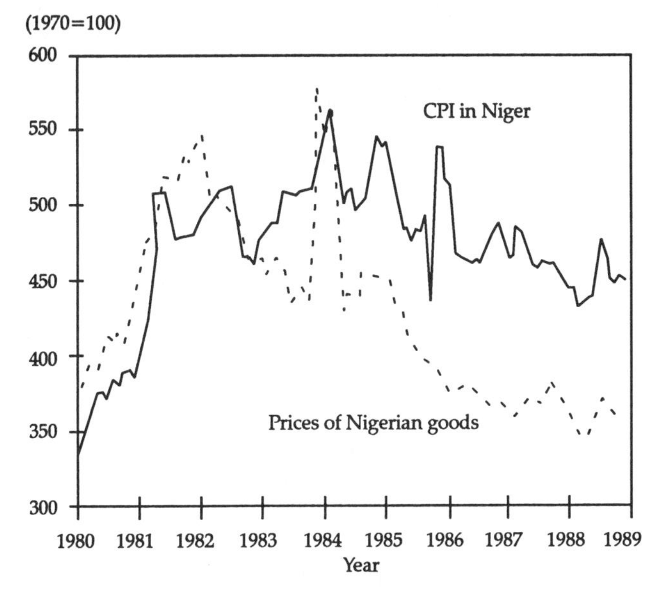

Source: BCEAO, Ministry of Planning and World Bank.

Figure 7-4. Index of Herd Size and Cash Balances, 1980-89

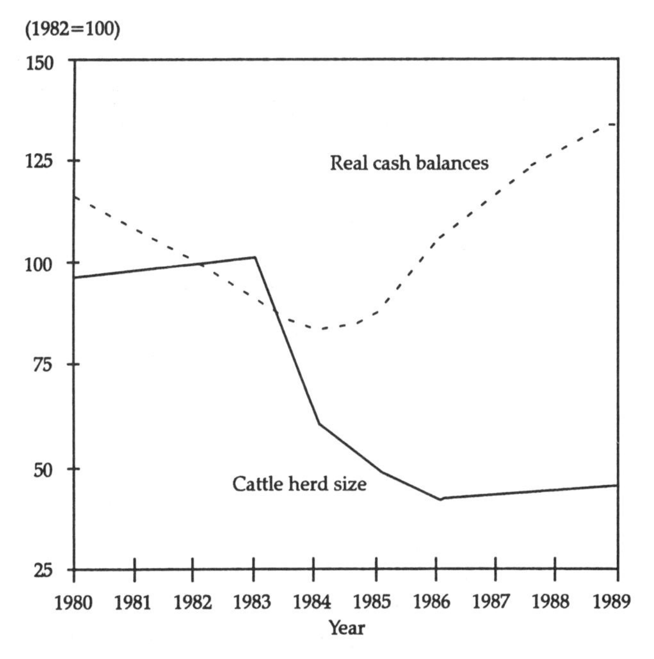

Source: BCEAO, and Ministry of Agriculture.

Probably the former is affected by the price of goods which are not traded with Nigeria, but with Europe and the rest of the world. The prices of European goods have not fallen as have those of the goods traded with Nigeria.

Hence, another approach is required. In the developed countries' literature, the elasticities approach has been replaced by the asset approach. The latter seems capable of being adapted to the cases of less developed countries. However, one must take into account some specific aspects of these countries, especially in Africa. The next section is devoted to the development of an asset approach adapted to the structural features of the economy of Niger.

An Adapted Asset Approach

In the model presented in this section we attempt to capture as simply as possible a specific feature of the Niger economy which must be taken into account for adapting the asset approach to the balance of payments in Niger. Cross-border trade with Nigeria involves to a large extent storable goods, like cattle and grains, which we refer to as African goods. These African goods can be held as assets; granaries and herds are the bulk of the portfolios of most African people. Therefore, cross-border trade may be regarded as a channel whereby agents can swap physical assets like cattle against cash balances and vice versa.

Niger also trades other goods with the rest of the world by official channels. We refer to these goods as European goods. Hence, we analyze a simple model in which there are two goods. E denotes the quantity of European goods consumed and is meant to capture the flow of official imports, and A denotes the quantity of African goods consumed. African goods are assumed to be storable; their total inventory level is denoted by H. We assume this asset to be productive, like cattle, so that the output of African goods is equal to $F(H)$, $F'0$, $F"$

S denotes the amount of African goods sold on the market, which can be negative in the case of an acquisition. The accumulation equation for this asset can be written:

$$dH/dt = F(H) - A - S. \quad (7\text{–}9)$$

The alternative asset is real cash balances. As we have seen above, the market for the naira can be regarded as an indirect channel for acquiring convertible currency. M denotes the nominal quantity of money held, and p^E and p^A are the money prices of the European goods and African goods, respectively. The following equation:

$$p = p(p^E, p^A), \quad (7\text{–}10)$$

defines the consumer price index $p(.)$, which is increasing, differentiable, and homogeneous of degree one. Accumulation of real cash balances (M/p) is governed by the following equation:

$$(7\text{–}11) \quad d(M/p)/dt = (p^A{}_s - p^E E - (d \log p/dt)M)/p.$$

We assume that the representative agent is an unconstrained price taker in a large market, and so we treat H and M/p as "jump variables", i.e. as right-hand differentiable functions of time. Equations (7–9) and (7–11) thus refer to the right-hand derivatives with respect to time for these two variables. Given these two differential equations and neglecting population growth, we assume that the representative agent seeks to maximize:

$$(7\text{–}12) \qquad \int_0^{\infty} (U(E,A)+W(M/p))e^{-rt}dt,$$

in which the utility function $U(.)$ is twice differentiable, increasing, concave and homogeneous of degree one, and the wealth function $W(.)$ is increasing, twice differentiable and strictly concave. Including real cash balances in the objective function, with additive separability, is an assumption which dates back to Sidrausky (1967) (See Blanchard and Fischer, 1989, for a discussion of the Sidrausky model). The constant rate of time preference is denoted r.

Applying Pontryagin's Maximum Principle to this problem and rearranging terms, we get three marginal conditions. First, optimization of the consumption bundle implies:

$$(7\text{–}13) \qquad U_A/U_E = p^A/p^E,$$

where U_A and U_E are the partial derivatives of $U(.)$ with respect to A and E, respectively. Because of the homogeneity assumption, from (7–13) we can derive a negative relation linking the ratio of the quantity of the European goods to the quantity of the African goods to their relative price:

$$(7\text{–}14) \qquad E = A f(p^A/p^E), f' \; 0.$$

In the utility function, $U(.)$, the elasticity of substitution between E and A is defined by $s = (p^A/p^E)f'/f \; 0$. Second, we have the fundamental arbitrage equation of this portfolio model which reads:

$$(7\text{–}15) \qquad F'(H) + d \log(p^A/p)/dt = (p^A/p)(W'/U_A) - d \log p/dt.$$

This relation equates the real rate of return on holding African goods, on the left-hand side, to the real rate of return on holding real cash balances, on the right-hand side. The former is made of the marginal product of the stock of African goods plus the expected capital gains entailed by an increase in their real price. The latter is made of the subjective value of the services provided by the cash holding, less the expected capital loss due to inflation.

Lastly, we get the standard Keynes-Ramsey formula, which equates the psychological rate of interest to the real rate of interest. Here, it reads:

$$(7\text{–}16) \qquad r - d \log U_A/dt = F'(H).$$

Using again the homogeneity assumption, and (7–14), we can invert (7–16) to get the optimal inventory of African goods as:

$$(7\text{–}17) \qquad H^* (d \log(p^A/p^E)/dt) = F'^{-1}(r-h \, d \log(p^A/p^E)/dt), \; H^{*\prime} \; 0,$$

where $h = -s(AU_{AA}/U_A) \; 0$. This result is again intuitive, and describes a simple speculative behavior (assuming perfect foresight): an increase in the expected rate of growth of the relative price of African goods leads to an increase in the optimal inventory level (herd size, for example). The mechanism involved here is not asset substitution. It is based on the relation between saving and the return on assets, as the argument of the function $F'^{-1}(.)$ measures the real rate of interest. It can alternatively be expressed as the positive effect of an expected improvement of the terms of trade on investment in productive assets.

Similarly, using (7–16) and the arbitrage equation (7–15), one can write the optimal cash balance as:

$$(7\text{–}18) \quad (M/p)^* = W'^{-1}((p^U{}_A/p^A)(r-hd \log(p^A/p^E)/dt + d \log p^A/dt))$$

Given the real rate of interest, which has a negative impact, this equation says that real cash balances will be lower, the higher the rate of inflation, which is here measured by $d\log p^A/dt$. The argument in the second parenthesis may be regarded as a kind of monetary rate of interest.

We can modify this equation to make it more suitable for the econometric application presented below, given our data limitations. Using the definition of the price index $p(.)$ given in (7–10), we can change (7–18) into:

$$(7\text{–}19) \quad (M/p)^* = W'^{-1}((p^U{}_A/p^A)(r-(h/g)d \log(p^A/p) /dt+d \log pA/dt))$$

where $0g=p^E p_p E/p$ is the share of European goods in the price index.

We have been able to test a log-linear approximation of equation (7–10), using monthly data from 1984.04 to 1988.10. We use again circulation fiduciaire as our monetary aggregate. The price variable representing the Nigerian CPI is the rural price index, which is multiplied by the parallel exchange rate to obtain the CFA franc price of Nigerian goods. Therefore ep* is a good indicator of the price of African goods. On the other hand, since our price index for Nigerien goods is the CPI for African consumption in Niamey, it is relatively more influenced by the prices of European goods. Then *(ep*)* corresponds to p^A and *(ep*/p)* to (p^A/p).

The following equation is based on this convention:

$$\text{(7–20)} \quad \log(M/p) = \underset{(4.74)}{-3.5} - \underset{(2.60)}{0.35}\, D\log(ep^*) + \underset{(1.26)}{0.17}\, D\log(ep^*/p) + \underset{(2.46)}{0.27}\log(ep^*/p) - \underset{(2.98)}{0.001}\, RF + \underset{(34.34)}{0.98}\, AR(1)$$

$N=67,\ R^{2a}= 0.93,\ DW = 2.08,\ F = 165.6.$

Here, *D* log(.) represents the rate of growth of the corresponding variable computed over 12 months. RF is an average of the monthly rainfall in the towns of Maradi, Tahoua and Zinder, which should be representative of the climatic situation in the relevant geographic zone. Its significant impact and negative sign may be regarded as capturing the negative impact of rainfall on the perceived uncertainty of the near future. This shifts the *W'*(.) curve downwards, by reducing the precautionary demand for money. *AR*(1) is the coefficient of first-order residuals autoregression, estimated by the Cochrane-Orcutt method. Its value suggests that the residuals are generated by a random walk process, i.e. that shocks tend to be permanent. As for equations (7–7) and (7–8) above, we have represented by N the number of observations used, by R^{2a} the usual R^2 adjusted for degrees of freedom, etc. The t-ratios are in parentheses below the estimated coefficients.

The rate of growth of the relative price of African goods is not significant at the 20% (two-tailed) confidence level, although it has the correct sign. This might simply be due to the strong correlation existing between *D* log*(ep*)* and *D* log*(ep*/p)*, but in terms of our theoretical model, it can be rationalized as showing that (h/g) is close to zero, which is the case if either AU_{AA}/U_A is close to zero, i.e. if U_A is nearly constant, or if the elasticity of substitution between European goods *E* and African goods *A* (s) is not significantly different from zero. Moreover, if these two elasticities are significantly different from zero, but small, their product *h* is even smaller and may not be significantly different from zero. Dropping this variable does not alter much the other features of the equation:

$$\text{(7–21)} \quad \log(M/p) = \underset{(4.56)}{-3.67} - \underset{(2.77)}{0.21}\, D\log(ep^*) + \underset{(3.09)}{0.31}\log(ep^*/p) - \underset{(3.16)}{0.001}\, RF + \underset{(36.01)}{0.98}\, AR(1)$$

$N=67,\ R^{2a} = 0.93,\ DW = 2.10,\ F=204.67$

The fit remains impressive and there is no trace of residuals autocorrelation. However, the coefficient of *D* log*(ep*)* is reduced, as this variable now captures to some extent the positive effect of Dlog(ep*/p). This mongrel coefficient remains significant, but this suggest that (7–20) is preferable, as it separates clearly the two effects.

These equations bring out quite forcefully the role of the price index for Nigerian goods, and hence of the parallel exchange rate of the naira, in determining the level of real cash balances in Niger. We have used *D* log*(ep*)* and *D* log*(ep*/p)* as explanatory variables, assuming implicitly static expectations. Various attempts have been made at applying the rational expectations hypothesis with various leads, using the "error in variable" method advocated by Wickens (1982). The results were much less satisfactory.

As we have seen above, the price of African goods, as defined here, decreased over most of the period analyzed. This would explain the increase in real cash balances which has taken place. The increase in real cash balances has probably played an active role in sustaining the surplus of the parallel balance of payments which we have presented above. However we have not been able to provide a direct test of this assumption, for lack of appropriate data.

Further Evidence

Figure 7- 4 presents indexes of the real cash balances series and of the estimated herd size for cattle. Data for the latter are known to be very inaccurate, but are the only evidence available on this topic. The estimated herd size series shows a sharp drop in 1984, corresponding to the drought, but no recovery took place after that date.

The real cash balances series shows sustained growth in the second half of the period. Before the drought, herd size was increasing and real cash balances were decreasing. After the drought, herd size decreased or remained at a low level, and cash balances increased in real terms. The econometric exercise presented above suggests that the continuing depreciation of the naira on the parallel market, pulling the price index of African goods downwards both in nominal and in real terms, played a large part in determining this change in portfolio composition.

The parallel exchange rate of the naira stabilized with respect to the CFA franc in 1989. If this break in the downward trend which dominated the 1980s would be confirmed in the future, it would remove the incentive to hold large and increasing real cash balances. There may be a sudden drop in the amount of real cash balances and this might result in the parallel balance of payments turning to a deficit position.

Conclusion

In this paper we have shown first how a small economy with a convertible currency like Niger can acquire foreign exchange via its cross-border trade with Nigeria, despite the latter having an inconvertible currency. The reason for this result is that the naira market is well unified by arbitrage and must therefore be cleared globally, by adjustment of the parallel exchange rate, and not on a bilateral basis with all the countries participating in this market. Hence Niger can run a surplus in its parallel balance of payments, provided the rest of the participants in this market run a net deficit of the same magnitude. Hence, cross-border trade should be given some attention in the design of macroeconomic policy in Nigeria's neighboring countries.

We have found an indicator of such a surplus of the parallel balance of payments by looking at the series of banknotes issued in the BEAC monetary union and purchased by the BCEAO in Niger. It happens that most of these notes are bought in Maradi and Zinder, which are known as the main centers for cross-border trade with Nigeria. This indicator shows that Niger has had a sustained surplus since 1983.

We have shown that this surplus cannot be explained by applying the elasticities approach and referring to the competitiveness of Nigerien goods. A sort of purchasing power parity relationship passed econometric tests. This suggests that the price level in Niger is mainly determined exogenously by the price of the goods which are tradeable with Nigeria. The price series for Nigerian goods is characterized by a downward slope, which explains the price deflation which took place in Niger in the second half of the 1980s. This downward trend may play a part in explaining the above mentioned surplus, within an asset approach framework.

We have analyzed a simple model of asset accumulation adapted to the structural features of the Niger economy. The main point is that cross-border trade in this area involves mainly storable goods like cattle and grains, and can thus be analyzed as a means for exchanging physical assets ("African goods") against cash balances. We have derived from this model an econometric equation to explain the behavior of real cash balances, based on monthly data. The results confirm that the increase in real cash balances which occurred during the 1980s can be explained by the decrease in the price of African goods. Further evidence was added to strengthen this point by showing that cattle herd size did not recover after the drought, but continued to decline.

We have not been able to provide a direct test of the relationship between the improvement in the parallel balance of payments and the increase in real cash balances induced by price deflation. Such a test would require a more complicated model with specification of the balance of official payments. In the case of a convertible currency like the CFA franc, a surplus in the parallel balance of payments can be used either to increase cash balances or to transfer funds from the parallel market to the official one. In Niger during the 1980s, the accumulation of cash balances certainly played a part in the improvement in the parallel balance of payments.

Notes

1. "Nigerien" means "from Niger", while "Nigerian" means "from Nigeria".

2. The validity of this assumption depends on the institutional features of the commercial sector. It would not be appropriate in its simple form for the smuggling of cocoa out of Ghana, for example, where the parastatal Cocobod has a monopoly on the cocoa trade and a private agent transporting cocoa is smuggling. Any attempt at cloaking this trade must be more indirect, using official transportation of other goods as a cover. There is no such monopoly now in Niger.

3. This is at variance with the assumption tested by Azam and Besley (1989) for the case of Ghana, where the supply curve of the smuggling industry is upward sloping.

4. One should qualify this statement by remarking that the African CPI in Niamey might not be very representative. Its weights are based on a survey done in

1961-62, with a sample of 317 households. In particular, rents are not included, whereas a recent survey has shown that they account for more than 15 percent of the expenditures of urban households in Niamey. But it seems likely that the corresponding figures for rural households and households living in smaller towns is much smaller.

5. We neglect other goods, such as European goods like Benson & Hedges cigarettes, which are smuggled into Nigeria via Niger, and Nigeria-made Peugeot cars which are smuggled into Niger, which play a not-so-negligible part in the real world (see Azam 1990a).

6. This homogeneity assumption is meant to simplify the presentation as a shortcut through some dynamic complexities. It gives the model "bang-bang" reactions. It should not be pushed too far, as it removes from the model the well known consumption smoothing behavior which is found in most standard theories of consumption and saving (see footnotes 8 and 9 below).

7. If the homogeneity assumption (see footnote 7) was given up, the growth rate of A or E would show up in equation (17). But it could be substituted out in the full blown dynamic analysis, to get (17).

8. This residuals dynamic may be regarded as a simplified way to take into account the dynamics squeezed out of the theoretical model by our simplifying assumption.

References

Arnoult, E.J. 1983. "Cross-Border Trade Between Niger and Nigeria." In Elliot Berg and Associates, *Joint Program Assessment of Grain Marketing in Niger.* 2, 1-12, Niamey, Niger: USAID.

Azam, J.-P. 1990a. "The Balance of Cross Border Trade between Niger and Nigeria (1980-87): Evidence from the Parallel Market for the Naira." In M. Roemer and C. Jones, eds., *Markets in Developing Countries: Parallel, Fragmented and Black.* San Francisco: International Center for Economic Growth. Forthcoming.

Azam, J.-P. 1990b. "Marchés Paralléles et Convertibilité: Analyse théorique avec références aux économies africaines." *Revue Economique.* Forthcoming.

Azam, J.-P. and T. Besley. 1989. "General Equilibrium with Parallel Markets for Goods and Foreign Exchange: Theory and Application to Ghana." *World Development* 17, 1921-1930.

Azam, J.-P. and C. Vernhes. 1990. "La détermination des taux de change parallèles en Afrique: modèle macroéconomique et test économétrique (Ghana, Nigeria, Zaire)." *Economie et Prévision.* Forthcoming.

Blanchard, O.J. and S. Fischer. 1989. *Lectures on Macroeconomics,* Cambridge, Massachusetts: MIT Press.

Devarajan, S., C. Jones and M. Roemer. 1989. "Markets under Price Controls in Partial and General Equilibrium." *World Development* 17, 1881-1893.

Grégoire, E. 1986. *Les Alhazai de Maradi (Niger).* Paris: Editions de l'Orstom.

Guillaumont, P. et al. 1988. "La politique d'ajustement au Niger 1980-1987." Report to the Government of Niger, forthcoming in an abridged version as "Ajustement structurel, ajustement informel: le cas de Niger."

McAleer, M., A.R. Pagan and P.A. Volker. 1985. "What Will Take the Con out of Econometrics?" *American Economic Review* 75, 293-307.

Pitt, M.M. 1981. "Smuggling and Price Disparity." *Journal of International Economics* 11, 447-458.

Sidrausky, M. 1967. "Rational Choice and Patterns of Growth in a Monetary Economy." *American Economic Review* 57, 534-544.

Wickens, M.R. 1982. "The Efficient Estimation of Econometric Models with Rational Expectations." *Review of Economic Studies* 49, 817-838.

8

Comments on Parallel Markets

Paul Collier and Vikram Nehru

Paul Collier

I found these to be three rather convincing papers. But I want to step back from them and ask about the framework which might be used to look at the relationships between changing the exchange rate and inflation, which is the core of two of the papers.

We come to this problem with quite a lot of intellectual baggage, a lot of which is inevitable, but unfortunately dysfunctional. Let me sketch two of the models that we bring along. One is the orthodox monetarist model as applied to developed countries where we have purchasing power parity. What happens when we devalue? The price of all tradable goods rises in domestic currency and that pushes up the demand for money, so for a given money supply, the balance of payments improves. That is what I would call a balance of payments improving devaluation. I think it is not relevant in the present context.

The second model, which is a structuralist model, is particularly relevant to the Latin American context. The main focus is that a devaluation causes cost plus inflation, partly through pushing up import costs and partly through pushing up wages in response to import costs, which somehow causes an endogenous increase in the money supply.

I think that neither one of these models is a very good framework for analyzing the devaluations in Africa over the last decade. The simple view of the exchange rate premium ignores two potentially important effects. One is that parallel market transactions involving foreign exchange are illegal, and illegality sometimes, but not always, involves cost. The other effect that is ignored in the simple quota model is that foreign exchange is an asset, so there will be asset demand effects. All three papers are really going beyond the simple endogenous trade policy model to incorporate either asset effects or illegality effects or both.

The Effect of Relative Prices on Devaluation

Within the endogenous trade policy model, what happens when we devalue? There are relative price effects and there are price level effects. What happens to relative prices? The devaluation permits a trade liberalization. That is the first effect, so that the domestic price of exportables rises relative to importables. There are two relative prices in this endogenous trade policy model. One is the domestic price of exportables to importables. The other is the domestic price of importables to nontradeables.

We are told that the IMF measure of the real exchange rate is a proxy for the relative price of importables to nontradeables. But the effect of a devaluation in the endogenous trade policy model is that the relative price of nontradeables to importables rises. In the typical African devaluation, the IMF measure of the real exchange rate goes down a lot, and the relative price of nontradeables to importables actually goes up. That is one reason why the IMF measure of the real exchange rate is so extremely confusing in the African context. The concept of the real exchange rate (the IMF measure) predates the whole structural period. It was a concept devised basically for developed economies, and I think it is being misapplied in the African context.

In developed countries trade policy does not matter very much. It can be ignored in the first approximation. There is only one relative price—the price of tradeables to nontradeables. In Africa there are two relative prices and one damn term—the real exchange rate. And there is only one measure, which is the IMF looking at these CPIs in Africa and wholesale price indexes in the US. The concept and its measure are not up to the job.

The Inflationary Effect and Effects on the Price Level of a Devaluation

When the trade policy is endogenous, we would expect that the price level would fall as long as the money supply is constant. This is because trade liberalization leads to some real income gains. If the money supply is constant and real income goes up, the price level goes down. So we would expect from our simple model that devaluation in the African context would be counterinflationary, as long as the money supply is constant.

Another rue to the devaluation is how it changes the money supply, which is a matter of how it changes the budget and the politics of the off-setting effects. Tax revenue from exports goes down because the overvaluation of the exchange rate is an implicit tax on exports. But off-setting that, the implicit subsidy on imports, which is the quasi-expenditure of the government, also goes down. Usually in Africa, because imports exceed exports, the budgetary effect of the reduction in the implicit subsidy exceeds the budgetary effect of the reduction in the implicit tax. You can usually expect the budget to improve. There are two reasons why we would expect devaluation to be counterinflationary. The first is the effect of raising real income for a given money supply. The second is the effect of the improvement in the budget and reduction in the money supply.

The Chhibber-Shafik paper takes a structuralist perspective on the relationship between devaluation and inflation. The authors say that they get counterintuitive results in that devaluation in Ghana has actually reduced inflation. I am not at all surprised by the result, which to me is entirely intuitive. It is not counterintuitive if you start from a cost plus framework. In my opinion this just shows that the framework is not right for Africa. The O'Connell-Kaufmann paper comes to exactly the same result—that the devaluation of the official exchange rate in Tanzania has been counterinflationary. It also confirms the endogeneity of trade policy in response to terms-of-trade shocks.

Illegality and the Effect of Assets

Two modifications of the simple endogenous trade policy model involve illegality and the effect of assets. These can be discussed in the context of exchange rate unification—unification of the parallel and the official exchange rate.

Illegality

The effects of illegality on the premium cut both ways. Illegality of exports tends to raise the premium, which has to cover the costs of getting dollars. Illegality of imports tends to lower the premium. This is where Kaufmann and O'Connell's work on the own-funded imports scheme in Tanzania is particularly interesting. I think they show that the own-funded import scheme has indeed tended to raise the premium. A policy inference from that might appear to be that we jump straight to unification of an exchange rate and forget about an intermediate stage where we have an own-funded import scheme because an own-funded import scheme pushes the premium up relative to what would happen with unification. I think one qualification to that is that what Kaufmann and O'Connell are doing is true in a flow sense, but certainly in Tanzania (and I think most countries), prior to reforms there had been quite a lot of build-up of illegally held dollars, so there is a stock of dollars held abroad. On introduction of the own-funded import scheme, that stock of dollars tends to be repatriated. That repatriation of dollars tends to depress the premium. It is not a sustainable thing, but something that will happen over two or three years. So we might cash in on this short-run repatriation effect without having the premium overshoot the level it would reach were we to reunify the rate.

Asset Effects

All three papers stress asset effects. The Kaufmann-O'Connell paper and Chhibber-Shafik paper stress interest rate effects. The Azam paper ingeniously adds the real economy effects; cattle in the end are the real determining feature. I have one problem with the interest rates effects on the premium. The policy conclusion seems to be to raise nominal interest rates. I am worried by that because certainly in the Tanzanian context, and I think perhaps in the Ghanaian context, the government is heavily in debt to the banking system. In Tanzania the banking system is basically there to lend money to the government and parastatal organizations which are acquiring the bank loans. The private sector is just making bank deposits. So when we raise nominal interest rates, there is a direct pass-through onto government expenditure and so an increase in the budget deficit. Thus raising nominal interest rates just increases the inflation tax. The holders of money pay the inflation tax. So increasing the nominal interest rate does not increase the interest rate on money; it is purely a transfer between some holders of money and other holders of money. The real return goes up in Tanzania for that quarter of money holders who are holding their money in interest bearing money. The counterpart is that the real interest rate goes down for three-quarters of money holders who are holding non-interest bearing money. I worry that we are getting these interest rates effects in the model. I do not think we should.

Vikram Nehru

The three papers presented—although all on the subject of parallel foreign markets—were written with different objectives in mind. The Chhibber-Shafik paper on Ghana examines the consequences of these markets for inflation; the paper by Azam traces the cases of large purchases of BEAC bank notes in Niamey, Niger; and the Kaufmann-O'Connell paper focuses on the effect of Tanzania's macroeconomic policies on the premium of the parallel foreign exchange market. I would first like to make two general observations and then proceed to examine some specific issues in the papers presented.

My first general observation is that, on the basis of these papers, it is clear that channels for capital flows have become permanently etched in the African landscape, well concealed to be sure, from the eyes of the law, but a reality nevertheless that no policymaker can ignore. These channels have served two functions: a defense by the private sector to maintain its real wealth in the face of efforts by the State to redistribute national resources to the public sector; and a way to cope with uncertainties ensuing from genuine concerns of economic and political instability. All three papers have observed that a critical factor explaining parallel foreign exchange rate movements is private capital flows arising as a result of portfolio adjustments or through such means as over and under invoicing. What are the implications of this important observation? The first is that the marginal rate for foreign exchange is no longer an exogenous variable but an endogenous one that policymakers need to target for several reasons—the most important being that the parallel rate effectively sets the relative price of tradables in the economy. The second is that aggregate demand management is central for stability of the exchange rate. And the third is the elevation of monetary policy as one of the tools for achieving external balance, given that domestic deposit rates adjusted for expected depreciation need to be related to return on financial assets abroad. In other words, quite apart from its other virtues, the existence of parallel foreign exchange markets imposes a discipline on macroeconomic management that can only improve the domestic environment for investment and growth.

My second general observation is on the key issue of foreign exchange market unification mentioned in both the Ghana and Tanzania papers. The evolution of Tanzania's official and parallel foreign exchange markets appear to be identical to those of Ghana, but with a two-to-three year time lag. In this regard, the papers on Tanzania and Ghana have struck the right cautionary note with regard to the question of "unification" of the official and parallel markets. Both papers are appropriately circumspect on the question of capital controls remaining an essential feature in the official foreign exchange market. In moving from its present system of foreign exchange allocation through import licensing, Tanzania could usefully study Ghana's experience with the foreign exchange auction and its gradual liberalization since September 1986. Only five weeks ago, Ghana introduced an extended interbank foreign exchange market supported by weekly wholesale auctions with a system of capital controls which, to my mind, strikes a fine balance between caution and liberalization.

There is also a more practical and mundane purpose behind maintaining a monitoring system for the use of official foreign exchange—that of ensuring the documentation support for disbursements of program aid. Once all current transactions and amortization payments are covered by the official market, the effective difference in coverage of the official and parallel foreign markets will then be entirely due to capital flows, which the analysis in these papers shows us to be highly influenced by the level of domestic deposit rates adjusted for expected depreciation of the exchange rate. The focus of policy would then shift to monetary management as a means to bringing about the "unification" of the two markets that most economists consider so desirable.

The Azam Paper

I found the paper by Azam on cross-border trade between Niger and Nigeria intriguing and fascinating. Mr. Azam's hypothesis starts with the observation that a real appreciation of the Nigerien CFA franc vis-a-vis the Nigerian naira coincided with a cross-border surplus in favor of Niger as indicated by purchases of BEAC banknotes by the BCEAO in Niger. The reason, according to Mr. Azam, is the reduction in the price of all rural goods in Niger following a similar decline in the rural Nigerian CPI—including the price of cattle, which is a part of the Hausas' savings portfolio. We are then told that Niger's bilateral balance of payments surplus with Nigeria can be attributed to sales of this "bovine portfolio" by the Nigerien Hausas to their Nigerian brethren across the border, because of the relatively higher return to holding cash balances than to owning cattle. This is an ingenious explanation, but one that warrants some further examination.

First of all, we should interpret with some caution, the results of an analysis which uses data on the ownership of cattle for countries that share a thousand mile common border with little meaning to Hausa cattle herders. But even were this data to be accurate, figure 4 in the paper does not appear to support the hypothesis. Although herd sizes stabilized after 1986, the rise of real cash balances appears to continue unabated. An alternative hypothesis, and one that deseves some examination, is that the real depreciation of the Naira led to capital outflows from Northern Nigeria to Niger. With a fixed exchange rate, the consequences would be an increase in real cash balances, which, presumably, were not drawn down because of continuing uncertainties in Nigeria. Finally, on the Azam paper, I felt that the analysis could have benefitted from a more comprehensive treatment of the realtionship between the parallel exchange rate, domestic inflation, and the money supply in Niger. The statement that nontraded goods do not seem to play a significant part in the CPI does not appear to borne out by the results in equation 6. I would suspect that adding money to equations 6 and 19 would provide more convincing results and a better understanding of what exactly is going on in Niger.

The Kaufmann-O'Connell Paper

I found little to disagree with in the analysis of the relationship between the parallel market premium and key macroeconomic variables in the Kaufmann-O'Connell paper. However, the reduction of the large set of interrelated variables into the two reduced form equations given in tables 3–1 and 3–3 left one somewhat confused. I was also puzzled as to why the domestic money stock is expressed in terms of foreign exchange after being converted at the official exchange rate. Finally, I agree with the authors' judgment that an important lacuna in the model is the absence of a fiscal module. A key question would be how changes in the official rate would affect the budget deficit and in turn, through the money supply, have a bearing on the parallel premium. Finally, I did not get a sense of what is being done in monetary management and financial sector reform in Tanzania, and how this may have a bearing on the interest parity differential. In this context, it would be useful if the paper ran some "what-if" simulations to highlight key policy options.

The Chhibber-Shafik Paper

I am at somewhat of a loss to comment on the paper by Chhibber and Shafik because the authors graciously took on board some of the comments I had made on earlier drafts. I therefore look forward to a more objective assessment by Paul Collier. Nevertheless, permit me to ask one, perhaps simple, question. Why is it that in the simulations, the general course of inflation appears largely unaffected despite significant changes in key assumptions? Can the authors point to certain fundamental factors driving inflation that appear impervious to large variations in the official exchange rate or production?

9

Issues in Public Expenditure Policy in Africa: Evidence from Tanzania's Experience

Laurean W. Rutayisire

In an effort to counter Tanzania's economic crisis, the authorities launched a structural adjustment program in 1982 and an economic recovery program in 1986. These programs articulated the new objectives of economic policy which the country would pursue and the measures by which the objectives would be achieved. The objectives have been the following: to reduce the rate of inflation; to achieve balance of payments adjustment so as to alleviate the existing extreme foreign exchange scarcity and the consequent underutilization of domestic production; to achieve an increase in the productivity of parastatal enterprises and improvement in public sector management; to secure increases in food and cash crop production by providing adequate producer incentives, marketing and distribution systems, and resources available to agriculture; and to maintain the already achieved equity in income distribution, including the provision of social services and other basic needs to the majority of the population.

According to the structural adjustment and economic recovery programs, the implementation of the above objectives involved the following measures. First, the government has constrained expenditure to reduce money supply growth and inflation. This has involved decreasing government's share in total domestic credit, in order to minimize its crowding out impact on the availability of bank credit to the rest of the economy. In order to reduce government expenditure, various budget subsidies have been abolished. The second set of measures adopted has consisted of the rationalization of prod-uction in the public sector to increase capacity utilization, improve labor utilization, and reduce unpro-ductive activities. Put into practice, this has involved reducing public sector employment, closing some public corporations, shelving some development pr-ojects, and decentralizing some central government functions to local governments. Complementary to these measures, the authorities have further resolved to increase the role of the market in allocation. Thus, interest rate and producer pricing policies have been used to improve efficiency in the allocation of domestic resources. Exchange rate policy has been used to improve the country's international export competitiveness and to provide export producer incentives. There have also been trade liberalization measures, which have included removal of restrictions on a wide range of imports and reduction in the number of commodities under price control.

The economy of Tanzania is beginning to experience signs of recovery, including the reversal of output decline. However, the country's economic crisis seems to have acquired an additional dimension. There has been an increasing deterioration in the quality of the country's basic economic and social infrastructure, which is apparently causing a constraint on subsequent recovery efforts. There are two major perspectives on explaining the causes of such deterioration in the quality of economic and social infrastructure. One line of argument is that the fiscal and monetary policy measures adopted under the structural adjustment and economic recovery programs have been perversely restrictive. The result has been inadequate funds for infrastructural rehabilitation and maintenance.[1] An alternative view similarly maintains that the deterioration in the quality of the country's economic and social infrastructure is attributable to underfinancing. However, it relates the underfinancing of provisions to the overgrown size of government expenditure.In other words, the underfinancing has been the result of a thin distribution of government expenditure over a wide range of items.[2]

This paper attempts to explain the decline in the quality of Tanzania's basic economic and social infrastructure. In order to identify the appropriate perspective, the literature on optimal and sustainable level of government expenditure is first surveyed in the second section. The question of what can be regarded as an optimal and sustainable level of government expenditure is also examined in this section. Section three proceeds to ascertain the optimal and sustainable level of government expenditure in Tanzania and to what extent the current size of government expenditure in the country has departed from such a level. During the structural adjustment and economic recovery programs, the level of government expenditure has remained in excess of the optimal and sustainable level for the country. Therefore the factors which explain such a persistently high level of government expenditure are explored. Some of the effects of increased government deficit spending are examined in the fourth section. Suggestions and conclusions are presented in section five.

On the Optimality and Sustainability of Government Expenditure

An optimal and sustainable level of government expenditure maximizes the government's welfare improvement program subject to the constraints imposed by the government's permanent income, its target of reducing inflation, and the objective of minimizing the burden and crowding out impact of government debt (Zee 1988, Buiter 1983). It has been argued that the inflation adjusted budget should be balanced.This would support government fiscal and monetary policy which is committed to bringing down inflation and to creating conditions in which sustainable economic growth can be achieved (Miller 1982). Balancing the government budget adjusted for inflation requires that government expenditure be held constant in real terms; it should be allowed to increase only in nominal terms.

An optimum and sustainable government expenditure level can be ascertained as in the following equation (Zee 1988, Buiter 1983b, Miller 1982):

$$(9\text{–}1)\ D = G - \overline{T} - \dot{K} - (\dot{P}/P)\,G_{R-1} - (\dot{e}/e\,G^{ex} - nPSFD)$$

The unsustainable or permanent deficit is denoted by D and government expenditure is denoted by G. Government revenue, is expressed in permanent income terms as a three year moving average; $\dot{K}$ denotes nominal government capital expenditure; $\dot{P}/P$ is the actual inflation rate; $\dot{e}/e$ represents change in the nominal exchange rate; $G_{R\text{-}1}$ is government recurrent expenditure lagged one period; G^{ex} is net foreign obligations denominated in local currency; and $nPSFD$ is a growth adjustment term. $PSFD$ denotes the budget deficit and n denotes real GDP growth.

First government permanent income ($\overline{T}$) is subtracted from overall government expenditure (G). Government permanent income should be the income which the government expects to earn on average and over a long period. This definition considers current and discounted future streams of revenue from taxes, government property, and capital stock. Government permanent income has been proxied by a three year moving average of government tax revenue, royalties on government natural property rights, dividends from public enterprises, and other government revenue. The approximation of permanent income by a moving average of income series has been used in a number of studies.[3]

Government capital expenditure ($\dot{K}$) is subtracted from overall government expenditure less government permanent income. The residual constitutes government recurrent expenditure, which takes into account the productivity gain which may derive from government capital projects. The sale of government projects can also generate revenue with which the government budget deficit may be bridged. Government recurrent expenditure is partly on account of servicing capital projects. However, it is largely made up of general recurrent expenditure (consumption spending) and debt service. Thus the equation emphasizes the need to balance government recurrent expenditure and government permanent income.

In the existing literature,[4] the need to account for the impact of inflation on the government budget deficit has been discussed mainly in connection with the government debt. The argument has been that government deficit spending is liable to increase due to inflation indexed interest payments. However, it has also been argued that inflation leads to real decline in the outstanding stock of government debt. The excess of the latter impact over the former should be considered as additional revenue due to the inflation tax. The converse would cause a net increase in the government budget deficit as a result of inflation.

Although the inflation tax should be taken into account, in practice government accounting is done on a cash basis rather than on an accrual basis. Therefore, the adjustment in the above equation is for nominal increases in government expenditure due to inflation. This is in view of the fact that the inflation tax on outstanding debt is only an accrual item. However, cost price interactions affect the cost of public provisions as well. The adjustment for inflation which is provided for in equation (9–1) thus considers both debt and nondebt government recurrent expenditure.

The issue behind inflation adjustment of the government budget deficit is the need to hold government expenditure at a constant level in real terms and only provide for nominal increases in government expenditure due to inflation. Thus, in the above equation, a nominal increase in government expenditure due to inflation is provided for on government expenditure in the preceding period, which, according to the above argument, should be held constant. Holding the government expenditure at a constant level in real terms is consistent with the policy target of bringing down inflation. Moreover, inflation indexation can enhance price stabilization efforts to the extent that it will institute inertia in cost price interaction.[5]

The equation also provides for the impact of exchange rate change $\dot{e}/e$ on the government budget deficit. In particular, the adjustment indicated provides for the net direct impact of exchange rate change on the government budget through foreign exchange denominated receipts and payments, including the import content of the budget G^{ex}. The equation provides for real GDP growth n. Increases in real GDP entail increases in government capability through increased revenue with which to sustain a higher level of expenditure. Thus, in summary, the balancing of government recurrent expenditure and government permanent income is adjusted for inflation, exchange rate change, and real GDP growth.

However, the equation does not include a cyclical adjustment term. During the later years of the structural adjustment program and the economic recovery program, the authorities have abolished a wide range of budget subsidies and increased discretionary taxes. These measures have tended to minimize the cyclical sensitivity of the government budget. Hence, a provision for this adjustment has not been made in the equation.

Other studies have suggested that a sustainable government expenditure program should provide for the government's major role in income redistribution or the social contract deficit (Bossons 1986). Considering the economic policy measures which were summarized in the introduction, it is evident that during the structural adjustment and economic recovery programs emphasis has been on securing functional income redistribution by adopting appropriate producer pricing and marketing reforms. Hence a provision for the government budget's income support has not been made in the equation.

In addition to these qualifying remarks, this study has concentrated on the sustainability of total government expenditure rather than its composition. This perspective has been maintained in macroeconomic stabilization policies. A treatment of the composition of government expenditure in Tanzania has been made elsewhere.

Table 9–1. *The sustainability of government expenditure in Tanzania: fiscal years 1982/83 to 1987/88*
(millions of shillings)

Item		*1982/83*	*1983/84*	*1984/85*	*1985/86*	*1986/87*	*1987/1988*
1	Overall government expenditure	19,276	23,918	26,728	33,220	55,481	77,326
2	Government permanent income[a]	11,919	14,644	17,662	23,392	32,940	49,400
3	=1-2	7,357	9,274	9,066	9,827	22,542	27,926
4	Capital account expenditure	4,404	5,736	5,391	5,817	15,091	17,255
5	= 3-4	2,953	3,538	3,675	4,010	7,451	10,671
6	Inflation adjustment[b]	1,960	2,231	2,727	3,201	4,110	6,059
7	= 5-6	992	1,307	947	810	3,340	4,613
8	Exchange rate change adjustment (net)	153	449	0	179	-3,139	-3,313
9	= 7-8	839	858	947	630	6,479	7,925
10	Adjustment for GDP growth	45	-233	-211	297	723	1,198
11	= 9-10 surplus (-) and defecit (+)	794	1,091	1,158	333	5,756	6,727

Notes: [a] Permanent income is computed as a three year moving average of government tax and nontax revenues excluding external import support grants. [b] In the course of government expenditure estimates the authorities have directed a 15 percent inflation adjustment, which is also used here.

Source: The computations are based on the optmal and sustainable government expenditure level equation in the text (9–1). Data has been obtained from the *Economic Survey, Bank of Tanzania Reports, and Government Expenditures Supply Notes Vol. II.*

The Size of Government Expenditure in Tanzania

The above measurement of the sustainability of government expenditure has been applied to the case of Tanzania. The focus here is on the period covered by the structural adjustment and economic recovery programs. The results in table 9–1 proceed according to the above equation. As is indicated by the results in row 11, the level of government expenditure in Tanzania entails a budget deficit which is not sustainable within the constraints imposed by the country's economic stabilization and recovery targets and the constraint imposed by the government's permanent income.

In table 9–2, the results are presented for the same measurement process with a change in the inflation adjustment term. In this case, the inflation adjustment in row 6 of table 9–2 provides for a full inflation indexation of government expenditure. As a consequence of this change, the results in row 11 indicate a budget surplus during the period covered by the structural adjustment program (1982/83 to 1985/86). An unsustainable deficit is indicated during the period covered by the economic recovery program (1986/87 to 1987/88).

These results indicate the extent of the conflict in perspective which was reviewed in the introduction. The major difference is in the level of inflation indexation accommodated by the government budget. However, even after accommodating a full inflation indexation, the results still show that the current level of government expenditure in the country is unsustainable. Such a level cannot be sustainable without eventual recourse to raising seigniorage revenue or increasing reliance on foreign borrowing. The current level of government expenditure cannot be sustainable unless there is an increase in the current and expected future levels of government revenue and increased productivity in the public sector.

In an attempt to examine the factors which explain the level of government expenditure in a country, it is useful to begin by looking into the rationale for governmental intervention. Besides the standard market breakdown arguments and merit goods pro-vision, the control of the commanding heights of the economy argument was important in the case of Tan- zania. Whatever the justification for governmental intervention, concern has been expressed in the lit-erature about how governmental agencies have ifl-uenced government expenditure levels. Various oth-er factors have also been emphasised in the havioral explanation of high government expenditure.[8]

Figure 9–1 shows the trend in government expenditure during the period from 1970 to 1987 in Tanzania. There have been displacements[9] in government expenditure levels in 1977, following the break up of the East African Community; in 1979, following the Idi Amin war; and in 1981/82, following the onset of the country's current economic crisis and the failure of the National Economic Survival Programme (NESP). However, overall there has been steady growth in government expenditure.

The displacement effects are treated as random shocks. Our focus addresses the institutional perception of the tax price of government expenditure and

Table 9–2. *The sustainability of government expenditure in Tanzania with full inflation indexation: fiscal years 1982/83 to 1987/88*
(millions of shillings)

Item		1982/83	1983/84	1984/85	1985/86	1986/87	1987/1988
1	Overall government expenditure	19,276	23,918	26,728	33,220	55,481	77,326
2	Government permanent income[a]	11,919	14,644	17,662	23,392	32,940	49,400
3	=1-2	7,357	9,274	9,066	9,827	22,542	27,926
4	Capital account expenditure	4,404	5,736	5,391	5,817	15,091	17,255
5	= 3-4	2,953	3,538	3,675	3,936	7,451	10,671
6	Inflation adjustment[b]	3,780	4,026	6,569	7,101	8,887	12,097
7	= 5-6	-828	-488	-2,895	-3,165	-1,436	-1,425
8	Exchange rate change adjustment (net)	153	449	0	179	-3,139	-3,313
9	= 7-8	-981	-937	-2,895	-3,344	1,703	1,887
10	Adjustment for GDP growth	45	-233	-211	297	723	1,198
11	= 9-10 surplus (-) and defect (+)	-1,025.9	-704	-2,684	-3,641	980	689

Notes: [a] Permanent income is computed as a three year moving average of government tax and nontax revenues excluding external import support grants. [b] In the course of government expenditure estimates the authorities have directed a 15 percent inflation adjustment, which is also used here.

Source: The computations are based on the optmal and sistainable government expenditure level equation in the text (D+...). Data has been obtained from the *Economic Survey, Bank of Tanzania Reports, and Government Expenditures Supply Notes Vol. II.*

how that has influenced increases in government expenditure, the accounting procedures used for public expenditure, and the cost price impacts on government expenditure. According to the institutional perception of the tax price of government expenditure (Buchanan and Wagner),[10] deficit financing, among other consequences, reduces the perceived price of public provisions. This, in turn, induces higher demand for public provisions on the part of voters and parliamentary members.

The public agency behavior, or the incrementalist hypothesis, postulates a systematic upward bias in government expenditure determination (Wildavsky 1964). It asserts that in the budget making process only a few unusual departures of amounts in the budget from those allocated in preceding years will receive scrutiny, and that a large number of average increase items will be approved routinely. However, the hypothesis does not explain why this occurs. The experience of Tanzania seems to suggest that lack of proper procedures for costing of public provisions and distorted incentives which result in agency expenditure maximization are among the main factors behind the systematic public expenditure bias.

There are several input cost effects on government expenditure. One is the impact of population growth on government expenditure. The quality level of services offered is limited by deteriorating capacity. The presumption is that government responds in order to restore the quality of the services. Another cost impact is the input price impact. Our emphasis has been on import prices and their influence on the import content of the government budget and the domestic price level.[11]

Figure 9-1. Trends in Government Expenditure in Tanzania, 1970-87

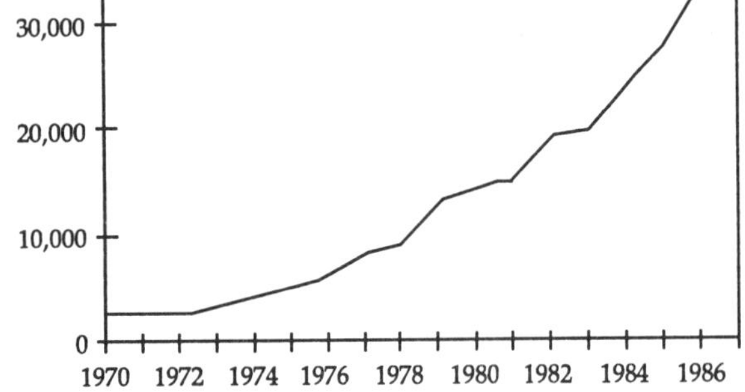

The above hypotheses have been tested for Tanzania. The results are shown in the following equation:

$$\text{(9–2)}\quad G_1 = 0.032 - 0.531T_1 + 1.394G_{1(-1)} + 0.438IMP_{(-1)} + 42.175POP_1$$

$$(0.119)\ (-1.118)\ (10.555^{*})\ (2.280^{**})\ (1.508)$$

$R^2 = 0.963$; $\bar{R}^2 = 0.950$; $DW = 1.88$;
F-Stat. = 77.9; $d.f. = 12$
*, ** = significant at 2% and 5% respectively

Here G_1 is overall government expenditure as a percentage of GDP. The ratio of tax revenue to total financing of government expenditure, T_1, has been used as a measurement of the perception of the tax price of government expenditure. G_1 lagged one period $G_{1(-1)}$ represents the incrementalist hypothesis. The percentage change in the import price index $\dot{IMP}_{(-1)}$ measures the input cost impact. The ratio of total population to total government expenditure $POP1$ is a further measure of the cost price impact.

The above results indicate that the incrementalist hypothesis and the input price impact have significantly influenced government expenditure increases in Tanzania. The tax price impact hypothesis argues that a reduction in the ratio T_1"or increased reliance on deficit financing"should lead to a reduction in the perceived tax cost and conversely to a rise in government expenditure. In the above results, the coefficient estimate has the expected negative sign but is insignificant. This could indicate the extent to which the tax price of government expenditure has not had any significant price rationing impact on the demand for government expenditure increases.

The coefficient of the ratio of population to government expenditure is positive as expected, but is also insignificant. The lack of expenditure response to counter the impact of population pressure on the quality of public service provisions would support the contention that the quality of services has deteriorated. However, evidence of unsustainably high government expenditure may indicate that the decline in the quality of services reflects a thin allocation of government finances over a broad expenditure commitment.

Some Implications of Government Deficit Spending Increases

A number of studies have shown the extent to which the monetization of deficit spending has been among the main factors which have accounted for Tanzania's inflation.[12] This section focuses on the implications of government deficit spending and its possible crowding-out impact.

Excess government borrowing from the banking system has resulted in quantitative credit rationing in the rest of the economy. The testing of the above hypothesis has involved specifying and estimating the bank credit supply function. This includes a variable for government borrowing from the banking system. A significant negative coefficient on this variable should indicate crowding out, while a significant positive coefficient should indicate crowding in.

On the basis of the bank credit supply function specified by Bryan and Carleton,[13] the following equation has been adopted:

(9–3) $CDR_t = a_0 + a_1TDDR_t + a_2GBDR_t + a_3RL_t + U_t.$

The credit to total deposits ratio is denoted by CDR_t, for time period t. The ratio of time deposits to total deposits, $TDDR_t$, is introduced as a measure of the extent to which banks do not need to hold excess reserves for the purpose of meeting the demand for cash. Given other factors, an increase in the ratio of time deposits to total deposits should facilitate increases in credit advances by banks. Government borrowing from the banking system as a percentage of total deposits, $GBDR_t$, has been introduced as a measure of the extent to which credit ceilings are imposed on banks' credit to the rest of the economy as a result of excess government borrowing from the banking system. The real lending rate, RL_t, is included in the model as a measure of returns on bank lending operations. An increase in the lending rate should be expected to induce banks to provide more loan advances.

The data for the variables have been computed from the Bank of Tanzania reports. A weighted average of lending rates on loans of various maturities has been computed. Hence, the real lending rate has been obtained as the weighted average of the nominal lending rates less expected inflation. The above credit equation has been fitted on bank credit to agricultural production, agricultural marketing, manufacturing, public enterprises, and the private sector (table 9–3).

The estimation of the crowding out of credit to the rest of the economy can be regarded as one facet of the impact. A second facet relates to the impact of such a crowding out of credit on output. In order to estimate the consequent output implication, the Lucas supply function has been adopted. The function postulates that, given a situation in which production and factor supplying entities base their decisions on relative prices only, then change in real output will vary with perceived relative prices and with its own lagged value.

Lucas' specification of this assertion is the following:

(9–4) $Y_t = b_1(P_t - E(P_t/I_{t-1})) + b_2Y_{t-1}.$

The variables denote real output, Y_t; the actual inflation rate, P_t; the expected inflation rate, $E(P_t/I_{t-1})$; and lagged real output, Y_{t-1}.Lucas' conclusion is that any policy response which purports to increase output and leads to inflation, will only lead to output increases as long as the actual inflation rate exceeds the expected inflation rate. This function has been fitted on data in developed and less developed countries.[15]

The Lucas supply function specified in the above equation has been extended to include a proxy for credit rationing as:

(9–5) $Y_t = b_1(P_t - \hat{E}(P_t/I_{t-1})) + b_2 Y_{t-1} + b_3B^*_t$

Credit rationing in the rest of the economy, B^*_t, has been measured by the amount by which credit to the rest of the economy has decreased as a result of government bank borrowing. The variable has been generated as a fitted value from the regression of credit to the rest of the economy on government borrowing from the banking system.

During the structural adjustment and economic recovery programs, restraints on government borrowing from the banking system have succeeded in reducing such borrowing to a level below the targets set by the programs. Hence the period before the structural adjustment program, from 1966 to 1982, has been considered to be more relevant in the analysis of the crowding out effect. Results obtained using the ordinary least squares estimator are presented in table 9–3.

As shown in equation (9–1), which refers credit to agricultural production, the coefficient on government borrowing from the banking system is significantly negative. In the case of credit to agricultural marketing, which is shown in equation (9–2), the coefficient on government borrowing is significantly positive. Thus, while credit to agricultural production has been crowded out, credit to agricultural marketing has been crowded in.

Equation (9–3) indicates estimates of credit to manufacturing.The coefficient on government borrowing is negative but insignificant. A crowding in of the public enterprise investment would be ex-

Table 9-3. *The quantitative credit rationing impact of government borrowing from the banking system in Tanzania, 1966 to 1982*

(i) Dependent variable: $CADR_t$ (credit to agricultural production as a percentage of bank deposits)

	Constant	$TDDR_t$	$GBDR_t$	RL_t
Coefficient	0.072	0.001	-0.121	-0.0001
T-statistic	(2.196)	(0.01)	(-3.075*)	(-0.087)

$R^2 = 0.437$ $\bar{R}^2 = 0.307$ DW = 1.7 F-Stat = 3.4 d.f. = 13

(ii) Dependent variable: $CAMDR_t$ (credit to agricultural marketing as a percentage of bank deposits)

	Constant	$TDDR_t$	$GBDR_t$	RL_t
Coefficient	0.264	0.401	0.373	-0.004
T-statistic	(3.470)	(-1.584)	(3.016*)	(-1.750)

$R^2 = 0.469$ $\bar{R}^2 = 0.347$ DW = 1.2 F-Stat = 3.53 d.f. = 13

(iii) Dependent variable: $CAMDR_t$ (credit tothe manufacturing sector as a percentage of bank deposits)

	Constant	$TDDR_t$	$GBDR_t$	RL_t
Coefficient	0.145	0.079	-0.196	-0.001
T-statistic	(1.074)	(0.178)	(-0.849)	(-0.155)

$R^2 = 0.45$ DW = 1.63 F-Stat = 0.2 d.f. = 13

(iv) Dependent variable: $CAMDR_t$ (credit to the private sector measured by commercial bank assets held in the private sector)

	Constant	$TDDR_t$	$GBDR_t$	RL_t
Coefficient	0.667	-0.487	-0.633	0.003
T-statistic	(2.280)	(-0.969)	(-1.832***)	(0.632)

$R^2 = 0.382$ $\bar{R}^2 = 0.239$ DW = 1.64 F-Stat = 2.5 d.f. = 13

(v) Dependent variable: $CAMDR_t$ (credit to public enterprises measured by commercial bank assets held in public enterprises)

	Constant	$TDDR_t$	$GBDR_t$	RL_t
Coefficient	0.098	0.274	1.094	-0.007
T-statistic	(0.430)	(0.375)	(3.956*)	(-0.704)

$R^2 = 0.549$ $\bar{R}^2 = 0.445$ DW = 0.8 F-Stat = 5.3 d.f. = 13

Table 9–3. ***(continued)***

(vi) Dependent variable: LQA_t (natural log of output of the agricultural sector)

	Constant	*$LCPI_t$*	*LQA_{t-1}*	*LB^*_t*
Coefficient	-0.008	-0.280	0.880	-0.042
T-statistic	(-0.714)	(-2.773*)	(2.767*)	(-0.843)

$R^2 = 0.653$ $\bar{R}^2 = 0.513$ DW = 1.83 F-Stat = 6.33 d.f. = 12

(vii) Dependent variable: $LQMF_t$ (natural log of output of the manufacturing sector)

	Constant	*$LCPI_t$*	*$LQMF_{t-1}$*	*LB^*_t*
Coefficient	-0.026	-1.045	0.888	-0.508
T-statistic	(-1.555)	(-6.382*)	(7.653*)	(-0.826)

$R^2 = 0.923$ $\bar{R}^2 = 0.904$ DW = 2.07 F-Stat = 48.05 d.f. = 12

Note: The asterisks, * and ***, indicate significance at 2 percent and 10 percent levels respectively. TDDR is the ratio of time deposits to total deposits; GBDR is government borrowing from the banking system as a percentage of total deposits; RL is the real lending rate; CPI is the consumer price index; B* is a proxy for the crowding out impact of government borrowing on the rest of the economy's credit; and L indicates natural logs.

pected, since manufacturing public enterprises have accounted for a large share of the country's investment through the period in question. The estimation of manufacturing sector credit in equation (9–3) aggregates public and private manufacturing undertakings; a disaggregation is made in equations (9–4) and (9–5). Equation (9–4) presents commercial bank credit to private enterprises. The coefficient on government borrowing is significantly negative. Equation (9–5) relates to public enterprises' credit, and the coefficient on government borrowing is significantly positive. These results indicate the extent to which government borrowing from the banking system has crowded out credit to agricultural production and private enterprises. The same borrowing has crowded in credit to agricultural marketing and public enterprises.

In equations (9–6) and (9–7), the Lucas supply function has been fitted on agricultural sector output and manufacturing sector output respectively. A proxy of the crowding out impact of government borrowing on the rest of the economy's credit has been included in the equation. As shown, however, the coefficient on the variable, though negative as expected, is insignificant. Ambiguity as to the output effect of excess government borrowing from the banking system may reflect the crowding in and crowding out effects which are evidenced in the above results. Evidently, credit to agricultural marketing is as critical in importance as credit to agricultural production, because peasant producers are motivated by the ability to sell their crops. Similarly, credit to public enterprises has a positive impact on their output, due to liquidity constraints on working capital. As for the output effect of the crowding out of credit to private enterprises, existing evidence indicates that a large fraction of private enterprises' investment finance has derived from alternative sources other than credit from the National Bank of Commerce.

Conclusion

This paper has addressed the size of government expenditure in Tanzania. Two conflicting concerns over the current level of government expenditure in the country have been identified. One is the concern that the current measures to bring about reduction in the size of government expenditure have been unduly contractionary. The other is the view that government expenditure is still too high and overstretched.

The model of optimal government deficit and debt has been applied to Tanzania. It has shown that the current level of government expenditure is unsustainable within the constraints of the government's average income and its economic stabilization targets under the structural adjustment and economic recovery programs. Hence, the prevalence of the underfinancing of government provisions despite the unsustainably high level of government expenditure corroborates the broadbased nature of government commitments.

Welfare improvement motivations and unexpected shocks partially explain the high level of government expenditure. However, the government's expenditure estimation procedure (costing) and its tolerance of deficit financing have caused the government budget constraint to be less binding on the level of government expenditure. The cost price impact has also been evidenced to have had a significant influence on government expenditure increases.

Government deficit spending has affected investment in the rest of the economy. Agricultural production and private enterprise credit have been crowded out. Credit to agricultural marketing and public enterprises has been crowded in.The net effect on output has been shown to be indeterminate.

There are several policy suggestions based on these findings. First, the sustainability of government expenditure raises a case for further control of government expenditure in the country. Overall restraints on government expenditure, such as those in the current macroeconomic stabilization policy measures, need to be emphasized. This is especially important in relation to eventual monetization and the debt service burden, and to increasing the binding nature of the government budget constraint and inducing prudence in the government budget.

Second, unless there is a dramatic increase in the government's current and expected future income, the sustainability of government expenditure in Tanzania will require reduction in government expenditure. Measures to increase income would include a dramatic reversal of the inadequate performance of public enterprises.

Third and finally, according to the results presented, government borrowing from the banking system can be used to some extent to obtain needed crowding in effects. In this respect, government guarantees of credit to peasant producers and agricultural marketing should be emphasized.

Notes

1. See ERP (June 1986); Finance Minister's Budget Speech (June 1987); and Tibaijuka (1988).
2. See Finance Minister's Budget Speech (June 1988).
3. See Mikesell and Zinser (1973) for a review of the literature.
4. See for instance Miller, op cit.
5. For the debate see Begg (1982), ch.6.
6. See Buiter 1983 and Rutayisire 1987.
7. See Rutayisire 1988.
8. For a review of the literature see Atkinson and Stiglitz (1980).
9. The thrusting of the level of government expenditure to a higher platau of spending as a result of such a shock as war without reverting to thepre-war level has been propounded by Peacock and Wiseman (1961) as the displacement effect.
10. See Buchanan and Wagner (1977); also the review in Shibata and Kimura (1986).
11. For evidence see Rutayyisire and Mgonja (1989).
12. Among others see Rutayisire and Mgonja (1989)
13. Bryan and Carelton (1967).
14. See Lucas (1973).
15. Ibid.
16. See Rutayisire (1990).

References

Atkinson, A.B. and J.E. Stiglitz (1980); *Lectures on Public Economics* (McGraw-Hill).

Begg, D.K.H. (1982); *The Rational Expectations Revolution in Macroeconomics* (Philip Allan).

Bryan, W.R. and W.T. Carleton (1967); Short-run Adjustment of an Individual Bank, *Econometrica*, Vol. 35, No. 2, (April), pp. 321-347.

Bossons, J. (1986); Measuring the Viability of Implicit Intergenerational Social Contracts in *Public Finance and Public Debt, Proceedings of the 40th Congress of the International Institute of Public Finance* (Wayne State University).

Buchanan, J.M. and R.E. Wagner (1977); *Democracy in Deficit: The Political Legacy of Lord Keynes* (New York: Academic Press, Inc.).

Buiter, H.W. (1983); The Theory of Optimum Deficits and Debt, *NBER Working Paper Series*, Working Paper No. 1232 (Cambridge, Mass: NBER).

Lucas, R.E. Jr. (1973); Some International Evidence on Output - Inflation Trade-Offs, *American Economic Review*, Vol. 68, pp. 326-334.

Mikesell, R.F. and J.E. Zinser (1973); The Nature of the Savings Function in Developing Countries: A Survey of the Theoretical and Empirical Literature, *Journal of Economic Literature pp. 1-26.*

Miller, M. (1982); Inflation-adjusting the Public Sector Financial Deficit, in J. Kay *The 1982 Budget*(Oxford Basil Blackwell).

Peacock, A.T. and J. Wiseman (1961); *The Growth of Public Expenditure in the United Kingdom* (Princeton University Press).

Rutayisire, L.W. (1990); Issues in Monetary and Credit Allocation Policies in the African Context: Evidence from

Tanzania's Experience. A paper presented at the African Centre for Monetary Studies and National Bank of Ethiopia Seminar on Experience with Instruments of Economic Policy, Addis Ababa.

Rutayisire, L.W. (1988); Ways and Means of Improving the Quality and Effectiveness of Public Expenditure Programming: A Case Study of Tanzania in *Ways and Means of Improving the Quality of Public Expenditure Programming,* ECA/PHSB/BUD 188/7 (2.1) (a) (Addis Ababa: United Nations Economic Commission for Africa, October).

Rutayisire, L.W. (1987); Measurement of Government Budget Deficit and Fiscal Stance in a Less Developed Economy: The Case of Tanzania, 1966-198, *World Development,* Vol. 15, No. 10/11 (Oxford: Pergamon Press), pp. 1337-1351.

Rutayisire, L.W. and G.S. Mgonja (1989); The Effect of Devaluation on Government Budget Deficit: A Case Study of Tanzania. A paper presented at a workshop organized by the African Economic Research Consortium (AERC), Harare, Zimbabwe.

Shibata, H. and Y. Kimura (1986); Are Budget Deficits the Cause of Growth in Government Expenditures? *Public Finance and Debt.* Proceedings of the 40th Congress of the International Institute of Public Finance (Detroit, Michigan: Wayne University Press).

Tanzania, Economic Recovery Programme (June 1986)., Finance Minister's Budget Speech (1987 & 1988).

Tibaijuka, A.K., The Need to Monitor the Welfare Implications of Structural Adjustment Programmes in Tanzania. A paper presented at the 5th Economic Policy Workshop (Dar es Salaam, May 1988).

Wildavsky, A. (1964); *The Politics of the Budgetary Process* (Boston: Little Brown and Co).

Zee, H.H. (1988); The Sustainability and Optimality of Government Expenditure, *IMF Staff Papers,* Vol. 35, No. 4, pp.658-685.

10

Poverty Concious Restructuring of Public Expenditures

Marco Ferroni and Ravi Kanbur

Public expenditure choices play a fundamental role in poverty alleviation through their effects on the supply response of adjusting economies and their contribution to human capital formation. Public investment and recurrent spending are important determinants of the quality and quantity of economic and social infrastructure which, in turn, affect human resources and labor productivity, as well as producers' ability to take advantage of adjustment-related changes in relative prices. But government budgets have become tighter during the 1980s in sub-saharan Africa because of stagnating economic growth (particularly in the more easily taxed formal sector), declining commodity prices and a more restricted international credit environment. Resource constraints and the need to control inflation by means of more conservative monetary and fiscal policy are motivating public finance reform in many countries today. The objective of these processes of reform is to bring spending more closely in line with revenue and to raise the efficiency and effectiveness of the government's participation in the economy in support of a specified development path and redistributive (including poverty alleviation) goals. The objective of this paper is to provide an overview of public expenditure patterns in sub-saharan Africa during the adjustment decade of the 1980s and to develop a theoretical framework for retargeting of public spending toward poverty alleviation. The second section is devoted to an analysis of trends in public expenditures, paying particular attention to the extent to which expenditures focus on, and reach, the poor. In the third section, the links between public expenditure restructuring and poverty alleviation are explored in a framework which takes into account human resource interactions and the multidimensionality of the standard of living.

The following section illustrates applications of the framework with African examples. Three areas of applications are explored. We first review a recent attempt at deriving weights for various components of the standard of living (i.e. UNDP's Human Development Index). The question asked is whether the social valuation of specified components of the standard of living, which is implicit in the Human Development Index, can be used as a guide to prioritize public expenditures. The second application uses a production function approach to human resources. The question here is how basic needs inputs (for example, health programs, education services) and the cost thereof in terms of public expenditure are linked to specified achievements in human development such as levels of child mortality, life expectancy, and adult literacy. The third application focuses on the analysis of consumer budgets of the poor. This is combined with data on public spending and used to determine the degree to which government subsidies are reaching the poor in one particular country, Côte d'Ivoire.

The final section concludes the paper and suggests priority areas of research and data collection to support poverty conscious public expenditure reform.

Public Expenditure Patterns in Africa During the Adjustment Decade

In this section, recent trends in public expenditure patterns and, in particular, in spending benefitting the poor are analyzed. This is a necessary step toward the formulation of recommendations, later in the paper, regarding poverty conscious public expenditure reform. We look at aggregate expenditures, the intersectoral and intrasectoral composition of expenditures and the quality and effectiveness of programs and services delivered by the government.

Aggregate and Sectoral Expenditures

The study of public expenditure patterns in sub-saharan Africa is hampered by data deficiencies. There are no comprehensive data on program expenditures (as opposed to broad sectoral aggregates) and on regional and local expenditures. The IMF Government Finance Statistics (GFS) are the only comprehensive data source on central government spending. Even though the GFS data are in reality largely budget figures rather than actual expenditure data, they are used in this paper for want of a published alternative. The GFS figures are, however, complemented by evidence from the World Banks' recent public expenditure reviews in Africa, and this resource, as well as other knowledge makes it possible to develop at least preliminary conclusions regarding the poverty focus and poverty alleviation effect of public spending.

It is often assumed that central government expenditures in Africa and spending on social services (health, education) have plummeted during the debt and adjustment decade of the 1980s. GFS figures indicate that, on average, total real government spending (including interest payments) has grown during the 1980s with a flattening of the trend between 1982 and 1984 (see table 10–1, which was developed using data for 20 African countries chosen on the basis of data availability for all spending categories and years considered in the table). The 1982-84 period was one of drought in the Region and particular resource constraints which were later relaxed to a degree by an increase in donor finance and debt rescheduling in support of the adjustment process. Real per capita expenditures declined in 1983 and 1984, but grew in all other years and regained in 1987 the level attained in 1982. Real discretionary spending (i.e. government spending exclusive of debt service payments) increased in the early 1980s, stagnated in the mid-80s (it declined in 1983 and 1984 in per capita terms) and recovered thereafter. Discretionary spending as a fraction of GDP declined in 1983 and 1984, stagnated during the following two years and recovered in 1987. The overall picture, then, is one of a set-back during 1983 and 1984 and moderate growth in the other years of the decade.

As might be expected, GFS data point to considerable between-country variation around these average trends. In a thorough, GFS-based study of fiscal policy change in Africa, Sahn (1990: 27-31) notes that discretionary government expenditures in low-income and oil importing countries expanded at a much slower pace than in middle-income and oil exporting countries during the 1980s. The author's country-specific analysis also indicates that there has been little growth in discretionary expenditures in CFA countries and east Africa, the mean trend of the latter group of countries being lowered by the decline in Tanzanian real expenditures observed during the 1980s. On the other hand, growth has been more rapid in central and southern Africa (extreme cases in the latter group of countries are Botswana, which exhibits rapid expenditure growth, and Zambia, where real expenditures declined, but began to recover during the second half of the 1980s). The countries where discretionary public expenditures per capita (expressed in constant 1980 US $) declined during the 1980s include Kenya, Liberia, Madagascar, Malawi, Mauritius, Niger, Nigeria, Sierra Leone, and Togo. They increased, according to GFS data, in Burkina Faso, Cameroon, Ethiopia, Ghana, Mali, Somalia and Zimbabwe, among other countries.

Looking at the decade as a whole, rather than its first half, GFS data indicate, in Sahn's words (p. 29), that there have been no "across-the-board reductions" in real discretionary government spending and that for most countries expenditures were "on the rise, or at least steady ... despite the proliferation of IMF and World Bank loans that often carry with them conditions involving budgetary austerity." In view of the 1983-84 set-back in real spending referred to above, it seems, in fact, that adjustment programs, which really proliferated after 1984, have helped restore real spending by lowering the effective rate of inflation and raising foreign resource transfers and (through debt-rescheduling) discretionary budgets.

Table 10–1 shows that social and economic sector spending tended to increase during the decade under review, both as fractions of total discretionary expenditure and in real per capita terms. (There was a decline in average economic service expenditures between 1986 and 1987, but not in spending on the agricultural component thereof.) Trends in social sector spending are closely correlated with trends in discretionary governmental spending in the data used. While social expenditures fluctuated between years, they were higher in 1987 (the last year reported in the table) than at the beginning of the decade. This, of course, should not be taken to mean that essential social services were adequately funded. But there is no evidence, in table 10–1, of a decline in real resources devoted to social services. This expenditure category does not seem to have been vulnerable to cut-backs in government expenditures in Africa or to have borne the brunt of reductions in aggregate expenditures where they have occurred.

This supports the hypothesis formulated by Hicks and Kubisch (1984) that governments tend to protect social expenditures in times of economic difficulty. In their study, social expenditures emerged as the most protected of five expenditure categories examined in 32 developing countries for the period 1972-80. (The authors consider their conclusions preliminary, because their study is limited to consolidated central government accounts, excluding expenditures of state and local governments.) Updating this analysis for 1979-84, Pinstrup-Andersen et al. (1987: 75-81) found similar aggregate results for Africa but considerable reduction in social expenditure in Latin America. Sahn (1990: 56) found the elasticity of health (education) expenditures with respect to total government expenditures in sub-saharan Africa to be rather high, viz. 0.96 (1.08, respectively) for the 1985-87 period. Elasticities were estimated for three periods during 1974-87. The estimated values of elasticities increased with time, indicating that health and education expenditures were given progressively higher priority as time passed. The elasticities with respect to GDP were found to be above unity for both health and education, indicating that increases in aggregate economic growth have led to more than proportionate increases in government spending on the social sectors in Africa.

These trends do not permit conclusions regarding the redistributive or poverty alleviation effect of public spending. They actually appear to indicate that if poverty alleviation effects were weak (as will be argued below), and to the extent that GFS data can be believed, resource scarcity cannot, *a priori,* be invoked as the reason. Inequitable intrasectoral expenditure patterns, low effectiveness in program delivery and inadequate ratios of recurrent to capital spending are the reasons why the poor capture a disproportionately small share of public subsidies and services, as explained below. While increases in financial resources to support anti-poverty programs are needed in Africa, raising the poverty focus of governmental expenditures is also likely to require changes in the within and between sector composition of public spending, among other measures, as well as improvements in the factors which constrain the effectiveness of program delivery.

Intrasectoral Resource Allocation

Anti-poverty spending encompasses economic sector spending to raise the income opportunities of the poor, "essential" social services, and transfer payments or "safety nets" to bolster the consumption of the chronically poor and, transitorily, the needy lo ers under the adjustment process. Economic sector expenditures to raise the incomes of the poor and to increase their contribution to growth include projects and programs to build infrastructure and institutions in support of labor-intensive sectors and patterns of production. Credit programs and interventions to improve poor people's access to inputs and services and to improve the functioning and transparency of the markets in which they trade are items of high priority. Agriculture (including smallholder agriculture) is a key sector to be promoted, despite the fact that agricultural supply response takes time, since no other sector is capable of comparably participatory growth. Yet, macro price policy and development spending have tended to favor industry, in keeping with the African post-independence tradition of inward-looking, industry-based thinking about development. Within agriculture, public spending is often found to be biased in favor of state farms as opposed to private holdings and commercial farming as opposed to the smallholder sector. Rural infrastructure development and agricultural research and extension are deemed to be underfunded in many African countries (see, for example, World Bank 1989a, 1989b, 1989c).

Essential social services geared toward the poor are normally said to include primary and preventive health care (as opposed to hospital-based care, for example) and primary, rather than higher-level, education. The focus on primary education is explained by the fact that this is the foundation of schooling, that primary education is known to yield high private and social rates of return relative to higher-level education, and that primary enrollment rates have stagnated in sub-saharan Africa during the 1980s, and the quality of schooling has declined (World Bank, 1987). To judge the impact of public spending on the poor it is, therefore, necessary to study the record regarding the intrasectoral allocation of resources and the effectiveness of program targeting and delivery. This task cannot be carried out with reference to the aggregate GFS, and there is no comparably comprehensive, disaggregated source. The

Table 10–1. ***Avaerage trends in public expenditures, 20 African countries (1980-87)***
(1980 U.S. dollars, indexed: 1980=100)

	Aggregate government expenditures				
Year	*Total*	*Per capita*	*Discretionary*	*Per capita discretionary*	*Discretionary as a percentage of GDP*
1980	100.0	100.0	100.0	100.0	24.7
1981	108.0	104.8	107.7	104.6	25.3
1982	117.7	111.0	115.7	109.1	25.5
1983	117.8	107.9	113.2	103.6	24.0
1984	114.0	101.2	107.3	95.2	22.8
1985	122.5	105.5	113.8	98.0	22.8
1986	129.5	108.1	117.5	98.2	22.5

Sectoral expenditures

	Per capita	*Share of discretionary budget (percentage)*		*Per capita*	*Share of discretionary budget (percentage)*
Health			**Economic services**		
1980	8.4	5.5	1980	41.6	5.5
1981	8.8	5.6	1981	38.2	5.6
1982	9.5	5.8	1982	39.7	5.8
1983	9.0	6.0	1983	37.7	6.0
1984	8.7	6.0	1984	37.6	6.0
1985	8.7	6.0	1985	39.2	6.0
1986	9.1	6.0	1986	43.0	6.0
1987	9.9	5.8	1987	39.7	5.8
Education			**of which agricultural services**		
1980	24.0	15.4	1980	13.5	8.1
1981	25.3	15.1	1981	11.9	7.6
1982	26.0	16.3	1982	13.9	8.4
1983	25.4	17.0	1983	11.7	8.7
1984	24.8	16.6	1984	12.6	8.4
1985	24.2	16.4	1985	12.5	8.4
1986	24.6	16.7	1986	13.3	9.7
1987	25.3	15.4	1987	14.7	9.5
Social services (health and education combined)					
1980	32.4	19.9			
1981	34.1	20.7			
1982	35.5	22.1			
1983	34.4	23.0			
1984	33.5	22.6			
1985	32.9	22.4			
1986	33.7	22.7			
1987	35.2	21.2			

* See text for explanation. The countries are: Botswana, Burkina Faso, Cameroon, Ethiopia, Ghana, Kenya, Liberia, Madagascar, Malawi, Mali, Mauritius, Niger, Nigeria, Sierra Leone, Somalia, Swaziland, Togo, Uganda, Zambia, Zimbabwe.
Source: Caluculated on the casis of data from IMF, Government Finance Statistics (1989 Yearbook, Washington, D.C., 1989.

evidence must be pieced together, and we do this on the basis of World Bank Sector Reports and recent Public Expenditure Reviews, focusing in particular on social sector spending and food-linked transfer payments. It is worth noting, however, that while Public Expenditure Reviews contain much that is relevant to our purpose, they do not focus explicitly on poverty, but on sector strategies and the implications of expenditure levels and patterns for the overall monetary and fiscal stance.

In contrast with the identified priority areas of anti-poverty social spending, public expenditures on health and education in Africa tend to be characterized by disproportionate resource allocation to higher order services which benefit the wealthier and, in particular, urban-based groups of society. For example, the budget of Ghana's Ministry of Health is directed predominantly at curative care for the one-third of the population living in urban areas (World Bank, 1989d: 44). In Tanzania, hospital services account for 68% in the 1988/89 budget of the central Ministry of Health, whereas preventive services claim 5.9% (World Bank, 1989b: 86). While this underestimates the actual expenditure on preventive care, because hospitals also offer this kind of care (for example, immunizations) and because the main responsibility for preventive care is reported to be at the district level and reflected in the recurrent budget of the Ministry of Local Government, this budget, too, shows only a small percentage of expenditures (4%) devoted to preventive care. Of total government recurrent expenditures for health, including those channelled to district councils, only 8.8% was allocated to primary health facilities in 1987/88 (World Bank, 1989e: 86-87).

In education, this pattern of discriminatory public spending finds expression in the fact that subsidies per user increase dramatically with the level of schooling. Thus, public recurrent expenditure per primary pupil in sub-saharan Africa has been estimated at $48 in 1983 (15% of GNP per capita). It amounted to $223 per secondary student (62% of GNP per capita) and $2710 per tertiary student, or eight times GNP per capita (World Bank, 1987: 141-143). While post-primary education must, of course, be promoted, it would appear to be appropriate, on equity grounds, to increase students' participation in the financing of costs, combined with the devel- opment of an educational credit system and a sch-olarship fund to enable students from poor families to enroll. Public funds thus released could be div- erted to primary education where there is normal ly believed to be little scope for cost recovery, but where there is great need to increase coverage and quality.

Turning to the issue of transfers, the last of the three broad areas of anti-poverty public spending identified above, and focussing on food subsidies, one finds that governments do not normally publish data regarding explicit financial subsidies. Real spending on food subsidies is assumed to have declined in many African countries during the 1980s (for Zambia, this is documented in Pinstrup-Andersen et al., 1987: 86), but the impact of this trend on the poor is not straightforward. It depends on the distribution of the incidence of subsidies across the income spectrum, adaptive responses of the poor to price increases (both on the income and the consumption side), and whether policies to compensate the poor for price increases are put into effect. Since, as is well known, African food subsidies are not well targeted on the poor, declines in real financial subsidies may not be particularly harmful to this group. Malnutrition is, of course, wide-spread (see Alderman, 1990, for evidence regarding the incidence of malnutrition in selected African countries, and an analysis of the prevalence and determinants of malnutrition in Ghana), calling for remedial action capable of improving both income and human resource factors (for example, mothers' education) which determine nutritional status. Food subsidies clearly have a role to play among the envelope of policy options to improve nutrition (Behrman, 1989: 101), but they should be narrowly targeted to the poor--a condition which, for administrative and political reasons, it may be difficult to fulfill.

A method to determine the poverty consequences of food price changes is proposed and illustrated later. It hinges on the recognition that, in order to reach the poor and to significantly affect their real incomes, government subsidies must be directed at goods and services which are widely consumed by this group. To keep leakages of benefits to nonpoor groups small, the poor must account for a large proportion of the total national consumption of subsidized goods and services. The described pattern of intrasectoral resource allocation and the analysis of consumer budgets suggest that neither of these conditions is usually fulfilled.

Effectiveness of Public Expenditures

The efficiency of resource use and the effectiveness of public service delivery have been found to decline when government budgets contract (Gallagher and Ogbu, 1989). This is in large measure due to the circumstance that, in times of resource scarcity, governments tend to maintain the number of workers on their pay roll, and nominal salaries, at the expense of non-personnel inputs. Nevertheless, real civil servant pay has declined drastically in some African countries (in Tanzania it declined by a factor of five between the early 1970s and the late 1980s; World Bank, 1989f: Vol. I, p. 3), leading to a crisis of motivation and ability to deliver on the part of the civil service and prompting many competent officials to move to the private sector or engage in moonlighting activities. The frequently observed insufficiency of funds for materials, operations and maintenance, and mobility of personnel has had the effect of undermining government delivery systems in such sectors as transport, agricultural extension, and education and health. Inadequate operations and maintenance expenditures have diminished the productivity of past capital spending and have ushered in a period (which began in the 1970s in many countries) of degradation of economic and social infrastructure. The implementation of donor-assisted development projects has been slowed down by the lack of matching recurrent expenditures from governments.

In countries where data are available, wages and salaries have typically absorbed from one-half to three-quarters of recurrent expenditures. In Ghana's health budget, for example, wages took 62% and 76% in 1985 and 1986, respectively (Vogel, 1988: 26). The share of wages and personnel emoluments in Rwanda averaged 48% of recurrent spending in the health sector during 1982-88 and took over 80% in education. Only 3-4% of recurrent spending in education was left for materials and less than 1% for maintenance (World Bank, 1989b). This pattern of budgetary allocation under adverse conditions, unjustifiable on economic grounds, but understandable for political and social reasons, has all but crippled many government services in the Region due to shortages of drugs and disposable supplies in health, shortages of text books and school supplies in education, and immobility of service delivery personnel.

The effectiveness (and the poverty alleviation impact) of a given evolution of real discretionary expenditures can be very different depending on the differential trend of recurrent and investment expenditures within that envelope. For example, an expansion in discretionary expenditures coupled with a rapid decline in the ratio of recurrent to capital spending could exacerbate the shortage of funds for nonwage recurrent inputs referred to above. This was probably the case in Cameroon in the latter part of the 1980s since the average ratio of recurrent to capital expenditures fell from 2.3 to 1.3 between 1978-80 and 1986-87 in that country, while real wage expenditures rose (Sahn, 1990; Tables 7 and 8). In most African countries, however, the ratio of recurrent to investment expenditures appears to have increased during at least part of the 1980s according to GFS data, although the ratio varied considerably between countries and years (Sahn, 1990; Table 7). The analysis of the implications of given functional compositions of expenditure and of the decline in capital spending in terms of future growth and the ability to deliver services requires a careful country by country evaluation of the trade-offs involved. This is beyond the scope of this paper. Where investment expenditures have declined in the face of growing discretionary expenditures, additional resources to defray recurrent costs have become available. Nevertheless, the Bank's recent Public Expenditure Reviews which were analyzed as part of the research undertaken for this paper consistently point to a lack of funds to cover the recurrent expenditures associated with essential services. This would indicate a need to raise the budgetary resources of sectoral agencies providing essential services and infrastructure, while at the same time undertaking measures to correct imbalances in the ratio of wage to nonwage recurrent expenditure. Raising the budget of the identified agencies for reasons of poverty alleviation is likely to require resource reallocation, and decision criteria to guide this process are developed later in this paper.

Beyond inappropriate input combinations and ratios of recurrent to investment spending, a crucial factor constraining the effectiveness and efficiency of public spending is the lack of adequate planning and management and, hence, of coherent public expenditure strategies. Efforts to raise the productivity of budgetary resources include the setting of realistic goals (arguably a difficult task in the light of political pressures and rapidly expanding populations), the explicit acknowledgment of recurrent cost implications of new projects and past investments, the exploration of opportunities to raise new revenue, the promotion of manpower competence in the civil service through training and management improvements, and the rationalization of donor assistance. It is now widely recognized that projects are often included in public spending portfolios because of the availability of donor funding, rather than their agreement with articulated sectoral priorities. Similarly, ministerial management capabilities are taxed to their limit by the diversity of projects and partners. In Zambia's agriculture sector, for example, there are about 150 projects supported by more than 20 donor agencies, each maintaining its own procedures and requirements. "Overlaps and conflict between the objectives of different projects are commonplace" (World Bank, 1989a: Vol. II, p. 29). The solution to the problems arising from donor proliferation is, of course, not to suggest a cut-back in donor assistance, but its progressive subordinat ion to government priorities in the 1990s, although it is recognized that these priorities may not yet be fully spelled out in some countries.

Conclusion

Our discussion suggests that the poverty allevition impact of public expenditures in sub-saharan Africa is low because recurrent expenditures are under-funded, government subsidies tend to be directed to higher order services and commodities not widely consumed by the poor, and because management problems and suboptimal input combinations are diminishing the efficiency and effectiveness of service delivery. This is also the conclusion of the World Development Report 1990, on poverty, which finds that government spending "tends to be skewed away from the people who need it most--the poor" (World Bank 1990a, p. 76). An important assumption implicit in this conclusion is that the "poor" are separated from the "nonpoor" by a poverty line defined as the level of income which cuts off, say, the lower 30%-40% of the income distribution. While this is not the place to discuss the measurement of poverty or the conceptual and practical difficulties inherent in the notions of relative and absolute poverty, it is obv- ious that if everybody, or a vast majority of the population, were classified as being poor (it is sometimes suggested in Africa that this be done), our conclus ion regarding the poverty focus of public expen- ditures would not hold. The definition of a poverty line which helps policy makers focus on the lower end of the distribution of income is a prerequisite for the formulation of public policy, including public expenditure reform, seeking to alleviate poverty.

If poverty alleviation is accepted as an explicit objective of public expenditure strategies, it is necessary to review and redefine the sectoral, intrasectoral and functional distribution of funds in the three areas of essential economic and social services and infrastructure, and transfer payments, identified earlier. It is, however, not obvious how this may be best done, because human capital inputs interact with each other and with directly income-enhancing economic sector expenditures, and these interactions are difficult to quantify. For example, what is the poverty alleviation effect of an additional dollar spent on agricultural extension for women farmers compared with an additional dollar spent on primary health care? A framework to analyze this problem is developed in the next section.

Restructuring of Public Expenditures for Poverty Alleviation: a Theoretical Framework

Our analysis of public expenditure patterns in the previous section focused on two policy variables, viz. the flow of funds to certain sectors and programs, and the factors which determine the efficiency and effectiveness of expenditures (i.e. the functional composition of expenditures and public expenditure management). In this section we will focus on the first of these variables. We hope to contribute to the formulation of poverty sensitive public expenditure strategies through an analysis of public expenditure choices and effects. This is complicated, among other reasons, by the fact that the various dimensions of the standard of living of households and individuals are affected individually and jointly by the different components of a governmental spending program. Thus, public expenditure restructuring will have complex effects with many interactions, and difficult choices will have to be made from the various trade-offs. The purpose of the analysis that follows is to develop a framework in which some of these trade-offs can be clarified.

The multi-dimensionality of the standard of living has to be faced head on in assessing the consequences of public expenditure restructuring. The standard focus on income or expenditure based measures of welfare must be complemented by the concept of "basic needs" which was introduced in the 1970s. The "basic needs" literature stresses a number of indicators (in particular, life expectancy, literacy, health, nutrition, and housing) as being complementary, if not superior, to the usual income/expenditure indicator, and argues for a strategy to increase the values of these indicators. A closer reading reveals several arguments (sometimes not clearly separated) for a basic needs strategy:

(i) Thinking of basic needs requirements as entering the standard of living directly, it is argued that the standard of living of the poor can be raised more efficiently by focussing on basic needs. There are, in turn, two sub-arguments here. One is that for any poor person a dollar spent directly on basic needs will be better than a dollar spent directly on income raising (which will then, indirectly, influence basic needs). Another is that basic needs spending could be better targeted toward the poor.

(ii) Thinking of basic needs achievements as being inputs into income generation, it is argued that the rate of return to such investment is higher than that, for example, in directly productive physical capital. Thus, even if income/expenditure were the ultimate aim, a basic needs strategy is superior. There is once again a targeting argument to supplement and bolster the basic case.

The "social sectors" in a governmental expenditure program display all of these considerations, but they lie along a spectrum. In the case of housing we come closest to the pure consumption end of the spectrum (at least, we have not seen studies that argue for housing in terms of its productivity enhancing properties). In fact, many studies on poverty do indeed monetize housing consumption by imputation through hedonic regressions and the like. Education is perhaps at the other end of the spectrum, where the literature concentrates primarily on its productivity aspects rather than on the consumption value. In principle, the decision rule for public resource allocation is straight forward—it depends on rates of return to different levels of education.

Health and nutrition are located toward the middle of the spectrum. They enter the standard of living directly as well as indirectly through productivity effects. The well-known controversy about the relative merits of growth oriented versus basic needs oriented public action in Sri Lanka (Bhalla and Glewwe, 1986; Sen, 1981; Isenman, 1980; Anand and Kanbur, 1990) can be seen to take the first route. All participants in the debate have agreed that infant mortality, for example, should be reduced because this is a desirable goal by itself. The only question is whether this is best achieved through income growth or through basic needs intervention. An issue which is not addressed in the literature just cited is the feedback of improved health and nutrition on income growth. Another question which is not raised, although it is implicit in the discussion, is whether direct intervention (i.e. "social sector" spending) can be better targeted towards the poor. In this context, however, it should be noted that the entire discussion on Sri Lanka is based on national average indicators of basic needs fulfillment—the question of basic needs fulfillment of the *poor* can therefore be addressed only indirectly, if at all.

There seem, then, to be (at least) four complicating aspects of public expenditure restructuring: (i) One component of public expenditure can affect several aspects of the standard of living of a typical household; (ii) Each component of the standard of living can affect other components of the standard of living for a typical household; (iii) The different components of the standard of living have to be valued relative to each other for a typical household; and (iv) Some components of public expenditure when passed through (i) - (iii) above, are more effective than others in raising the standard of living of *poor* households.

Let us start by highlighting (i) - (iii) and focus, therefore, on a typical (not necessarily poor) household or individual. To put some of these issues into a common framework, consider the following simple model. Basic needs achievement B is a function of social expenditure E and income Y as follows:

$$B = \alpha_0 + \alpha_E E + \alpha_Y Y \quad (10\text{–}1)$$

Income on the other hand is a function of basic needs achievement and "productive" expenditure I:

$$Y = \beta_0 + \beta_B B + \beta_I I \quad (10\text{–}2)$$

There is a budget constraint

$$G = E + I \quad (10\text{–}3)$$

and the valuation of B and Y to give the "true" standard of living W is,

$$W = \gamma_0 + \gamma_B B + \gamma_Y Y \quad (10\text{–}4)$$

The government faces the choice of restructuring by changing the balance between E and I. In which direction should it move? In order to answer this question let us first solve (10–1) and (10–2) to give the values of B and Y for any given values of E and I:

$$B^* = \frac{\alpha_0 + \alpha_Y \beta_0 + \alpha_E E + \alpha_Y \beta_I I}{1 - \alpha_Y \beta_B} \quad (10\text{–}5)$$

$$Y^* = \frac{\beta_0 + \beta_B \alpha_0 + \beta_B \alpha_E E + \beta_I I}{1 - \alpha_Y \beta_B} \quad (10\text{–}6)$$

Substituting these into (10–4) we get

$$W^* = \frac{\gamma_0 + \gamma_B (\alpha_0 + \alpha_Y \beta_0) + \gamma_Y(\beta_B \alpha_0)}{1 - \alpha_Y \beta_B} + \frac{[(\gamma_B \alpha_B + \gamma_Y \beta_B \alpha_E) E + (\gamma_B \alpha_Y \beta_I + \gamma_Y \beta_I) I]}{1 - \alpha_Y \beta_B} \quad (10\text{–}7)$$

At the margin, therefore, the choice between putting one more dollar in E versus I depends on the comparison

$$\alpha_E (\gamma + \gamma_Y \beta_B) >< \beta_I (\gamma_Y + \gamma_B \alpha_Y) \quad (10\text{–}8)$$

The comparison depends on a combination of productivity and valuation considerations. For example, suppose that we were interested only in basic needs achievement, so that $\gamma_Y = 0$. Then the choice between E and I depends on

$$\alpha_E >< \beta_I \alpha_Y \quad (10\text{–}9)$$

In other words, it depends on a comparison of direct versus indirect effects. This comparison is at the heart of the debate on Sri Lanka's policies in the 1960s and 1970s. Anand and Kanbur (1990) have estimated the relationship (1) for Sri Lanka on time series data, and conclude that for the infant mortality rate the direct effects of social expenditure greatly outweigh the indirect effects operating through investment in income earning opportunities. But this is a country specific finding which may not apply to Africa.

If basic needs had no productivity effects, so that $\beta_B = 0$ then (10–8) collapses to

$$(10\text{–}10) \qquad \alpha_E \gamma_B >< \beta_I [\gamma_Y + \gamma_B \alpha_Y] \quad \text{or} \quad (\alpha_E - \beta_I \alpha_Y)\gamma_B >< \beta_I \gamma_Y$$

Thus a prerequisite for basic needs expenditure to be worth considering at all is that direct expenditure leads to a bigger effect (α_E) on basic needs than the indirect effect through income increase ($\beta_I \gamma_Y$). This is the Sri Lanka controversy again, but now there are extra considerations in γ_B and γ_Y. Even if the direct effect was greater, the social valuation of basic needs achievement would have to be sufficiently high for basic needs expenditure to be worthwhile.

Finally, if basic needs had only productivity effects and there were no feedbacks via income to basic needs, so that $\alpha_Y = 0$, and basic needs were not valued for themselves but only for the income they generate, so that $\gamma_B = 0$, then (10–8) collapses to

$$(10\text{–}11) \qquad \alpha_E \beta_B >< \beta_I$$

This is a straight-forward marginal productivity comparison between expenditure on education, say, and expenditure on other production activities. A unit of expenditure diverted from these other activities has opportunity cost β_I in terms of income foregone. But it leads to an increase in education, which in turn leads to an increase $\alpha_E \beta_B$ in income.

Thus we see that many of the strands of the arguments surrounding basic needs fall out as special cases of (10–8). Moreover, the above analysis can be applied to the choice between any two categories of public expenditure regardless of the level of sectoral or intrasectoral aggregation—health and education, health and economic infrastructure, primary and higher levels of eduction. While the technical parameters α and β are necessary and not easy to estimate, what seems to stand out is the need to arrive at clear social weights γ between different dimensions of the standard of living.

The above discussion highlights some of the interactions between different categories of public expenditure in their impact on the standard of living of a *typical* household or individual. We now turn to the extent to which the policy actions affect the poor. As noted earlier, an implicit argument in some of the literature is that certain types of expenditure are to be preferred because they are or can be better targeted. This argument needs to be made explicit, to be made precise, and to be quantified.

Suppose that there are n units in the economy, indexed by $i = 1, 2, \ldots, n$. Out of the total expenditures E and I, let E_i and I_i be the amounts that reach unit i. Clearly,

$$(10\text{–}12) \qquad \sum_{i=1}^{n} E_i = E \, ; \sum_{i=1}^{n} I_i = I$$

The individual counterpart to (10–7) is thus given by

$$(10\text{–}13) \quad W_i^* = \frac{\gamma_o + \gamma_B(\alpha_o + \alpha_Y \beta_o) + \gamma_Y(\beta_o + \beta_B \alpha_o)}{1 - \alpha_Y \beta_B} + \left(\frac{\gamma_B \alpha_E + \gamma_Y \beta_B \alpha_E}{1 - \alpha_Y \beta_b}\right) E_i + \left(\frac{\gamma_B \alpha_Y \beta_I + \gamma_Y \beta_I}{1 - \alpha_Y \beta_B}\right) I_i$$

Equation (10–13) gives the effect on individual income of expenditures E_i and I_i reaching individual i. We now need to formalize a focus on poor units, so as to gauge the impact of restructuring on them. This is not the place to discuss in detail the drawing of poverty lines or the formulation of poverty indices. There is now a large literature on this. Suffice it to say that given a poverty line z which delineates poor from nonpoor, one index that is becoming quite commonly used is that put forward by Foster, Greer and Thorbecke (1984). This is defined by:

$$(10\text{–}14) \qquad P_\alpha = \frac{1}{n} \sum_{i=1}^{q} [(z - W^*i)/z]^\alpha \; ; \; \alpha \geq 0$$

When $\alpha = 0$, this index turns into the commonly used head count ratio or incidence of poverty. When $\alpha = 1$, it measures the normalized "poverty gap." As the value of increases, more and more weight is given to the poorest of the poor. This family of measures has proved useful in operationalising poverty measurement, while allowing us to represent a range of value judgements through the ability to vary the parameter.

Suppose now that *a marginal* budgetary shift from E to I occurs in the aggregate. How does this feed through to individuals? One might entertain different possibilities. One is that individuals gain or loose in proportion to their current levels of E and I. Let each E_i become

$E_i(1+\theta)$ and each I_i become $I_i(1-\sigma)$. Clearly, from the budget constraint (10–3):

$$(10\text{–}15) \qquad \theta E = \sigma I$$

Totally differentiating (10–14) and using (10–15) we get

$$(10\text{–}16)\ dP_\alpha = -\frac{1}{n}\sum_{i=1}^{q} (\alpha/z)\,[(z-W^{\bullet})/z]^{\alpha-1}\{C_E\, E_i d\theta - C_I I_i\, d\sigma]$$

$$= -\frac{E}{n}\sum_{i=1}^{q} (\alpha/z)\,[(z-W^{\bullet})/z]^{\alpha^{-1}}\,[C_E\,(E_i/E) - C_I\,(I_i/I)\,]$$

Where,

$$C_E = \frac{\gamma_B\,\alpha_E + \gamma_Y\,\beta_B\,\alpha_E}{1-\alpha_Y\,\beta_B}$$

$$C_E = \frac{\gamma_B\,\alpha_Y\,\beta_I + \gamma_Y\,\beta I}{1-\alpha_Y\,\beta_B}$$

As might be expected, the impact on poverty depends on the current shares of total expenditure of each type reaching the poor. Thus, when $\alpha = 1$, we get

$$(10\text{–}17) \qquad dP_1 = -(E/nz)\,[C_E(E^P/E) - C_I(I^P/I)]$$

Where E^P and I^P are expenditures of each type reaching the poor. Thus, with this framework the "targeting" case for certain categories of expenditure relies on high values of ratios of the type Ep/E and I^P/I. These are in principle verifiable and quantifiable.

Toward Implementation: Some African Illustrations

It is convenient to think of the problem of poverty conscious public expenditure restructuring as follows. For a given total of public expenditure, the policy instrument available to us is to alter the composition of expenditure between *relatively broad* sectors and programs within sectors. While the answer to the question "how broad?" is country specific, we would like to retain the sense that we are not here discussing very fine micro management of individual programs. While this is important, our task here is to change the pattern of resource flows at a more aggregative sectoral or intrasectoral level. Given the current structure of utilization of these resources, we wish to determine how best to allocate expenditure across different categories. The theoretical analysis highlights three sets of parameters that are crucial: (i) those that quantify the importance of one dimension as opposed to another in the social valuation of the standard of living, (ii) those that quantify the links between public expenditure and achievements along several dimensions of the standard of living, and (iii) an assessment of what fraction of public expenditure in any given category reaches the poor. We now consider each of these parameters in turn.

Social Weights on Different Dimensions of the Standard of Living

Attaching weights to the various dimensions of the standard of living is a *normative* question. The topic of weighting different outcomes in, say, health, education and income needs considerable thought, not least from policy makers, so as to arrive at coherent ways of assessing the normative evaluation of alternative policies. Otherwise, every policy can be justified on some weighting or the other.

Developing such a weighting is not an easy task, and attempts to do so are fraught with danger. One recent attempt is that of UNDP (1990), whose Human Development Report advances a Human Development Index (HDI) as a measure of achievement that incorporates income and non-income factors. The HDI is defined as follows. First, we specify three relevant indicators at the national level as its components—life expectancy (X_1), literacy (X_2), and the logarithm of real GDP per capita (X_3). Looking across a range of countries, the maximum and minimum value for each indicator is established. A "deprivation" index for the ith indicator and the jth country is then defined as

$$I_{ij} = \frac{(max_j\,X_{ij} - x_{ij})}{max_j\,X_{ij} - min j\,X_{ij})}$$

Clearly, I_{ij} lies between 0 and 1. UNDP (1990) then defines the deprivation index for country j as a simple average of the three deprivation indexes for the country:

$$I_j = (1/3)\sum_{i=1}^{3} I_{ij}$$

The "Human Development Index" is then defined as:

$$(HDI)_j = 1 - I_j$$

The reader is referred to the original source, UNDP (1990), for the calculated values of the *HDI*. It is worth noting that, in the report, 44 countries are defined as having a "low" level of human development or an *HDI* of less than 0.5. Of these, 32 are in sub-saharan Africa. However, while the *HDI* is an interesting attempt at arriving at a unidimensional measure of achievement, it has to be viewed with caution. First, the normalization assumption seems problematic. In effect the index views achievement relative to the best country in the sample. Thus, if Japan's life expectancy were to fall, Kenya's *HDI* would go up! That is an odd sort of index to have. Second, even if we set aside this normalization problem, the index essentially gives equal weight to achievement along the three dimensions. Is this an accurate reflection of value judgements, and is it even clear what value judgements this implies?

To see how one might set about developing an index from a coherent set of value judgements, and only as an illustration, let an individual's income at time t be y_t and his utility of income $U(y_t)$. Then his/her expected lifetime utility at birth is:

$$W = E \int_0^T U(y_t) e^{-\delta t} dt$$

where $\int$ is a discount rate, T is lifetime (a random variable) and E is the expectation operator. Then, if T is exponentially distributed with parameter λ,

$$W = \int_0^\infty U(y_t) e^{-(\delta + \lambda)t} dt$$

Moreover, if $y_t = y$ for all t and $U(y) = 1n\, y$, then

$$W = \frac{L}{1 + \delta L} \; [1ny]$$

where $L = 1/\lambda$ is the expected lifetime. Thus, if $= 0$, we get

$$W = L. [\ln y]$$

The above expression is related to (at least two elements of) the *HDI*, but is nowhere near it in actuality—country rankings could be vastly different as between the two. However, the index just derived at least has the virtue of making its value judgements clear and transparent. With the *HDI* it is not clear what the judgements underlying the very precise implied tradeoffs are or where they came from. Thus, there does not seem to be a basis for recommending the *HDI*'s weighting of welfare components as a guide to public expenditure reform.

Public Expenditures and Welfare Achievements

The second consideration mentioned above (i.e. the quantification of the links between public spending and welfare achievements) calls for an evaluation of complementarities (or substitutabilities) between different types of interventions in affecting relevant outcomes. For example, it is often argued that improving mothers' education is the best investment one can make for children's health. But *quantitative* estimates of the effect are rare. Would a dollar diverted from primary health care to primary education for girls really lead to a long term decline in child mortality? If so, by how much? Within the health sector, would a dollar diverted from drugs in rural health centers to improving quality of health personnel in these same health centers lead to an improvement in measured health indicators? If so, by how much? Equally difficult questions lurk in the nutrition-health area. The complementarities here are well recognized but it is their quantification that is problematic and lagging behind. If the dreadful choice between maintaining a nutrition program or a vaccination program has to be made, what are the tradeoffs? A badly nourished person is more likely to develop complications from a disease. But how much more likely?

These are not questions to which we have answers, but we doubt that policy analysts have often enough posed the question in this way, so as to force the limited econometric analysis that exists to speak to these issues. One notable exception is a recent Bank study on "Health Care Cost, Financing and Utilization" in Nigeria (World Bank, 1990b). This study concentrates on the problems and prospects for cost recovery in health care, using as an example Ogun State in Nigeria. Based on a household survey, the study first estimates the demand for health care in its various dimensions as a function of individual variables (income, education, etc.), price of care, as well as quality variables (for example, the availability of drugs and the physical condition of facilities). In this health production function (or reduced form demand analysis) the frequency of visits to different types of health facilities is taken as the dependent variable.

Such studies, on the intermediate or proximate determinants of achievements are common enough. Two recent studies of this type on the determinants of nutritional achievements in Africa are by Alderman (1990) and Sahn (1990). The explanatory variables in these studies include parental education and its effect on children's nutritional status as measured by anthropometric achievements. What is unknown in this approach is the influence of (past) public expenditure strategies on today's level of parental education.

Taking frequency of visits to modern (private and public) health centers as the policy variable of interest, the Nigeria study then simulates the impact of various cost-recovery related policy changes on the outcome. Examples of this might be *(i)* an increase in the price of health services provided by the public sector, (ii) an increase in the price plus an improvement in quality of facilities (suitably defined), and (iii) improvements in drug availability, and so on. For 19 such policy options the study simulates the impact on frequency of visits and on the public budget. The study illustrates what is required before expenditure restructuring analysis can be done even on one quite narrowly defined outcome—frequency of visits to health centers.

While the simulation of the impact of these policy changes on the frequency of visits uses the estimated demand for health functions discussed above, their impact on the public budget requires a different type of analysis: in particular, the costs of quality and drug availability improvements have to be estimated. This requires a detailed cost analysis of the different components of the public health system, including personnel, drugs, and physical infrastructure. While this is presented in the Nigeria study, this type of analysis is typically not available for African countries. It is precisely this direction of research which is needed to complement the growing literature on estimates of household level health production functions.

Public Expenditures and the Consumption Patterns of the Poor

The third component of the framework developed in Section 3 is the assessment of what fraction of public expenditure of a particular type is reaching the poor. It may be surmised that this is more easily analyzed than the questions asked in the two preceding subsections. This is in fact correct, but (like the estimation of the production function of basic needs satisfaction referred to above) it relies on the availability of household survey data of a type that is not yet widely available in African countries. We will illustrate the analysis with three examples from Côte d'Ivoire, a country which has undergone dramatic economic difficulties in the 1980s, and has been forced to reconsider the level and composition of its public expenditure.

Let us start the discussion with a consideration of consumer price subsidies and their impact on the public budget and on the poor. As between subsidies on one commodity or the other, it can be shown that in expression (10–17) $C_E = C_I$, so that in a poverty conscious restructuring of subsidies the government should move towards commodities that have a high value of the ratio of poor people's consumption to national consumption. This is shown formally in Besley and Kanbur (1988) but is intuitively obvious. A unit reduction in the national subsidy on a commodity saves the budget an amount proportional to the total amount consumed in the economy. The impact on the reduction of the poverty gap is (to a first order of approximation) proportional to the total amount consumed by the poor. Hence the critical ratio is that of the latter to the former--a poverty conscious, and *balanced budget*, restructuring would mean spending more on commodities that had a higher value of this ratio.

The above analysis can be complicated considerably, but the rule of thumb developed here has the virtue of simplicity. It can be applied to household income/expenditure surveys that disaggregate consumption by commodities and permit the establishment of a poverty line. Let us start by analyzing each household's accounts in terms of sources of income and destinations of expenditure. In fact, total expenditure (per capita) will be our measure of individual welfare. The destinations of expenditure can be broken up into several basic categories: for example, consumer expenditure on food, consumption of home produced food, consumption of home produced nonfood items, other consumption expenditure including nonfood items, and remittances paid out. Each of these can in turn be disaggregated further to any level that the data will allow and the policy analysis requires. Thus, if the price of kerosene is an important policy instrument this should appear as a separate category in the disaggregation of consumer expenditure on nonfood items. Similarly, if the price of rice looms large in policy discussions of consumer subsidies, these should appear as a separate category. In any event, given a mutually exclusive and exhaustive categorization of expenditures, and given a poverty line, an expenditure decomposition matrix can be drawn up .

Under the total in each cell should be the percentages it represents of the column total and of the row total. These percentages are highly relevant for the analysis of the poverty impact of food subsidy changes. But they have to be used with care. Often one finds analysts using column percentages to claim that the impact of a food subsidy reduction will or will not be large on the poor. Thus, if a particular commodity accounts for a small fraction of poor households' total consumption (i.e. a low column percentage), it is argued that this is an attractive commodity for a subsidy cut. However, this does not take into account what the saving will be on the fiscal deficit account—which is why the subsidy is being cut in the first place. An appropriate question to ask is: What is the poverty impact per unit of deficit reduction? For this it is the row percentage that is relevant, as argued above. Of course one can make the analysis much more complicated, but in the operational context a matrix such as the one suggested here should prove useful as a first cut.

Often a price change will have an effect on producers of the commodity also. A similar strategy can be applied to the sources of income. Starting off from very broad categories (employment income, agricultural income, nonfarm self-employment income, remittances received), we can disaggregate down as far as the data will allow and the analysis requires. Employment income can be further broken down by production sectors and agricultural income by crop. Nonfarm self-employment income can also be further disaggregated to production sectors where relative price changes are occurring.

An illustration of the above analysis is the case of the price of rice in Côte d'Ivoire. From the Living Standards Measurement Survey (1985) for Côte d'Ivoire, Kanbur (1990) calculated the indicators in table 10–2. The details of the calculations are given in the original source and are not our concern at the moment. In the third Côte d'Ivoire Structural Adjustment Loan Agreement (1986) explicit mention is made of the need to bring domestic prices of rice more in line with international prices. The efficiency based welfare economics of such policy reform is well known--the sum of producer and consumer surplus increases relative to the distorted equilibrium. However, there are distributional implications depending on how prices are changed and in what direction.

Table 10–2 shows that the incidence of poverty among rice farmers is 35.7% (compared to 30% for all Ivorians). Rice producers thus have a special claim in a policy of poverty alleviation. The mean area farmed by poor rice farmers is lower than that for all rice farmers. Combining 1 and 2 we find that the ratio of land farmed by poor farmers to total land farmed is 28.6 percent. While the relevant ratio suggested by theory is that of rice production by the poor to total rice production, the above is an adequate proxy. This ratio is to be compared with ratio 3 in table 10–2. It is seen from these that rice is not really the poor man's food. Only 8.7 percent of total consumption is accounted for by the poor. Thus from the point of view of poverty targeting, if the choice is between reducing producer price or increasing consumer price, the indicators suggest that it is the latter which will do least damage to poverty at the aggregate national level. Similarly, if the choice is between increasing the producer price or reducing the consumer price, the former has priority in a poverty focussed strategy.

Finally, compare 10–3 and 10–4. Relative to food on average, rice is decidedly the consumption of the non-poor. Thus, if the choice is between increasing

Table 10–2. *Rice and poverty in Côte d'Ivoire*

1. Incidence of poverty among farmers	35.7%
2. Ratio of mean area of poor to mean area of all	80.0%
3. Ratio of rice consumption by poor to total	8.7%
4. Ratio of food consumption by poor tp total food consumption	13.4%

Source: Kanbur (1990)

the consumer price of rice and the price of food in general, rice would be the prime candidate. Likewise, where price reductions are concerned, food in general is preferred to rice. Similar indicators can be calculated for other foods as policy issues emerge around them during adjustment. The theory developed in Besley and Kanbur (1988), and implemented here for rice, which is highly relevant to the Côte d'Ivoire policy dialogue, has wide applicability and can be utilized for other countries as comparable or better data begin to become available.

Another illustration of how household survey data can be used in assessing public expenditure restructuring priorities is in the area of housing in Côte d'Ivoire. In the Second Structural Adjustment Loan Agreement (1983), the government announced a major disengagement from the housing sector, noting explicitly that it would pass on costs to public sector tenants in rental housing. What might be the conseqquences of this for the poor? Table 10–3 below provides some information on this. Using data from the LSMS, it gives an indication of how much of the public expenditure on rent was in fact going to the poor.

The table shows that while out of all individuals in the sample who rented, 27.3 percent rented from SICOGI/SOGEFIHA or some other public agency, only 6.9 percent of the poor did so. In fact, out of the 730 individuals in the sample who rented from these agencies, only 14 were below the poverty line. This seems fairly conclusive evidence that the poor have not benefited from the operation of SOCOGI/SOGEFIHA; and table 10–3 goes on to confirm this by showing that of those for whom rent is subsidized or paid by someone else, this subsidy comes from a public agency 80.5% of the time, but there were no poor who fell into this category in the sample.

Thus, the disengagement of the government from its present activities in the housing sector is unlikely to be detrimental to the poor given that current intervention is largely in favor of the top 70 percent of the population and not the bottom 30 percent. However, a caveat is in order. It remains true that for the vast

Table 10–3. *Rental housing characteristics by poverty group in Côte d'Ivoire*

	Poor	*All*
1. Own house (%)	91.9	74.4
2. Of thosw who rent, rental from SICOGI/SOGEFIHA/ public agency	6.9	27.3
3. Of those who rent, those for whom rent is paid by someone else (%)	6.9	12.8
4. Of those for whom rent is paid by someone else, payment by SICOGI/SOGEFIHA/other public agency	0.0	80.5

Source: Glewwe and de Tray (1988)

majority of Ivorians, even more so for poor Ivorians in the rural sector, rental housing is not a concern. It is the quality of housing and amenities which matter, and it is to these that we now turn. Glewwe (1986) has analyzed these in some detail. He notes that the amenities of the poor are relatively worse on every count. Relatively more of them have no toilet access and of the people who do have such access, very few of the poor have access to a flush toilet or toilet inside the house. Similarly, the source of drinking water is a well (with or without pump) for most poor people and the next most important source is river, lake, spring, or marsh. These figures lay out the need for a restructuring of expenditure away from rental support to programs of improvement of amenities to the poor.

Finally, we look at the example of education scholarships in Côte d'Ivoire. Table 10–4, adapted from Glewwe and de Tray (1988), illustrates the clear case for restructuring public subsidies away from generalized scholarships.

Table 10–4. *School attendance and scholarships in Côte d'Ivoire*

	Poorest 30%	*Wealthiest 70%*	*All*
Household members attending school			
Primary	11.7	19.7	17.3
Secondary	2.4	8.4	6.6
University	0.1	0.7	0.5
Scholarship money			
Household expenditure	0.8	0.6	0.6
Per household member (CFAF/year)	467.3	1417.6	1137.9

Source: Glewwe and de Tray (1988)

On the basis of the figures in table 10–5, Glewwe and de Tray (1988) conclude as follows:

"... reductions in funding for university education will have very little effect on the poor. If these funds were instead used to improve the quality and availability of primary education instruction, the poor would likely receive substantial benefits. In terms of scholarships received, the impact is roughly the same among the poor and the non-poor measured relative to household expenditure levels. However, since the wealthier households have much higher expenditure levels in per capita terms, much more scholarship money is going to wealthier households (whose members are more likely to be in school at high levels of education) than to poor households. Cutting scholarship money across the board will not disproportionately affect the poor, while targeting that money to improve the quality of primary education is clearly to their advantage. Overall, policies for funding changes in education are more likely to benefit the poor than hurt them."

The above three examples, in the area of food subsidies, housing and education, illustrate the role that comprehensive household surveys in Africa could play in identifying poverty conscious public expenditure restructuring possibilities. The collection of such data should clearly be a priority.

Summary and Conclusion

In the foregoing pages, recent patterns of public expenditures in sub-saharan Africa were reviewed, an analytical framework to guide marginal resource allocation between spending alternatives was developed, and the scope and some limitations of poverty conscious public expenditure restructuring analysis were illustrated by means of selected applications. It was concluded, that the poverty alleviation record of past public expenditure strategies is rather limited. However, decisions regarding the reallocation of resources to directly poverty-reducing spending programs are complicated by the fact that the various human capital inputs interact with each other and with productive sector expenditures targeted on the poor in producing welfare outcomes. How is one to prioritize the use of scarce resources for poverty reduction?

The answer suggested was that, given the multidimensionality of the standard of living, the choice between social and economic sector resource allocation at the margin (or indeed between any spending alternatives regardless of their sectoral affiliation and level of aggregation) depends on the comparison of total expected welfare effects working through both basic needs and income mechanisms. A distinction was made between direct and indirect expenditures, the former taking place in the primary sector in which one seeks to obtain improvements (for example, health), the latter occurring in other (for example, income-enhancing) sectors which will lead to improvements in health through interactive effects. The analysis implied that there is normally a case for a combination of direct and indirect expenditures. Approaches to the analysis of the poverty alleviation impact of public expenditures were discussed. Three pieces of information were identified as needed, viz. the weight to be attached to various components of the standard of living, estimations of the linkages between expenditures and achievements, and knowledge regarding the fraction of expenditures reaching the poor. The social valuation of the various components of the standard of living is a normative matter. We have therefore recommended caution in the use of weighting schemes, such as UNDP's Human Development Index, as a basis for decisions reg- arding changes in the composition of public expenditure.

However, much is to be gained from production function approaches to the analysis of welfare achievements and from analyses which link public expenditures and the consumption patterns of the poor. For example, an individual's health production function can be said to include, among other arguments, a set of health inputs (medical consultations, preventive care, availability of health facilities) and a set of household "public" goods such as sanitation facilities, water quality and the like (World Bank, 1990c: 91). These inputs are a function of public expenditure levels and patterns (unless they are provided privately), and their contribution to health outcomes can be estimated on the basis of appropriately designed household surveys. The availability and prices of medical inputs and household "public goods" are determinants of the demand for health. Changes in public policy regarding these determinants (e.g. pricing or cost recovery policy; the expansion of health facilities) will thus affect health care choices as governed by the parameters of the health production function. If combined with studies of the cost of public services and their components, this demand side analysis allows one to trace changes in welfare achievements (or proxies thereof) to changes in public expenditures as exemplified in World Bank (1990b).

In conclusion, then, it is noted that household surveys are a prerequisite for policy research on the determinants of basic needs satisfaction and the degree of overlap between the pattern of government subsidies and the consumption habits of the poor. The collection of household budget and consumption data should therefore be promoted. The reader is referred to World Bank (1991) for the description of a prototype household survey which is designed to capture the information needed to perform the types of analysis identified above. On the supply side cost data on specified services and their input components, and data on service utilization, are required. Cost data are needed both to link expenditures and achievements and as a basis for the improvement of public expenditure management. The collection and analysis of these demand and supply side data is a demanding task, but (as we hope to have demonstrated) well worth the effort in terms of the insights gained to guide the restructuring of public expenditures from the point of view of poverty reduction.

References

Alderman, H. (1990): "Nutritional Status in Ghana and Its Determinants," Social Dimensions of Adjustment Working Paper No. 3. World Bank.

Anand, S. and R. Kanbur (1990): "Public Policy and Basic Needs Provision: Intervention and Achievement in Sri Lanka," forthcoming in J. Dieze and A.K. Sen (eds) *Hunger and Public Policy*, Oxford University Press.

Behrman, J. (1989): "Interactions Among Human Resources and Poverty: What We Know and What We Don't Know," PHR Department, Processed (February) World Bank.

Besley, T. and R. Kanbur (1988): "Food Subsidies and Poverty Alleviation," *Economic Journal*.

Bhalla, S.S. and Glewwe, P. (1986): "Growth and Equity in Developing Countries: A Reinterpretation of the Sri Lankan Experience," *World Bank Economic Review*, Vol. 1, No. 1.

Foster, J., J. Greer and E. Thorbecke (1984): "A Class of Decomposable Poverty Measures," *Econometrica*.

Gallagher, M. and O.M. Ogbu, 1989. Public Expenditures Resource Use and Social Services in Sub-Saharan Africa. Draft, Africa Region Technical Department, The World Bank: Washington DC.

Gertler, P. and J. van der Gaag (1988): "Measuring the Willingness to Pay for Social Services in Developing Countries," Living Standards Measurement Study, Working Paper No. 45. Washington DC: World Bank.

Gertler, P., L. Locay and W. Sanderson (1987): "Are User Fees Regressive? The Welfare Implications of Health Care Financing Proposals in Peru," *Journal Econometrics* 36, pp. 67-80.

Glewwe, P. and D. de Tray (1988): "The Poor During Adjustment: A Case Study of Côte d'Ivoire," Living Standards Measurement Study, Working Paper no. 47, World Bank.

Hicks, N. and A. Kubisch (1984): "Cutting Government Expenditures in LDCs," *Finance and Development*, Sept.

Isenman, P. (1980): "Basic Needs: The Case of Sri Lanka," *World Development*

Jaeger, W. and C. Humphreys (1988): "The Effect of Policy Reforms on Agricultural Incentives in Sub-Saharan Africa," *American Journal of Agricultural Economics* 70, pp. 1036-43.

Kanbur, R. (1990): "Poverty and the Social Dimensions of Adjustment in Côte d'Ivoire," Social Dimensions of Adjustment Working Paper No. 2, World Bank.

Pinstrup-Anderson, P., M. Jaramillo and F. Stewart (1987): "The Impact of Government Expenditure," in Cornia, A., Jolly, R. and Stewart, F. (eds) *Adjustment with a Human Face: Protecting the Vulnerable and Promoting Growth* (A UNICEF Study). Clarendon Press, Oxford.

Sahn, D. (1990) "Fiscal and Exchange Rate Reforms in Africa: "Considering the Impact on the Poor," Cornell University, Food and Nutrition Policy Program, Processed (March), Washington DC.

Sahn, D. (1990): "Malnutrition in Côte d'Ivoire," Social Dimensions of Adjustment Working Paper No. 4, World Bank.

Sen, A.K. (1981): "Public Action and the Quality of Life in Developing Countries" *Oxford Bulletin of Economics and Statistics.*

UNDP (1990)" *Human Development Report*, Oxford University Press.

Vogel, R.J. (1988): "Cost Recovery in the Health Care Sector: Selected Country Studies in West Africa," World Bank Technical Paper Number 82, Washington DC: World Bank.

World Bank (1987): "Education in Sub-Saharan Africa: Policies for Adjustment, Revitalization, and Expansion," A World Bank Policy Study. Washington DC: World Bank.

__, (1989a): Zambia: Public Expenditure Review, Washington DC: World Bank.

__, (1989b): Rwanda: Public Expenditure Program: An Investment in Economic Strategy. Report No. 7717-RW, Washington DC: World Bank.

__, (1989c) Mozambique: Public Expenditure Review, Washington DC: World Bank.

__, (1989d) Ghana: Population, Health and Nutrition Sector Review. Report No. 7597-GH. Washington DC: World Bank.

__, (1989e) Tanzania: Population, Health and Nutrition Sector Review. Report No. 7495-TA. Washington DC: World Bank.

__, (1989f) Tanzania: Public Expenditure Review, Washington DC: World Bank.

__, (1989g) Malawi: Public Expenditure Review: Report No. 7281-MAI. Washington DC: World Bank.

__, (1990a): World Development Report 1990, Washington DC: World Bank.

__, (1990b): "Health Care Cost, Financing and Utilization in Nigeria," In preparation, Washington DC: World Bank.

____, (1990c): "Analysis Plans for Understanding the Social Dimensions of Adjustment," Social Dimensions of Adjustment Unit, Report No. 8393-AFR, Washington DC: World Bank.

____, (1991): "The SDA Integrated Survey," Social Dimensions of Adjustment Unit, Washington DC: World Bank.22

11

Manufacturers' Responses to Infrastructure Deficiencies in Nigeria: Private Alternatives and Policy Options

Kyu Sik Lee and Alex Anas

In many countries in Africa, infrastructure provision suffers from two kinds of inefficiencies. The first is the presence of a public sector with a relatively high level of capital investment in place but which remains non-performing or unable to provide steady and reliable infrastructural services. The second, a consequence of the first, is that many users of the public infrastructural services find it necessary to provide their own facilities in whole or in part by incurring the much higher costs of private provision. These two extremes, (i) the non-performing public sector and (ii) the private provision responses of firms, are well known to exist in Nigeria.

The solution to this problem of infrastructural deficiencies in Nigeria and other African countries is not likely to be a technological one. It is generally understood that in these countries large additional capital outlays or extensive rehabilitation programs cannot be fully effective without progress in improving institutional organization, logistical support services, and administration. Yet, it is these areas which are the least well understood and where progress has remained elusive, difficult and unpredictable. Thus, it is realistic to assume that the public sector will continue to remain non-performing for some time to come and that the infrastructural deficiency problem will need to be addressed in a way which minimizes the social cost in a shorter timeframe. This would require fine-tuning regulatory regimes and the existing institutional structure, and coming up with more efficient pricing policies in order to induce active private sector participation in infrastructure service provisions.

The purpose of this paper is threefold. First, we document how infrastructural deficiencies affect manufacturing firms of different sizes in different regions. Second, we describe how firms respond to the deficiencies, identify the costs of these responses, and estimate the extent of private cost as a measure of the willingness of firms to pay for reliable services. Third, based on these observations, we offer alternative policy options for improving infrastructure provisions in Nigeria. These policy options provide alternatives between the two extremes of the non-performing public sector and the private provision by individual manufacturers. The policy options discussed in this paper include: (i) regulatory changes for enabling fuller utilization of existing private provision capacities, for example, by allowing the sale of excess private power supply; (ii) private sector participation in selected infrastructure support activities, such as production, distribution, maintenance, metering or revenue collection; and (iii) alternative pricing and tariff strategies which exploit observed variations in private provisions by firm size and location.

Our analyses in this paper are based on the empirical results from the survey of manufacturing establishments conducted for this research project. The questionnaire was developed by the World Bank (supported by then the West Africa Regional Research Fund) in collaboration with the Nigerian Industrial Development Bank. The field survey was implemented by Arthur Andersen & Co., Lagos. A stratified random sample was drawn from the sample frame of manufacturing establishments provided by the Nigerian Federal Office of Statistics. The sample covered five states: Lagos, Anambra, Imo, Kaduna and Kano. The survey consisted of 36 pages with 349 computer readable variables, and was completed in late 1988 for 179 manufacturing establishments. The sample firms covered all manufacturing industries (at the two-digit level of the Standard Industrial Classification) and a continuum of firm sizes as measured by employment. Infrastructural deficiencies and firms' private provision responses are covered for five subsectors: electricity, water supply, transportation of freight and personnel, telecommunications, and waste disposal. (The Nigerian Industrial Development Bank has completed the survey for an additional 66 establishments among its client firms. This data is still being processed and not included in our analysis in this paper.)

The paper is organized as follows. The second section documents and discusses the extent, apparent causes, and incidence of infrastructural deficiencies in Nigeria. We have drawn from World Bank project reports, and institutional and other qualitative information on the state and causes of deficiencies in selected infrastructural subsectors in order to complement the information from the establishment survey data collected. The third section focuses on the alternative private provision responses of manufacturers. Prior to the survey of establishments, our knowledge of private response options of the firms was based on rough aggregate figures, anecdotal descriptions of selected cases, or specific field interviews of several firms. The survey findings clearly document the presence of a wide range of responses, and the frequency of their occurrences, and also their incidence and costs by region, size of firm, and other characteristics. The fourth section presents the estimates of capital costs of various private provisions and analyzes the private cost as a measure of the willingness of firms to pay for reliable services. The fifth section discusses several policy options developed and their economic rationale. To the extent possible, we use the survey results to give a preliminary empirical justification of the potential benefits of the policy options considered.

To make quantitative estimates of the benefits of the suggested policies it will be necessary to implement empirically the analytical framework developed in Anas and Lee (1988), by estimating the degree by which individual firms in the sample will respond to changes in policies such as tariffs or regulatory constraints. The current paper sets the stage for such an analysis by identifying the appropriate response patterns. The plan for an econometric analysis and the associated measurement needs are briefly mentioned in the concluding section.

The Extent and Causes of Infrastructure Deficiencies

It is common knowledge that Nigerian manufacturers suffer from frequent interruptions of publicly provided services such as electricity, water, telecommunications, transport, and waste disposal and by the poor quality of these services when and where they are available. A detailed discussion of these problems for each infrastructural sector is given in Anas and Lee (1988; pp.3-8).

The Nigerian Industrial Development Bank (NIDB), the collaborating local institution of this study, has been particularly concerned about these problems, since financing industrial projects has been its main activity. According to NIDB's staff, frequent power cuts and voltage fluctuations have forced almost every industrial establishment in the country to undertake extra investments in generators in order to avoid production losses as well as damage to machinery and equipment. For similar reasons, extra investments are also made in sinking boreholes and installing water treatment plants. Such extra investments raise industrial costs and make it difficult for local industrial products to compete in price with their imported counterparts. By unduly enlarging the overhead and running costs, they lengthen the gestation period of industrial projects.

State monopoly enterprises such as the Nigerian Electric Power Authority (NEPA) or the Nigerian Telephone Company (NITEL) have a large amount of capital investment already in place but fail to deliver their services at the level required to meet the demand. Such failures not only result in the waste of scarce resources but also significantly affect manufacturing and other productive activities in the Nigerian economy. Therefore, it is important to emphasize that infrastructure services are intermediate inputs used in producing final goods and services and that the inadequate supply of these services will adversely affect the productivity growth of industries and economic development in general.

The causes of infrastructural failures may be grouped into two kinds. The first is relatively well understood and relates to shortcomings of the technology used by the public sector, including problems in the day to day management, and operation and maintenance of the facilities. The second is more complex in nature and less well controlled, and relates to general problems with administration, bureaucracy, planning, metering, billing for services delivered, revenue collection, personnel training in the public sector, and lack of appropriate incentives for management and personnel in part because of civil service pay ceilings. This second set of factors has remained the key problem over the years because further investments in additional facilities is easily rendered ineffective if the institutional organization and logistical support systems are lacking.

Assessing the actual burdens imposed by the current inadequacies and the costs of ongoing adjustments will be useful as the government continues to make strategic investment choices which involve the following types of trade-offs:

- among different users of the infrastructural services such as residential versus manufacturing;
- between additional capital investments versus maintenance and rehabilitation of existing facilities or the training and recruitment of personnel;
- among different infrastructural subsectors such as electric power versus telecommunications; (iv) between as well as within regions and cities;
- between alternative pricing and tariff structures;
- between assisting the private sector in its self-provision efforts versus supporting further the public infrastructure sectors; and
- between different organizational and structural reforms focused on deregulation, commercialization, and the partial privatization of selected infrastructure related functions in individual subsectors.

Causes of the Deficiencies

World Bank studies and project work in the last decade have documented the extent and causes of infrastructural failures in each sector. Taking electricity generation as one example, the current situation can be gleaned from two project appraisal reports which are eight years apart (World Bank 1981 and 1989). The basic types and causes of failure remain essentially unchanged over the entire decade of the 1980s. Technological causes of failure in this sector are primarily related to transmission and distribution. The ratio of the available capacity to that of the installed capacity is generally low and as much as 50 percent of installed capacity may be essentially inoperable at any given time. However, operable generating capacity is still considered substantial and essentially adequate. Most power interruptions (nearly two thirds) are a result of bottlenecks on the transmission and distribution networks. These recurring transmission problems are believed to be due to the lack of spare parts or the delays in obtaining them. In addition, shortages of materials, vehicles and foreign exchange have been the key factors which have constrained the expansion of the distribution system. In recent years these factors have been aggravated by the sharp fall in the price of oil which has reduced the public budget, as well as by the sharp devaluation in the value of the naira which makes imported spare parts even more expensive. A persistent problem has been the frequent overloading of transformers. The fact that only 400 to 500 of NEPA's fleet of 3,000 vehicles are operational has systematically hampered routine maintenance of the distribution network. Similarly, lack of properly trained personnel is the apparent cause of failures to maintain circuit breakers on the transmission network. Another area which contributes to these problems is the inadequacy of NEPA's monitoring facilities in its National System Control Center which is supposed to track and quickly service failing components on the national network (World Bank 1989).

Most recent studies have paid attention to the nontechnological factors contributing to power interruptions and failures, and the current government efforts to partially commercialize a number of parastatals have also included NEPA (World Bank 1989). NEPA will therefore have more autonomy in wage and compensation policy, tariff setting and in determining its own capital expenditure program. It is generally recognized that current NEPA tariffs for electricity have essentially no relationship to economic opportunity costs. For example, electricity tariffs remain unchanged since 1979 when they were raised to 7 kobos per kWh. At this level, it is estimated that they are about one seventh of the long run marginal cost of supply and do not even cover the cash operating costs of generation, transmission and distribution. It is known that in most of developed and advanced developing countries such as Korea, the tariff per kWh is about 7 US cents (52.5 kobos). With the already established commercialization of the Nigerian National Petroleum Company (NNPC) which supplies gas fuel to NEPA, gas prices are going up and NEPA would have to raise its electricity tariff soon, as NEPA becomes subject to a higher degree of market discipline. In addition, it has been recommended that the tariff be raised in stages in the next several years (World Bank 1989).

The problems of underpricing are also observed in water supply. The Lagos State Water Commission (LSWC) since 1986 is operating under a new tariff which raised water prices for manufacturers by about 40 percent and a further increase of 270 percent was due for approval. A vendor licensing system authorized to levy direct charges for water sold at public standposts is under discussion (World Bank 1988). Since industrial water use is beginning to exhaust the groundwater supplies of the Lagos State region, it is reasonable to expect that tariff increases for industrial use of water may become more feasible in the future. (More detailed discussions of the causes of deficiencies appear in Lee, Stein, and Lorentzen 1989.)

The Incidence of the Deficiencies by Firm Size and Region

Our data reveal that there are large variations in the availability and quality of public infrastructure services and in firms' private provision responses across regions and firm sizes. Such observations imply an important role in government strategy regarding infrastructural policy reform. Variations in private provision patterns can be summarized as follows. Table 11–1 shows that only 14 out of the 179 firms, or 7.8 percent do not have their own electricity generators. Twelve of these firms are in Anambra and Imo and two are in Lagos while all firms in Kaduna and Kano have their own generators. For the firms that do not have their own generators (or "captive firms"), the supply is not 100 percent reliable. Table 11–2 shows that the captive firms are generally small ones. Moreover, table 11–3 shows that the smaller firms are subject to the bulk of the power failure incidents. Some small firms do not have their own generating equipment, not because the burden of poor electricity supplies is less per unit of output for them, but rather because the generation cost per unit of electricity is higher for them because of economies of scale in electricity generation.

Table 11–1. ***Distribution of manufacturing establishments by region and source of electricity***

Region	Source of electricity for production operation: NEPA only	NEPA main	Own gen. main	Own gen. only	Total
Lagos					
frequency	2.00	68.00	10.00	2.00	82.00
row%	2.44	82.93	12.20	2.44	100.00
column%	14.29	48.57	50.00	40.00	45.81
Anambra/Imo					
frequency	12.00	22.00	1.00	1.00	36.00
row%	33.33	61.11	2.78	2.78	100.00
column%	85.71	15.71	5.00	20.00	20.11
Kaduna/Kano					
frequency	0.00	50.00	9.00	2.00	61.00
row%	0.00	81.97	14.75	3.28	100.00
column%	0.00	35.71	45.00	40.00	34.08
Total					
frequency	14.00	140.00	20.00	5.00	179.00
row%	7.82	78.21	11.17	2.79	100.00
column%	100.00	100.00	100.00	100.00	100.00

Note: a. NEPA only = using 100% from NEPA; NEPA main = NEPA as the main source and own generators as standby; Own gen. main = NEPA as standby; Own gen. only = 100% from own generators.

Source: NIBD/IBRD Establishment Survey, 1988.

Small firms are the ones that cannot afford capital investments for boreholes, vehicles for the shipment of products, motor cycles and couriers, and radio equipment. Compared to the other two regions, Anambra-Imo has a higher concentration of small firms and the burden of inadequate infrastructure seems to be more serious there.

The heavy incidence of infrastructural failures among small firms has an implication for the growth of industries and the generation of employment. According to the "incubator hypothesis" that was tested in the earlier World Bank research on industrial location in Bogota (Lee 1989a) and in Seoul (Lee 1985), it was observed that small new firms spend their early years near the city center or in an old industrial area with easy access to good utilities and other essential services. They do so because it is prohibitively expensive for small firms to operate in outlying areas where infrastructure services are poor. As they grow and become more independent, they tend to move out of the central area for more space. In Nigeria and perhaps in most African countries, large cities with poor infrastructure cannot offer the incubator function for small new firms. Since small firms cannot afford their own generators and boreholes and other facilities, the burdens of inadequate public infrastructure services are especially severe for the small firms which start and grow in those cities. This has a serious negative implication for the birth and growth of small firms and for the generation of employment and income. The studies mentioned above (Lee 1985, 1989a) showed that small new firms generate between 60 to 80 percent of the new jobs created in large cities in Asia and Latin America. This implies high returns in Nigeria to selectively improving infrastructure service provisions for particular users at particular locations, since the observed service reli-

Table 11–2. *Distribution of manufacturing estalishments' source of electricity by firm size*

	Firm size							
Source of electricity	*1-19*	*20-49*	*50-99*	*100-199*	*200-499*	*500-999*	*1,000 & over*	*Total*
NEPA only								
frequency	11.00	3.00	0.00	0.00	0.00	0.00	0.00	14.00
row%	78.57	21.43	0.00	0.00	0.00	0.00	0.00	100.00
column%	68.75	8.57	0.00	0.00	0.00	0.00	0.00	7.82
NEPA main								
frequency	3.00	26.00	35.00	30.00	25.00	13.00	8.00	140.00
row%	2.14	18.57	25.00	21.43	17.86	9.29	5.71	100.00
column%	18.75	74.29	79.55	85.71	96.15	86.67	100.00	78.21
Own gen. main								
frequency	2.00	4.00	8.00	5.00	0.00	1.00	0.00	20.00
row%	10.00	20.00	40.00	25.00	0.00	5.00	0.00	100.00
column%	12.50	11.43	18.18	14.29	0.00	6.67	0.00	2.79
Own gen. only								
frequency	0.00	2.00	1.00	0.00	1.00	1.00	0.00	5.00
row%	0.00	40.00	20.00	0.00	20.00	20.00	0.00	100.00
column%	0.00	5.71	2.27	0.00	3.85	6.67	0.00	2.79
Total								
frequency	16.00	35.00	44.00	35.00	26.00	15.00	8.00	179.00
row%	8.94	19.55	24.58	19.55	14.53	8.38	4.47	100.00
column%	100.00	100.00	100.00	100.00	100.00	100.00	100.00	100.00

Source: NDIB/IBRD Establishment Survey, 1988

Table 11–3. *Distribution of manufacturing estalishments' power outages by firm size*

	Firm size							
Average number of power outages per week	*1-19*	*20-49*	*50-99*	*100-199*	*200-499*	*500-999*	*1,000 & over*	*Total*
Less than 5/week								
frequency	7.00	11.00	15.00	8.00	7.00	7.00	3.00	58.00
row%	78.57	21.43	0.00	0.00	0.00	0.00	0.00	100.00
column%	68.75	8.57	0.00	0.00	0.00	0.00	0.00	32.40
5-10/week								
frequency	9.00	26.00	22.00	19.00	15.00	4.00	2.00	91.00
row%	2.14	18.57	25.00	21.43	17.86	9.29	5.71	100.00
column%	18.75	74.29	79.55	85.71	96.15	86.67	100.00	50.84
More than 10/week								
frequency	2.00	4.00	7.00	8.00	4.00	4.00	3.00	30.00
row%	10.00	20.00	40.00	25.00	0.00	5.00	0.00	100.00
column%	12.50	11.43	18.18	14.29	0.00	6.67	0.00	16.76
Total								
frequency	160.00	35.00	44.00	35.00	26.00	15.00	8.00	179.00
row%	8.94	19.55	24.58	19.55	14.53	8.38	4.47	100.00
column%	100.00	100.00	100.00	100.00	100.00	100.00	100.00	100.00

Source: NDIB/IBRD Establishment Survey, 1988

ability problems are to an extent location and user specific.

Alternative Responses of Manufacturers

There are essentially four ways in which firms might respond to infrastructural deficiencies. These are:

- relocation,
- factor substitution,
- private provision, and
- output reduction.

Below we discuss the economic rationale behind each of these responses, and why they are or are not observed in Nigeria.

Relocation

The firm may relocate to a site with better infrastructure services. Such relocation can occur within a city or from one region to another. Our survey results show that Nigerian firms do not move to other locations from the initial site. Even though 50 percent of the firms had been at their present location since 1980, only two out of the 179 sample firms indicated that they had relocated from another location. This absence of mobility is striking considering that the average annual moving rate observed in large cities in other developing countries is about 5 percent. The relative immobility of Nigerian firms is consistent with the fact that the capacity, regularity and quality of infrastructure vary from bad to worse within and across Nigerian cities. This tends to limit the gains in infrastructure quality that can be achieved by moving to new locations. Nigerian firms instead undertake their own extensive capital expenditures and incur regular operations and maintenance outlays to provide their own services. The high setup cost with a large amount of initial capital investment for own service provision would make it difficult for the firms to move.

Another problem with relocation is that it often involves trading one infrastructural deficiency for another. For example, a firm that moves into the Lagos area because it is much cheaper to sink boreholes there (since the water table is high), might better its water supply, but the firm may face new problems such as losses in production time due to the commuting delays of employees.

Factor Substitution

The firm may substitute away from the use of the poorly provided service by adjusting its mode of production in favor of those inputs and raw materials which are less infrastructure intensive. For example, if a firm has a choice between a labor intensive and a capital intensive process and if the labor intensive process relies less on infrastructure than the capital intensive one, the firm's strategy would be to substitute labor for capital thus reducing the quantity of infrastructure inputs. The various private provision activities with large capital expenditures undertaken by the Nigerian firms indicate that their ability to adjust to the relative prices of labor, machines, materials, or various infrastructure service inputs is rather constrained by the current technologies in use. Since such input substitution possibilities are limited, the firms operate inefficiently by providing their own infrastructure services when these are crucial for their operations. In case of a milk processing plant, for example, even if the public power supply were available at proper voltage for as much as 90 percent of the time, the firm could not afford to eliminate its own generators with 100 percent capacity because any voltage surges and drops at a critical time would threaten key equipment in the production process and result in much waste.

Private Provision

As already mentioned, numerous strategies are available for the firms to provide their own infrastructure services. The fact that the vast majority of firms do so even when the publicly provided infrastructure services are extremely inexpensive, indicates the importance of having reliable infrastructural inputs. Private provision as a strategy is not entirely separate from factor substitution. In fact, by providing their own infrastructural services, firms are substituting internal capital in the form of equipment, machinery as well as labor in the form of maintenance personnel for the publicly provided infrastructure services which are not forthcoming. As documented in Anas and Lee (1988), Nigerian firms are observed to pursue four different private response strategies. These are:

- Self-sufficiency: The firm provides its own infrastructural services to the point where it does not need any public inputs. For example, table 11–1 shows that only 5 out of the 179 surveyed firms are in this mode with respect to electricity generation.
- Standby private provision: The firm has its own infrastructural facilities in place and switches to these facilities when the quality or reliability of the public services falls below a critical level. From table 11–1, 140 firms or 78 perc-

ent of those surveyed are in this situation with respect to power supply.

- Public source as standby: The firm relies primarily on its own facilities but switches to the public supply during those times of the day when the public source delivers a high quality service. Again, from table 11–1, twenty firms or about 11 percent of the surveyed firms reported such behavior.
- Captivity: The firm continues to rely on the public source exclusively despite the very low reliability of such a service. It is reasonable to expect that captivity will be the dominant mode among the very small firms who cannot afford infrastructural capital investments. Only 14 or 7.8 percent of the surveyed firms reported such behavior in the case of electricity.

Anas and Lee (1988) argued that there are economic incentives for three additional regimes of private provision which are not observed in Nigeria because of government regulations on the supply and trading of infrastructure services by private entities. These regimes are:

- joint production;
- satellite behavior; and
- shared production.

"Joint production" refers to the case where a firm, typically a large one, which has already made a substantial investment in infrastructural capital finds it profitable to sell part of its infrastructural output to other firms. With few exceptions, this has not been possible in electricity production in Nigeria, because private producers of electricity are not normally allowed to sell surplus power to other firms or even back to NEPA. "Satellite behavior" is the other side of the coin with respect to joint production. A satellite firm is one which purchases infrastructure services from another firm that has surplus infrastructure services to sell. At times of power interruption, for example, a satellite firm would switch from NEPA to the generators of a nearby private producer. "Shared production" refers to the possibility of firms coming together in a club type of arrangement called "utility pool" to share the cost of infrastructural capital inputs by building their own facilities. (A theoretical framework for the club type arrangement is discussed in McGuire 1974.) The above typology of private provision alternatives is applicable to all five infrastructure subsectors considered in this study.

Table 11–4. *Values of private infrastructure provision as a percentage of the total value of machinery and equipment*

Private provision	*Small firms*[a]	*Large firms*	*Total*
Generators	24.78	10.06	10.42
Boreholes	2.81	1.91	1.91
Vehicles for workers	5.49	2.84	2.86
Vehicles for shipments of goods	10.95	4.47	4.62
Vehicles for garbage disposal	0.15	0.48	0.48
Radio equipment	1.48	0.59	0.59

Note: a Establishments with less than 50 employees. The values of generators, borehole, and radio equipment are included in the total values of machinery amd equipment, but those vehicles are not included.

Source: Anas and Lee (1988).

Output Reduction

This response to infrastructural deficiencies is also common. Firms which are captive or use their own standby equipment are subject to output reduction either on a regular basis or when their own equipment fails to operate properly. However, the chief impact of output reduction necessarily falls on small firms which find it too expensive to pursue another response, or on very large power intensive firms which cannot find appropriate size equipment (e.g. generators) to meet their service needs. It is difficult to observe, but it undoubtedly happens that many small firms in Nigeria have either shut down or have failed to grow to any critical size because of infrastructural deficiencies. Also, births of new firms will be reduced if many must shut down soon after birth because of infrastructural inadequacies.

Costs of Private Provision

Capital Costs and their Incidence

The firms that we have surveyed provide a telling story of the incidence of private provision which is by far the most dominant response among Nigerian manufacturers. Table 11–4 summarizes the data on average current market values of various equipment and facilities used for own service provisions and their share of the total value of the firm's machinery and equipment for production. We find that the capital value of generators and support facilities such as the switches and transformers is on the average 25 percent of the total value of machinery and equipment for small firms (with less than 50 employees) and 10 percent for large firms. This share varies widely across the five states and by firm sizes, from 4 percent for large firms in Imo to 36 percent for small firms in Anambra. The average value of capital for electricity generation including all firms is 954,000 naira (about 130,000 U.S. dollars). This value is almost four times larger than the share of capital for boreholes and treatment facilities. The average value is 260,000 naira for all firms with boreholes, which is about 2 percent of the total value of machinery and equipment. This share value varies from 0.5 percent in Kano to 2.1 percent, or six times higher, in Lagos. Although water supply takes up a much smaller share of equipment and machinery than does electricity, the share is again higher for small firms than it is for large ones, by about 50 percent.

Although only about 15 percent of the firms provide transport for their workers, the share of these vehicles in total capital equipment is 5.5 percent for small firms and just under 2.8 percent for large firms. The low ratio of self-provision observed in transporting one's own workers mean that, at least in Lagos, a great deal of production time is lost because of the late arrival of workers. When firms choose not to make capital expenditures for their own provision of certain services, they often incur comparable costs in other forms such as in lost production time. In Lagos, long commuting time is not due to the distances between residences and workplaces but due to long waiting times for buses. Savings from employing workers with lower wages are limited by the firms' inability to get them to the factory on time. In the shipment of goods, 63 percent of the surveyed firms had their own vehicles. These vehicles make up 11 percent of total capital equipment for small firms but only slightly more than 4 percent for large firms. The average capital value of these vehicles was 387,000 naira for each firm. Capital expenditures such as radio equipment and motorcycles for couriers are small compared to generators and boreholes, but returns to these investments are extremely high. About 37 percent of the firms have radio equipment and its share in the total value of machinery and equipment is nearly three times higher for small firms. On the average, managers spend more than 10 hours per week on the road to deliver messages or hold conversations that could be handled in moments over a working phone line.

Figure 11-1. Proportion of Electric Power Supply from Own Generators, 1987

Firm size	Percent
1-9	35.46
10-19	44.94
20-49	45.15
50-99	41.45
100-199	32.33
200-499	23.61
500-999	29.75
1000 +	14.81

Source: Based on data from NIDB/IBRD Project Establishment Survey, 1988.

The Private Cost as a Measure of Willingness to Pay for Reliable Services

As documented in the above section, manufacturers incur high capital cost in installing own facilities for providing their own services. In the case of electric power generation, the survey reveals that nearly all standby firms have installed capacity sufficient to run the entire plant during a period of NEPA power interruption. The data also indicates that the sample firms as a whole 25 percent of all power used by them during 1987 came from their own generators and 75 percent from NEPA. (The breakdown by firm size is shown in figure 11–1) Because the typical installed private generation capacity is approximately sufficient to run the entire plant (with some reserve for maintenance), this means that about 75 percent of the generation capacity remains idle. This idle capacity results in extremely high total average cost of private power generation as shown below. The high cost of private provision sustained by the firms is the implicit value of service reliability that the firms are willing to pay for. A precise measure of willingness to pay can be determined by calculating the average cost (per kWh) of electricity produced by the firm's own generators (as the lower bound). When the average cost of the privately produced power is higher than the price charged by NEPA, the difference between these two gives the premium which manufacturers are willing to incur in order to insure themselves of an uninterrupted power supply at all times.

Tables 11–5 and 11–6 show two such sets of computations on the average cost of private power generation. Table 11–5 shows the average cost computed using each firm's reported power consumption from own generators during 1987 for different firm size categories (25.48 percent of the total consumption for the sample firms as a whole). Thus, these figures reflect the cost of holding idle generating capacity. Table 11–6 shows the average cost of electric power generation for different firm size categories assuming that 100 percent of power supply comes from own generators. In both tables 11–5 and 11–6, the capital recovery cost is computed by annualizing the current market value of the firm's generators and accessories using the remaining service life. The recurring costs of fuel, maintenance, and labor, are added to the capital cost. In table 11–6, these reported recurring costs are appropriately adjusted for the full utilization case as explained in footnote (a) to table 11–6. The average cost schedule by firm size has been calculated with different sets of assumptions on (i) the real rate of interest and (ii) the exchange rate. In our discussion below, we refer to the average cost schedule computed with the 10 percent real interest rate and the current exchange rate of 7.5 Naira per US dollars. (During the 1980s the average inflation rate was about 12 percent and the current commercial lending rate is about 20 percent.)

Table 11–5 shows that at the actual average utilization rate of 25 percent of the generating capacity, the average cost per kWh is 4.61 Naira, which is 66 times the present NEPA price of 7 kobos. Suppose that the NEPA tariff were to be adjusted to 30 kobos, a rate currently charged by a private supplier in Lagos. The average firm would still be incurring 15 times the new NEPA price at the actual utilization rate of 25 percent. Even under the assumption of 100 percent supply from own generators, the average cost of 1.41 Naira for all firms (table 11–6) will be five times higher than NEPA's 30 kobos. The premium is highest for the 20-49 person firm size category with a factor of 6 while for the largest size category of 1,000 or more persons the premium is a factor of only 1.3. Small firms pay for a higher premium because of economies of scale in electric power generation.

Table 11–5. *Average cost of electric power generation by firm size: underutilization case*[a]

Firm size	Interest rate[b]	Average cost (Naira/kWh) 5%	10%	15%
		(1987 exchange rate US$1-4.0 Naira)		
All firms		2.540	2.834	3.150
0-9		0.374	0.426	0.483
10-19		0.698	0.781	0.871
20-49		3.336	3.740	4.171
50-99		2.698	3.009	3.346
100-199		2.573	2.936	3.328
200-499		2.357	2.564	2.780
500-999		1.442	1.611	1.793
1,000 +		2.327	2.439	2.556
		(1989 exchange rate US$1 = 7.5 Naira)[c]		
All firms		4.061	4.612	5.204
0-9		0.634	0.732	0.838
10-19		1.086	1.243	1.412
20-49		5.701	6,457	7.267
50-99		4.191	4.775	5.407
100-199		4.196	4.876	5.611
200-499		3.718	4.106	4.512
500-999		2.063	2.379	2.721
1,000 +		3.105	3.315	3.534

Note: [a] The average utilization of installed geberating capacity was 25.48%. [b] Interest rates represent hypothetical real rates. [c] Adjusted for the values of the generators and accessories only.

Source: NIDB/IBRD Project Establishment Survey, 1988.

Table 11–6. *Average cost of electric power generation by firm size: full utilization case*[a]

Firm size	Interest rate[b]	Average cost (Naira/kWh) 5%	10%	15%
		(1987 exchange rate US$1-4.0 Naira)		
All firms		0.959	1.021	1.086
0-9		0.143	0.155	0.169
10-19		0.453	0.463	0.493
20-49		1.101	1.180	1.263
50-99		1.045	1.122	1.206
100-199		1.023	1.091	1.163
200-499		1.018	1.060	1.104
500-999		0.675	0.712	0.752
1,000 +		0.314	0.326	0.339
		(1989 exchange rate US$1 = 7.5 Naira)[c]		
All firms		1.291	1.407	1.530
0-9		0.205	0.228	0.252
10-19		0.568	0.621	0.677
20-49		1.106	1.752	1.908
50-99		1.380	1.525	1.682
100-199		1.444	1.572	1.708
200-499		1.243	1.322	1.406
500-999		0.821	0.890	0.996
1,000 +		0.314	0.339	0.423

Note: [a] Assumed 100% of electric power supply comes from own generators. Fuel consumption and maintenence cost are adjusted accordingly. Fuel by a factor of 4 and maintenence and parts by 3, when utilization rate increases from 25% to 100%. [b] Interest rates represent hypothetical real rates. [c] Adjusted for the values of the generators and accessories only.

Source: NIDB/IBRD Project Establishment Survey, 1988.

From the 20-49 person firm size category the average cost declines exponentially with the firm size. The cost schedules in tables 11–5 and 11–6 have been fitted to semi-log and double-log regressions as reported in table 11–7. The slope coefficients are all statistically significant. The average cost values shown in tables 11–5 and 11–6 are plotted in figure 11–2 (excluding the values for firms with less than 20 employees).

The premium paid by firms varies with firm size. Such variation should be a central concern in the design of appropriate policies for both efficiency and equity reasons. The smallest group with less than 20 employees shows an average cost that is lower than the sample mean. This is not because they can generate electric power at lower cost however. Rather it is because they cannot afford to make the expensive capital investment to meet the required power need. They may be able to generate enough power to support the lighting and other critical elements.

The evidence of the presence of economies of scale in electric power generation is clear from the 20-49 size category as mentioned above. The cost of producing 100 percent of power supply from the installed generating capacity falls by a factor of 4.4 (from 1.752 to 0.399 in table 6) as firm size increases from "20-49" to "1,000 and over." When the cost of idle capacity is included, the average cost in the same range of firm sizes falls by only a factor of 1.9 (from 6.457 to 3.315 in table 11–5). Since large firms can achieve great scale economies when their capital intensive equipment is fully utilized, the fall in average cost is higher in the case of fuller utilization. From the above analysis, we can conclude that the premium over the NEPA price declines with an increase in firm size and that even after a hypothetical tariff increase to 30 kobos per kWh, the fuller utilization case premiums would still be larger than the NEPA price for all firm sizes. Of the average total cost of 4.61 naira in the case of underutilization, the average variable cost is 80 kobos for the sample firms as a whole (table 11–8). A NEPA price of 30 kobos will be only about a third of the average variable cost of self-generation. In some developed countries, gas turbine generators are widely used and they do not manifest economies of scale. This technology is seldom used in Nigeria as yet. The minimum size for gas turbine generators however is likely to be too large for the need of most individual firms.

Developing Policy Options for Improving Service Provision

As explained in the introduction, in Nigeria two extreme cases of inefficiency in the provision of infrastructural services are observed: first, the non-performing public sector with heavy capital investments; second, the costly provision of services by individual firms themselves. The self-provision response has developed over the years because of nonperformance in the public sector. Without the extensive private provision responses, the total welfare

Table 11–7. *Regression of the average cost of electric power generation firm size.*

	Semi-log			Double-log		
	5%[a]	10%	15%	5%	10%	15%
			Fuel utilization case[b]			
Constant	-0.172	-0.095	-0.005	1.153	1.323	1.432
	(1.50)	(0.83)	(0.04)	(2.41)	(2.77)	(3.06)
Slope	-0.000619	-0.000625	-0.000636	-0.306	-0.325	-0.328
	(3.25)	(3.39)	(3.39)	(3.19)	(3.39)	(3.49)
R^2	0.0669	0.0670	0.0720	0.0649	0.0721	0.0761
N	149.0[c]	150.0	150.0	149.0	150.0	150.0
			Underutilization case[d]			
Constant	0.646	0.773	0.870	1.629	1.830	1.921
	(4.89)	(5.86)	(6.46)	(2.95)	(3.33)	(3.41)
Slope	-0.000503	-0.000522	-0.000513	-0.229	-0.246	-0.240
	(2.30)	(2.39)	(2.29)	(2.07)	(2.23)	(2.16)
R^2	0.0355	0.0379	0.0349	0.0289	0.0332	0.0310
N	146.0	147.0	148.0	146.0	147.0	148.0

Note: a Interest rates represent hypothetical real rates. b Assumed 100% of electric power supply comes from own generators. Fuel consumption and maintenence cost are adjusted accordingly. Fuel by a factor of 4 and maintenence and parts by 3, when utilization rate increases from 25% to 100%. [b] Interest rates represent hypothetical real rates. [c] The total number of observations may not be the same because the log of negative values is not defined and they are treated as missing values. [d] The average utilization rate of installed generating capacity was 25.48%.

Source: NIDB/IBRD Project Establishment Survey, 1988.

loss resulting from the public sector failures would have been much higher in Nigeria.

At best, public sector performance is likely to improve very gradually. In addition, improvements in public sector performance will be accompanied by considerable upward adjustment in pricing and tariffs. Such adjustments which are necessary for long run efficiency however are bound to create hardships in the short run, as firms of all sizes and in all sectors and regions make their own adjustments. For these reasons, the correct policy perspective for Nigeria is not to stress improvements in public sector performance to the exclusion of private sector incentives. Rather, the challenge is to find feasible intermediate term policy options which bridge the gap between the above mentioned two extreme cases of inefficiency, namely, the nonperforming public sector and costly private provision by individual firms.

As discussed in Anas and Lee (1988), there are numerous opportunities that can be exploited for strengthening those already existing markets for the private supply of infrastructure services or creating new ones such that the costs of private provision are significantly reduced and more efficient private provision alternatives are offered. Policy options can be grouped into three categories: (i) regulatory changes which will induce fuller utilization of existing private provision capacities; (ii) private sector participation in selected subactivities; and (iii) changes in pricing and tariff structures. We will discuss below each of these policy areas as illustrations for possible policy options drawing on the survey results. More definitive policy recommendations will be made later in the study based on formal empirical analyses to be conducted.

Figure 11-2. Average Cost of Electric Power Generation, 1987

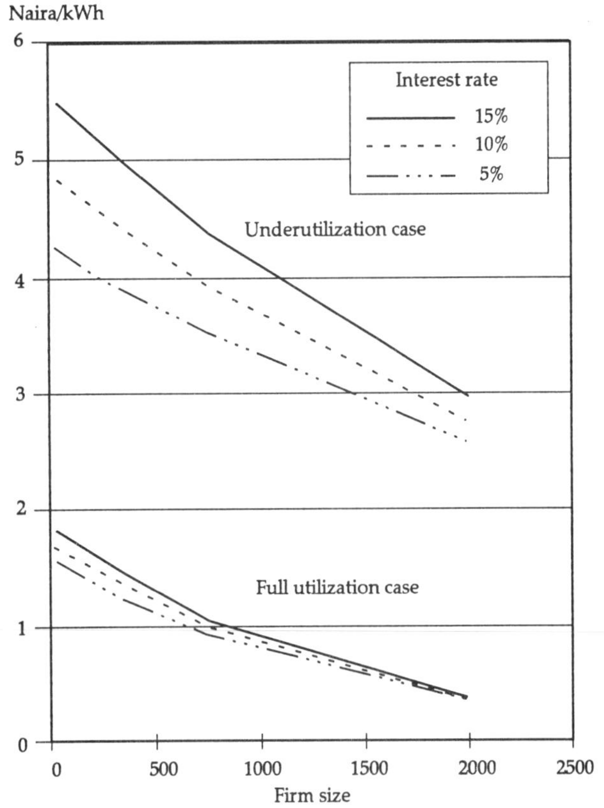

Note: The plotted values are from the regression of the values in Tables 11.2 and 11.3 excluding establishments with fewer than 20 employees (using the exchange rate of US$1.00+7.5 naira).

Source: Based on data from NIDB/IBRD Project Establishment Survey, 1988.

Regulatory Changes for Fuller Utilization of Private Provision Capacities

Some minor regulatory changes can generate significant benefits to individual firms. As noted earlier, most firms have standby generators which stay idle about 75 percent of the time. These firms however are not allowed to sell the excess power they produce to NEPA or to other firms. The potential cost savings from allowing such transactions can be large. The current regulations inhibit the regimes of "joint production", "satellite behavior", and "shared production." Indeed, the efficiency gains of allowing large firms with a high level of installed capacity to exploit fully their scale economies and to compete with NEPA by supplying smaller satellite firms could be significant. The presence of economies of scale was shown in the case of electric power generation.

Such regulatory changes could also motivate "shared production" whereby private manufacturers join forces to form certain types of "utility pools" to exploit economies of scale in the provision of each type of infrastructural service and economies of scope in the provision of several different infrastructural services at the same time. Utility pools should be quite feasible in the existing industrial estates or in areas with a relatively high concentration of industries. The participation of large firms in the infrastructure production process and competition with the public suppliers broadens the choices available to small firms and especially the "captive firms." Small firms in such an environment can become satellite firms or can join in utility pools. Small firms have very high willingness to pay for reliable electric power supply. Thus, they would be motivated to join a utility pool or to become satellites to larger firms.

A good example of a "utility pool" in place is the central effluent collection and treatment facility in the Agbara Industrial Estate which was established by a private developer. This central facility is operated by a management company. As the government attempts to tighten industrial pollution control, treating the effluent within individual firms would be prohibitively expensive, especially for small firms. Similarly, in industrial layouts in Lagos, the central collection and treatment of effluent by the management board should be technically feasible and will induce economies of scale. Such a management board could be further empowered to operate and manage "utility pools" which include a wide range of services such as electric power generation, garbage collection, and the shipment of goods. Another example of a central facility in place is the six megawatt standby generator of the University of Ibadan which serves the entire campus. A note prepared for a recent Industrial Sector Study (Lee 1989b) further discusses such possibilities for the existing industrial areas in Nigeria.

Table 11–8. *Average fixed and variable costs of owning own electric power generation per kWh*

Firm size	*Fixed cost*	*Variable cost*	*Total*
All firms			
Naira	3.180	0.803	4.612
percent	82.60	17.40	100.00
0-9			
Naira	0.655	0.077	0.732
percent	89.50	10.50	100.00
10-19			
Naira	0.990	0.253	1.243
percent	79.62	20.38	100.00
20-49			
Naira	5.824	0.634	6.457
percent	90.19	9.81	100.00
50-99			
Niara	3.784	0.991	4.775
percent	79.24	20.76	100.00
100-199			
Naira	4.157	0.719	4.876
percent	85.26	14.74	100.00
200-499			
Naira	3.305	0.801	4.106
percent	80.49	19.51	100.00
500-999			
Naira	1.646	0.733	2.379
percent	69.19	30.81	100.00
1,000 +			
Niara	1.877	1.438	3.315
percent	56.62	43.38	100.00

Note: [a] Annualized capital value of generators and accessories. [b] Include fuel, maintenence, parts and labor. For the sample firms as a whole, 25.48% of electric supply came from own generators.

Source: Anas and Lee (1988).

Private Sector Participation in Contestable Markets for the Supply of Infrastructure Related Services

Although more efficient pricing systems combined with appropriate relaxation of regulatory constraints can be introduced to induce improved public sector performance and to minimize the adverse impacts of infrastructural deficiencies on manufacturers, these strategies alone are unlikely to significantly improve the current situation in the short run or even in the medium term, because of the various inefficiencies in administration, financial management, and the operation and maintenance practices of the public agencies. Based on what is observed in Nigeria, a sensible way of breaking this inertia seems to be the encouragement of private sector participation in various infrastructure related functions and subactivities.

Indeed, in Nigeria we observe that some private firms are already engaged in certain types of infrastructure related subactivities. Recently, NEPA began subcontracting certain segments of its operations, such as maintenance for a power station and transmission facilities, to private firms. Many foreign firms including Siemens and ITT have already had maintenance contracts with the Nigerian Telephone Company (NITEL). The government allowed a private firm, DHL, to operate in Nigeria. DHL charges a much higher fee than the Nigerian Postal Service, but it is faster and more reliable, and thriving with good business. This is additional evidence that users of services have high degrees of willingness to pay when reliable services become available. This was also observed in the Maroko low income area in Victoria Island. This area, which NEPA never included in its network, has been served by a private entrepreneur who charges 30 kobos per kWh, four times higher than NEPA's 7 kobos. But this rate of 30 kobos is still many times lower than the average cost of own power generation. Another example is the air freight and passenger transport sector. In this area, a number of small privately owned domestic airlines provide stiff competition to Nigerian Airways because they supply more reliable service. Railroads, where the high sunk costs associated with the capital facilities make the industry less contestable, cannot as easily benefit from such private competition, but trucking has emerged as a very viable alternate transport mode.

In Nigeria, a broad continuum of options exist between the two extremes of inefficiency characterized above. These options amount to providing incentives for private entrepreneurs to engage in the supply of certain infrastructure services, thus creating appropriate market mechanisms. Such markets can be specialized to infrastructural services in the areas of production, distribution, maintenance, administration, metering and monitoring, or bill collection. The feasibility of creating and expanding such markets for the supply of these services by the private sector lies in the fact that the government fails to provide adequate services whereas the users are willing to pay for more reliable services when such are available, for example in the case of electric power supply.

A recent Bank case study by Whittington, Lauria and Mu (1989) documents how high willingness to pay for water has led to the emergence of a complex web of private market mechanisms for water distribution in Onitsha, a Nigerian town of 700,000. In this town, the private sector operates about 275 tanker trucks which purchase water from about 20 privately owned boreholes and sell it to businesses and households with storage tanks. Many of the households purchasing such water in turn sell it to individuals who are not equipped to store in large quantities, or to thousands of small mobile private vendors. The private vendors provide two times more water on the aggregate compared to the public utility and collect 10 times the revenue in rainy season and 24 times the revenue in dry season. Households pay these private vendors over twice the operations and maintenance costs of piped water, a strong indication of the willingness to pay for reliability, and clear evidence of the private sector's ability to compete with the public sector.

To operationalize a workable framework for promoting private participation in the infrastructure subsectors, the following strategies in three key areas need to be considered.

Regulatory regimes and market mechanisms. The first step is to improve the present regulatory regimes to provide a more favorable environment for private investors so that they can enter the market for a specific service and offer alternative sources of supply. Many of the public sector failures in Nigeria stem from the fact that most infrastructure services are provided by strongly centralized government monopolies. As discussed in Anas and Lee (1988), however, even some services which have the characteristics of public goods can be supplied with the participation of the private sector (also see Roth 1985). To the extent that the markets for certain infrastructural services are contestable (Baumol, Panzar and Willig 1982) because there are no large sunk costs involved in capital facilities, it should be feasible to liberalize restrictions against the setup and operation (entry and exit) of private firms.

There are a number of situations where such a strategy can be successful. A good example is the utility pool already discussed above. Individual firms in a pool may prefer to have a private infrastructure provider who will manage and operate a pool with shared facilities such as vehicles and waste collection equipment. This would allow the pool to take advantage of the economies of scale and scope, as well as to pass the transaction costs of administration and management to the private entrepreneur who would be self-financing by levying charges on the pool members.

As mentioned earlier, power generation is an area where private participation can be greatly increased by allowing private entrepreneurs to set up power plants which compete with NEPA. A successful arrangement exists at Jos where a privately owned power plant which was setup in colonial times has been allowed to operate. This firm supplies much of the local power needs and sells its excess power to NEPA. Additional private power providers are likely to emerge throughout Nigeria if the existing regulatory constraints were relaxed. If this were to happen, NEPA could stiffen its tariff structure since users would have the freedom to switch to the private suppliers. NEPA's transmission and distribution grids should be made accessible to such private power companies which can be required to pay appropriate access fees which reflect the marginal costs of serving them. Allowing access to the grids makes the generation of power a contestable activity which greatly increases the incentive for private participation. The levying of efficient access fees by NEPA would provide a source of revenue which aids in cost recovery while reducing some of NEPA's own power generation costs. This approach has been followed in Britain with respect to both the power authority (Henney 1987) and British Telecom (Beesley 1981). A wide range of options for private sector participation have also been considered in the past. These include, for example, farming out distribution functions to private firms (World Bank 1983b, Coyaud 1986).

Organizational and institutional mechanisms. To induce the development of appropriate market mechanisms for private sector participation in infrastructure supply, it will be necessary to allow appropriate institutional arrangements such as subcontracting or franchising to carry out a particular type of infrastructural service. Such mechanisms will tend to vary from sector to sector and will depend on the strength of incentives which are needed and the efficiency gains which will occur from private sector participation.

A good illustration is available in the waste collection and disposal subsector in Nigeria where a number of alternative institutional responses have been observed in recent years (Sulu 1987). While Lagos approached the problem of solid waste disposal by authorizing large capital expenditures (World Bank 1985), Ibadan implemented a citizen participation procedure in which private firms haul their garbage to designated points to be picked up by private licensed subcontractors or by the public sector. In Owerri the solution was to enter into a subcontract with the German firm SULO A.G., which made an unsolicited offer.

Luger (1989) in a recent World Bank discussion brief argues for more private sector participation in solid waste collection in the Lagos area to increase its share of industrially generated waste up from the current 7 percent. Luger breaks down solid waste collection into the following subactivities:

- pickup at the source and delivery to processing plant or transfer station;
- pickup at processing plant or transfer station and delivery to tipping site or resource recovery facilities;
- transfer points, tipping sites, processing facilities, or incinerators;
- maintenance of various facilities; and
- administration including bill collection.

While the Lagos State Waste Disposal Board (LSWDB) could continue to maintain control over regulation, the remaining subactivities are candidates for various forms of privatization on a case by case basis. For example, the private sector could be induced to set up landfill sites or resource recovery facilities if they are allowed to produce gas, energy, or compost which can be sold profitably. In finance, bill collection can be contracted out, where the contractor's payment is based on the percentage of outstanding revenues that are collected. Such a private collecting entity would be more motivated than the existing bureaucracy to achieve full revenue accrual. In the areas of pickup and delivery to intermediate points, there is a variety of available options including direct delivery by the manufacturing establishment's own vehicles, pickup by private entrepreneurs on a demand activated basis, or pickup by a private entity licensed to operate as a spatial monopoly within a particular district.

Monitoring mechanisms for market operations and service quality. As various infrastructure related functions currently under government control are decentralized and privatized, it will be important to redefine the role of the government for appropriate monitoring and supervision of efficient market operations. For example, if a subactivity such as bill collection or garbage pickup is contracted out to private firms, it will be necessary to monitor their success with revenue collection or quality of service in garbage pickup. Their contract renewal could be determined by a periodic competitive bidding process.

In sum, the government will play an important role in implementing the new institutional setups resulting from the policy options and reforms that might be adopted. More systematic analyses of economic and institutional feasibility will follow in this study.

Congestion, System Failures, and Pricing Policy

The fluctuations in the quality of public infrastructure services observed in Nigeria are, in part, a result of congestion in the use of the system. While the demand for the service from a public agency such as NEPA is a function of quality, the quality itself is a declining function of the quantity demanded due to congestion effects. The public agencies must consider the trade-offs between the quantity supplied and the quality (and reliability) of services in determining the pricing policy, especially in the short run when the ability to expand the system is limited. Treating congestion as endogenous is common in transportation and other urban infrastructure systems, and congested situations require the levying of an optimally set congestion toll which will reduce the load and congestion to a socially optimal level.

As an illustration, consider the electric power pricing by NEPA. Most power interruptions (nearly two thirds) are a result of bottlenecks on the transmission and distribution networks. It is commonly observed that in the industrial areas in Lagos when large energy intensive manufacturing plants such as steel mills start operating, the resulting voltage surge often damages machinery and equipment of smaller firms located in the vicinity. Large energy intensive firms place heavier loads on the system, thus tying up more operable transmission capacity. However, these large firms are the ones which can afford to have own generators, have a greater amount of unused generating capacity, and can produce electric power at a much lower average cost than small firms.

In the case of NEPA, the congestion is so severe that the system tends to fail completely resulting in frequent power outages. In such a situation, it would be desirable to raise the tariff to a sufficiently high level to clear the market. For example, at a NEPA tariff of

50 kobos per kWh, large firms may find self-generation cheaper and use their own generators more fully, thereby reducing congestion. Deregulation, to allow those firms to sell excess power, should provide added incentives to own generation of power. Small firms will then have better access to the system. Public supply quality is expected to improve at the higher NEPA price which smaller users may find still lower than the cost of self-generation. We have requested NEPA to provide us with the necessary data to document statistically the correlation between loads on the transmission network and the frequency of power failures, in order to measure the quality improvements that can be expected from inducing firms with different private provision capacities to reduce their use of the public supply in response to higher prices. A more comprehensive study of the market structure, including NEPA's costs and variations in demand by user types and locations is needed to determine the order of the price that will remove congestion.

Producing specific tariff systems for individual subsectors such as electric power, water, and telecommunications is beyond the scope of this study. In this research project, however, we intend to quantify relative efficiencies of alternative pricing regimes by simulating the responses of different types of firms to such regimes that reflect particular types of market structures. Possibilities for considering variations of the "two-part tariff", for example, were discussed in detail in the framework paper (Anas and Lee 1988). Bahl and Linn (1989) present an excellent review of pricing urban services. A recent paper by Heady (1988) stresses the role of public sector prices as instruments of cost recovery and explains the Bank's two-step practice in setting public sector prices. The first step calculates the marginal cost; the second step adjusts marginal cost to take account of other factors such as revenue shortfalls, market distortions, and distributional effects. All these factors are relevant to the Nigerian situation. Another recent paper by Julius and Alicbusan (1988) documents the two-step approach in more detail and surveys the use of such pricing policies in many countries and various public sectors. A clear discussion of short-run marginal cost pricing, economic user charges, and budget deficits is given in Meier (1983, pp.192-203) which is reprinted from Walters (1968) and Bennathan and Walters (1979) who also discuss nonlinear "two-part tariff" pricing.

Conclusions and Further Empirical Study

The main objectives of this paper were to document the extent, causes, and incidence of infrastructural deficiencies as they affect Nigerian manufacturers; to observe the responses of the manufacturers to these deficiencies; and to develop viable policy options based on the observations from the data collected. The results of the establishment survey revealed general patterns of deficiencies and self-provision responses by manufacturers which cut across all five infrastructure subsectors included in the study. In particular, in nearly all infrastructural activities, small firms face higher unit costs than larger firms do and the patterns of self-provision by firms differ a great deal by region within the country as well as by type of firm.

Our main thrust in developing policy options is that the ongoing structural adjustments in Nigeria, including changes in pricing, regulation, and institutional structure in most sectors, need to be extended to managing and accommodating the costs of the widespread private provision of infrastructure services resulting from public sector failures. Because improvements in public sector performance are likely to remain slow in the short and intermediate terms, manufacturers and especially small firms will continue to bear the costs of self-provision. Furthermore, with the ongoing upward adjustments in tariffs the burdens of the deficiencies which are borne by small firms will increase. To ease these private burdens and to improve the overall infrastructural provision in Nigeria, we have considered plausible policy options in the following three areas:

- Regulatory reforms such as the relaxation of regulatory restrictions against the trading of infrastructural services among manufacturers.
- Private sector participation in contestable markets for the supply of infrastructural services, wherever appropriate for selected subactivities such as production, delivery, maintenance, revenue collection, and finance, by means of various institutional mechanisms such as subcontracting, franchising, and districting.
- Alternative pricing policies taking into account the capacity limitation and congestion effects on the service facilities.

A set of more definite policy recommendations will be provided later in the study on the basis of the formal empirical analysis to be conducted with the establishment survey data. In particular, econometric work outlined in the framework paper (Anas and Lee 1988) and Verma and Lee (1988) will enable us to estimate key production and cost function parameters which will provide firm quantitative bases for policy analyses. Such econometric models can be used to simulate the responses of selected firms to various policy changes. Such simulations are essential in ord er to obtain better insight about the probable econ omic benefits that are likely to result from the policy options and the implementation strategies which we have discussed.

References

Anas, A., and K.S. Lee. (1988), "Infrastructure Investment and Productivity: The Case of Nigerian Manufacturing, A Framework for Policy Study." Infrastructure and Urban Development Department, Discussion Paper, Report INU 14. The World Bank.

Bahl, R.W., and J.F. Linn. (1989), *Urban Public Finance in Developing Countries*. The World Bank. (mimeo)

Baumol, W.J., and D.F. Bradford. (1979), "Optimal Departure from Marginal Cost Pricing." *American Economic Review*, 256-283.

Baumol, W.J., J.C. Panzar, and R.D. Willig. (1982), *Contestable Markets and the Theory of Industrial Structure*. Harcourt Brace Jovanovich.

Beesley, M.E. (1981), *Liberalization of the Use of British Telecommunications* Network. Department of Industry, Report to the Secretary of State, Her Majesty's Stationery Office, London.

Bennathan, E., and A.A. Walters. (1979), *Port Pricing and Investment Policy for Developing Countries*. Oxford University Press.

Coyaud, D. (1986), "Private and Public Alternatives for Providing Water Supply and Sewerage Services," in *Management Options for Urban Services*. The World Bank. (Report on Seminar held at Cesme, Turkey, November, 1985.)

Heady, C. (1988), "Public Sector Pricing in a Fiscal Context." Policy, Planning and Research Working Papers, WPS 179, The World Bank, April 1989.

Henney, A. (1987), "The Operation of a Power Market." (A paper prepared for the government and the electricity supply industry as a contribution to the debate on how to promote competitive power generation), London.

Julius, D. and A.P. Alicbusan(1988), "Public Sector Pricing Policies: Areviewof Bank Policy and Practice." Policy, Planning and Research Working Papers, WPS 49, The World Bank.

Lee, K.S. (1985), "An Evaluation of Decentralization Policies in Light of Changing Location Patterns of Employment in The Seoul Region." Urban Development Discussion Paper, UDD-60. The World Bank.

Lee, K.S. (1989a), *The Location of Jobs in a Developing Metropolis: Patterns of Growth in Bogota and Cali, Colombia*. Oxford University Press.

Lee, K.S. (1989b), "Infrastructure Constraints on Industrial Growth in Nigeria." The World Bank, (mimeo).

Lee, K.S., J. Stein, and J. Lorentzen. (1989), "Urban Infrastructure and Productivity: Issues for Investment and Operations and Maintenance." The World Bank, (mimeo).

Luger, M.I. (1989), "Privatization Options for the Solid Waste Sector in Nigeria." Discussion Brief, Infrastructure and Urban Development Department, The World Bank.

McGuire, M. (1974), "Group Segregation and Optimal Jurisdictions." *Journal of Political Economy*, Vol.82: 112-132.

Meier, G.M. (1983), *Pricing Policy for Development Management*. The Johns Hopkins Press. (Published for The Economic Development Institute of the World Bank.)

Meyer, J.R., and M.R. Straszheim. (1971), *Pricing and Project Evaluation*. The Brookings Institution.

Munasinghe, M. (1979a). *Economics of Power System Reliability and Planning*. Johns Hopkins Press.

Munasinghe, M. (1979b), "Electric Power Pricing Policy," *World Bank Staff Working Paper. No. 340*. The World Bank.

Munasinghe, M. (1988), "Energy Economics in Developing Countries: Analytical Framework and Problems of Application." *The Energy Journal*, Vol. 9, No. 1. (Available as *World Bank Reprint Series*: Number 421.)

Roth, G. (1987), Private Provision of Public Services in Developing Countries. Oxford University Press.

Sulu, S.A. (1987), "Development of Maintenance Strategy for Urban Solid Waste Collection." A Paper Presented at the 1987 National Workshop on Engineering Maintenance Organized by the Nigerian Society of Engineers.

Verma, S.K. and K.S. Lee. (1988), "Measuring the Productivity of Infrastructure Services in Nigerian Manufacturing: A Theoretical Framework." The World Bank (mimeo).

Walters, A.A. (1968), *The Economics of Road User Charges*. The Johns Hopkins Press.

Whittington, D., D.T. Lauria, and X. Mu. (1989), "Paying for Urban Services: A Study of Water Vending and Willingness to Pay for Water in Onitsha, Nigeria." Infrastructure and Urban Development Department, Case Study, Report INU 40.

World Bank. (1981). *Nigeria: The Power Transmission and Distribution Project*. Report No. 3041-UNI.

World Bank. (1983a). *Nigeria: Macroeconomic Policies for Structural Change*. Report No. 4506-UNI.

World Bank. (1983b). *Nigeria: Issues and Options in the Energy Sector*. Report No. 4440 UNI.

World Bank. (1983c). *Nigeria: Lagos Urban Sector Review*. Report No. 4479-UNI.

World Bank. (1985). *Nigeria: Lagos Solid Waste and Storm drainage Project*. Report No. 6375-UNI.

World Bank. (1986). *Trade Policy and Export Development Loan*. Report No. P-4358-UNI.

World Bank. (1988). *The Nigerian Structural Adjustment Program: Policies, Impact and Prospects*. Report No. 6716-UNI.

World Bank. (1988). *Nigeria: Lagos Water Supply Project*. Report No. 6375-UNI.

World Bank. (1989). *Nigeria: Power System Maintenance and Rehabilitation Project*. Report No. 7607-UNI.

12

Comments on Fiscal Deficit and Expenditure Policy

Yao Kouadio and Nii Kwaku Sowa

Yao Kouadio

The Ferroni and Kanbur Paper

The main objectives of Ferroni and Kanbur's interesting paper are to present an overview of public expenditure patterns in Sub-Saharan Africa, to develop a theoretical framework for retargeting public spending toward poverty alleviation, and to undertake some case studies based on the rules derived from the theoretical section.

I would like to indicate at the outset that I was a bit surprised by the finding according to which adjustment programs in Sub-Saharan Africa have not really led to a reduction in real public expenditures. Likewise, it seems odd that at the sectoral level, in opposition to what might be expected, social and economic sector public expenditures have not declined despite the need to adjust most economies through a reduction in domestic absorption.

Whatever the actual trend in overall spending, the authors point out that intrasectoral expenditure patterns remain inadequate in the sense that there is a disproportionate resource allocation toward higher-order services, that is those going to the wealthy. Thus, the poverty alleviation impact of public expenditures is less than it could be.

Following this overview of public expenditure patterns is a very fine section on retargeting of public expenditures, which raises a number of challenging questions of a normative nature. Recognizing that the standard of living is a multidimensional concept, the authors build a very simple, yet instructive, model in which the standard of living is a linear function of social expenditures and the level of income, and income is a function of basic needs and productive expenditures. The interesting feature of the model is that it allows for feedback effects as far as basic needs and income are concerned.

Given the budget constraint, the authors attempt to provide some simple decision rules as regards public expenditure restructuring. The major conclusion that emerges is that a necessary condition for favoring basic needs expenditures hinges on whether or not the direct effects of those expenditures exceed the indirect effects that are induced by increases in income. As correctly emphasized by the authors, this is clearly not a sufficient condition for the social valuation of basic needs and income.

Obviously, one could question the linearity assumption that is put forward in the theoretical model. But this is not the crux of the matter, since most of the implications remain valid even under nonlinearity. What is more important is related to the issue of estimating social weights among the different dimensions of the standard of living. Carrying the analysis one step further, Ferroni and Kanbur consider the effect of restructuring public expenditures on poverty using Foster, Greer and Thoebecke's poverty index. The implications of their analysis are contained in two formulae that are to be compared. The basic finding here is that the impact of public expenditure restructuring on poverty depends on the current shares of total expenditures reaching the poor. This insight is probably the major contribution of this paper.

After some digression on the Human Development Index, the usefulness of which is surely debatable, the paper introduces some empirical evidence based on some African countries. Thus, regarding Côte d'Ivoire and the issue of rice pricing policy, Ferroni and Kanbur suggest, in line with the theoretical prescriptions, that increasing the consumer price would be preferred to reducing the producer price if poverty alleviation is a priority for policymakers. Still, regar- ding housing policy in that country, the conclusion reached by the authors is that public disengagement is unlikely to be detrimental to the poor and is therefore to be adopted given the current size of the public finance deficit.

I do not have any quarrel with the above analysis if ne accepts, as a primary goal, poverty alleviation instead of some other objective. It should be made clear, however, that the conclusions do depend on the data set and that one should perhaps undertake sensitivity analysis, particularly as far as the poverty line is concerned.

On the whole, I find this paper very stimulating, although one feels frustrated at the difficulty of constructing true standard of living indices in order to facilitate the assessment of public expenditure restructuring.

Nii Kwaku Sowa

My comments on the three papers in this section have to do with the lessons that African policymakers can derive from them to help shape up their public expenditures.

Most African governments, soon after independence, embarked on various programs aimed at modernizing or developing their economies. A number of them adopted industrialization as a development goal and, in the process, established several import substituting industries. These industries, which were mainly state owned, were supported by weak infrastructural bases and have weak forward and backward linkages. Revenue generated from most of these enterprises could not support them and some had to rely on government (subvension) to pay their workers.

The state, in most African countries, also was the provider of basic utility services such as electricity, water and waste disposal. These services were provided at very subsidized rates to make them affordable to the poor.

In the face of mounting expenditures, African governments have had to contend with very narrow and weak tax bases, which have led to budgetary problems. By the beginning of the 1980s, all countries in Sub-Saharan Africa, except a few like Botswana, were saddled with budget deficits which had to be financed by governments printing money. These led the countries into other economic problems like inflation and balance of payments difficulties.

The three papers by Rutayisire, Ferroni and Kanbur, and Lee and Anas look at the fiscal problems of Africa from different angles. Rutayisire looks at the problem of budgeting, Lee and Anas examine the issue of how inefficiencies in infrastructural services get translated into unnecessary costs for manufacturers, and Ferroni and Kanbur provide solutions to expenditure management.

The Rutayisire Paper

In the paper on Tanzania, Rutayisire sets out to resolve what he thought was a conflict of opinion on government expenditure in that country. On the one hand is the concern about the reductions in the size of government expenditure, and on the other hand is the view that the level of government expenditure is still too high. In my opinion, there is no conflict of opinions: one can be cutting down on government expenditure, but so long as it has not come down to desirable levels, it can be said to be too high.

Rutayisire sets out to determine the optimal and sustainable level of government expenditure for Tanzania. His results show that the years of structural adjustment and economic recovery (1981 to 1988) have been years of unsustainable government spending. He unfortunately did not extend his analysis to pre-structural adjustment and economic recovery program years for us to compare. However, this result by itself resolves what he thought was a conflict of opinion. Structural adjustment and economic recovery programs call for restraint on government spending and yet, still public expenditure in Tanzania remains unsustainably too high. Thus, either Tanzania expands its tax base or there must be further cut-back of government spending with proper targeting (Ferroni and Kanbur).

The issue of sustainability of government expenditure will, however, have to be looked at carefully since it depends on certain factors which may not be properly measured. Budgetary procedures in most African countries are haphazard. Estimates of government spending do not follow proper costing procedures and rely mostly on percentage increases. This can either overestimate or underestimate government spending. In addition, inflationary expectations, if not well anticipated, may distort calculations for sustainability. In any case, even if government spending is unsustainable and calls for budgetary cuts, care should be taken not to cut essential services. Cuts in sectors which provide infrastructural support services can lead to serious repercussions in other sectors of the economy.

The Lee and Anas Paper

Lee and Anas, with Nigeria as a case study, look at the effect of inefficiencies in infrastructure provision on manufacturers and the latter's response to such deficiencies. This study is particularly interesting in the way it unearths a problem which can be found in most African countries. The paper considers the deficiencies in the provision of services such as electricity, water, telecommunication, transport, and waste disposal. It then examines the cost of these externalities to the manufacturers. It observes that because of frequent interruptions and the inefficiencies in the provision of these services, most manufacturers have stand-by units to support them. Firms which do not have any stand-by units suffer production loss not through their own inefficiencies, but by picking up external diseconomies from public corporations.

Policies recommended by Lee and Anas include changes in the regulation for the provision of these services to allow private entrepreneurs to enter the market. This move is already being implemented in most countries under structural adjustment programs. However, the nature of these services sometimes dictates the existence of a natural monopolist—in Africa the State.

There is also the suggestion that the market be deregulated to allow firms with stand-by units to sell their services to firms without stand-by units. This can only be a short-run solution, since a private company with a generator will sooner or later run into worse problems than the public company. There was a hint that sharing of generators by firms will ease congestion on the public lines. A better and longer term solution will be for proper layout planning before industries are allowed to set up.

Lee and Anas' calculations also show that firms are already sharing high costs in trying to supply infrastructural services themselves. It follows that firms will be prepared to pay higher prices now for better public service. Governments are, however, reluctant to increase the price of these utility services which they consider as part of the basic needs package for the citizenry.

The Ferroni and Kanbur Paper

If poverty conscious expenditure is what the government is thinking of, then Ferroni and Kanbur's tools become useful. A closer examination may show that the proportion of the poor who use these utilities—electricity, water (pipe-borne) and so on—is quite small. Thus subsidies on any of these utilities may be favoring the rich rather than the poor.

Some allusion has been made already to the usefulness of Ferroni and Kanbur's paper. The following comments are directed to specific issues in the paper.

- The first part makes some generalizations about public spending in Africa. A closer look at some countries will reveal that, contrary to what was expressed in the paper, real government expenditure has fallen in some countries. With current efforts at cost recovery in both health and education, certainly expenditure on these sectors would have to be dropped.
- It is true, as the paper observes, that anti-poverty spending must encompass economic sector spending as well as provision of basic needs. But for most people in Africa who live on the brink of survival, direct expenditure on basic needs leads to a bigger effect on their standard of living than an indirect effect through income increase.
- The paper attempts to develop a social welfare achievement index which is intended as an improvement on the UNDP's Human Development Index (HDI). The criticism against the HDI is that it weights each of the social indicators equally. The Ferroni-Kanbur indicator has a stronger theoretical foundation, but ends up with equal weights for the social indicators. Moreover, whereas the HDI has three social indicators, the Ferroni-Kanbur index has only two.

13

The Informal Financial Sector and Markets in Africa: an Empirical Study

Ernest Aryeetey and Mukwanason Hyuha

Difficulties in comprehending the way the informal sector operates have led to inaccurate speculation about how informal financial activity influences the impact of monetary and other economic policies in many African economies. The potential of the informal sector is usually underestimated; this sometimes leads to the pretence among policymakers that this activity is inconsequential. Even where it is acknowledged that a substantial informal financial sector exists, policymakers are usually at a loss with regard to capturing its activity within a policy framework. The sector thrives and affects growth processes in several ways; however, it is either unaccounted for or only improperly accounted for in the national accounts.

Here informal financial market refers to all financial transactions that take place beyond the functional scope of various countries' banking and other financial regulations. Those that are covered by these regulations belong to what are called here the formal financial sector and these are "by definition, bound by the laws and regulations that governments have enacted for the control, supervision, or facilitation of that sector" (USAID, 1989, page 7). Thus the formal sector includes the commercial banking system, near-banks, insurance companies, housing development banks, investment banks, etc. The informal financial sector includes a wide range of financial activity whose extent may differ from country to country. Institutions and activists in the informal sector include moneylenders, rotating savings and credit associations (ROSCAs), mobile bankers and susu operators, mutual assistance groups, landlords, neighbors, friends and family members, etc.

Ideally, analysis would be based on a disaggregation of informal financial activities in order to capture more accurately the functioning of informal groups, and their separate impacts on the economy. However, this is not possible presently in view of the limited information available. To amend the situation, even if partially, most of the illustrations here are based on dominant activities and groups. Thus, for example, the study of informal savings mobilization is Ghana dwells mostly on the susu system using the single collector. This system involves more than 95 percent of the market women who save informally (i.e. 76 percent of the total sample), and the market women are considered to be the largest group of informal savers in Ghana (Aryeetey and Gockel, 1990).

Current interest in what role the informal financial sector can play in the economic transformation of many African countries is derived from the dearth of financial resources for the capitalization process. Evidence of the inadequacy of existing resource allocation and resource allocating mechanisms is apparent in the severe economic crises the region has suffered since the mid 1970s. The crises reached alarming levels in the early part of the 1980s when the economies of most African countries were epitomized by severe stagnation and decline. These problems arose from deep-seated structural imbalances in most of the economies of the region.

To arrest the situation, many African countries are now pursuing various structural adjustment programs aimed at economic recovery and growth. In view of the austerity posed by many of these programs, governments have had to follow tight fiscal and monetary policies. These are reflected partly in reforms of the formal financial sector and public expenditure policy. These reforms invariably imply substantial real reductions in the amount of credit available to the private corporate and household sectors, which thereby forces more and more people to finance various economic activities informally. But relatively little is known of the mechanisms inherent in this activity.

Recent studies (Aryeetey and Gockel 1989; Hyuha, Ndanshau and Kipokola 1989; Chipeta and Mkandawire 1989) in Ghana, Tanzania, and Malawi have helped to throw some light on the internal workings of informal markets, as well as on some of the relationships between them and the established or formal financial market. This paper attempts to streamline the various country studies and map out common areas of operations and similarities in the various informal markets. The information obtained from the primary studies in the three countries is supplemented with information from recent studies in Zaire and Senegal by the US Agency for International Development (1989).

It is argued in this paper that the informal financial sector in Africa is relatively large and plays a crucial role in savings and investment. Thus, for optimal financial resource allocation, informal financial institutions and markets need to be given due consideration in policy design. The paper presents the characteristics of the informal financial sector in five African countries. It then looks at some of the ideas expressed about the nature and workings of the informal system. An attempt is then made to hypothesize within a very static framework the relationship between the formal and informal segments of the financial market in the fourth section. A summary and concluding remarks appear in the final section.

Characteristics of the Informal Financial Sector in Five Selected African Countries

This section presents a review of the informal financial sector in five African countries—Ghana, Malawi, Senegal, Tanzania, and Zaire (see table 13-1). The five cases address the following issues:

- the existence, spread, and estimated size of the sector;
- the types of institutions operating for savings mobilization, credit extension, and safekeeping functions;
- the loans extended by the informal financial sector and the average operating costs of the sector;
- interest rates and other charges on loans;
- the relationship between the informal and formal financial sectors.

Ghana

Under the original forms of land tenure systems in Ghana, rich landowners allocated parcels of their land to squatters or migrants for farming, with the condition that one-third or one-half of the total produce would be given to the landowner. Following the monetization of the economy with modern administration, this transaction was increasingly made in monetary terms. This practice eventually developed into the modern money lending business in rural and urban communities. Aside from money

Table 13-1. *Some macroeconomic features of the five countries*

Country	*Population (millions) 1987*	*GNP per capita (dollars)*	*Average annual inflation rate (percent) 1980-87*	*Gross national savings as a percentage of GNP 1980-87*	*Gross domestic savings as a percentage of GDP 1987*	*Gross domestic investment as a percentage of GNP 1987*	*Gross domestic investment as a percentage of GDP 1987*	*M2 money supply as a percentage of GDP 1987*
Ghana	13.6	390	48.3	2.0	4	7.1	11	11.7
Malawi	7.9	160	12.4	7.0	12	19.0	14	25.0
Senegal	7.0	520	9.1	-2.8	6	15.5	13	23.5
Tanzania	23.9	180	24.9	9.0	-6	18.7	17	25.9
Zaire	32.6	150	53.5	4.9	10	14.3	13	8.8

Source: World Bank, World Development Report 1989, Oxford University Press.

lending, other forms of informal financial activity include the single collector susu system or mobile banking, the rotating susu system or ROSCAs, mutual assistance groups or clubs, and credit unions.

The most widespread of these forms of saving and borrowing is the single collector susu system, which is believed to have begun in Ghana with migrant Nigerian traders (IPC 1988). With this activity, a collector (usually male) visits his clients at their work places, market stalls, shops, and homes and collects funds towards a savings plan. Savers deposit fixed amounts for an agreed period of time, after which their savings, less an agreed sum, are returned.

Over 76 percent of 1,000 market women interviewed by Aryeetey and Gockel (op cit) in Accra, Kumasi, and Takoradi (the three largest centers in Ghana) saved in this manner. The study of rural households by IPC (op cit) also showed that this was the most prevalent form of financial savings in those areas. What is interesting about this form of savings is the fact that no interest is earned on deposits. Rather, the saver pays a service charge through the savings retained by the operator at the end of the agreed period of the savings plan, usually one month. Thus, for a thirty-day plan, a day's savings is retained.

The susu system permits access to credit for various purposes. For as many as 31 percent of the market women saving with susu collectors, access to credit on flexible terms was the most important reason for saving. A major characteristic of these loans, however, is the fact that they are usually of a short term nature. They often must be repaid within three months. In 25 percent of the cases studied, they had to be repaid within one month. In most cases no interest was levied on credit acquired this way.

If the credit facility should attract interest at all, interest payments in this widespread informal activity are as in the following example. A woman who normally deposits one hundred cedis (C100) daily towards a 90-day savings plan would accrue C9,000. She may be permitted a loan of C4,500 in addition to her bulk sum of C9,000 (less the service charge). She then pays C120 daily to the collector for the next three months. Half of her normal deposits (C4,500 less service charge) is returned to her; the other half plus C1,800 are used to repay the principal and interest on the loan. The market women have very little problem with this arrangement since the 40 percent interest rate over three months is paid daily in relatively small amounts (C20 in this example). This may be compared to an average lending rate of 30 percent per annum charged by banks.

Moneylenders often make credit available for between 3 and 12 months. The average maturity period is 6 months at an interest rate of 50 percent. They are patronized in spite of the relatively high interest rates because those who require them usually have no other choice, and it is relatively easy for borrowers to acquire funds. In many cases, the borrower's expectations of a specific income within an agreed period, an introduction by someone known to the lender, and a written undertaking are enough to secure the borrower credit. The default rate is said to be quite low, but is believed to have gone up in the last decade, when personal incomes suffered greatly. Credit unions and mutual assistance groups often have longer term facilities.

The ratio of deposit money to M2 money supply and the ratio of the banking sector's claims on the private sector to GDP have been used to measure the size and relative significance of the informal market (Aryeetey and Gockel op cit and Wai 1957). They have both shown a relative decline in the strength of the formal banking system and therefore significant growth in the informal sector. For rural Ghana, IPC (op cit) estimated that 60 percent of all financial savings would be done informally. It was estimated that an association of 30 susu collectors in Tamale mobilized about C30 million in savings in a month. This contrasted with the mid-year total savings mobilized by the banks in the area of C11 million in 1988. The informal sector was more significant for private household sector savings mobilization and credit granting than was the formal financial sector. The fact that a substantial portion of savings accumulated in the susu system is used to finance working capital makes financing through the sector essential for small-scale investment.

There are significant links between the formal and informal financial sectors on the deposit and lending sides. It has been noted that, "most of the 'susu' operators deposit at the end of each day the total savings mobilized from the markets in one of the banks near the market" (Aryeetey and Gockel, op cit). This behavior was attributed to the security it affords the operators. Most of the bank deposits made by the susu operators are of the demand deposit type since the operators need to stay liquid at all times to make payments at short notice. Similarly, it is observed that many moneylenders operate bank accounts. They sometimes go as far as borrowing from banks, if possible, and relending to their customers. In view of these links between the two sectors, it is indeed likely that, through a proper assessment of the actual cost and demand structures of the two sectors, the informal sector could be made completely subject to various monetary and fiscal policies of the government, without losing its character.

Malawi

According to Chipeta and Mkandawire (1989), in Malawi the informal financial sector is larger than the formal financial sector. Using the ratio of informal to formal lending to the private sector, they estimated that the informal financial sector was about three times as large as the formal.

Institutions in the informal sector range from moneylenders (Katapila) to friends and neighbors. Table 13-2 shows the types of informal financial institutions that exist in Malawi and some of the interest rate charges on loans extended by the sector. Estate owners, friends, and employers are significant institutions in the informal financial sector (Chipeta and Mkandawire 1989).

Informal sector loans are extended for consumption and investment purposes. Consumption loans usually come from Rotating Savings and Credit Associations (ROSCAs or Chiperegarii) and may be spent on consumer durables. The loans are given without much bureaucracy and promptly, with minimal or no paperwork. Except for the moneylenders, interest rates on loans are either nonexistent or very low compared to those in the formal sector, as table 13-2 shows. The loan repayment rate is quite high despite the absence of formal collateral.

On the low interest rates, Chipeta and Mkandawire (1989) write the following:

"Most of the traders, farmers, friends, relatives and neighbors do not seem to charge interest on the loans that they grant. Traders may be looking at free loans as a means of maintaining and improving relations with their customers, while the rest may be looking at it as a means of maintaining solidarity."

The 'Katapila' charge high interest rates basically because they evolved customarily within such a tradition. The scarcity of funds and expectation of high default rates should also explain the high rates payable to the moneylenders.

The formal and informal sectors are closely linked. Chipeta and Mkandawire (1989) describe this relationship as follows:

"In general, the activities of the informal financial sector complement those of the regulated formal financial sector. The informal financial sector finances business enterprises that the formal financial sector does not finance. It also finances consumption of households that do not have access to formal financial institutions. But despite its success, the informal financial sector does not meet all the demand for loans."

It is also suggested by the two researchers that in view of its size, the informal financial sector plays a major role in the savings and investment process in Malawi. As such the formulation and implementation of policy in the country should take account of the informal financial sector.

Senegal

The United States Agency for International Development (1989) study of Senegal covered mainly urban areas. The relatively rich groundnut region and a few other rural areas were also included. The report maintains that:

"The informal sector, defined as that portion of economic activity which is for the most part outside the domain of the formal legal system and government, is very large...The team believes that approx-

Table 13-2. *The informal financial sector in Malawi: structure and interest rates*

	Nominal rate (annual percentage)		*Real rate (annual percentage)*	
Institution	*Deposit*	*Lending*	*Deposit*	*Lending*
Moneylenders	n.a.	300-1200	n.a.	285-1185
Estate owners	n.a.	2-25	n.a.	-13-10
Traders	n.a.	n.a.	n.a.	n.a.
Grain millers	n.a.	n.a.	n.a.	n.a.
Smallholder farmers	n.a.	n.a.	n.a.	n.a.
Savings & credit associations	94-100	24-720	79-85	9-705
Cooperative savings associations	0	0	0*	0*
Community funds	n.a.	12-600	n.a.	-3-85
Employers	n.a.	4.5-14	n.a.	-10.5-1
Friends	n.a.	0	n.a.	0*
Relatives	n.a.	0	n.a.	0*
Neighbors	n.a.	0	n.a.	0*

* Real rate is zero minus rate of inflation.
Source: Chipeta and Mkandawire (1989), Annex table 4, p. 30.

imately 60 percent of the population in Senegal...earn their living primarily in this sector of the economy. Information regarding the amount of income or full-time employment generated by the informal sector is not available, but it is clear that most people participate in the informal sector at least part of the time. Informal sector activities touch virtually everyone."

Table 13–3 shows the types of informal financial institutions that exist in Senegal. Loans from the informal financial sector are usually of a short term nature and in general the interest rates applied are very low. As in Ghana and Malawi, the operating costs per unit are relatively low, thus making the informal financial sector units low cost producers of financial services. This is attributed to the absence of red tape. The possession of collateral security is not required and literacy does not play an important role in loan acquisition. Credit from the sector is available for investment and consumption purposes.

In Senegal the informal financial sector is generally regarded as a necessary complement to formal services. There is a direct link between the two sectors in that members of the market savings group actually bank their savings with the formal system. It is therefore likely that policies initiated for formal financial development will have some effect on the informal sector. In 1989 it was reported that the informal sector was experiencing a credit squeeze and a liquidity problem following structural adjustment policies initiated by the government to affect primarily the formal system (USAID, 1989).

Table 13–3. *Major informal units and the uses of their credit in Senegal*

Institution	*Uses of its credit*
Rural	
Village savings associations (Caises Villageois)	Religious feast Pilgrimage to Mecca Mutual aid funds Communal projects, etc.
Family savings	Family use (self-finance, etc.)
ROSCAs (tontines)	Purchase of durables Other consumption Business finance
Urban	
ROSCAS (tontines)	Purchase of durables Other business finance Consumption
Market rotating savings	Investment Consumption
Market savings groups (banks)	Mainly investment

Source: USAID; Informal Financial Markets: Senegal and Zaire (1989).

Tanzania

Hyuha et al (1989) generated information on the informal financial sector in Tanzania in a survey covering 262 households in seven regions during the last half of 1989.[1] There is no estimation of the size of Tanzania's informal financial sector; however, it is likely to be as large as the formal financial sector. In general, it appears that there is complementarity between the two financial sectors. The major link between them is the semiformal cooperative societies, which obtain credit from the formal sector and lend to the informal sector.

Table 13–4 shows the types of informal financial units in Tanzania and the share of total households using each. The most common informal financial unit is the semi-official cooperative society. Table 13–5 shows the major purposes of informal sector borrowing in Tanzania, by share of the total sample. Informal sector loans are used for consumption and investment. According to the sample evidence, the largest amounts of credit go to crop finance, food, and business finance. These loans are of a short term nature. The informal financial sector appears to derive at least some of its patronage from borrowers who find the formal sector too complex:

> ...the most impending factors can be summarized as follows. First, the formal financial institutions are very complex and amorphous in the eyes of the peasants... The second factor is the time involved in the process of dealing with the formal financial institutions. In fact, up to 32.4% of the respondents reported that it is too time-consuming to deal with the institutions. Third, peasants and small-scale businesses normally do not have the right type of collateral security demanded by the formal financial institutions. Fourth, the process is not only complex and time-consuming but also bureaucratic. Lastly, the formal financial institutions are urban-based... (Hyuha and Ndanshau, 1990, p. 11-12).

Interpersonal relationships (friend, ethnicity, neighbor, etc.) explain the nature of interest rates, which are usually very low in the informal financial sector. Further, a lot of saving and lending goes on in kind, rather than in money form.

Zaire

The size of the informal sector in Zaire has not yet been ascertained. It is estimated to be considerably large since about "85 percent of the population in

Table 13-4. *Types of informal financial units in Tanzania by households*

Type of informal unit	*Share of total households (percentage)*
Moneylender	1.6
Cooperative society	35.7
Landlord	2.2
Shopkeeper/trader	4.9
Family member	19.9
Neighbor	11.1
Household in the village	9.5
Household outside the village	1.5
Church/mosque	0.6
Other	10.6

Source: Hyuha et al (1989)

Table 13-5. *Purpose of informal sector borrowing in Tanzania*

Expenditure item	*Share of total sample (percentage)*
Food consumption	20.2
Crop production*	24.6
Purchase of farm machinery	13.4
Purchase of livestock	3.7
Business/trade finance	17.1
Building of house	5.0
Children's education	9.1
Weddings	6.1

Note: Includes financing weeding and harvesting activities and machinery (ox-ploughs, tractors, etc.).
Source: Hyuha and Ndanshau (1990), Table 5, p.22.

Zaire earn their living primarily in this sector" (USAID 1989, p. 95).

The informal financial sector in Zaire is dominated by semiformal cooperative societies:

"Zaire's semiformal financial markets are dominated by cooperatives which lend primarily to the national government by purchasing its treasury bills. The small portion lend for commercial and personal purposes is shrinking further in relative importance. Cooperatives receive substantial deposits from members, mainly for safe-keeping. Where they exits the co-ops are well-located and acc- essible to small urban savers, but pay heavily negat- ive real interest rates, are poorly supervised and sub- ject to frequent fraud and theft (USAID 1989, p. 92).

Other units include mutual aid societies, ristournes (tontines), private banking agents, cooperative credit unions, mutual savings associations, and private banks. They are utilized mainly for safekeeping purposes; real interest rates are highly negative.

The USAID study covered only the Shaba region of Zaire. Nevertheless, it is evident that loans are of a short term nature and interest rates charged on the loans are very low. The loans are for consumption and investment purposes.

There is evidence of some interrelationship between the formal and informal sectors. The cooperatives sprang up and have grown tremendously in the last five years, mainly because of the near collapse of the formal sector. As in Senegal, structural adjustment policies have not checked inflation and have led to a serious cash scarcity (liquidity squeeze) which favors urban over rural areas. Thus, macroeconomic policies directed at the formal institutions and activities also affect the informal market.

Review of the Literature and Some Theoretical Underpinnings for the Existence of Informal Financial Markets

Many economists interested in African development are beginning to realize that the informal financial sector has a very important role to play in the growth and development process of the continent. This is especially true in rural areas (Rahman 1989). Several propositions have been made in explaining why informal financial markets continue to thrive despite the official encouragement and support of a more modern and sophisticated formal financial market.

Economic Dualism

Ever since Wai's (1957) pioneering study of the informal financial markets, most of what has been done in this area has been mainly descriptive. In most of the descriptions of informal financial markets, however, their links with formal financial institutions have been portrayed as weak (Ghatak 1981). Some of these suggestions of weak links find their roots in earlier theories of economic dualism. These suggest that, in many African and other developing countries, the modern sector fails to absorb the traditional sector completely largely due to inherent sociocultural reasons. It is difficult to accept this argument, however, in view of the fact that modern banking facilities cannot be found throughout the economy in most African countries.

The earlier theories of economic dualism are linked to the expectation that, as the economies of less developed countries grew and became modernized, the formal financial sector would tend to absorb the informal sector (Ghatak 1981). The contrary is suggested by evidence from Ghana. In Ghana, the informal financial sector has continued to grow even after reforms to improve efficiency and strengthen capitalization levels were introduced into the formal sector by the government in 1985. Operators in the informal sector have simultaneously introduced several innovations in order to keep up with the competition.

Inadequacy of Formal Institutions

The orthodox explanation of the need for credit by small-scale entrepreneurs and peasant farmers hinges on poverty. Therefore, many formal financial institutions have been set up by African governments to provide the required external finance and to pool together the meager domestic financial resources in the countries. However, these development banks have not adequately satisfied the needs of large sections of the intended clientele (Mauri 1987). The bureaucratic, complex, amorphous, and unfamiliar aspects of these institutions has distanced them from the targeted clientele. They are thus seen to be beyond the reach of peasant farmers, small-scale entrepreneurs, and ordinary household savers (see Ghatak 1981). Formal financial institutions are regarded as benefiting only some urban dwellers and large-scale farmers and entrepreneurs, particularly those in the commercial and trade sectors. In any case available formal credit is inadequate, difficult to obtain, and urban-biased.

In view of the difficulties associated with formal credit arrangements, the credit starved small peasant farmers, small-scale entrepreneurs, and households have had to make alternative arrangements for financing their expenditures. Many people have had to resort to the informal financial sector. Over time the informal financial sector has grown in size and importance. It is believed to be larger than the formal sector in savings mobilization and credit allocation to households in some countries (as observed in Malawi and Ghana).

Substitution

Aryeetey and Gockel (1990) observed a strong correlation between the loss of confidence in the banking system in Ghana and the expansion of the informal financial sector. They suggest that the informal sector thrives on the deficiencies in the formal financial sector, which would support the substitution or competition argument. It is often argued that in areas where no banks can be found or where the banking system does not function adequately, as in rural areas, the informal sector provides a very much appreciated substitute. Many people continue to patronize the facilities of the informal sector, especially in rural areas, in view of the monopoly power enjoyed by single units of the sector. Miracle et al (1980) think the use of formal banks is not as widespread as is usually imagined.

"The great bulk of the African population makes little or no use of formal savings and lending institutions. There are few banks in most areas,... (In) African countries over 70 percent of the population is rural, and banks are found almost exclusively in the larger urban centers. However location is only part of the problem. Even the portion of the population near urban centers makes strikingly limited use of formal financial institutions." (Miracle et al 1980, p. 701-2).

Complementarity

The Aryeetey and Gockel work (1990) makes it clear that the informal financial sector has managed to develop self-sustaining mechanisms that have helped it to achieve a greater level of product differentiation under term conditions. These have been made possible through innovations in savings mobilization techniques. Also, a lot of the recent literature based on empirical studies of informal financial markets appears to support the view that these markets often complement the activities of formal financial institutions. This is especially true for commercial banks. In areas where such banks exist, people demanding credit use both markets simultaneously (Hyuha, Ndanshau, and Kipokola 1989, and Chipeta and Mkandawire 1989). The informal financial sector's self-sustaining mechanisms, it is argued, assist it in performing this complementary role.

Seibel (1986) is also of the opinion that both sectors have their own peculiar strengths which guarantee them distinct markets. This may be corroborated by Mauri's (op cit) study of the financial markets in Ethiopia. What comes out clearly from the literature is the existence of two sectors, which are propelled by different forces. They are not entirely exclusive, yet manage to benefit from each other's inadequacies.

New Structuralist Argument

Some of the works of the new structuralists contest the financial liberalization propositions of McKinnon (1973) and Shaw (1973). The new structuralist attempts to introduce considerations of the informal sector into the analysis have assumed a significant relationship between the two markets (see Van Wijnbergen 1982, 1985; Taylor 1983). As Owen and Solis-Fallas (1989, p. 342), who are very critical of the new structuralist analysis, point out, "a key feature of the new structuralist critique is the emphasis on informal credit markets as an important source of residual financing."

The new structuralist argument is that increases in interest rates will push upward interest costs incurred in financing working capital. This will lead to a cost-push effect on prices which may then lower output under monopolistic conditions (Owen and Solis-Fallas, ibid). The growth expected by McKinnon and Shaw will not be achieved in the short run and a form of stagflation may be experienced.

Van Wijnbergen (1982, 1985) and others argue that the effects of increasing real bank deposit rates on short run portfolio allocation will depend on the extent to which households can substitute between bank deposits, loans on the informal financial market (curb markets), and unproductive assets (i.e. cash, gold, and commodity stocks). Households (which are assumed to be surplus units) allocate their real wealth, *W*, in such a manner that lending on the informal market, *L*, is characterized by the demand function:

$$L = f(P, i, r_{td}, y)\, W \qquad (13\text{–}1)$$

where *P* is the inflation rate, *i* is the nominal informal lending rate, r_{td} is the nominal time deposit rate, and y is real income.

There is very little empirical evidence of informal lending or other portfolio allocation being dependent on formal nominal deposit rates in Africa. The link between the formal and informal financial sectors is not via the deposit and lending rates; it is via other factors, such as security in the case of susu collectors. Aryeetey et al (1990), in testing the relevance of McKinnon-Shaw postulates for Ghana, observed that real deposit rates were insignificant in explaining changes in domestic savings in Ghana. In an inflationary environment, potential savers will reallocate their savings to the formal sector only if they can be sure of obtaining credit within a set period. Rising nominal interest rates do not guarantee this.

The new structuralists conclude that the informal sector helps to sustain the money market by reducing the contractionary effects of increased formal nominal rates. This appears to support the empirical evidence from various African country studies. However, Owen and Solis-Fallas (op cit) disagree with the assumptions underlying the new structuralist models, such as that of higher allocative efficiency. Owen and Solis-Fallas only see the activities of individual moneylenders and their clients as making up the entire informal financial market when they insist that "loans are of relatively short maturity... (and that) moneylenders operate under quasi-monopoly conditions with each lender active in a small-scale spatially defined submarket in which there are limited opportunities for maturity transformation or economies of scale in risk pooling, administration of loans, etc." (Owen and Solis-Fallas, op cit, p. 347). They overlook the existence of cooperatives and many other actors on the informal scene who may be more interested in term loans than earlier literature would suggest.

Some Hypotheses on the Relationship Between the Formal and Informal Financial Sectors: A Microeconomic Consideration

This section offers a tentative explanation for the higher prices in the informal market. To determine the demand and supply structures for the two sectors, and hence their relationships, it must be determined a priori whether the two sectors operate within a single financial market or separate markets. This is related to the issue of whether the two sectors offer a homogeneous product or a heterogeneous set of financial products, and whether their products are complements or substitutes. There are ambiguities in the internal workings of the informal sector across countries, and wide variations within countries for the different operators, such as mobile bankers and moneylenders. The tendency usually has been to make the joint activities of moneylenders representative of informal financial sector activities.

It is useful to compare the financial products offered by the formal and informal sectors on the asset side. The formal financial sector is dominated by commercial and development banks in many African countries. These offer short and long term credit, although evidence from most countries suggests that term lending is not very significant in view of internal management problems (see Aryeetey et al 1989 and Ghatak, 1981). In the informal financial sector there is greater emphasis on short term credit for working capital and consumption. The informal and formal sectors offer a similar product, which is probably not entirely homogeneous. Based on this criterion of product differentiation on the asset side, one may assume an oligopolistic relationship with some minor product differentiation. Within the informal

Figure 13-1. Marginal Cost Curves for the Formal and Informal Financial Sectors

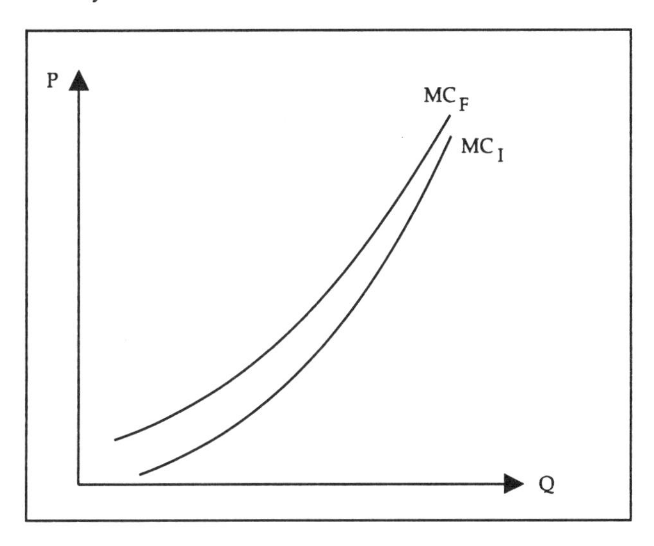

Figure 13-2. Specific, Nonspecific, and Potential Total Demand

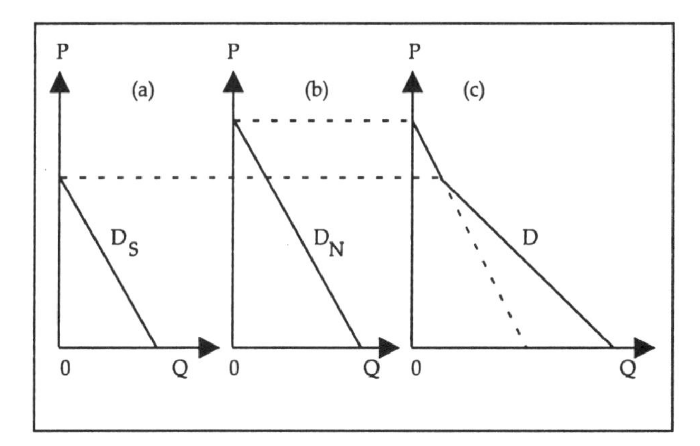

Source: Hemmer and Mannel, (1989).

sector itself there are monopolistic tendencies, as in the case of some moneylenders, and nearly perfectly competitive practices, as in the case of some susu operators.

On the supply side, the two sectors face different cost structures. This is important in explaining the different prices for their products in an oligopolistic situation. It is usually suggested that the operational costs of the individual actors in the informal sector are much less than those in the formal sector (USAID 1989). If costs were aggregated, however, it is likely that the difference is not large because the informal sector does not gain from economies of scale. Thus the marginal cost curves of the two sectors would tend to converge as the activity expands (see figure 13-1).

The closeness of the two marginal cost curves depends partly on the estimation of the risk element involved in informal sector lending. The marginal cost curve for the informal sector depends on the relative strengths of the monopolistic and perfectly competitive elements. It is, however, likely that a huge chunk of monopoly profit goes into individual estimations, especially in rural areas.

On the demand side, the two sectors cater to the credit needs of different target groups. Banks satisfy the demand of large established firms engaged in specific economic activities. Credit from the informal sector goes to individuals, small-scale businesses, and the agricultural sector. The two types of demand are not exclusive, however, as demand for the products of each sector has specific and non-specific elements. The two sectors have different demand curves in any case[2] (see figure 13-2). In most rural areas where the demand for credit cannot be satisfied by formal channels, the demand from the informal sector is specific or exclusive and this facility is an obvious substitute for what the formal sector would have offered. In urban areas much of the demand for credit is non-specific: many people interact relatively freely in both sectors and the issues of complementarity and substitution are not clear.

For example, the owner of a small processing plant may obtain credit from a commercial bank to expand his plant size. He may then borrow from friends to pay for new raw materials. In this case the products of the two sectors complement each other. If the bank had rejected his application for credit, and he had to invest in the expansion with credit from friends or a moneylender, such credit would be seen as a substitute for the unavailable formal credit. The issue then becomes which of the two sectors is predominant in urban and rural areas.

The greater the element of substitution in the demand, the more specific that demand is. Similarly, the greater the element of complementarity the more significant is the element of non-specificity of demand. It would seem that demand in the formal market would have a greater proportion of non-specificity (in view of the ease with which urban dwellers and other patrons of the sector can switch from one sector to the other) than demand in the informal market. Most of our studies show that it is easier for normally formal credit users to move from formal borrowing to informal borrowing than it is for normally informal borrowers to switch around. In

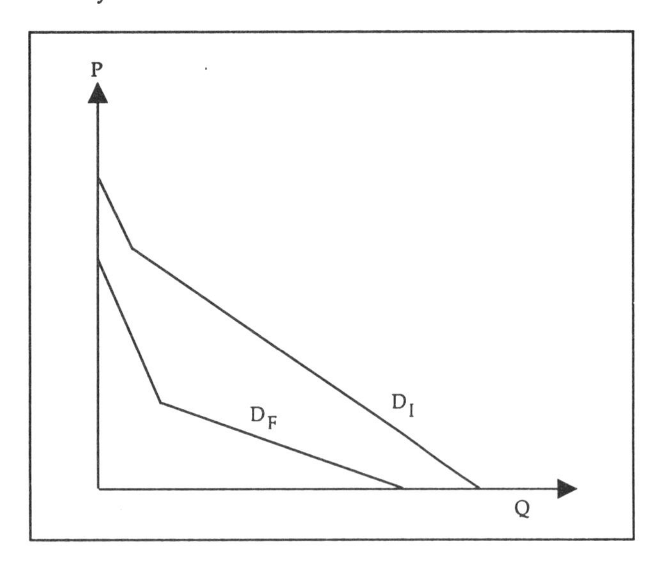

Figure 13-3. Total Demand Curves for the Formal and Informal Financial Sectors

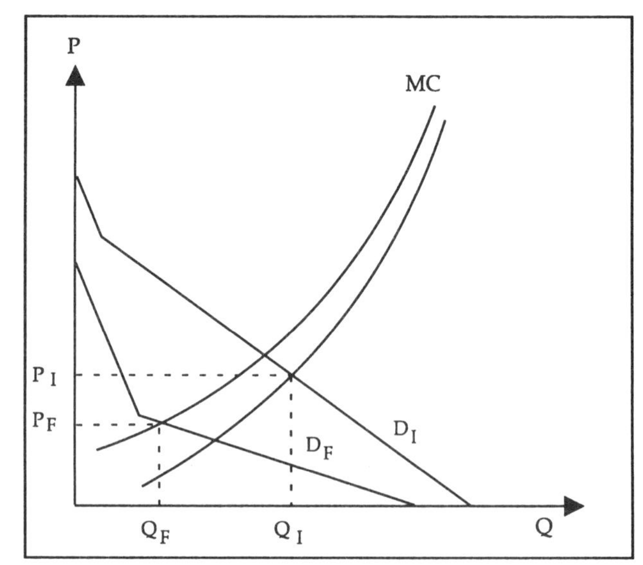

Figure 13-4. Pricing in the Formal and Informal Financial Markets

the determination of prices (interest rates) in both markets, therefore, the structure of the demand curves will play some role. The link between the two markets is established by the non-specific element in both markets. The interest rate at any given time in either sector would, if determined freely, depend on the proportions of specific and non-specific elements in the demand they face since these proportions determine elasticities.

The evidence gathered so far suggests that demand for informal credit may outweigh demand in the formal sector for several reasons. The former, in an extreme case, may involve total rural demand and a large part of urban demand. The existence of credit rationing in the formal sector is important in the analysis. These observations lead us to suggest demand curves for the sectors as in figure 13–3. The link between the two markets by way of the non-specific element in both demands is not portrayed here in view of the fact that this link has not yet been studied closely. Making allowances for this anomaly, superimposing figure 13–1 on figure 13–3 yields figure 13–4, which indicates the prices (interest rates) at which credit would be obtained from both markets at any given time. The higher interest rates in the informal sector may be explained more by supply side conditions following the relatively rapid rise of average and marginal costs in that sector, illustrated by MC_i approaching MC_f. This is also a trend we intend to investigate more closely.

In this static presentation, P_i cannot approach P_f despite our earlier assertion that there are non-specific elements in both demands. This can only be explained by the existence of strong institutional barriers vis-à-vis the demand for collateral and other institutional barriers. Thus, barriers to entry into the formal market would always lead to segmentation and probably provide an additional motivation for the existence and growth of an informal financial market, either in a complementary or competitive role.

Summary and Concluding Remarks

A large and significant informal financial sector exists alongside the formal financial sector in many African countries. Evidence from Ghana, Malawi, Senegal, Tanzania, and Zaire has been examined in this paper. It supports the argument that the informal sector is too large in these countries to be ignored in policy-making. This is even more so since actions taken in one sector have repercussions in the other.

The informal financial sector has similar characteristics in the five countries studied. This was by way of the observed types of operators and activists, relative size, relative operating costs, and relative pricing. Differences arise on the question of which informal units dominate the market. Thus, for example, while the susu appears dominant in Ghana, co-operative societies dominate informal financial activity in Tanzania.

Only recently has the importance of the informal sector to growth and development been highlighted in the literature, as per the new structuralists. However, some of the assumptions underlying the relationship between the two sectors—such as the assumption that they may relate via interest rates—appear overstretched. That assumption is not supported by the available empirical evidence.

It has also been shown here that the informal financial sector acts both as a complement and a substitute to the formal financial sector. It is this substitution effect that makes a relatively greater proportion of demand in this sector specific. A higher level of non-specificity in demand would work to push down prices in the sector in the long run in the absence of institutional barriers. In view of this, attempts at forging a closer link between the two markets would have to include, among other things, a concentration on reducing the level of non-specificity in formal sector demand as a means of aligning demand in both sectors and reducing the wide gap in interest rates. This may come about through expansion and improvements in the facilities which formal financial institutions have to offer.

Notes

The authors would like to express their appreciation for the financial and technical assistance provided by the African Economic Research Consortium (AERC), particularly Jeff Fine and Benno Ndulu, towards the preparatory work for this paper.

The use of material made available by our colleagues C. Chipeta and M.L.C. Mkandawire (Malawi), M. O. Ndanshau and J. P. Kipokola (Tanzania) and F. Gockel (Ghana) is also acknowledged. This became possible by virtue of the high level of cooperation within the AERC network on Informal Financial Markets in Africa.

1. See also Lundahl et al (1987), Msambichaka, Ndulu, and Amani (1986), and Due (1980).

2. We borrow here the illustration of specific and non-specific demand in the formal and informal sectors from Hemmer and Mannel.

References

Aryeetey, E. and F. Gockel (1990); *Mobilizing Domestic Resources for Capital Formation: The Role of Informal Financial Markets in Ghana,* paper presented at a workshop of the African Economic Research Consortium (AERC), Nairobi, 26-30 May.

Aryeetey, E., Y. Asante, F. Gockel, and A.Y. Kyei (1990); *Mobilizing Domestic Savings for African Development and Diversification: A Ghanaian Case,* paper presented at a workshop of the International Development Centre, University of Oxford, Oxford, 16-20 July.

Chipeta, C. and M.L.C. Mkandawire (1989); *The Informal Financial Sector in Malawi: Scope, Size and Role,* an interim research report presented at an AERC workshop, Harare, 4-8 December.

Due, J.N. (1980); *Costs, Returns and Repayment Experience in Ujamaa Villages in Tanzania,* 1973-1976, University Press of America, Washington.

Ghatak, S. (1981); *Monetary Economics in Developing Countries,* the MacMillan Press Ltd., London and Basingstoke.

Hemmer, H-R and C. Mannel (1989); "On the Economic Analysis of the Urban Informal Sector," *World Development,* Vol. 17, No. 10, pp. 1543-1552.

Hyuha, M., M.O. Ndanshau, and J.P. Kipokola (1989); *Informal Financial Markets in Tanzania: Scope, Structure and Policy Implications: Evidence from a Pilot Survey,* paper presented at an AERC workshop, Harare, 4-8 December.

Hyuha, M. and M.O. Ndanshau (1990); *Economic Adjustment Packages in Africa: The Informal Financial Sector as the Forgotten Half in Financial Reform in Tanzania,* mimeo., Department of Economics, University of Dar es Salaam.

IPC (1988); Rural Finance in Ghana, a research study prepared for Bank of Ghana on behalf of the Ministry of Economic Cooperation of the Federal Republic of Germany, Bonn.

Lundahl, M. et al (1987); "Agricultural Credit in Tanzania: A Peasant Perspective," *Savings and Development,* Vol. 11, No. 4.

Mauri, A. (1987); *The Role of Financial Intermediation in the Mobilization and Allocation of Household Savings in Developing Countries: Interlinks between Organized and Informal Circuits, The Case of Ethiopia,* paper presented at the International Experts' meeting on "Domestic Savings Mobilization through Formal and Informal Circuits: Comparative Experiences in Asian and African Developing Countries," Honolulu, 2-4 June.

Miracle, M.P., D.S. Miracle, and L. Cohen (1980); "Informal Savings Mobilization in Africa," *Economic Development and Cultural Change,* Vol. 28.

Msambichaka, L.A., B.J. Ndulu and H.K.R. Amani (1986); *Agricultural Development in Tanzania: Policy Evolution, Performance and Evaluation,* Friedrich Ebert Foundation, Bonn.

McKinnon, R.I. (1973); *Money and Capital in Economic Development,* The Brookings Institute, Washington.

Owen, P.D. and O. Solis-Fallas (1989); "Unorganized Money Markets and 'Unproductive' Assets in the New Structuralist Critique of Financial Liberalization," *Journal of Development Economics,* 31 p. 341-355.

Rahman, F.H. (1989); "Rural Savings: A Neglected Dimension of Rural Development," *The Courier,* Journal of the ACP-EEC, No. 115, May-June, pp. 70.

Seibel, H-D (1988); *Financial Innovations for Microenterprises:*

Linking Formal and Informal Financial Institutions in Africa and Asia, paper presented at the World Conference on "Support for Microenterprises," Washington, 6-9 June.

Shaw, E.S. (1973); *Financial Deepening in Economic Development*, Oxford University Press, New York.

Taylor, L. (1983); *Structuralist Macroeconomics: Applicable Models for the Third World*, Basic Books, New York.

van Wijnbergen, S. (1982); "Stagflationary Effects of Monetary Stabilization Policies: A Quantitative Analysis of South Korea," *Journal of Development Economics*, 10, April, pp. 133-169.

van Wijnbergen, S. (1985); "Macroeconomic Effects of Changes in Bank Interest Rates: Simulation Results for South Korea, *Journal of Development Economics*, 18, August, pp. 541-554.

Wai, U Tun (1957); "Interest Rates Outside the Organized Money Markets of Underdeveloped Countries," *IMF Staff Papers*, Vol. 6, No. 1.

14

Mobilizing Domestic Resources for African Development and Diversification: Structural Impediments to Financial Intermediation

Machiko Nissanke

As the deep economic crisis which crippled Sub-Saharan Africa throughout the 1980s wears on, official donors and African recipients are increasingly concerned about the prospect of foreign aid becoming structurally embedded in the economies of the region. After several years of policy reforms tied to structural adjustment loans, it is apparent that the foundations necessary for self-sustained growth have yet to be laid down. Many countries have sadly experienced institutional disintegration, infrastructural erosion, and the accelerated deterioration of systematic capacity during the crisis.

The financial policies embodied in structural adjustment programs derive most of their intellectual content from the financial repression hypothesis.[1] It is argued that a strategy of financial liberation will deepen the financial structure and renew economic development. However, little is known about how the dynamic mechanism of liberalization impacts on the financial structures of fragile low-income countries. Institution building is needed to achieve sufficient market depth to enable market forces to operate beneficially. The sequence and pace of liberalization must be very carefully worked out and must take into account the speed of market deepening and changing macroeconomic conditions.

Moreover, in low-income economies institutional and motivational bottlenecks might be more binding than the policy bottlenecks which are singled out by the financial repression hypothesis. Viable financial policies must be based on improved understanding of the structural characteristics and impediments in Sub-Saharan economies. It is imperative to examine in detail the prevailing structure of the financial system and the economies at large, in order to design carefully the transition process from repression to liberalization. It is also important to assess the effects of prevailing macroeconomic and financial policy regimes on domestic savings mobilization, and to grasp the financial dynamics of structural adjustment programs.

This paper tries to identify key motivational, institutional, and policy bottlenecks inhibiting the mobilization of domestic savings. It explores means to develop the financialization of domestic savings and the intermediation between savings and investment for economic diversification and development. The discussion presented here is preliminary and is based largely on studies of Ghana (Aryeety et al 1989) and Malawi (Chipeta and Mkandawire, 1989). The second section summarizes the patterns of financial and savings behavior of the household sector, corporate and small-scale enterprises, and the public sector. The third section evaluates the structure and functioning of formal financial institutions and markets. Several policy implications are discussed in the concluding section.

The Pattern of Saving and Financing

The analysis presented in this section is based partly on recently conducted field work. Unfortunately, lack of data hinders the analysis of the pattern of enterprise financing. Some interesting findings regarding the household sector are emerging from current research. The savings and financing patterns of the public sector are discussed here mainly in relation to those of the private sector.

The Household Sector

Detailed and reliable estimates of savings rates by region and type of household are hard to obtain. However, the available sample survey and previous studies[2] indicate, for example, that total savings in rural households in Ghana could constitute as much as half of estimated annual income. It is estimated that only 20 percent of these savings is accounted for by financial savings. While various forms of asset holding as stores of value indicate a large scale of efficiency loss to the economy,[3] this is undoubtedly also a reflection of the insufficient degree of monetization of the economy. Hence, analysis of rural savings should take into account the fact that saving behavior in a nonmonetized peasant economy can be distinctly different from that in a monetized economy (see, for example, Alamgir 1976 and Ghosh 1986). The rate of monetization of economic activities and the degree of financialization of savings are considerably higher in urban areas.

Most of the financial savings (between 80 and 95 percent) of the household sector have been mobilized through informal financial institutions and markets. In Ghana, informal financial savings (excluding personal loans) are estimated to constitute about two percent of GDP. In Malawi the total credit extended by the informal financial sector (including estate owners and personal loans among friends and families) has been estimated at 6.5 percent of GDP;[4] informal sector lending to the private sector was at least three times as great as that of the formal sector.[5]

Many of the traditional forms of saving and informal financial activities have long existed in African countries. The informal financial sector includes credit unions, which could be described as semi-formal institutions; the system of rotating savings; savings and credit associations; cooperative savings associations; the single collector system; and moneylenders. The nominal interest rate charged on credit varies enormously across these heterogeneous operations. In Ghana and Malawi informal credit is often granted with very flexible repayment terms in light of debtors' personal circumstances.

Numerous factors inhibit the majority of the population from using formal institutions. The institutional bottlenecks and weakness of the formal sector, and the lack of access to its credit facilities in the absence of required collateral have created a vibrant informal financial sector. In the informal sector the cost of credit can be much higher, but even low-income groups can have access to liquidity.[6]

The striking degree of malfunctioning of the formal financial institutions and the system as a whole[7] acts as a structural hindrance to more effective savings mobilization through the formal financial system. Commercial banks do not have the motivation to overcome their operational bottlenecks and make efforts at savings mobilization. This is due largely to the problem of excess liquidity. This has led to the striking situation that "in a rural area of Ghana (Tamale), one association of 'susu' collectors was reported to have mobilized C30 million in a month, amounting to C360 million ($2.45 million) in a year" (Bentil et al).[8] This compares with the average total savings deposits mobilized by commercial banks in that area of C11 million ($75,000) by mid-year in 1988.

Funds mobilized through one informal sector group are seldom used to finance other groups in the informal sector. Thus, the potential use of savings mobilized through the informal financial sector for economy-wide diversification remains unrealized. A process of industrialization and development based on informal small production units has not yet taken off the ground.[9] Few informal financial institutions provide long term finance in either country studied.

Evaluation of the current and potential links between the formal and informal sectors can be pertinent for future policy formation on savings mobilization. The two country studies emphasize the complementarity of relationships between the two sectors. However the facilities directly available to the population through the informal sector are to some extent presently substituting for the operations of formal institutions. A sizeable proportion of the demand deposits of banks is mobilized through informal sector operations. This is not reflected in the growth of bank liabilities because informal sector operators keep deposits in current demand and savings accounts, i.e. in highly liquid forms. As the income level remains the dominant variable in determining the personal savings rate, the terms of trade between the agricultural and other sectors has a crucial role to play in rural savings mobilization. In addition, our study indicates the need for recycling rural savings in the local vicinities on a visible scale. Savings mobilization and credit granting facilities must be made accessible and utilized actively within local communities.

The Pattern of Enterprise Financing

Unfortunately, few comprehensive data are available on the financing pattern of small-scale enterprises in Ghana and Malawi. It is, however, clear that most private entrepreneurs have financed their activities with loans from relatives and friends, informal sector loans, and their own savings. This has limited the scale of activities and investment.

Aryeetey et al analyzed the available balance sheets submitted by a limited number of established corporations in Ghana. The data show that, on average, equity financing is more dominant for public corporations than for private corporations, and the ratio of external finance to internal funding and the loan/equity ratio are higher for private corporations. The most common source of finance for the private sector is short term credit such as overdraft facilities, trade credit, and supplier's credit.

In Malawi there has been virtually no commercial bank lending to small- and medium-scale businesses. Within the private sector, the banks prefer to deal with the estate sector. At any rate, the commercial banks do not provide long term or development loans. In order to address these problems and to facilitate the development of small and medium enterprises, a number of specialized institutions have been established in Malawi during the past 10 years. These specialized institutions suffer acutely from inadequate funds to carry out their mandate, and very limited administrative capacity to manage an extensive credit network. The agro-industry, which uses the agricultural base as a source of raw materials, is regarded as an especially promising area for diversification. At this point, the supply of credit would be a binding constraint on the expansion of this sector which is vital for Malawi's future diversification.[10]

Savings and Public Sector Finance

This section presents data on public and private savings and public sector finance in Ghana and Malawi.

Public and Private Savings. The savings rates for Ghana and Malawi during 1969 to 1988 are prestend in table 14–1. Table 14–1 presents an approximate breakdown of gross national savings into public sector savings and private sector savings. Here public savings refers to government savings defined as the net excess of total revenue and grants over current expenditure, and private savings is calculated a residual after public savings have been deducted from gross national savings.

In Ghana, the gross domestic savings and gross national savings rates have been at a very depressed level (one of the lowest levels in the region). During 1972 to 1983, the Ghanaian government ran a persistent current budget deficit (see table 14–2), which further depressed the aggregate savings rates. The current account balance stayed in deficit throughout the period, despite continuous downward adjustments in current expenditure (see table 14–2). The root cause of the chronic deficits should be found in the rapidly deteriorating tax base, which declined at a rate much faster than that of expenditure. As a percentage of GDP, revenue dropped steadily from over 16 percent in 1975 to 6 percent in 1983.

In 1983, Ghana began reducing the deficit in its current budget and in 1985 began to generate a surplus with increased external grants. Current revenue,

Table 14–1. *Public and private savings (percentages of GNP)*

	1972-73	*1974-78*	*1979-83*	*1984-88*
Gross national savings rate				
Ghana	12.1	8.1	3.4	5.2
Malawi	8.7	12.1	2.4	6.1[c]
Public Savings rate[a]				
Ghana	-1.2	-2.7	-3.3	1.8
Malawi	0.5	3.1	3.4	1.9[c]
Private savings rate[b]				
Ghana	13.3	10.8	6.7	3.4
Malawi	8.2	9.0	-1.0	4.2[c]

Notes:
a. This is the current account balance.
b. This is the gross national savings rate less the public savings rate.
c. For 1984-86 only.
Source: IMF International Financial Statistics Yearbook and Government *Finance Statistics Yearbook.*

including grants, increased sharply from 8 percent of GDP in 1984 to 20 percent of GDP in 1987, whereas expenditure was increased from 9 to 21 percent of GDP in the same period. The increase in public savings since 1984 implies that the private sector savings rate, calculated as a residual, has continually followed a dwindling trend.

In Malawi, the trend in gross savings rates has been characterized by a distinct break between the 1970s and the 1980s, and by large yearly fluctuations, as shown in table 14–1. Savings rates increased during the 1970s and deteriorated significantly after 1979. The government ran a surplus on its current budget account on average throughout the 1970s and 1980s. In contrast to Ghana, this was achieved while keeping the level of current expenditure between 15 and 20 percent of GDP throughout the period. The difference in the fiscal situations in the two countries is explained by the revenue side: in Malawi, there was no deterioration in the domestically generated revenue base, and grants of considerable size consolidated the situation.

Public Finance. Tables 14–2 and 14–3 show summary data on public finance and its relationship to the domestic financial system in Ghana and Malawi.[11] The size of the deficit is much larger in both countries when capital expenditure is taken into account. Although there are interesting divergences between the two countries, the question of crowding out is likely to take a central place in the discussion of public sector finance, given the predominant share of the public sector in domestic credit allocation.

The government of Ghana financed its deficit almost exclusively by borrowing from the central bank and the banking institutions up to 1983, in the near absence of external finance. Capital expenditure increased after 1984, when the government accepted a structural adjustment loan. The mode of financing has changed dramatically since 1984 with a large inflow of external funds. The share of the public sector in total domestic credit increased enormously in the 1970s; since 1984 it has been decreasing slowly.

In Malawi, the government has kept its current budget balance in surplus. However, the overall budget deficit has been as large as 10 percent of GDP, which indicates the consistently high level of capital expenditure. Hence, the government has continually been faced with borrowing requirements of considerable size. In contrast to the Ghanaian case, the governm-

Table 14–2. *Public finance*

	1972-73	*1974-78*	*1979-83*	*1984-88*
Percentage of GDP:				
Current account blance				
Ghana	-1.6[a]	-3.2	-3.7	3.0[c]
Malawi	0.2[b]	3.0	3.2	1.7[c]
Overall budget surplus/defecit				
Ghana	-6.0[a]	-9.8	-5.8	-1.1[c]
Malawi	-6.0[b]	-6.9	-10.3	-7.6[c]
Government borrowing from the banking system				
Ghana	4.2[a]	9.6	5.3	0.5[d]
Malawi	1.3[b]	1.6	4.8	3.2[c]
Domestic credit				
Ghana	27.7	31.9	21.8	22.9
Malawi	12.1	23.2	37.7	31.8
Percentage:				
Growth in credit to public sector**				
Ghana	10.2	56.9	41.0	43.9
Malawi	37.5	64.7	29.2	8.2

Notes:
a. This is the current account balance.
b. This is the gross national savings rate less the public savings rate.
c. For 1984-86 only.
*Credit extended by the Central Bank and deposit money banks, i.e. those included in the monetary survey.
**In Ghana it includes the central government and non-financial public enterprises only. In Malawi it includes the central government and all official entities.
Source: IMF International Financial Statistics Yearbook and Government *Finance Statistics Yearbook.*

ent resorted to heavy external financing throughout the period. While this strategy invited its own burdensome problem of external debt management, it implies that government has a less dominant position as borrower in the domestic financial system in Malawi.

The Formal Financial System

Aggregate financial indicators, the structure of the formal financial system, the operational characteristics of financial institutions, and the constraints and scope of financial markets are discussed in this section.

Aggregate Financial Indicators

The time series data for M1, M2, and M3 as percentages of GDP in Ghana and Malawi during 1969 to 1987 are used to show the extent of financial deepening. In Ghana all the ratios reached a peak in 1976 and thereafter declined sharply. Currency accounted for 54 percent of M2 in 1960, between 35 and 38 percent of M2 during the mid 1960s to the mid 1970s, and increased gradually after 1976 to 54 percent during 1983 to 1986. These statistics reveal the process whereby movement toward financial deepening was reversed during 1976 to 1983. According to a recent study undertaken by the World Bank (Neal 1988), Ghana ranks second only to Zaire as having the lowest ratio of M2 to GDP in the world.

In Malawi, the ratio of currency as a percentage of M2 was 36 percent in 1965, declined to 20 percent in 1980, and remained at just under 20 percent throughout the 1980s. The currency-to-M2 ratio in Malawi is about one-third of that in Ghana. This reflects the difference in the stages of financial development in the two countries. There has been a steady increase in financial deepening in the Malawian economy during the 1967 to 1988 period.

The Structure of the Formal Financial System

Ghana's formal financial system is comprised of the Central Bank (the Bank of Ghana), four large commercial banks, three development banks (DFIs), two merchant banks, two smaller banks, and 20 small rural banks. The total assets of the banking system (excluding the central bank) were 231 billion cedis at the end of 1988. This was about a quarter of GDP. Of total assets, some 71 percent was accounted for by the four large commercial banks, 20 percent by DFIs, and 6 percent by merchant banks.

Bank density in Ghana increased from 1.9 branches per 100,000 inhabitants in 1976 to 3.4 in 1988. This

Table 14-3. ***Percentage of public sector in domestic credit***

	1972-73	*1974-78*	*1979-83*	*1984-88*
By the central banks:				
Total				
Ghana	95.9	95.1	99.0	89.8
Malawi	77.1	92.3	94.7	100.0
Claims on government				
Ghana	71.3	85.2	81.0	71.3
Malawi	77.1	47.1	72.7	83.8
Claims on public enterprises				
Ghana	24.6	9.9	18.0	18.5
Malawi	—	56.5	22.0	16.2
By the banking institution:*				
Total				
Ghana	65.2	80.3	90.0	83.2
Malawi	20.2	45.8	51.3	68.4
Claims on government				
Ghana	46.0	62.3	69.8	71.0
Malawi	17.7	24.4	36.4	53.6
Claims on public enterprises				
Ghana	19.2	18.0	20.2	12.1
Malawi	2.5	21.4	14.9	14.8

Notes: a. 1975-78. *Includes central banks and deposit money banks.
Source: *IMF International Financial Statistics Yearbook* and Government *Finance Statistics Yearbook.*

was due to 25 percent growth in the number of primary commercial bank branches between 1976 and 1988 and the establishment of 120 rural banks since 1976. There are currently 466 bank branches spread unevenly over the country.

Malawi's financial system comprises the Central Bank (the Reserve Bank of Malawi, RBM); two commercial banks; three development finance institutions; two finance houses; the New Building Society (NBS); the Post Office Savings Bank (POSB); a number of insurance companies; and the Malawi Union of Savings and Credit Cooperatives (MUSCCO), which is the umbrella organization for 58 credit unions. The financial system's total assets at the end of 1987, excluding RBM, amounted to 39 percent of GDP. Of the total assets, 68 percent was in the commercial banks, 12 percent in the insurance companies, 10 percent in the POSB, 4 percent in the New Building Society, 3 percent in the DFIs, and 2 percent in the financial houses.

The number of branches and state agencies of commercial banks and the POSB increased from 185 in 1967 to 331 in 1988. However, because of Malawi's high population growth rate, the ratio of people to banks deteriorated from 22,300 in 1967 to 25,000 in 1988. (The comparable number for Ghana in 1988 was 29,200.) The bank density is much higher in urban centers than in rural areas.

Operational Characteristics of Financial Institutions

This section examines the functions and operational characteristics of the formal financial system's main subsectors—commercial banks, development finance institutions, rural banks in Ghana, and the Post Office Savings Bank in Malawi.[12]

Commercial Banks. Analysis of the balance sheets of commercial banks in Ghana and Malawi reveals two strikingly similar features of their operational characteristics: excess liquidity held above the required level on a voluntary basis, and a heavy concentration of loan portfolios on the short end of the market. These banks hold assets far above the required statutory levels in cash with no return, and in treasury bills and government stocks with very low returns. The absence of viable projects to which to lend is cited as a prime reason for the excess of loanable funds. Faced with excess liquidity and limited lending opportunities, the banks lack incentive to mobilize additional savings.

Given the short term liability structure and the weak capital base of both countries, the portfolio management of commercial banks is geared to matching the maturity structure of assets with the existing maturities of liabilities. The banks argue that medium and long term lending on a greater scale would be improper because they are forced to maintain a sizeable proportion of their assets in highly liquid form. Instead they lend to a small number of well-established and financially strong private and parastatal companies, which have been the main beneficiaries of institutional credits offered at subsidized interest rates. The credit ceilings imposed for the purpose of macroeconomic stabilization by the International Monetary Fund programs are singularly ineffective in an environment in which portfolio management is conservative and undynamic.

In Ghana there are four large commercial banks: the Ghana Commercial Bank (GCB), Barclay Bank of Ghana (BBG), Standard Chartered Bank of Ghana (SBG), and the Social Security Bank (SSB). At the end of 1988, these banks accounted for about 77 percent of total deposit liabilities of financial institutions, 67 percent of the loans market, and 76 percent of private bills and securities. They held about 15 percent of their total deposit liabilities as excess reserves in 1988; the share of liquid assets was 60 percent of total deposit liabilities.

Excess liquidity and the high frequency savings cycles of the private sector have resulted in a very short term liability structure. About 70 percent of liabilities and between 85 and 90 percent of deposit liabilities originate from demand and savings deposits, the latter of which are operated almost like current accounts. The balance sheets also reveal that the commercial banks in Ghana are typically undercapitalized.

Commercial bank loans in Ghana are predominately of short term maturity, and usually are made by extending overdraft facilities. It was estimated that in 1987 only 15 percent of commercial bank loans had a maturity of over three years. Only through continuous rollovers do the banks, in effect, provide some long term finance to their more credit worthy customers. Commerce and the construction sector are favored by the commercial banks, although they are required to hold at least 20 percent of their total loan portfolio as credit to the agricultural sector.

In Malawi commercial banking activity is undertaken by two banks, the National Bank of Malawi (NBM) and the Commercial Bank of Malawi (CBM). The former has 76 percent of commercial bank deposits and 60 percent of assets. The two banks have accounted for around 80 percent of the total deposit liabilities of financial institutions throughout the post-independence years.

Excess liquid assets held by Malawi's commercial banks amounted to 20 percent of total deposits in

1975. In 1988 excess liquid assets were 38 percent above the minimum reserve ratio and accounted for 68 percent of total deposits.[13] A salient shift in the composition of deposits appears to have taken place in favor of time deposits since the early 1970s. At the National Bank of Malawi, the share of time deposits was over 50 percent in the mid 1980s. However, more than 80 percent of the time deposits have a maturity of less than one year (Chipeta and Mkandawire).

Due to increasing excess liquidity, the commercial banks have actively discouraged people from depositing money in interest bearing time deposit accounts with maturities of more than one year. The cost of keeping the deposits in long maturities has been considered too high because of the absence of good investment opportunities and the wide margin between deposit and lending rates. This suggests that interest rate liberalization alone is ineffectual for efficient savings mobilization when the financial intermediaries are awash with excess liquidity.

The commercial banks in Malawi are characterized by their extreme risk aversion. Banks are known to extend credit only to well-established large enterprises. The large proportion of commercial bank lending is for capital over short periods and is in the form of fluctuating overdrafts.

In the sectoral credit distribution of the commercial banks in Malawi, agriculture has accounted for the largest proportion (over 50 percent throughout the 1980s). Most agricultural credit has been absorbed in financing the tobacco sector. The wholesale and retail trade has the next largest share in credit distribution.

While the private sector in Malawi remains the dominant borrower (with between 60 and 70 percent of bank claims), the central government's share of claims has increased significantly in recent years and accounts for between 25 and 30 percent of bank credit. Smallholders, smallholder tenants on estates, nontraditional exporters, and small and medium enterprises have received very little direct credit from commercial banks in Malawi.

Development Finance Institutions. The inadequacies of conventional commercial banking practice and the dearth of medium and long term credit to the priority sectors of the economy necessitated the establishment of development finance institutions in both countries. They were typically founded on term loans from external sources and have been continuously financed either directly by external sources or indirectly through the government, often through refinancing schemes evolved by the central bank.

Three development finance institutions were established by the central bank in Ghana: the National Investment Bank, in 1963; the Agricultural Development Bank, in 1965; and the Bank for Housing and Construction, in 1973. In 1988 the three banks had 26 percent of the market share of loans and advances in the banking system.

The development finance institutions were dependent on term loans from external sources involving exchange and interest rate risks. They assumed large financial risks by specializing in long term lending in high risk areas such as agricultural and industrial finance. They have experienced acute illiquidity and insolvency problems and have suffered from a substantial number of nonperforming portfolios with high default rates and bad debts. They are often not able to meet the Central Bank's reserve requirements and have had to be frequently refinanced and recapitalized by the Central Bank or the government.

Due to the insufficiency of funds available from external and government sources, Ghana's development finance institutions have mobilized savings. At the end of 1988, they had 13 percent of the market share of total deposits in the banking system. Faced with chronic insolvency problems, they emphasized short term self-liquidating loans and advances. Consequently, even the development finance institutions have begun to shift away from the manufacturing and nontraditional agricultural sectors.

There are three development finance institutions in Malawi: the Industrial and Development Bank (INDEBANK), the Industrial and Development Fund (INDEFUND), and the Small Enterprise Development Organization of Malawi (SEDOM). These development finance institutions rely totally on foreign resources and do not mobilize domestic savings.[14] INDEFUND and SEDOM have difficulties in carrying out their functions due to inadequate resources and staff shortages. In-house technical and administrative capacity is reported to be grossly inadequate. It is reported that INDEFUND has consistently made losses in its operations. SEDOM is more effici- ent in its operations, but also faces a rising level of arrears in its overall portfolio. In view of the vital importance of their functions for the country's diversification prospects, it is critical to reduce the financial constraints on the development finance institutions and to improve their resource base and operational efficiency.

Rural Banks in Ghana. The rural banks in Ghana are unit banks that have been established since 1976 to mobilize resources and extend credit locally. They are owned and managed by their local communities and assist rural based industrialization through loans to cottage industries. The central bank holds preference shares equivalent to one-third of the initial share capital. However, the major source of funds is deposit liabilities, which made up 70 percent of total resources in 1988. The rural banks serve small borrowers and depositors and have been quite successful at savings mobilization.

Between 1977 and 1988, the share of agriculture in total credit from rural banks was halved. Credit to agriculture amounted to 54 percent of total credit in 1985, but had dropped to 34 percent by 1988. The share of trading activities increased from 13 percent in 1985 to 26 percent in 1988.

Moreover, the rural banks have accumulated excess liquidity. They had a primary reserve ratio of 25 percent and a secondary reserve ratio of 26 percent as of mid-June 1988, which were far in excess of the mandatory requirements of 10 and 20 percent in the respective reserve categories. It is reported that managers of the rural banks have resorted to low risk investment in government. Meanwhile an increasing number of households are obtaining credit from the informal financial sector.

Post Office Savings Bank in Malawi. Nearly 90 percent of Malawi's population lives in rural areas. The network of the Post Office Savings Bank (POSB) has the most extensive coverage for savings mobilization among all formal institutions in the rural areas. POSB has 286 deposit collecting points and has mobilized over 12 percent of total savings by formal financial institutions. The interest earned by its depositors is tax-exempt, which gives POSB a competitive edge in mobilizing resources. However, because it primarily serves low-income groups, it has a low minimum deposit balance requirement and a ceiling is imposed on deposits for individuals and corporations. Its deposit liabilities are noticeably affected by changes in rural cash income and by remittances of deferred payments by Malawian migrant workers.

The POSB lends exclusively to the government, as its statutes require, through the purchase of government securities. Chipeta and Mkandawire argue that the lack of lending facilities discourages depositors, despite the attraction of nearness to their homes, and drives small depositors toward the informal financial market. Drawing upon other countries' experiences, they suggest granting management and financial autonomy to make the POSB a more important instrument for resource mobilization and credit allocation to small-scale rural traders and manufacturing enterprises.

Financial Markets: Constraints and Scope

Ghana and Malawi have not been able to generate long term credit and loan provisions for self-sustainable diversification. In spite of high liquidity in the overall banking system, the potentially productive sectors in the economy are not able to receive adequate institutional credit. In view of the acute need for export diversification, the lack of availability of export financing for nontraditional exports should be regarded as a critical bottleneck.

Segmentation of the formal financial system in these countries has become structurally embedded into their economies. There is very little liquidity management involving interbank credit arrangements, whereby resources could be shifted from the surplus sector to the deficit sector among financial intermediaries. This section presents the current state of, and scope for, developing money and capital markets in Ghana and Malawi.

Money Markets. The mismatch of liquidity positions within the banking system calls for the creation of an active interbank money market and the establishment of a discount house that trades in a variety of short maturity financial products.

In Ghana the Consolidated Discount House Limited (CDHL) was formally institutionalized in 1988 as an interbank intermediary for short term assets. It is jointly owned by eight banks and six insurance companies. However, large portions of the banking system are still characterized by idle funds, while the other financial institutions and the industrial and commercial sectors face severe liquidity problems. The CDHL's principal portfolios are in Bank of Ghana bills, treasury bills, cocoa bills, grain bills, bankers acceptances, negotiable certificates of deposit, and government stocks. Commercial papers are not yet featured. The CDHL has functioned as an important source of short term borrowing for the government.

In Malawi, no formal money market exists. The interfinancial institution claims are at present mainly in the form of deposits of NFBIs at commercial banks and at the Central Bank. The existing range of non-government short term paper is very limited. Three-month treasury bills remain the major short term debt paper in the economy. By far the largest proportion of treasury bills has usually been held by commercial banks. Hence, the development of the money market for short term assets in Malawi appears to be presently hampered by the very limited number of financial products on offer and prospective transactors on the demand and supply sides.

Capital Markets. The acute shortage of term loans with long maturity in both countries requires the urgent development of an active and broad based capital market. A capital market could transform and lengthen credit maturities and considerably reduce the credit risks inherent in transactions among individuals. It would enable corporate issuers to access long term funds provided by investors and to offer liquidity in the secondary market for shares and bonds.

In Ghana there is no capital market despite many attempts to establish one and the presence of some 18 public companies. The National Trust Holding Company was established in 1976 as the nucleus of a capital market to provide brokerage services to the investing public and shareholders. Very few shares have been issued, however, despite the abundant potential supply of funds from the commercial banks and the Social Security and National Insurance Trust.

In Malawi, in the absence of an active stock exchange, the capital market is extremely shallow. It only deals in securities involving central government, locally registered stocks. The major institutions that have so far participated in these dealings are the financial intermediaries and the three holding organizations, namely MDC, ADMARC, and the Press Corporation.

Thoughts on Policy Implications

The findings presented in this paper have led us tentatively to conclude that the widely accepted wisdom that general capital shortages are a core problem for development may be an illusion. The real bottlenecks and binding constraints must be determined for each country at each state of economic development.

One of the core problems in Africa is the fragmented state of domestic resource mobilization. The system has been segmented into formal and informal sectors, and little interaction has taken place among the formal institutions. There is great potential for savings mobilization for small-scale productive activities through the development of the informal sector and the strengthening of potential links between the formal and informal sectors. However, the limitations of the present state of resource mobilization and intermediation are critical.

Savings mobilized in the household sector through informal agents and organizations are small in size per unit of transaction as well as short term in frequency. This must be in part a reflection of the nature of saving in low-income developing countries, which has been pointed to by Deaton (1989). Such saving is high frequency and aimed at intertemporal consumption smoothing, in contrast to intergenerational or life-cycle hump saving. It provides a buffer between uncertain and unpredictable income and an already low level of consumption. In addition, the combination of high risks and high transaction costs acts as a severe constraint on viable large-scale financial intermediation by the informal financial sector. The informal sector currently plays an important role only in savings mobilization and not in efficient actual intermediation.

In view of the extent of the segmentation of the formal system, it is necessary to reorganize some of the financial institutions and to encourage the creation of effective money and capital markets for channelling funds into productive activities. However, the observed mismatch of liquidity positions within the formal financial system in these countries is a structural phenomenon. While short term money abounds in the form of excess liquidity in one segment of the banking system, the economy as a whole is characterized by an endemic shortage of long term loan provisions for productive investment and diversification. The financial system currently lacks the ability to take on the risks associated with the maturity term transformation which is indispensable for the active intermediation of financial flows between surplus and deficit units of the economy.

The fundamental problem of high risk in these fragile African economies should be addressed in a general program of financial development. Such a program would take into consideration the close correlation between the motivation for efficient savings mobilization and the investment climate. Evidence presented here suggests that interest rate adjustment, which is a key instrument of financial liberalization programs, has had little effect in improving savings mobilization and the efficiency of capital allocation.[15] In Africa, the financial market is still too shallow to make interest rates a key equilibriating variable between savings and investment. Under existing conditions, the freeing of interest rates is not likely to induce financial intermediaries to assume greater risks.

Measures to be taken should include efforts to build up the local capabilities in risk and asset and liability management, and in the government's macroeconomic management. Risk sharing or credit risk insurance schemes of some sort are urgently required.[16] It must also be explicitly recognized that in a monocultural economy risk covariance is extremely high. Financial intermediaries cannot function efficiently without appropriate devices and policy instruments in place to deal with this problem. The mechanism of reducing the general financial risk inherent in commodity based economies includes, at the national level, improvements in the design and administration of macroeconomic and financial polic ies. A stable macroeconomic environment is a fundamental precondition if the interest rate is to function as an efficient instrument of savings mobilization.

Issues related to the improvement of the international monetary, financial, and trade systems should be reexamined. External official and private financial flows and foreign investment might assist in lessening the risks which are so detrimental to domestic savings mobilization and hence to long term, sustainable development in low-income developing countries. Above all, unless the risks associated with foreign exchange, interest rates, and commodity prices exogenous to these economies are greatly reduced, problems in domestic finance will remain.

Notes

This paper is part of our ongoing research project on "Domestic Resource Mobilization for African Development and Diversification" in which comparative empirical studies have been undertaken of the financial system of five African countries -- Ghana, Kenya, Malawi, Tanzania and Zambia. The project is set to look systematically and in a substantial way into the micro-macro links of the issues on finance and development. Due to the incompleteness of the study at this stage, this paper is very much preliminary and based largely on the interim reports on Ghana and Malawi.

The author wishes to acknowledge the contribution of Ernest Aryeetey, Yaw Asante, Alexander Kyei, Fritz Gockel (Ghanaian team) and C. Chipeta and M.L.C. Mkandawire (Malawian team) whose research findings have been drawn on extensively in the writing of this paper. The author would also like to thank the Gatsby Charitable Foundation, SIDA and DANIDA for their generous financial support.

1. For a critical review of the recent theoretical debate on finance and development, see Nissanke (1990).

2. Aryeetey et al (1989) cite estimates made by Bentil (1988).

3. Aryeetey et al (1989) suggest that what may appear as an investment may actually be saving as a hedge against inflation.

4. Since the coverage of the informal financial sector differs between the two country studies, the size of the informal sector presented here is not strictly comparable. It can be noted that both studies exclude credit unions.

5. As Chandavarkar (1985) argues, the existence of informal finance of this size may have significant implications for the efficacy of monetary policy.

6. Aryeetey et al (1989) note that the liquidity problems of the rural population stem partially from their holding their savings in illiquid real assets. While they appear to be sensitive to the cost of holding financial assets, they often are less sensitive to the cost of borrowing to ease liquidity problems.

7. As Germidis et al (1989) note, financial dualism can be ascribed as much to the shortcomings of the formal sector as to the intrinsic dualism of economic and social structures in developing countries and the population's attachment to traditional modes of behavior.

8. Due to the large seasonal fluctuations characteristic of rural household income, however, it is necessary to apply some caution in interpreting the annual estimate presented by Bentil.

9. An exception to this can be found in the case of Malawi, where it has been estimated that one quarter of the total credit generated by the informal sector, K63.1 million, was used to finance trade and manufacturing in 1988 (Chipeta and Mkandawire, 1989a).

10. Chipeta and Mkandawire (1989b) report that the government of Malawi has recently considered a number of new measures to expand credit resources for smallholders and small and medium enterprises. In view of the scope for the expansion of informal financing of trade and diversification, Chipeta and Mkandawire argue for including the informal financial entities in the country's plan for promoting small and medium enterprises.

11. While in Table 1 the current account balance is expressed as a percentage of GNP to calculate the public savings rate, in Table 2 it is expressed as a percentage of GDP in order to be compatible with the rest of the indicators in the table.

12. Due to limited space, central banks, monetary policy, and other financial institutions are not included in this discussion.

13. It is reported that the trade liberalization measures taken in 1989 subsequently eased the banks' excess liquidity problems to some extent.

14. In order to improve the situation, Chipeta and Mkandawire (1989b) suggest that SEDOM and INDEFUND should be allowed to participate in domestic resource mobilization by issuing long term paper on which higher interest rates could be paid than those on bank deposits.

15. Cho reports findings which suggest that "in the presence of capital market imperfections, elimination of interest rate ceilings is no longer thought beneficial" (Cho 1986). Anderson and Khambata (1985) argue that "the right combination of policies is a relaxation of the controls combined with interventions to address shortcomings, essentially of an institutional nature, in the capital markets" (p. 350).

16. However, as Meyer and Cuevas (1990) note, many conventional devices of loan insurance and guaranteed

funds have failed to live up to expectations in terms of additionality in lending. This points to the need for more innovative schemes. Ragazzi (1981) suggests establishing stabilization funds or other devices to cushion the impact of adverse circumstances on small and medium enterprises and farmers.

References

Alamgir, Mohiuddin (1976), "Rural Savings and Investment in Developing Countries: Some Conceptual and Empirical Issues", *Bangladesh Development Studies*, January 1976, pp. 1-48.

Anderson, Dennis and Fardida Khambata (1985), "Financing Small-Scale Industry and Agriculture in Developing Countries: The Merits and Limitations of 'Commercial' Policies", *Economic Development and Cultural Change*, pp. 349-371.

Aryeetey, Ernest, Eric Asante, Fritz Gockel and Alexander Kyei (1989a), "Mobilizing Domestic Savings for African Development and Industrialization: A Ghanaian Case", an interim country report presented at Workshop on Domestic Resource Mobilization in Harare, December 1989.

Aryeetey, Ernest and Fritz Gockel (1989b), "Mobilizing Domestic Resources for Capital Formation: The Role of Informal Financial Markets in Ghana", preliminary report presented at Workshop of the AERC, Harare, December 1989.

Bentil, B. et al (1988), "Rural Finance in Ghana", a research study prepared for the Bank of Ghana on behalf of the Ministry of Economic Co-operation of the Federal Republic of Germany.

Chandavarkar, Anand G. (1985), "The Non-Institutional Financial Sector in Developing Countries: Macroeconomic Implications for Savings Policies", *Savings and Development*, no. 2.

Chipeta, C. and M.L.C. Mkandawire (1989a), "Mobilizing Domestic Savings for African Development and Industrialization: A Case Study of Malawi", an interim country report presented at Workshop on Domestic Resource Mobilization in Harare, December 1989.

Chipeta, C. and M.L.C. Mkandawire (1989b), "The Informal Financial Sector in Malawi: Scope, Size and Role", progress report presented at Workshop of the AERC, Harare, December 1989.

Cho, Yoon Je (1986), "Inefficiencies from Financial Liberalization in the Absence of Well-Functioning Equity Markets", *Journal of Money, Credit and Banking*, vol. 18, no. 2, pp. 191-199.

Deaton, Angus (1989), "Saving in Developing Countries: theory and review", paper presented at the First Annual World Bank Conference on Economic Development, Washington, D.C., April 1989.

Germidis, Dimitri, Dennis Kessler and Rachel Meghir (1989), "Mobilizing Domestic Savings for Development: What Role for the Formal and Informal Financial Sectors?", OECD, summary note on the Development Centre Study, Paris.

Ghosh, Dipak, (1986), "Savings Behaviour in the Non-Monetized Sector and Its Implications", Savings and Development, no. 2.

Malawi/USAID (1987), "New Directions for Promoting Small and Medium Scale Enterprises in Malawi: Constraints and Prospects for Growth".

Meyer, Richard L. and Carlos E. Cuevas (1990), "Reducing the Transaction Costs of Financial Intermediation: Theory and Innovations", paper prepared for the International Conference on Savings and Credit for Development, Copenhagen, May 28-31.

Neal, Craig R. (1988), "Macro-Financial Indicators for One Hundred Seventeen Developing and Industrial Countries", December 1988. The World Bank, Washington, D.C.

Nissanke, Machiko (1990), "Theoretical Issues on Finance and Development: a critical literature survey", Queen Elizabeth House, Oxford University discussion paper.

Ragazzi, Giorgio (1981), "Savings Mobilization in Africa", *Savings and Development*, no. 1.

15

Monetary Cooperation in the CFA Zone

Patrick Honohan

There has been considerable discussion over the past few years on the merits for small countries of joining a monetary union. For the developing world, this debate has centered on the experience of the CFA or franc zone in Africa. This zone is the largest and most enduring of currency blocs. Most of the recent policy discussion on the CFA zone has focussed on three types of question: competitiveness, the mechanics of bank restructuring and the comparative growth performance of CFA versus non-CFA countries.[1] The focus of this paper is rather different. It is designed as a guide to understanding how the zone works as a mechanism for monetary cooperation.

When considering the merits of joining a monetary union small countries naturally value the credible commitment to exchange rate and price stability which membership represents and which would be hard to sustain through unilaterally pegging one's currency.[2] There is also a potential saving through pooling of the external reserves of member currencies.

But there might also be other advantages. Capital might flow more freely to where it was most needed within a monetary union. Provided the distribution of benefits of the union was reasonable, this could be to the advantage of all members, even those whose low capital productivity resulted in their becoming net lenders within the union. Furthermore the operation of monetary policy and the prudential supervision of the banking system might be more effective if the resources of several small countries were pooled in a strong and independent central bank.

This paper examines whether such additional advantages have been realized in the CFA zone. Unfortunately, the experience which is described in this paper is not an encouraging one. Despite the fixed exchange rate and an elaborate set of rules for avoiding excessive credit expansion, the CFA zone has almost foundered in widespread bank insolvency. Although the zone's institutional set-up seems equitable, in practice the burden of paying for the losses involved is likely to fall disproportionately on the poorer countries, while many of the non-performing credits have been made in some of the more prosperous countries of the zone.

The paper is organized as follows. Section 2 describes the member countries of the zone and points out in particular the diversity not only in their economic structure but in the state of the financial system. Some countries are net borrowers within the zone, some net lenders. This might suggest that some efficient mechanism is in place to channel investable resources to where they are most needed, but this is not the case. Section 3 shows that there is no effective regional money market and considers how this might be remedied. In the absence of a market to determine the regional distribution of credit, this must be done administratively. Section 4 discusses how credit has been distributed between the member countries and highlights the asymmetries which have arisen. While temporary fluctuations in export receipts have influenced part of the distribution, the bulk of available credit has not been used as a revolving fund to meet such temporary needs, but has increasingly gone to a few favored countries. Again this cannot be explained as a flow to where funds are best rewarded, as much of the credit has gone to banks which have subsequently failed. The costs of these bad debts affects the distribution of the seignorage benefits of the zone; section 5 explains how seignorage is divided in the unions and shows how the losses may fall disproportionately on non-favored countries. Section 6 concludes by considering how the recent banking crisis has revealed the deficiencies in the zone's arrangements.

The CFA Countries

The CFA zone is made up of two separate unions, the West African Monetary Union (UMOA) and the countries which have formed the Bank of the Central African States (BEAC).[3] Though not all of the former French colonies in Sub-Saharan Africa (SSA) joined one or other of the two unions, and though there have been defections[4] the current situation is that thirteen countries with a total population of some seventy million share just two central banks and two currencies. The currencies, each of which is known as the CFA franc, are both valued at one-fiftieth of the French franc, a peg they have retained for forty years. Along with the fixed exchange rate, the members of the CFA zone have traditionally retained an open international capital market, with statutory freedom of capital movements among the member countries and with France.

The monetary mechanism in the two zones has been characterized by:

- an annual monetary programming exercise which determines the planned growth in domestic credit in each member country.
- implementation of this monetary program through credit ceilings to each government and, in some cases, to the private sector, or through ceilings on central bank refinancing of private sector credit.
- administered interest rates, including preferential rates for priority sectors.

The total land area of the CFA zone amounts to 31% of SSA, though its population is only 16% and its total GNP about 22% of that in SSA. The members include some of the poorest countries in the world (chief economic indicators for the members are shown in table 15–1). All of them rely heavily on exports of a variety of primary products for the foreign earnings needed for imports of necessities. Most are also heavily dependent on official transfers and concessional lending.[5]

It is striking how varied the CFA countries are. Some are open to the sea, some are landlocked. In terms of surface area, each of the largest countries (Chad, Niger and Mali) is almost fifty times the size of the smallest, Equatorial Guinea. The density of population in Togo is almost fourteen times that in

Table 15–1. *CFA Zone—general indicators*

Country	Population (millions) mid. 1986	Area ('000s sq. km.)	GDP (1986 $ million)	GNP (per capita 1986-dollars)	GNP (per capita growth 1965 1980	GDP growth 1965-1980	GDP growth 1980-1986	Avg. inflation 1965-1980	Avg. inflation 1980-1986	Life expectancy years 1986	M2/GDP 1965	M2/GDP 1985
UMOA												
Benin	4.2	113	1,320	270	0.2	2.3	3.6	7.4	8.6	50	10.6	22.8
Burkina Faso	8.1	274	930	150	1.3	3.5	2.5	6.2	6.3	47	9.3	22.1
Côte d'Ivoire	10.7	323	7,320	730	1.2	6.8	-0.3	9.3	8.3	52	21.8	29.4
Mali	7.6	1,240	1,650	180	1.1	4.1	0.4	—	7.4	47	—	23.0
Niger	6.6	1,267	2,080	260	-2.2	0.3	-2.6	7.5	6.6	44	3.8	15.9
Senegal	6.8	196	3,470	420	-0.6	2.1	3.2	6.5	9.5	47	15.3	24.5
Togo	3.1	57	980	250	0.2	4.5	-1.1	6.9	6.7	53	10.9	45.3
UMOA	47.0	3,470	18,020	358	0.0	4.0	1.0	8.0	8.0	49	16.0	26.0
BEAC												
C.A.R.	2.7	623	900	290	-0.6	2.6	1.1	8.5	11.5	50	13.5	17.4
Chad	5.1	1,284	—	—	—	0.1	—	6.3	—	45	9.3	25.5
Cameroon	10.5	475	11,280	910	3.9	5.1	8.2	9.0	11.0	56	12.5	19.4
Congo	2.0	342	2,000	990	3.6	5.9	5.1	7.1	7.5	58	16.5	20.1
Gabon	1.0	268	3,190	3,080	1.9	9.5	1.5	12.7	4.8	52	16.5	26.3
BEAC	21.0	2,992	17,370	950	3.0	5.0	6.0	9.0	9.0	55	14.0	—
Sub-Saharan Africa	424.0	20,895	165,990	370	1.0	6.0	0.0	13.0	16.0	50	—	—

Sources: IMF and World Bank data

Chad. Per capita income in oil-rich Gabon is estimated at fifteen times the figure for Burkina Faso. Only in regard to population size and inflation rates does the range of CFA experience fail to encompass the Sub-Saharan Africa average (table 15–2).[6] Even here, there is a wide divergence within the CFA zone with individual country populations from less than a half a million to over 11 million; and, though long-term inflation trends converge around the French norm (see Honohan, 1990), cumulative 1980-86 inflation varied between 32% and 92%.

Exports of CFA countries are overwhelmingly (over 90 per cent in each of the countries) composed of primary products or of lightly processed primary products such as vegetable oils, cotton fabrics, wooden veneers and fertilizers. However, it is significant that different countries depend on different product mixes. For example, for the seven countries of the UMOA, a listing of the products which directly or indirectly account for at least five per cent of exports (to industrial countries) involves fourteen quite distinct products (table 15–3).

Since the member countries of both unions (but especially the UMOA) depend on different products for the bulk of their exports, and since the prices of these commodities are not very closely correlated,[7] there is substantial scope for pooling the risks of fluctuations in export markets.

Reflecting the wide differences in level of development, the range and number of financial institutions varies from country to country within the zone. The dominant financial institutions are banks; the non-bank financial sector is negligible. Most of the largest commercial banks are at least partly government-owned, often with key management and systems provided by French banks which are minority shareholders. There are also large government-owned development banks, some of which accept deposits. As has already been mentioned and is further discussed below, banks in most of the countries have been found to be insolvent. This applies to a substantial fraction of the banking system in Cameroon, Côte d'Ivoire, Senegal, Benin and Chad.

The monetary structure of the UMOA countries also displays wide variations from country to country. As shown in table 6, many of the CFA banking systems are heavily indebted to the central banks. The average indebtedness of the commercial banks to the central bank throughout the zone in recent years has been about one-third of their non-government lending. There are just two countries (Togo and Burkina Faso) at the other extreme, with the commercial banks accepting deposits so far in excess of what they can usefully lend that they place substantial deposits with the regional central bank. But in most of the countries, and especially in Côte d'Ivoire, Senegal and recently Cameroon, lending by the commercial banks far outstrips deposit resources, with the shortfall being made up through refinancing by the central bank. In all of the countries, direct lending by the banks to government is small.

The central bank in each of the two unions thus provides an intermediation function between the financial systems of the surplus and deficit countries. One way of looking at this is to observe that, while the central banks' total net claims on commercial banks is of the same order of magnitude as the total currency issue, three countries in UMOA and one in BEAC have commercial banking systems indebted to the central bank in an amount in excess of the national currency holdings. In effect, some of the resources mobilized through currency issue in the

Table 15–2. *The CFA countries in Sub-Saharan Africa*

	SSA average	*CFA lo*	*CFA hi*
Population density (pop./km.^{2})	20	4	36
Population (million)	12	<1	11
Area (km2)	597	28	1,284
Inflation average (1980-86)%	16	5	12
Life expectancy at birth (yrs)	50	44	58
Growth average (1980-86) %	0	-3	8
Percent share in GDP:			
Agriculture	36	10	50
Industry	25	13	54
Savings (GDS)	11	7	30
Investors	14	9	37
Export share	19	14	47
Government concerns	13	9	26
Primary products as a percentage of exports	57	21	96
Percent of manufactured exports to industrial centers	34	11	82
Persons per physician	25	3	56
School enrollment:			
Primary %	75	23	>100
Secondary %	23	5	25
Agricultural share in the workforce%	75	62	87
Urbanization %	23	8	4
Infant mortality %	113	74	144

Source: World Development Report 1988, The World Bank; Data refers to 1986 or a recent year.

Table 15–3. *UMOA: Commodity composition of exports*

Product	*Benin*	*Burkina Faso*	*Côte d'Ivoire*	*Mali*	*Niger*	*Senegal*	*Togo*
Cotton	21	35	—	68	—	—	5
Coffee	7	—	25	—	—	—	40
Cocoa	16	—	36	—	—	—	14
Uranium	5	—	—	—	97	—	—
Phosphates	—	—	—	—	—	16	—
Fruit and Vegetables	—	7	7	—	—	—	—
Groundnuts	—	—	—	—	—	44	—
Karite Almonds	—	36	—	—	—	—	—
Petroleum	35	—	—	—	—	—	—
Fish	—	—	—	—	—	28	—
Wood	—	—	15	—	—	—	—
Gold	—	—	—	14	—	—	—
Hides	—	7	—	—	—	—	—

Note: Table shows percentage of each country's exports to industrial countries of varios commodities and products manufactured from those commodities. Only commodities accounting for more than 5% of total exports are included.
Source: World Trade Statistics, UN.

other countries are being channeled to the banking systems in the other four countries. (The central banks also lend to the government, their overall intermediation role is discussed further below).

A Union-Wide Money Market?

In a single country the financial system is relied upon to intermediate financial flows from surplus to deficit entities. Bank and non-bank financial intermediaries and securities markets are the means by which savings flows reach those who need to borrow in order to invest in real resources or in order to meet a temporary excess of consumption needs over income. This is not always done with perfect efficiency. Various market and institutional obstacles stand in the way. For example informational asymmetries and the incentive for abuses mean that savers do not have perfect confidence in those who wish to use their funds. Standard financial contracts (such as the debt contract) do not always divide the risks between lender and borrower in an ideal manner; contracts which might be better from that point of view can prove to be unenforceable. All this is common to all countries. The challenge for the monetary union is to attempt to ensure that financial intermediation works as smoothly across international frontiers within the zone as it works within each member country.

National frontiers could create several types of obstacle to financial intermediation, including controls on foreign exchange movements, risk of changes in foreign exchange rates, lack of knowledge about counterparties associated with language differences and limited trading relations.[8] Some of these problems are well dealt with by the CFA unions, but some are not and the net result is that no effective union-wide regional financial market exists.

Where the CFA unions do function reasonably well is in the area of exchange controls and foreign exchange risk, though even here there are problems. In particular it has not always been as easy in practice to *transfer funds* from CFA countries to France in recent years as it has been in theory.[9] Transfers of funds between the CFA countries are much freer especially since there are only two currencies involved: transfers between two countries of the UMOA involve no foreign exchange transaction; likewise between two BEAC zone countries; nevertheless, there is always the residual risk that a member country might, at a later date, impose restrictions on foreigners attempting to repatriate funds. The fixed exchange rate reduces but does not eliminate the risk of *foreign exchange losses*: balance of payments difficulties have, from time to time, generated market fears of a change in the parity. For capital between UMOA or BEAC countries those risks are smaller: after all, one is dealing with a single currency and the scenarios under which repatriation of the funds to another African country within the union could entail an exchange loss are more remote.

Cotonou in Benin is about 2,500 km from Dakar in Senegal. This figure, together with the fact that the UMOA is larger than India, gives some idea of the geographical scale of the unions we are concerned

Table 15-4. UMOA: Variability of export prices

	Country						
Period	*Benin*	*Burkina Faso*	*Côte d'Ivoire*	*Mali*	*Niger*	*Senegal*	*Togo*
1980	21	35	—	68	—	—	5
1981	7	—	25	—	—	—	40
1982	1	—	36	—	—	—	14
1983	5	—	—	—	97	—	—
1984	—	—	—	—	—	16	—
1985	—	7	7	—	—	—	—
1986	—	—	—	—	—	44	—
1987	—	36	—	—	—	—	—
1988	35	—	—	—	—	—	—
Variance	—	—	—	—	—	28	—

Matrix of correlations

	Burkina Faso	*Côte d'Ivoire*	*Mali*	*Senegal*	*Togo*
Benin	0.067	0.415	0.667	0.625	0.711
Burkina Faso	—	0.927	0.206	0.130	0.474
Côte d'Ivoire	—	—	0.385	0.292	0.693
Mali	—	—	—	0.679	0.681
Senegal	—	—	—	—	0.608

Source: World Bank data

about. Over such large scales the information flows necessary to build sufficient confidence to allow borrowing and lending to take place efficiently require the use of trusted intermediaries. In the present stage of development of CFA financial markets this means that savings flows must be intermediated by the banking system. Thus a saver in Dakar may trust his local banker; a bank in Cotonou may have a creditworthy client. Unfortunately, international loans within the unions, even between banks, are virtually unknown. (Even within a single member country as developed as Senegal or Côte d'Ivoire the interbank market is very thin and confined to the leading banks only.)[10] Accordingly, it is unlikely, in the posited situation, that the Dakar funds will find their way to the borrower in Cotonou.

The BCEAO does operate what is known as the "Marche Monetaire", but this is little more than the combination of a deposit facility for banks, and a facility for secured borrowing by banks from it and at its discretion. Thus, if a bank in Benin for example, has a demand for liquid funds it may not be able to meet this demand from the marche monetaire even if it is able to provide adequate security, and even if a bank in Senegal, for example, has just placed surplus funds with the marche monetaire.

The BCEAO has not attempted to adjust interest rates to clear the domestic market for bank liquidity.[11] Furthermore, the gap between borrowing and deposit rates in the marche is as wide as one per cent per annum—far higher than interbank spreads in a developed money market.

The situation in the BEAC zone is no better; there is no "marche monetaire", though the BEAC does accept deposits and makes loans. Once again the loans are quite discretionary and depend on policy considerations other than pure banking prudence. The interest rates on loans bear no relation to market-clearing rates.

In short, it has to be said that there is no effective regional money market in the CFA zone. Under the present circumstances, the easiest way to encourage the development of one would be for the central banks to act as intermediaries matching the needs of sound deficit banks with the supply from surplus banks. In view of the fragile state of the banking system it is more than unlikely that any private intermediary would undertake this activity. After all, three large French banks have affiliates in most of the CFA countries, yet they do not undertake interbank transactions even between affiliates in different countries. Only the central bank has sufficient information to judge the soundness of participant banks. But each central bank must also manage monetary policy, and this is also done by means of influencing the quantities of bank liquidity in the different coun-

Table 15-5. *Matrix of trade paterns of CFA countries*
(Exports % of total exports of each country)

	Benin	Burkina Faso	Côte d'Ivoire	Mali	Niger	Senegal	Togo	Cameroon	C.A.R.	Chad	Congo	Gabon	Total CFA	French franc
Benin	0.0	0.0	0.0	0.1	1.1	0.1	0.0	0.0	0.0	0.0	0.1	0.0	0.1	0.0
Burkina Faso	0.1	0.0	0.0	0.1	0.3	0.0	0.2	0.0	0.0	0.0	0.0	0.0	0.1	0.0
Côte d'Ivoire	3.3	28.9	0.0	23.3	9.6	4.2	5.4	1.1	0.0	0.0	0.0	1.4	4.2	0.4
Mali	0.0	0.4	0.2	0.0	0.1	0.0	0.0	0.0	0.0	0.0	0.0	0.0	0.1	0.0
Niger	0.6	0.0	0.1	0.2	0.0	0.0	0.1	0.0	0.0	0.1	0.0	0.0	0.1	0.2
Senegal	0.9	1.2	1.6	6.4	0.8	0.0	0.9	0.9	0.2	0.2	0.8	0.2	1.1	0.1
Togo	0.1	0.3	0.0	0.1	0.6	0.0	0.0	0.0	0.0	0.1	0.0	0.0	0.1	0.0
Cameroon	0.0	0.0	2.8	0.0	0.0	0.0	0.0	0.0	0.0	19.5	0.0	0.8	1.0	0.4
C.A.R.	0.0	0.0	0.0	0.0	0.0	0.0	0.0	0.0	0.0	0.2	0.0	0.0	0.0	0.0
Chad	0.0	0.0	0.0	0.0	0.0	0.0	0.0	0.4	0.0	0.0	0.1	0.0	0.1	0.0
Congo	0.0	0.0	0.0	0.0	0.0	0.0	0.0	0.0	0.0	0.9	0.0	0.0	0.0	0.1
Gabon	0.0	0.0	1.4	0.0	0.0	0.5	0.0	0.0	0.0	0.0	0.0	0.0	0.4	0.4
CFA	4.9	31.5	6.3	30.2	12.5	4.7	6.6	2.4	0.3	21.0	1.0	2.4	0.0	1.7
French franc	21.6	34.3	35.3	29.3	39.3	34.9	27.5	43.0	59.3	38.0	53.4	55.0	38.7	0.0
CFA franc	26.4	41.6	41.6	59.9	51.9	39.6	34.2	45.4	59.6	59.0	54.4	57.4	45.9	1.7
World	100.0	100.0	100.0	100.0	100.0	100.0	100.0	100.0	100.0	100.0	100.0	100.0	100.0	100.0
Benin	0.0	0.3	0.7	0.0	0.3	0.9	0.2	0.0	0.0	0.0	0.0	0.0	0.3	0.1
Burkina Faso	0.1	0.0	4.3	0.2	0.7	0.8	2.0	0.0	0.0	0.0	0.0	0.0	1.3	0.1
Côte d'Ivoire	0.3	1.8	0.0	3.6	0.6	5.7	0.0	2.3	0.0	0.0	0.0	1.6	1.3	0.5
Mali	0.2	0.5	4.2	0.0	0.2	5.1	0.2	0.0	0.0	0.0	0.0	0.0	1.5	0.1
Niger	1.5	0.8	1.2	1.3	0.0	0.4	0.7	0.0	0.0	0.0	0.0	0.0	0.4	0.1
Senegal	0.0	0.0	1.8	0.3	0.0	0.0	0.0	0.9	0.0	0.0	0.0	0.3	0.6	0.3
Togo	0.2	0.5	0.3	0.1	0.1	0.7	0.0	0.0	0.0	0.0	0.0	0.0	0.1	0.1
Cameroon	0.0	0.0	0.8	0.0	0.0	2.7	0.0	0.0	0.0	10.6	0.0	0.1	0.5	0.6
C.A.R.	0.0	0.0	0.0	0.0	0.0	0.0	0.1	0.0	0.0	0.0	0.0	0.0	0.0	0.1
Chad	0.0	0.0	0.0	0.0	0.0	0.1	0.1	1.2	0.3	0.0	0.1	0.0	0.3	0.0
Congo	0.2	0.0	0.1	0.0	0.1	0.0	0.1	0.0	0.0	0.5	0.0	0.0	0.1	0.2
Gabon	0.0	0.0	0.5	0.0	0.0	0.3	0.0	0.3	0.0	0.0	0.0	0.0	0.2	0.4
CFA	2.6	3.8	13.9	6.4	2.0	17.7	3.5	3.9	0.3	11.2	0.1	2.0	0.0	2.5
French franc	10.8	38.1	24.9	14.1	88.0	34.8	17.0	26.4	18.1	12.6	12.8	35.9	38.7	0.0
CFA franc	13.4	42.0	38.8	20.5	90.0	52.5	20.4	30.3	18.3	23.8	12.9	37.9	45.9	2.5
World	100.0	100.0	100.0	100.0	100.0	100.0	100.0	100.0	100.0	100.0	100.0	100.0	100.0	100.0

Source: World Bank data

tries. An institutional framework for resolving such conflicts would need to be devised.

For example, in order to ensure a separation of the functions of monetary control and international intermediation, one arm of the central bank (call it the "monetary policy arm") would, as at present, be charged with planning the available amount of central bank credit for the union as a whole. Another arm (the "intermediation arm") would match supply of and demand for bank liquidity on the basis meeting any demands from sound banks presenting acceptable paper as security. The intermediation arm would accept deposits and would also borrow from the monetary policy arm. The latter would adjust the interest rate on its loans to ensure that the market cleared, i.e. that the intermediation arm was not borrowing more than the planned ceiling. Having a market-clearing interest rate is the key to satisfactory operation of international intermediation within the union. Sound banks will not over-borrow if the interest rate is not subsidized.[12]

To summarize, the international flow of funds within the UMOA and BEAC unions is not supported by an adequate institutional infrastructure. The prevailing risks are too great for private markets to perform this function given the information at their

Table 15–6. *CFA countries monetary survey*
(end-1987, in billions of CFA francs)

	Net external assets	*Domestic credit*	*Money (M2)*	*Currency*	*Claims of central bank on commercial bank*
Benin	-55	106	97	20	49
Burkina Faso	68	68	127	44	8
Côte d'Ivoire	-378	1,363	930	305	511
Mali	-43	189	130	61	32
Niger	19	104	114	36	30
Senegal	-206	539	332	101	156
Togo	65	97	164	48	5
UMOA	-530	2,466	1,894	615	780
Cameroon	-150	939	678	171	342
Central African Republic	18	48	60	41	13
Chad	9	78	76	47	39
Congo	-32	219	138	55	42
Equatorial Guinea	-4	11	9	7	4
Gabon	-33	384	238	49	29
BEAC	-192	1,679	1,199	370	467

Source: IMF Finance Yearbook

disposal. The central banks of each union are better positioned to channel funds from surplus to deficit countries. At present they do this without explicit reference to market criteria; they should move to a market-based system of allocating bank liquidity throughout their respective zones, while preserving monetary control.

Monetary Policy

Ideally, the available credit in the union should go to where it yields the highest return. This would presumably involve flows to support domestic expenditure in member countries faced with a transitory balance of payments difficulties resulting from adverse export shocks. On a longer term basis, there should also be a tendency for members where the marginal product of capital was low to be net lenders towards those where the marginal product of capital was high. In this section the process of credit allocation is reviewed to assess whether it flows in accordance with these criteria.

In order to protect the external position of the union, and ultimately to ensure the sustainability of the fixed parity, the monetary authorities avoid creating too much domestic credit. In practice, there is an annual credit planning exercise for each country separately, with the individual plans being simply added together to obtain the monetary plan for each union. The modelling underlying the plan is based on a simple standard fixed exchange rate quantity theory framework. Credit creation in excess of the increase in money demand will spill over into the balance of payments and deplete the union's reserves. Therefore the credit plan is in principle designed to limit such excesses, both for individual countries and for the union as a whole.

Broadly speaking, an expansion in central bank refinancing of bank credit leads to an expansion in overall bank credit, franc for franc. Although the money stock responds to such an increase at first, there is no sustained increase in money demand: over time the credit expansion leaks out into the balance of payments.[14]

Because there are no private banking flows between member countries, central bank refinancing in one country will result in an increase in bank credit in that country, and not in the other countries. This allows the central bank to pursue independent credit objectives for each member country, as is often required by IMF programs. (If a true international money market were established in the unions then union-wide liquidity might have to be adjusted in response to an emergent overshoot of the credit ceiling in one country).[15]

Among the factors apparently influencing the allocation of central bank credit between countries, at least in the short-run, appears to be each country's export receipts. In particular, UMOA countries who experienced a decline in the value of exports (relative to the average of the union experience) have tended to obtain a higher share of central bank credit. To this extent, the members have been able to benefit from the risk-pooling afforded by the diversity of export products in the union. However, efficient pooling of these risks requires the available credit to be used as a revolving fund, with borrowings to cover short-term disturbances being repaid in better times.

In fact, not all countries appear to treat central bank credit to the banking system as a revolving fund. Somehow, at least in the UMOA, the larger and more prosperous countries have gained a disproportionate share in the regional distribution of credit. The major shifts in the distribution of credit over the years in favor of these countries have shown no tendency to be reversed. Figure 15–1 shows lending to banks by the BCEAO expressed as a percentage of bread money. The upper panel of the figure shows the three countries, Côte d'Ivoire, Senegal and Benin, which have received the greatest proportionate share rising to more than 60 per cent of each country's wide money stock; the lower panel show the other three,[17] none of whom ever received more than about 40 percent of wide money (less than 20 per cent for

Figure 15-1. BCEAO lending to banks as a percentage of M_2, 1970-88

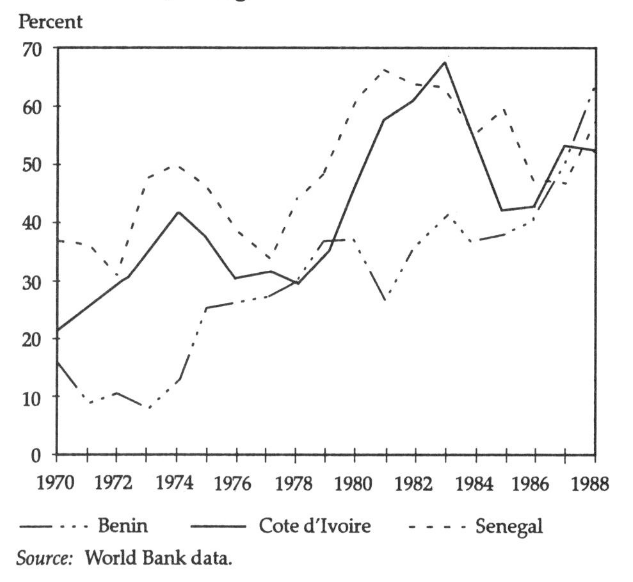

Source: World Bank data.

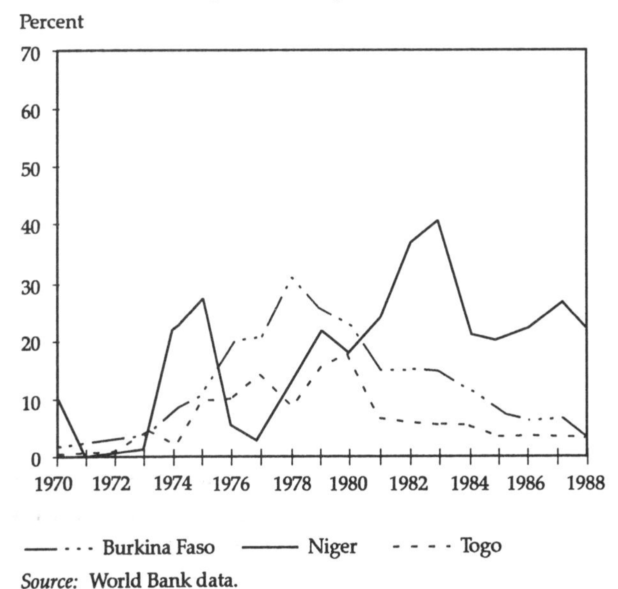

Source: World Bank data.

Figure 15-2. Net Foreign Assets as a percentage of M_2, 1970-88

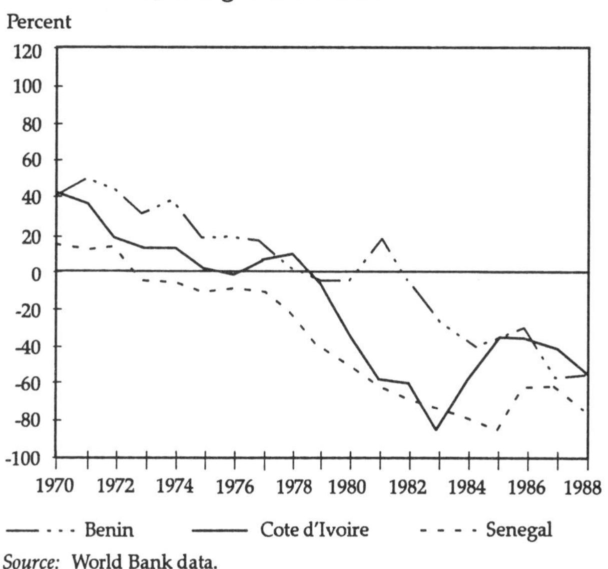

Source: World Bank data.

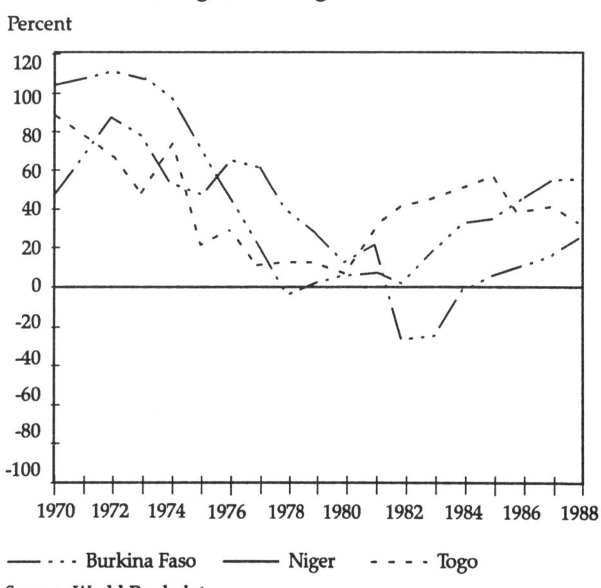

Source: World Bank data.

Togo) and whose credit has tended to revert to low levels.

BCEAO direct lending to governments has also been important, even when one leaves out borrowing from the IMF channeled through the Central Bank. The same country pattern is true of total BCEAO lending, including lending to governments. Since excessive credit expansions have leaked out through the balance of payments, this skewed distribution of credit has resulted in a skewed pattern of external deficits and external reserves, with the three larger countries holding disproportionately low gross external reserves.

Côte d'Ivoire's share of union GDP has fluctuated between about 40 per cent and 55 per cent in the period 1970-86 (with its share of union population growing from 23 per cent to over 27 per cent). Before 1980 Côte d'Ivoire's share in union external reserves fluct- uated between 22 per cent and 62 per cent, but in that year it fell below 7 per cent and has been as low as 1 percent, it share has never recovered to pre-1980 levels. A similar pattern is noted for *Senegal* and *Ben- in*. The fall in these three countries' shares of external reserves was taken up largely by *Togo*. Although Togo's share of union GDP has been falling from only about 7 per cent to about 5 per cent, and although its share in union external reserves never exceeded 20 per cent before 1980, since 1981 its share in union external reserves has never fallen below 40 percent.

This dramatic shift in relative shares of the union's gross external reserves corresponds to a disproportionate allocation of central bank credit to Cote d'Ivoire, Senegal and Benin. It is also reflected in the net external assets of the monetary system (figure 2). The excess credit leaked out into the balance of payments resulting effectively in these countries indirectly borrowing from (poorer) Togo to finance imports. This would be unsurprising on a temporary basis, but as it has been sustained for many years it suggests that there is no adequate mechanism for ensuring that the resources of the union are treated as a short-term revolving fund. The forces for repayment are extremely weak.[18]

As already suggested, a permanent pattern of persistent borrowers and patient lenders might be justified on the basis of a higher marginal product of capital in the borrowing countries, in which case the lending countries might achieve a better return than if they had increased domestic expenditure. However, the fact that much of the credit has gone to banks which have failed proves that credit has not sought out high-yielding activities. Indeed, the financial return to smaller members in the form of central bank dividends—or seignorage—will have been seriously eroded by these losses. It is to these financial returns that we now turn.

The Distribution of Seignorage

Membership of a monetary union is not a zero-sum game; some of its benefits, such as those of price stability, are public goods that can be enjoyed by all. The benefits of seignorage are divided among the members. This section explains how seignorage is divided and shows that the division is not as equitable as it may appear at first sight.

Broadly speaking, seignorage is the benefit which accrues to the issuer of money by virtue of being able to issue interest-free currency rather than having to incur interest charges on borrowing. The total value of the currency issue in the UMOA is about $2 billion, in the BEAC zone it is about $1.3 billion. Most of the benefit of the seignorage on this currency issue should flow to member governments through the distribution of BEAC profits, but some is transferred to those who are given subsidized credit from the Central bank.

The easiest way to trace the flow of seignorage is to consider the accounts of the central banks. Taking the BCEAO 1986 balance sheet (table 15–7) for example it can be seen that currency is the largest single item[19] representing over half of the BCEAO's liabilities. The principal corresponding assets are claims on banks and on the national treasuries, the latter being little more than one-third as important as the former. These claims are interest bearing, while the currency is not. This offers a margin which, after the operating costs of the BCEAO are deducted, represents most of the flow of benefits to member governments arising out of seignorage. The remainder of the seignorage is passed through the implicit interest subsidy on government borrowing and on some of the commercial bank borrowing from the BCEAO.[20]

The most natural distribution of seignorage would be in proportion to each member's national currency issue. In accordance with the UMOA statutes, however, the BCEAO's dividend is distributed equally among the member states. Therefore were it not for the subsidy on BCEAO lending rates it would seem that we could say that the seignorage is divided equally among the countries and thus represents a progressive redistribution from the large (and rich) to the small countries).[21] But there is an important caveat.

The caveat arises because a large part of the BCEAO's lending to banks has become non-performing. In Benin, Senegal and Côte d'Ivoire in particular, the BCEAO holds especially large claims on insolvent banks. Although it falls to each national government to meet BCEAO claims on insolvent banks, the BCEAO has already decided to consolidate these debts at very concessional rates. As the total of the claims in question are of the same order of magnitude as the currency issue, such concessions will severely dent the flow of seignorage benefits coming through the BCEAO dividends. In fact, by making some very rough calculations on the distribution of seignorage after retrospective account is taken of the annualized cost of these concessions, it appears that close to one hundred per cent of the seignorage of recent years has benefitted Côte d'Ivoire, Senegal and Benin.[22]

The full story about who benefits from the seignorage therefore becomes a combination of equal distribution of central bank profits, subsidies on treasury borrowings, and concessions made to individual countries in respect of bank insolvencies.[23] Though the smaller member countries have benefitted over the years by a larger distribution than they would have received had the distribution been proportionate to currency issue, they will now suffer by the equal distribution of losses on the BCEAO's portfolio. In the end it may be that they will effectively lose all of the seignorage.

For the BEAC, the seignorage analysis is more complex. The distribution of BEAC profits is not equal, but depends on a formula designed to take account of the actual contribution of each member country to the profits of the BEAC.[24] Whatever about the merits of the formula in normal circumstances, it is not clear how it will be applied if it becomes necessary to take account of heavy write-offs on BEAC loans to weak banks. Therefore it is too early to judge the likely distributional impact among countries of such losses.

The overall conclusion must be that the arrangements for distributing seignorage have been vitiated by the failure of the central banks to ensure that the resources mobilized by the currency issue were invested in a sound manner.

Concluding Remarks: the Banking Crisis

Previous sections have shown that, although the CFA zone has ensured a degree of price stability over the long-run, money and credit policy has not been without problems. The distribution of credit among different member states has not followed market principles but has been determined administratively. While the administrative arrangements are not obviously inequitable in themselves, the UMOA has slid into a very lop-sided situation where three countries —the three most prosperous—have taken the bulk of central bank credit and apparently squandered it. Between them, banks in these three countries have run up a net capital deficiency approximately equal to the average currency issue for the whole union—a sum not far short of $2 billion or ten per cent of the union GNP for one year. It is as if the central bank had printed the entire currency issue only to hand it to insouciant borrowers in these three countries.

How does it arise that the three countries with the largest bank insolvencies are also the countries which have received the lion's share of central bank credit? It is difficult to avoid the inference that, for all the talk of monetary programming, the inter-country allocation of credit has been driven by pressure to finance what has proved to be unsound lending by the banks in the most influential countries. Perhaps envisaging such an eventuality, the statutes of the union require each national government to make good any claims of the BCEAO on an insolvent bank.[25] Despite this, the governments involved have obtained generous rescheduling terms from the BCEAO and are also hoping for substantial additional concessionary assistance from bilateral and multilateral donors to finance the debts.

If one is to think in terms of winners and losers, so far the winners are those who benefitted from the

Table 15—7. ***BCEAO balance sheet***

(September 1986) In billions CFA Francs

Assets	
Gold	21
IMR reserve position and SDRs	16
Foreign exchange	357
(of which operations account)	(95)
Claims on banks and OFIs	596
Claims on national treasuries	431
(of which IMF and Trust Fund)	(399)
Fixed assets, etc.	47
Suspense accounts and miscellaneous	341
Total assets-total liabilities	1,809
Liabilities	
Current	528
Claims on foreign banks, etc.	57
Claims of local banks and OFIs	234
Claims of national treasuries, etc.	80
Transfers being executed	10
Obligations in foreign exchange	0
IMF, SDRs, and trust fund	426
Capital and reserves	107
Suspense accounts and miscellaneous	340

Source: BCEAO.

unwise and un-repaid loans made by the failed banks. The four poorest countries of the UMOA are, so far, losers; their claims on future dividends from the BCEAO are compromised by the uncertain value of its claims on the failing banks and the three less-poor governments. The tentative calculation above suggested that these countries have lost seignorage equivalent in capitalized terms to the value of currency in circulation.

Perhaps over-emphasis on the question of maintaining the parity led to preoccupation with macroeconomic aggregates, such as the total amount of central bank credit in the system to the neglect of the prudent distribution of this credit, and the adequate management of banks in the union. It is even possible that the device of a statutory ceiling on advances to government (designed to facilitate the credit restraint necessary to assure the exchange rate) provided a false sense of security and may have perversely led the central bank into a complacent approach to the growth of parastatal and government influenced credit.

On any reckoning this is a very poor performance of the union. It leads one to ask whether alternative statutory framework could have led to a more favorable outcome, with the resources of the union being distributed in a genuinely revolving fund to meet fluctuations in external needs of individual countries rather than being systematically channeled to a select few. The establishment of a more effective and market-based mechanism for channeling surplus funds would help: a true "marche monetaire" where solvency of the borrowing institution would be the sole criterion of whether it could have access to funds at a market-clearing rate.

In the BEAC there are also serious problems of bank insolvency, though in this case the evolution of central bank credit has changed sharply in the last couple of years, largely as a result of the fall in oil prices, so that it is too early to speak in terms of long-term structural inequalities between countries in the BEAC zone.

One must fear that the very unequal distribution of economic power of the member countries will still give Côte d'Ivoire, and to a lesser extent Senegal, in the UMOA, and Cameroon in the BEAC, the ability and incentive to continue to abuse the system when it has been recapitalized. Perhaps the power structure of the CFA zone would be more diffuse if the two unions were to merge. The dominance of Côte d'Ivoire would certainly be tempered, and the likelihood of a four-way coalition of the prosperous countries working to the disadvantage of the rest would seem somewhat remote. Each of Cameroon, Gabon, Côte d'Ivoire and Senegal might, in that wider grouping, provide a check on the others.

Notes

1. See, for example, the references by Devarajan and de Melo (1987a and b), Dittus (1987), Guillaumont & Guillaumont (1984 and 1988), Krumm (1985 and 1987), Ouattara (1986).

2. By embodying the exchange rate decision in a multicountry institutional set-up involving a joint central bank, each member country commits itself to an arrangement form which would be rather costly to exit. This pre-commitment enhances the credibility of the fixed exchange rate arrangement for each member. Pressures on any one member to adjust the exchange rate will be more easily resisted by the authorities because of the sunk costs of union membership. The credibility of the exchange rate commitment could then lead to a more stable monetary and inflation evolution. A companion paper (Honohan, 1990) provides evidence that franc zone inflation has been largely determined by French inflation.

3. While much of what is said in this paper applies to the whole CFA zone, some of the evidence provided is based only on experience in UMOA.

4. Guinea (Conkary) abandoned the French franc shortly after independence; Mauritania and Madagascar stayed with the French franc until 1973, though neither joined a union. Mali left the Franc zone after independence, only to rejoin it in 1967, and to become a member of the UMOA in 1984. Togo was not a founder member of UMOA either, but joined after a change of government in 1963. In 1985 Equatorial Guinea became both the smallest member of the CFA zone and the first member not to have been a former French colony. Monaco and the Comoros Islands are the remaining non-CFA members of the franc zone apart from France and its dependencies.

5. Official development assistance to the eight poorest members averaged over 12 percent of GNP in 1987.

6. And even for inflation the median of Sub-Saharan African countries is within the CFA range.

7. Table 15–4 shows that the variability of each UMOA country's typical export basket has not been strongly correlated across the countries in the 1980s.

8. Note, however that the reliability of official statistics which indicate a rather low level of intra-regional trade have been questioned by Yeats (1989b). In line with some studies of colonial relationships (cf. Austen (1987)), the substantial trade with France has prompted some to suggest that the existence of the zone provides extensive benefits to France. A recent study on these lines is Yeats (1989a).

9. There are some limitations which mean that capital is not fully free to flow to France. First, transfers are increas-

ingly being delated or even (in some countries) refused except where the bank can provide the funds abroad itself. This is probably the most important limitation. Second, banks are not permitted to hold more than working balances outside their own country. In some cases this has been stretched to imply that banks must always use to the full all foreign lines of credit available to them. Third, insurance companies are only allowed a very restricted range of investment possibilities outside their own country. Fourth, a transfer fee of 0.5% has been payable to the BCEAO on all transfers (for current or capital payments) made to France. This hardly discourages flows of a long-term character, and is said to be bypassed by many banks wherever possible. Fifth, export receipts must be repatriated to the exporters' country. (This serves to block one possible way of avoiding the transfer fee). Sixth, there have been requirements to declare export of CFA currency notes; these however are largely ineffective as the currency notes have been freely acceptable in Europe.

10. The main reason for this has been the uncertain solvency condition of several banks and the strained liquidity position. In recent months even some of the largest banks in Côte d'Ivoire have found themselves unable to meet their obligations on the due date, a development which has encouraged several banks to withdraw altogether from interbank lending.

11. For example, during a very tight period for the Ivorian banks in January 1988, the marche monetaire interest rates were actually lowered in line with interest rate movements in Paris, despite the persistence of excess demand for loans from the marche monetaire.

12. There is a long-standing debate over the appropriate relationship between CFA interest rates and those in Paris. As discussed in Honohan (1990) the policy-determined CFA rates have tended to track Paris rates, as would be the normal market tendency considering the fixed exchange rate and the relatively open capital market. However, because of the residual risks perceived by foreign banks, capital will not always flow in plentifully in response to a small interest differential. Higher interest rates may sometimes be necessary to allow the market to choose between competing uses for the limited funds available within the union.

13. In view of the small degree of economic interaction between the member countries, there are no substantial spillover effects of credit creation in one country on demand conditions in the others.

14. This process is documented in Honohan (1990). Other factors can also influence credit growth; the banks' borrowings from correspondents have been an important factor in the past.

15. Alternatively, if on or more country credit ceilings were being exceeded, country-specific reserve requirements could be imposed in those countries to choke off the expansion of credit. This might be the best solution if it were to prevail only for a short while. Sustained use of such devices would serve to recreate a wedge between national credit markets and impede the market allocation of credit.

16. As indicated by the negative correlation between country shares in monetary authorities claims on banks and government and shares in union-wide exports.

17. Mali is not shown; it joined the union in 1984.

18. In contrast, the union as a whole tends to repay drawings from its overdraft facility at the French Treasury.

19. Excluding the IMF-related items and the miscellaneous items "comptes d'ordre et divers."

20. There is not enough published information to make a precise calculation of each of the elements in the seignorage. Rough approximations based on 1987 balances and interest rates suggest that about one-third of the flow of UMOA seignorage benefits for that year cam through the interest subsidies. For the BEAC the proportion was more like three-fifths.

21. The rough calculations based on 1987 suggest that Côte d'Ivoire could have taken almost 30% of the UMOA seignorage flow and Senegal 16%. In each case the take was not greater than these countries' share (50% and 16%) of total currency in circulation; but it was greater than would be implied by equal distribution (14%). Benin's estimated take (12%) was the third largest, even though it had the lowest share of currency in circulation.

22. Adjusting the estimates mentioned in previous footnotes gives these three countries 46%, 26%, and 16% respectively. These calculations are very tentative as the scale of the losses is not yet fully determined.

23. The country-by-country assignment of interest charges on the BCEAO's operations account borrowing at the French treasury could also be relevant; but so long as these remain mostly as unpaid book entries between the members and the BCEAO they need not have any importance and are not therefore discussed here.

24. Some 70% of the profit distribution is divided on that basis. A further 15% is distributed in proportions to the currency issue and the remainder (15%) is distributed equally among the members. In calculating the contribution of each member country to profits, interest payments to the BEAC from a given country are regarded as part of that country's contribution to profits. Interest earnings from outside the union are allocated in proportion to each member country's external reserves. A detailed critique of the formula and its application is not attempted here, but it is clear that it does not successfully mimic the distribution of seignorage earnings which would prevail if each country operated with an independent currency issue. A rough calculation suggests that the Central African Republic and Chad (both of which have comparatively small borrowings and are among the poorest of the BEAC countries) are the most disadvantaged by the formula.

25. This provisions may actually have encouraged complacency on the part of the central bank in making loans to unsound or doubtful banks, as it divorced responsibility for losses from the decision to lend and meant that the central bank had no clear incentive to ensure that the banks to which it lent were sound. As a result, the central bank operated bank supervision as if it were a disinterested party, merely reporting its findings to the national authorities and taking no steps directly to curb bad management practices. On the other hand the provision it has certainly not been effective in discouraging governments from arranging for loans to made for political rather than sound economic purposes.

References

Austen, R. (1987), African Economic History: Internal Development and External Dependency, London, James Currie Publishers.

Bhatia, R.J. (1985), The West African Monetary Union: An Analytical Review, (Washington, D.C., International Monetary Fund).

Devarajan, S. And J. de Melo (1987a), "Evaluating Participation in African Monetary Unions: A Statistical Analysis of the CFA Zones", World Development, Vol 15, No. 4, pp. 483-96.

Devarajan, S. And J. de Melo (1987b), "Adjustment with a Fixed Exchange Rate: Cameroon, Côte d'Ivoire and Senegal", The World Bank Economic Review, Vol. 1 No. 3, pp. 447-87.

Dittus, P. (1987), "Structural adjustment in the franc zone - the case of Mali", World Bank CPD Discussion Paper 1987-11.

Guillaumont, P. And S. (1984), Zone franc et developpement africaine, (Paris, Economica).

Guillaumont, P. And S., eds. (1988), Strategies de developpement comparees: zone franc et hors zone franc, (Paris, Economica).

Guillaumont, P., S. Guillaumont and P. Plane (1988), "Participating in African Monetary Unions: An Alternative Evaluation", World Development, Vol 16, No. 5, pp. 569-76.

Honohan, P. (1983), "Measures of Exchange Rate Variability for One Hundred Countries", Applied Economics, Vol. 15, pp. 583-602.

Honohan, P. (1990), "Price and Monetary Convergence in Currency Unions: The Franc and Rand Zones", World Bank, Washington, D.C.

Krumm, K. (1985), "Adjustment policies in the Ivory Coast in the framework of UMOA", World Bank CPD Discussion paper 1985-6, Washington, D.C.

Krumm, K. L. (1987), "Adjustment in the franc zone: focus on the real exchange rate", World Bank CPD Discussion Paper 1987-7, Washington, D.C.

Ouattara, A. (1986), The Balance of Payments Adjustment Process in Developing Countries: The Experience of Ivory Coast, Report to G-24 (UNDP), UN, New York.

Yansane, K. (1984), Controle de l'activite bancaire dans les pays africains de la zone franc, Paris, Librarie Generale de droit a de Jurisprudence..

Yeats, A. (1989a), "Do African Countries Pay More For Imports? - Yes", PPR Working Paper 270, World Bank, Washington, D.C.

Yeats, A. (1989b), "On the accuracy of economic observations: Do Sub-Saharan Trade Statistics Mean Anything?", PPR Working Paper 307, World Bank, Washington, D.C.

16

Comments on Financial Sector Policy

F.M. Mwega and Diery Seck

F.M. Mwega

My discussion focuses on the paper presented by Earnest Aryeetey and Mukwanason Hyuha entitled *The Informal Financial Sector and Markets in Africa: An Empirical Study* and the paper by Machiko Nissanke entitled *Mobilizing Domestic Resources for African Development and Diversification: Structural Impediments to Financial Intermediation.*

The Aryeetey and Hyuha Paper

I found the paper by Aryeetey and Hyuha very stimulating and think it makes an important contribution to our understanding of the informal financial markets in Africa. My presentation is therefore in the nature of raising issues for discussion and further research.

The paper addresses itself to a number of issues. The first issue is the anatomy of the informal financial markets in Africa. The paper shows that the informal sector consists of a wide range of activities which vary from country to country. While the savings and credit societies are for example dominant in the informal financial markets of Tanzania and Zaire, the so-called 'susu' system is dominant in Ghana. In Malawi, estate owners, friends, relatives and employers are dominant. It is therefore very difficult to derive a general theory of how operators in the informal financial market behave, unless the sector is disaggregated by the type of economic activity. It is in this light that the models developed in the paper should be seen. One part of the paper, for instance, argues that interest rates in the informal sector are higher than in the formal financial sector and develops a model to explain this. Yet this argument is to a large extent only valid for money lenders who constitute a small component of the informal financial sector.

The second issue analyzed in the paper is the size of the informal sector. In a number of country case studies, such as those of Ghana and Malawi, some evidence is provided that shows the informal sector to be very large. Indeed, the paper argues that the informal financial sector is larger than the formal financial sector. In other country case studies however, this evidence is not convincing. In the case of Senegal and Zaire, the paper cites evidence that is interpreted to mean that 60 to 85 percent of the population earn their livelihood in the informal financial sector. This does not seem plausible. In these two cases (Senegal and Zaire), it is necessary to distinguish between the informal sector broadly defined and the informal financial sector. In general, there is need for more research on the size of this sector relative to the formal financial sector. This is because of the amount of policy attention the sector should receive is to a large extent, a function of its size. There is also need to study the internal dynamics of the sector, how it has evolved over time and what factors have influenced this.

A third issue that the paper analyzes is the linkages between the informal financial sector and the formal financial system. The paper concludes that the two sectors are, to a large extent, complementary. In the rural areas, informal institutions fill the gap created by inadequate formal financial institutions, while in urban areas the informal financial sector coexists side-by-side with the formal financial sector. Further issues however can be raised about this relationship. To what extent is it a dualistic relationship in which a modern sector coexists with a traditional sector?

Lastly, the paper tackles the policy implications of having a large informal financial sector. The basic argument by the paper is that since the sector is very large, it should be taken into account in policy formulation. Issues for further investigation then include the following. What kind of policy interventions are required? Does the formal financial sector reduce the effectiveness of monetary policy? Does the sector create money in the same way as commercial banks?

The sector may have a limited influence on the effectiveness of monetary policy because a number of the operators such as money lenders. 'susu' collectors, etc. in the informal financial sector deposit their funds with formal financial institutions. It is only when they need to make payments that they withdraw the money or write cheques. The activities of these operators can therefore be controlled for monetary policy purposes through the regulation of formal financial institutions. Indeed, the paper states that in the case of Senegal, a squeeze in the formal financial sector leads to a squeeze in the informal financial market.

The paper argues that interest rates in the formal financial sector do not influence those in the informal financial markets. If major operators in the informal financial sector deposit their funds with the formal financial institutions, and sometimes borrow from them to on-lend in the informal financial market, it is nor clear why the authors come to this conclusion.

The Nissanke Paper

I also found the paper by Machiko Nissanke very useful. It is quite true that African countries have to increase domestic savings if they are to reduce their dependence on net external capital inflows which in any case have been going down over time. The paper discusses the potential of mobilizing household, enterprise, and public savings and the shortcomings of the formal financial system in allocating the mobilized savings efficiently to investment. The paper argues persuasively on the need to take into account the structure and the institutions of a country.

According to the paper, household savings comprise a large proportion of total savings in Africa and a major problem in mobilizing household savings, especially in rural areas, is that a large component of this is in-kind, making it difficult to mobilize them through the financial system. At the same time, a large proportion of the financialized savings are mobilized by the informal financial system, which as stated above, may be larger than the formal system in terms of the mobilized deposits and the credit given.

Similarly, the paper cites evidence that the formal financial system does not seem to reach the small and medium sized firms so that they have mainly to utilize their own savings. These savings are however, not adequate. Formal sector credit tends to go to parastatals and to large firms while institutions that have been set up to provide credit to the small and medium sized firms do not seem to be effective for various reasons. The paper tells a similar story about the difficulties of mobilizing public savings in African countries using Ghana and Malawi as case studies.

The paper raises a number of issues. First, how do the various sources of savings interact with one another? There is for example, a large literature that shows that an increase in public savings that is achieved through a reduction in budget deficits may not increase total savings much if this depresses the economy and hence reduces private savings through a Keynesian transmission mechanism.

Second, what is the degree of market segmentation of the formal financial system? The paper argues that there is a high degree of segmentation of the formal financial system in which commercial banks keep excess reserves while the other institutions face a shortage of funds to lend. But to what extent are commercial banks' excess reserves really excessive? In discussions with bankers while doing a study of monetary policy in Kenya, they said the reserves they keep are not excessive and are approximately what they would keep if there were no minimum required reserves ratio.

Lastly, in the kind of situation described in the paper, what would be the effectiveness and the appropriate sequencing in the introduction of financial reforms?

Diery Seck

The Aryeetey and Hyuha Paper

The paper by Aryeetey and Hyuha can be seen as part of a growing literature on a topic that until recently had attracted little attention on the part of students of development economics—namely, the informal financial sector in Sub-Saharan Africa. In order to tackle this challenging issue, the authors set the following three goals in their paper: a) give a review of the empirical evidence on the size, nature and workings of the informal financial sector in five different Sub-Saharan African countries, b) discuss briefly the literature aimed at explaining the growth of informal financial markets in Third World countries in general and in Africa in particular, and c) propose a rationale that establishes a relationship between the formal and informal financial sectors, based on a micro-economic approach.

Research on this topic is timely and important for a number of reasons, some of which being:

- lack of success of most economic policies that have been implemented by African countries over the last thirty years and that had the common characteristic of ignoring the informal sector or taking its compliance with stated policies for granted;
- failure of the banking system in most African countries to perform its asset transformation function in a manner that is satisfactory even by the standards of policymakers and conducive to capital formation and growth;
- growing budget deficits and the realization that more than half the economic activity, being informal, may be beyond the reach of economic policies, at least of the tax collector.

Without a doubt, understanding the informal sector, financial or otherwise, is quite an intellectual challenge and may have become a prerequisite to the formulation of viable economic policies for Sub-Saharan Africa. Why? Because, as has been documented by Aryeetey and Hyuha, it is the economic sphere where more than half the people seek alternatives to what is being offered by the formal sector of the economy and often shield themselves from the effects of government policies and regulations. Anything that we learn about informal markets in Africa is likely to teach us something about the nature and extent of the imperfections that undermine the normal functioning of African markets and consequently point the direction for needed policy changes. In that sense the authors' effort should be commended.

The paper should be seen as a continuing effort by Mr. Aryeetey and others—see the paper's bibliography—at gathering data on and analyzing informal financial markets in Africa with the admitted desire to rationalize the body of knowledge on the topic. This partly explains the structure of the paper where the three major sections are loosely related and could have been developed into full papers in their own right if the authors had chosen to do so.

The findings of the paper can be summarized as follows. Empirical research in several African countries has shown than more than half the financial transactions are done in the informal financial sector; the informal financial sector is in fact a collection of different financial schemes with different types of participants and markedly different interest rates prevail at the same time in the same geographical markets although quoted by different types of market participants. It is also found that the informal financial market's success is at least partly due to the failure of the formal market to meet the needs of segments of the population and that interest rates are a weak channel if any for a possible link between the two markets. Whether the two markets are complementary or in competition with one another is still an open question.

The Size and Nature of Informal Financial Markets in Sub-Saharan Africa

Given the structure of the paper, the most appropriate way to discuss it is to examine its sections one at a time. The second section analyzes the informal financial sector in 5 countries. The definition of the informal financial sector is a catch all statement that is arbitrary and does not allow for an accurate measurement of the size of the market. Clearly, there is need for a definition of the informal financial sector that is better suited for rigorous theoretical analysis as well as tractable empirical work.

For instance, despite the marked differences in the activities of the sector in the uses of loaned funds, the types of contractual arrangements and above all in the prevailing interest rates, the authors make blanket statements such as ".. the informal and formal sectors offer a similar product, which is probably not entirely homogenous" (in the fourth section of the paper), or ".. through a proper assessment of the actual cost and demand structures of the two sectors, the informal sector could be made completely (my emphasis) subject to various monetary and fiscal policies of government" (in the section on Ghana).

The authors' definition also fails to give insights into the classification of the various segments of the informal financial sector. These segments can be categorized according to different criteria namely, by type of contractual arrangement, by type of underlying motives of the participants, by type of timing of cash flows, etc. Proper analysis of the informal financial sector must, as a prerequisite, acknowledge the diversity of the activities that are performed under the umbrella of the sector and perhaps be limited to the confines of each major segment of the market to be really useful.

While the literature on the informal financial sector is still in its infancy, there is a crying need for better techniques of measurement of the volume of transactions in the informal sector. Use of ratios such as Deposit money/M2 or Banking sector claims/GDP only result in doubtful comparisons. Appropriate measures should allow the direct assessment of the magnitude of each type of activity in the informal financial sector and the technique that consists in estimating the total amount saved by market women through the "susu" scheme by way of interviews is a step in the right direction.[1]

One issue that begs examination is why lending rates differ so much from one segment of the informal sector to another. Does the interest rate differential imply market segmentation within the informal sector and if so, how to account for the persistence of such segmentation? The authors mention that moneylenders in Ghana and co-operatives societies in Tanzania borrow from the bank to lend to the informal sector. Research into how widespread the phenomenon is and its potential for expanding the reach of the formal sector to hitherto unexploited segments of the population might be fruitful.

The Informal Sector: Growth and Links with the Formal Sector

In the third section of their paper Aryeetey and Hyuha address the issue of duality in African financial markets and that of complementarity or substitution between the informal and formal sectors. On the duality issue, they reject the view that socio-cultural attachments of participants to the informal sector help explain the durability of the informal market. Their reason is that "modern banking facilities cannot be found throughout the economy in most African countries".

I fail to see the logic of their argument. Moreover, Aryeetey and Gockel (1989) give empirical evidence that shows that for market women in Ghana, the proximity of a banking facility does not constitute an incentive to use the services of the formal banking sector; in fact proximity is found to be a disincentive! While the cultural attachment factor cannot be proved right or wrong based on available data, the counter-argument opposed by Aryeetey and Hyuha is certainly not valid.

To account for the duality of the financial sector, the authors use the historical evidence on the inability of the formal financial sector to satisfy the needs of portions of the population. The success of the informal financial sector, they explain, is based on the need for people who were left out of the formal sector to find alternative ways to satisfy their need for financial services. As a result, the authors implicitly side with proponents of substitution in the complementarity vs substitution controversy. This approach is biased and only invites counter-examples. Furthermore, it ignores the fact that some segments of the "modern" informal sector pre-date the era of the modern banking facilities, even if one readily admits that the informal sector may take advantage of the misfortunes of the banking sector. Some forms of activity in the informal sector would persist even if the formal sector performed ideally. In short, it takes more than historical accounts to settle the dispute that is raging between the complementarity camp and the substitution camp.

In their discussion of the possible relation between the formal and informal sectors, the authors exclude a link via interest rates, based on regression results obtained by Aryeetey and Gockel (1989). Their position is hasty and inappropriate because the empirical evidence provided by the regression estimates can be challenged on several grounds. First, although the authors acknowledge the presence of collinearity among the explanatory variables, its possible effect on the sign and significance of the regression coefficients is unaccounted for. Second, a single cross section regression in a single country is clearly not enough to reject the link via interest rates of the formal and informal sectors. Consequently, the issue is still unresolved and calls for further empirical evidence.

The third section of the paper has left unanswered several key questions in the understanding of the duality of African financial markets, and more importantly, the internal mechanisms of the informal sector. For example, what is the explanation for the lack of unified interest rates, given that customers have access to various segments of the informal market? What are the competitive pressures within the informal market and why do they fail to establish parity of the interest rates quoted by the different segments of the informal sector? Are some segments of the informal financial sector more closely related to the formal sector and what are the prospects for some of the segments of the informal sector being unified with the formal sector with respect to interest rates, especially in urban areas?

Relative Pricing and Cost Structures of the Two Sectors

In the fourth section of the paper, Aryeetey and Hyuha use a micro-economic approach to explain the higher pricing in the informal financial sector. The rationale that they propose is interesting and promising. However further clarification is needed with respect to the market structures of the informal and formal sectors. Their claim that "one could talk of an oligopolistic relationship with some minor product differentiation should be backed by a stronger theoretical demonstration. As for their admission that there are monopolistic tendencies and near-perfectly competitive practices in the informal financial sector, a statement that is not taken into account in the rest of the paper, it clearly indicates the urgency for more basic research directed at discovering the informal financial market.

The authors mention that the operating costs of the individual actors in the informal sector are lower than those of the formal sector. I do not see why simple aggregation of the participants in both sectors would result in closer operating costs for the two sectors given that actors in both markets operate on an individual basis. Moreover, the authors state that the informal sector does not gain from economies of scale which explains the convergence of the marginal costs of the two sectors as the activity expands. Such a view can be challenged because no in-depth analysis of the cost functions of the various segments of the informal sector has been undertaken yet.

The authors' claim that there is a tendency for the marginal cost curves of the two sectors to converge needs stronger substantiation. As for the authors' pessimistic note with respect to the likelihood of the two markets coming closer together because of barriers of entry into the formal sector, one can only hope that the recent developments in the financial markets and monetary policies of many African countries will irreversibly turn barriers of entry into a thing of the past.

Aryeetey and Hyuha use the concepts of specific and non-specific loans to analyze the issue of competitive asset pricing in the two financial sectors. While these concepts are helpful in understanding the pricing mechanism suggested in the paper, the authors have not used them to develop a full-fledged theoretical asset pricing model. An equally valid approach is to view the formal financial sector as segmented and the informal financial sector as integrated. Equilibrium asset pricing models for segmented and integrated financial markets are proposed by Eun and Janakiramanan (1986), Subrahmanyam (1975a,b) and Stapleton and Subrahmayam (1977). The authors may wish to make use of this literature in their future research on asset pricing in Sub-Saharan Africa's informal and formal financial sectors.

The Honohan Paper

Mr. Honohan claims that his paper is designed as a guide to understanding how the CFA Zone works as a mechanism for monetary cooperation. Taking a cost-benefit approach, he lists the potential gains that can accrue to small countries when they are part of a monetary union, namely:

- credible commitment to exchange rate and price stability,
- pooling of reserves
- larger capital market with free movement of funds within the zone and
- a more effective and independent monetary policy by a unified central bank.

The paper's intent is to determine if the last two potential gains are verified in the CFA Zone.

The conclusion of the paper can be summarized as an indictment of the way in which the BCEAO manages the money market in the Zone. In the author's opinion, despite an effective policy of fixed exchange rate and exchange control and an elaborate set of rules to avoid excessive credit expansion, the CFA Zone almost collapsed because of widespread bank insolvency. According to his rationale, the poorer countries of the union will pay more than their share of the losses sustained as a result of the failure of the Zone's banking system. He gives two reasons for that:

- According to the statutes of the UMOA, losses of the banking sector should be covered by the respective home governments of the banks. Yet, the BCEAO has decided to cover all the losses. Consequently, all the member countries of the monetary union will equally share the burden of those losses;
- The countries where the banking sector suffered most losses happen to be the three richer ones, i.e. Côte d'Ivoire, Senegal and Benin and coverage of losses of the banks by the central bank amounts to a skewed distribution of seignorage profiting those richer countries alone.

My discussion of the paper is organized as follows. First, a few general comments will underline my appreciation of Mr. Honohan's study. Second, the points about which the views expressed in the paper diverge from mine are discussed. Third, I will attempt a brief extension of the issues under discussion in the paper. A few suggestions for further research will serve as the concluding note to this discussion.

General Comments

The second section of the paper constitutes an excellent and well documented presentation of the CFA Zone and of its member countries. The similarities and differences of those countries among them on the one hand and with non-member countries of Sub-Saharan African on the other hand are expertly illustrated. While it is not a part of the main discussion of the article, Mr. Honohan's underscoring of the potential for export revenue diversification—because of low correlations between countries' export revenue indexes—through risk pooling is an interesting point. It might be an avenue for fruitful future research. The author gives ample statistical evidence to support his views.

In a way, Mr. Honohan describes how, in the case of the UMOA, the attempt by a central bank to play a major role in financial intermediation has failed miserably. The main reasons for the failure are:

- credit ceilings to governments and to the private sector;
- inefficient domestic inter-bank money market
- little or no inter-country capital flow within the Zone;
- passive supervision of commercial banks;
- failure by the central bank to use market clearing interest rates in the money market and in re-discounting.

In a word, the central bank has failed to achieve allocative efficiency—best loan prospects get funding at lowest rate—and operational efficiency—narrow spreads in inter-bank money market. As a corollary to the author's findings, the monetary union could possibly break down, not because it will no longer be able to manage the monetary affairs of the member countries, but because the poorer countries may justifiably feel that while the gains of the monetary union are equally shared, they pay a disproportionate share of the cost of the functioning of the monetary union.

A Few Notes of Dissent

The author claims that, with the exception of Togo and Burkina Faso, the commercial banks of the UMOA are heavily indebted to the central bank. I have used the same data and found that, given the role played by the central bank in capital allocation within the Zone and its consistent policy of negative real rates of interest, it is not the commercial banks but the central governments that are heavily indebted. Negative real deposit rates force commercial banks to operate with low volumes of deposits and consequently to borrow substantially from the central bank. Central governments of the UMOA borrow from the central bank to finance their growing budget deficits.

Furthermore, recent trends show that the central governments have engaged in a policy of crowding out that has been debilitating for the private sector. As is shown in the paper, the share of the central bank's claims on the economy going to the central government has increased dramatically between 1979 and 1988 for all the countries except Benin and Mali. The data series are not complete for those two countries. In addition to this direct form of crowding out, one must keep in mind that a considerable amount of the central bank's credit extended to commercial banks is loaned by the commercial banks to state firms. Overall, central bank's funding of the private sector through commercial banks is dwarfed by its direct and indirect funding of the public sector.

The author's focus on the shortcomings of the money market—he calls it by its French name: marché monétaire—has led him to advocate a major role for the central bank in that money market. Such a view can be challenged on several grounds. While the inter-bank money market is needed to recycle excess bank liquidity, what is more important is the emergence of an organized capital market. Failure to mention it may lead to the false assumption that an adequate capital allocation mechanism can be insured by the commercial banks or the central bank.

A separate and well-functioning organized capital market that is autonomous from monetary authorities is needed because:

- there is a situation of chronic crowding out;
- the banking sector is in crisis and may fail to accomplish its asset transformation function properly;
- there is a need for a long term substitute for external funding that is part of the countries' respective structural adjustment programs, and;
- there is a need to restore savings and investment to their pre-crisis levels.

There are other reasons for not entrusting the central bank with the organization of the money market. The first one is that, given the bureaucratic tradition of the BCEAO, its ability to run side by side a "monetary policy arm" and an "intermediation arm" that are independent from one another, as is suggested by Mr. Honohan, can be questioned. The second reason lies in the contradictory situation whereby the central bank would run a money market within the Zone,

Table 16-1. *The respective shares of the central government and commercial banks in the claims of the central bank*

	Benin		*Burkina Faso*	
	Government	*Banks*	*Government*	*Banks*
1979	n.a	n.a.	8.9	91.1
1980	n.a.	n.a.	11.7	88.3
1981	n.a.	n.a.	35.3	64.7
1982	n.a.	n.a.	41.4	58.6
1983	n.a.	n.a.	31.5	68.5
1984	n.a.	n.a.	48.6	51.4
1985	n.a.	n.a.	48.8	51.2
1986	21.8	78.2	56.1	43.9
1987	21.5	78.5	60.8	39.2
1988	16.5	83.5	74.8	25.2
	Côte d'Ivoire		*Mali*	
	Government	*Banks*	*Government*	*Bank*
1979	7.1	92.9	n.a.	n.a.
1980	24.9	75.1	n.a.	n.a.
1981	30.8	69.2	n.a.	n.a.
1982	28.1	71.9	n.a.	n.a.
1983	37.4	62.6	n.a.	n.a.
1984	42.0	58.0	87.8	12.2
1985	43.5	56.5	82.4	17.6
1986	42.0	58.0	70.7	29.3
1987	34.0	66.0	73.6	26.4
1988	35.8	64.2	74.8	25.2
	Niger		*Senegal*	
	Government	*Banks*	*Government*	*Banks*
1979	-85.7	185.7	16.0	84.0
1980	1.5	98.5	18.1	81.9
1981	-24.3	124.3	26.7	73.3
1982	23.0	77.0	36.2	63.8
1983	19.5	80.5	40.1	59.9
1984	40.8	59.2	45.5	54.5
1985	52.4	47.6	45.8	54.2
1986	42.4	57.6	49.	50.3
1987	25.1	74.9	50.7	49.3
1988	35.8	64.2	45.6	54.4
	Togo			
	Government	*Banks*		
1979	55.8	44.2		
1980	47.9	52.1		
1981	74.5	25.0		
1982	75.1	24.9		
1983	79.4	20.5		
1984	77.7	22.3		
1985	88.3	11.7		
1986	87.9	12.1		
1987	89.1	10.9		
1988	91.5	8.5		

Source: BCEAO

using market-determined rates and still enforce credit rationing schemes such as assigning credit ceilings to each bank in the banking system. While a money market operating under market clearing conditions fosters competition, credit ceilings undermine it because they tend to maintain banks' market shares fixed, irrespective of their relative performances.

Mr. Honohan dispels the possibility for private intermediaries to organize the money market because of the fragile state of the banking system and of the unacceptable prevailing levels of risk surrounding the banking sector. It can be argued that the perceived risk level is a direct consequence of the manner in which the central bank has been operating the money market, namely a deliberate policy of geographical segmentation across member countries, lack of active supervision of commercial banks, and failure to use market determined lending terms in the money market.

Extension to Mr. Honohan's Study

Mr. Honohan has shown that the CFA Zone has failed to provide member countries with an efficient and well organized money market with free movement of funds across countries of the Zone. He has also demonstrated that, while they use the credit facilities of the central bank the least, the poorer countries of the union pay a disproportionate share of the losses caused by the massive failure of the banking system. A natural extension of the author's analysis is to examine the gains that the private sector of the monetary union derived from monetary cooperation within the Zone.

In that respect, two issues pertaining to the effect of monetary policy of the private sector will be briefly discussed: the extent and impact of crowding out and of negative real deposit rates of interest on private firms. How well does the private sector fare in its cooperation with the public sector with respect to aggregate credit allocation by the central bank? As is shown in table16–1, except for Benin and Mali, which have incomplete data series, the historical evidence of the period from 1979 to 1988 points to a trend of increasing crowding out for all the countries of the UMOA.

In absolute terms, for Côte d'Ivoire the net credit granted by the central bank to the central government has been multiplied by a factor of 18.5 while that of the commercial banks is 2.5.[2] The corresponding numbers are for Senegal: 10.8 and 2.5, for Togo: 3.3 and 0.4, for Burkina Faso: 11.7 and 0.4 and the government of Niger changed from being a net lender to the central bank in 1979—6.3 billion francs—to being a net borrower in 1988—16.2 billion—while the claims on the commercial banks grew by a factor of 2.1.

Clearly, during the difficult years of 1982 to 1988, for most countries of the UMOA, the central bank chose to fund governments' budget deficits rather than the private sector with adverse consequences on the liquidity of the banking sector and on the level of real economic activity. The national private sectors that were penalized the most are those of Togo, Burkina Faso and Niger, those very countries that, Mr. Honohan shows, have paid a disproportionate share of the losses incurred by the union's banking system. To what extent the central bank has played a role in the current crisis of the banking sector is a research topic worth pursuing.

Negative real deposit rates of interest have complex implications because they determine the extent to which commercial banks seek funding from the central bank, which limits their operational autonomy. In other words, negative deposit interest rates give the central bank non statutory powers over the banks by way of conditionality on re-discounting operations e.g. credit ceilings, mandatory loan portfolio composition for banks etc.. They also allow the central bank to have a direct control on the level of growth of the productive sector since it can vary over time the minimum quality of the paper that it can accept for re-discounting.

Table 16–2. *Average real deposit rates of interest and differential between deposit rates of B CEAO and France*

Period	*Diff*[1]	*Burkina Faso*	*Côte d' Ivoire*	*Niger*	*Senegal*	*Togo*
1970-88	-0.32	-1.19	-3.45	-1.71	-3.02	-1.96
1985-88	0.35	4.48	1.83	9.01	2.63	5.38
1980-84	-0.87	-2.01	-1.21	-3.15	-4.09	-2.81
1975-79	-0.59	-6.74	-10.12	-8.68	-5.46	-6.02
1970-74	-0.04	0.66	-3.24	-1.86	-4.04	-2.92

[1] Difference between the nominal deposit rate of interest of the BCEAO and France.
Source: International Financial Statistics, IMF

Has the central bank sought to implement policies of low interest rates? Table 16–2 provides part of the answer. The column labelled DIFF displays the difference between the nominal deposit rate of interest of the BCEAO and that of France. The historical differences are not in UMOA's favor except for the last four years. The real deposit rates follow the same historical pattern in that they have been negative on average through 1984. They become positive starting 1985 because member countries experienced low or negative rates of inflation.

Capital flight or use of the informal financial sector are likely responses to consistently low deposit interest rates. For the productive sector, what seemed to be a good deal as far as borrowing rates are concerned turned out to be a bad deal with respect to the aggregate amount of credit supply. Thwarted growth and a higher incidence of bankruptcies may have been the inevitable consequence for firms that were in need of new funding. It is noteworthy that the countries that have had the lowest real deposit rates of interest are the richer and economically bigger countries of the monetary union, which gives an indication of the extent to which the central bank had to step in and fund the private sector.

A Few Concluding Suggestions

Mr. Honohan's paper has been very illuminating in terms of how the benefits and costs of being part of a monetary union can be inequitably distributed among the participants of the union. Several questions still remain unanswered and could be the focus of future research. In that regard three issues worth investigating are as follows.

What is the impact of crowding out on the private sector of a monetary union such as the UMOA and are there ways to mitigate it if the crowding out is to persist?

What is the potential contribution of using market determined interest rates in capital formation and growth whether the money market is organized by the central bank or by private intermediaries?

The UMOA has recently created a supranational supervising committee that will monitor the integrity and quality of the management of banking institutions in the zone. What will be its level of effectiveness compared to an insurance deposit scheme whereby insured banks pay premia that are commensurate with the quality of their respective portfolios as opposed to their levels of deposits or loans?

Notes

1. See Aryeetey, E. and Gockel, F., Mobilizing Domestic Resources for Capital Formation: The Role of Informal Financial Markets in Ghana, Paper presented at the Workshop of the African Economic Research Consortium (AERC), Harare, 4-8 December 1989.

2. Credit granted to other financial institutions is not included but is negligible.

References

Aryeetey, E. and Gockel, F., 1989, *Mobilizing Domestic Resources for Capital Formation: The Role of Informal Financial Markets in Ghana,* Paper presented at the Workshop of the African Economic Research Consortium (AERC), Harare, 4-8 December .

Eun, C. and Janakiraman, S., 1986, "A Model of International Asset Pricing With a Constraint on the Foreign Equity Ownership", *Journal of Finance, 41, 897-914.*

Stapleton, R. C. and Subrahmanyam, M., 1977, "Market Imperfections, Capital Market Equilibrium and Corporation Finance", *Journal of Finance, 32, 307-319.*

Subrahmanyam, M. 1975a, "On the Optimality of International Capital Market Integration", *Journal of Financial Economics, 2, 3-28.*

Subrahmanyam, M. 1975b, "International Capital Market Equilibrium and Investor Welfare With Unequal Interest Rates", in E. J. Elton and M. J. Gruber, eds.: *International Capital Markets (North Holland, Amsterdam).*

17

Commodity Exports and Real Income in Africa: a Preliminary Analysis

Arvind Panagariya and Maurice Schiff

This paper is part of a project whose principal objective is to address the frequent concern that a simultaneous expansion of exports by several developing countries is likely to lead to a decline in their terms of trade, export revenues and real income. This concern dominated the writings of development economists during the 1950s and has kept resurfacing in one context or another ever since. The essential argument as expounded in the early writings—e.g., Nurkse—was that exports of developing countries consisted mainly of primary products for which world demand was inelastic. Therefore, any productivity gains in exportables were likely to be passed on to importing countries via a change in the terms of trade favorable to the latter. There was little to be gained by relying on exports as the engine of growth.

A number of economists, including Krueger (1961), Cairncross (1962), Keesing (1967), Balassa (1978) and Bhagwati (1978, 1988), refuted the wisdom of this export pessimism. In particular, the recent paper by Bhagwati (1988) provides a comprehensive review of the controversy and makes a convincing case that the fears of elasticity pessimists, old as well as new, are ill founded. Among other things, he points to the phenomenal growth of exports *and* incomes of many East Asian countries to counter export pessimists. He also argues that the dramatic shift in the export composition of developing countries toward manufactures and the potential for intraindustry specialization provide further reasons why the world demand is unlikely to be a binding constraint in the future.

The general case for export pessimism has lost much of its force, at least for now. Concerns have remained very much alive, however, with respect to some specific countries and commodities. Thus, fears continue to be expressed that a simultaneous expansion of exports of certain commodities (e.g., cocoa, coffee and tea) by several African countries, most recently resulting from the adoption of structural adjustment programs, may lead to a decline in the real incomes and exports revenues of the countries. These fears have been sufficiently serious that since 1968 the World Bank has had special lending guidelines for commodities such as coffee, cocoa and tea. Bank policy is to deny lending for output expansion of these commodities, unless the country has no viable economic alternative or if the country has suffered a recent loss in production due to climatic or other reasons.

Concerns regarding possible harmful effects of commodity export expansion have reemerged recently in the context of structural adjustment both within and outside the Bank. For instance, such concerns were raised recently in the German parliament which, in turn, prompted the following query to the Chief Economist of the World Bank for Africa:

"In the framework of structural adjustment programs, many African countries endeavor to increase their exports of agricultural products. An increased supply of goods may soon lead to price declines of the correspondent products, so that additional revenues may not be realized. How does the World Bank justify its correspondent policy advice? What can be done to avoid the negative results?"

A similar sentiment has been expressed by Professor H.W. Singer in an exchange with Mr. C. Humphreys in the December 1989 issue of *Finance and Development*. Professor Singer asserts that the "adverse changes in terms of trade, far from being 'exogenous', are related to the export expansion. . . recommended simultaneously to indebted countries". In response, Mr. Humphreys writes that the argument of the negative effect on the terms of trade of export expansion is weak, and suggests Africa should raise productivity in its traditional exports, diversify its export base, and expand production of imported commodities. Perhaps there is some truth in both views but without systematic analysis we cannot reach a consensus.

The recent World Bank Board paper "Strengthening Trade Policy Reform" also emphasizes the need for studying the effects of increased commodity exports on the terms of trade and real incomes in Sub-Saharan Africa. Similarly, the study on Africa by Landell-Mills, Agarwala and Please, "Sub-Saharan Africa: From Crisis to Sustainable Growth" notes that some of the poorest African countries have been hit hardest by adverse terms of trade changes over the past 30 years.

These examples show that the fears raised by elasticity pessimists are very much alive in the context of African commodity exports. The problem is viewed as being especially serious for countries whose exports are concentrated in commodities that are said to exhibit low import demand elasticities (e.g., Cote d'Ivoire which exports coffee and cocoa). Rhetoric has been strong on the part of proponents as well as opponents of pessimism. But evidence provided to date on either side is sketchy. We believe that there is a real need to study the issue in depth and understand whether the export pessimists are justified and if so what can be done to maximize these countries' gains from exports. The present paper is a first effort in this direction.

A central premise behind the project is that domestic-policy instruments should be employed to promote efficiency at home while trade policy instruments should be used to deal with problems related to foreign demand.[2] Thus, the policy measures designed to correct domestic price distortions and real wage and exchange rate misalignment are well advised. The issue which deserves closer scrutiny, however, is whether the current trade taxes adequately handle the problems which arise from low demand elasticities in the world market.

There are two analytically distinct sources of terms of trade deterioration: exogenous deterioration due to low income elasticity of demand and endogenous deterioration resulting from increased productivity. In the latter case, increased productivity may be accompanied by a decline in income and export revenues if the price elasticity of demand is low. Our concern here is solely with the problems which arise from a low price elasticity of demand in the world market and their implications for trade policy. Problems which arise from low income elasticity do not require policy intervention and will not concern us here.[2]

Given a low price elasticity, problems can arise at both the national and international level. Thus, export expansion by one country affects not only its own income but that of the other countries as well. In particular, if export expansion is the result of increased productivity, the country expanding exports is likely to gain while the other exporters will lose. The project will study the effects at both levels.

In this vein, we would like to seek answers to the following questions in this project:

- What is the likelihood that export expansion resulting from better and fuller use of resources can lead to a decline in real incomes and export revenues in African and non-African countries? For which commodities is this outcome plausible? What are the key parameters determining the impact of export expansion on the terms of trade, export earnings and real incomes?
- Will further reductions in export taxes lead to an increase or decrease in real incomes? What will be the impact of tax reductions on tax revenues and output quantities?
- Empirically, how important is the issue of interdependence? For instance, while considering a further tax reduction on cocoa, should Ghana pay attention to policy changes in Côte d'Ivoire and Malaysia? In which commodities, if any, does interdependence play an important role?
- Can countries realize most of the gains from trade by choosing taxes optimally in an independent fashion?

In terms of development of theory, we should note that the issue of income- and revenue-maximizing exports taxes when two or more countries compete against each other in the world market for the same commodity has simply not been studied. The traditional literature deals with the situation in which two countries choose taxes on each other's exports so as to maximize income (welfare) or revenue.[3]

Relevant Commodities and Countries

Commodities that concern us must satisfy two important criteria. First, they must account for a significant share of exports of one or more African countries. Second, African countries must have a large share in the world market for those commodities. The six most important commodities based on these criteria are cocoa, coffee, tea, cotton, tobacco and groundnuts. Of these, the first three—cocoa, coffee and tea—are exported exclusively by developing countries while the last three are exported by both developing and developed countries. Developing countries do not import cocoa at all and account for only 7.9 percent of net imports of coffee. The remaining commodities are imported by developing countries in large volume both in absolute and relative terms.

Africa's share in the world market for cocoa is larger than that for any of the other commodities. Côte d'Ivoire and Ghana each has a large share in the world market and depends heavily on this commodity for export receipts. The two countries also raise substantial tax revenues from this commodity. Coffee is the next most important commodity for Africa. Once again, Côte d'Ivoire is the largest African exporter. In terms of export share for Africa, tea was not far behind coffee in 1986, with Kenya being the largest African exporter.

Our focus in this paper is on cocoa. Cocoa is not only the most important commodity export from Africa, but it also provides a clean case in that it is exported exclusively by developing countries and imported exclusively by developed countries. Tea and coffee, on the other hand, are also imported by some developing countries and raise complex distributional issues within developing countries. These commodities will be examined at a later stage.

Cocoa is exported in large amounts by non-African countries as well. We will incorporate explicitly these non-African countries into the analysis. This will allow us to examine the impact on African countries of changes occurring outside Africa, and vice versa.

The remainder of the paper is organized as follows. In Section II, we outline the model and its application to cocoa. In Section III, we present results of a number of simulations. Finally, in Section IV, we conclude the paper.

The Model

Our model consists of a multi-country demand-supply system for the commodity examined. The effects of adjustment programs are captured by parameters of the supply function. For instance, improvements in productive efficiency via fuller and better use of resources are represented by shifts in the intercept and/or slope parameters.[4]

Within a partial equilibrium framework, the principal theoretical issue is the determination of income —and revenue— maximizing export taxes when two or more countries compete against each other in the world market. Although there is a large body of literature on optimal trade taxes, the issue of how such taxes are determined when two or more countries compete against one another has simply not been addressed.

In this set-up strategic considerations inevitably come into play. Fortunately, there have been important developments in the area of strategic trade policy in recent years which allow us to address the problem in a reasonable way.[5]

The analytic solution for income — and revenue — maximizing export taxes for several countries who compete in the world market is provided in Panagariya and Schiff (1990). Below, we present a brief diagrammatic exposition of the basic model and its application to cocoa. In the following section, we present the simulation results with actual and income-maximizing taxes. Simulation exercises with revenue—maximizing taxes will be taken up at a later stage.

The basic structure of the problem can be explained conveniently with the help of a three-country setup. Denote the three countries by A, B and C and assume that the former two export cocoa to the latter. Assume that exporters do not consume and the importer does not produce cocoa. Individual producers and consumers are perfectly competitive. Each exporting country's government chooses the export tax so as to maximize the country's profits taking the other exporting country's tax rate as given. This behavioral assumption leads to what is called the Nash equilibrium in the game theory literature.[6]

Consider country A's problem. For a given export tax by B, say t_B, A's government must choose t_A so as to maximize profits from exports. We can obtain the excess demand facing A by subtracting B's supply from the world (i.e., country C's) demand. This excess demand curve is represented by $D_A(t_B)$ in figure 17–1. Coresponding to D_A, we can draw a marginal revenue curve, MR_A.

We assume that the supply curve, S_A, reflects the true marginal social costs of producing cocoa in A. Then the country's income-maximizing equilibrium will coincide with the profit-maximizing equilibrium. This equilibrium is given by the intersection of S_A with the MR_A curve. The corresponding world and domestic prices are given by P^W and P_A and the difference between them, $P^W - P_A = t_AP^W$, equals the *per-unit* export tax. Tax revenue and producers' surplus, respectively, are given by P_WNMP_A and P_AME. In this paper, we will refer to the sum of these areas as the country's profits from cocoa exports.

An increase in t_B shifts B's supply curve (not shown in figure 17–1) to the left and hence A's excess demand curve to the right. This causes the optimal t_A to change. It is easily shown that as t_B rises, the optimal t_A also rises. Moreover, A's profits associated with the higher (t_A, t_B) combination are higher as well.

In figure 17–2, curve R_A shows country A's optimal tax rate for different values of t_B. In Panagariya and Schiff (1990), we demonstrate that if demand and supply curves are linear, the shape of R_A must be as shown in figure 17–2. Analogously to R_A, we can derive R_B which shows the optimal values of t_B for different values of t_A. We will refer to R_A and R_B as reaction curves and to the point of their intersection, N, as the Nash equilibrium.

Suppose that B's tax happens to be t^1_B. The corresponding optimal tax for A, as shown by its reaction curve R_A, will be t^1_A. We refer to t^1_A as A's "myopic" optimal tax to emphasize that t^1_A is optimal only if A expects B to continue to hold its tax at t^1_B. As R_B shows, when A's tax is at t^1_A, B's income - maximizing tax is lower than t^1_B so that the latter is unlikely to hold its tax rate at t^1_B.

For any given value of t_A, we can also define a myopic optimal tax rate for country B. Thus, if $t_A = t^2_A$, B's myopic optimal tax rate is t^2_B. As in the previous case, if B fixes its tax at t^2_B, A will want to choose a lower tax than t^2_A. It is evident that only at N is A's tax rate optimal given B's tax rate and vice versa. In this sense, N is the Nash equilibrium.

A final point which deserves noting is that it is entirely possible that the profits of both A and B can be higher at arbitrarily chosen tax rates (t^2_A, t^1B) than at Nash equilibrium. The problem with these arbitrary rates is, of course, that at least one country can increase its income by changing the tax rate. Therefore, the arbitrary rates are not sustainable under the Nash assumption. The lower profits at N than at some higher rates is simply the result of "tax-competition" between the two countries.

In our simulations, we extend the above model to allow for nine exporters. These exporters are Cameroon, Côte d'Ivoire, Ghana and Nigeria in Africa, Brazil and Ecuador in Latin America, Indonesia and Malaysia in Asia, and Oceania. Markets are assumed to be competitive in each country, and domestic con-

Figure 17-1. Equilibrium Price Determination

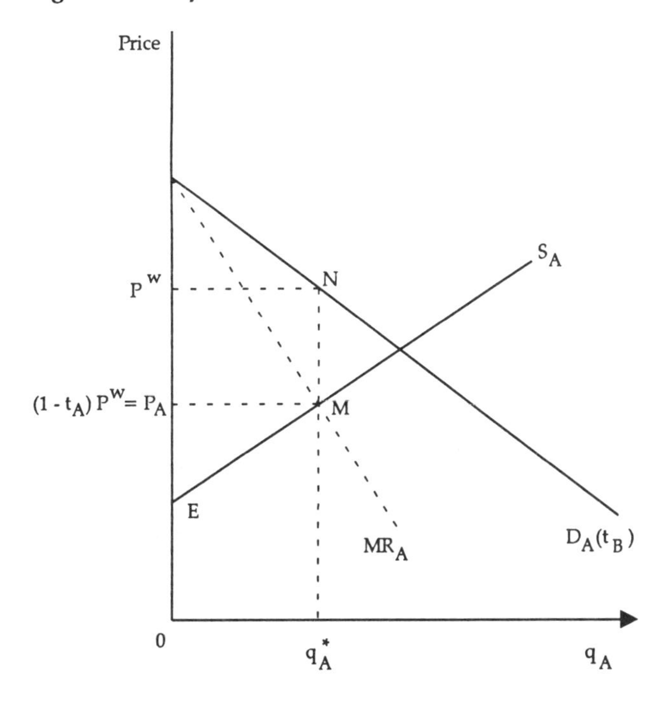

Figure 17-2. Optimal Tax for Country A Given Taxes in Country B

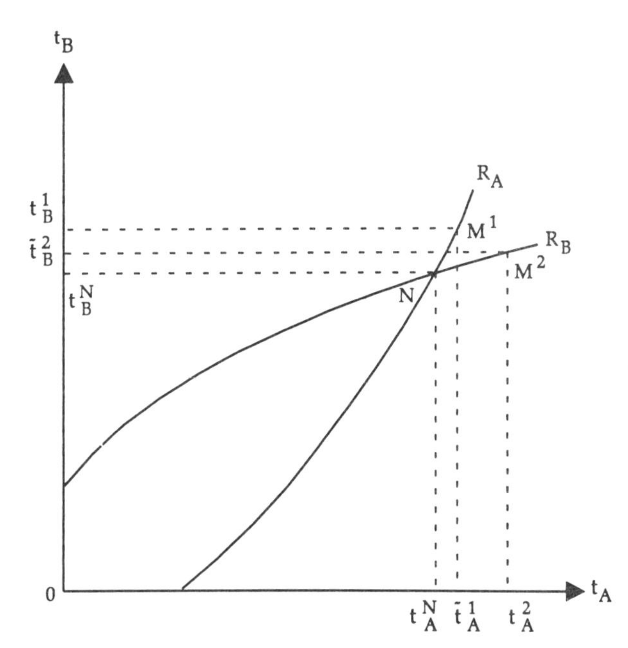

sumption is assumed to be zero. We derive both the myopic optimal export taxes and Nash export taxes for each country.

We also employ the model to perform several comparative statics exercises. We simulate the effects of improvements in production efficiency in African and non-African countries on real income and tax revenues in the nine countries. The simulations are done both when tax rates are held constant and when they are altered endogenously (Nash taxes).

The world demand and country supply curves have been linearized by using existing elasticity estimates and 1986 prices and quantities. Knowledge of the elasticities is not sufficient to solve for the tax rates. The exact form of the various functions also matters. For instance, several studies have estimated demand functions in log form, and have obtained elasticity estimates smaller than one. Clearly, the elasticity cannot be smaller than one along the entire demand curve as this would imply an infinitely high export tax at the world level.

The functional form is a matter which we plan to take up in the future. At this stage, we have assumed linear curves, which in the case of the demand function implies finite export taxes. Our results must thus be regarded as being preliminary. More definite results must await a more careful examination of the precise form of the demand and supply functions.

Trade policy interventions should be designed not to serve short-term stabilization objectives, but rather should be implemented in order to improve the long-term efficiency of resource allocation in those specific cases where the market fails to do so.[7] Hence, long-term elasticity estimates were used as a basis for linearizing the supply functions. Thus, the simulations performed in the following section should be understood as generating comparisons between alternative long-run equilibria after all relevant planting changes have taken place.

Moreover, since we are concerned with the long-run we abstract from the demand for stocks and focus on consumption demand. We assume for simplicity that the residual output exported by the small producers—those not included in our group of nine major producers—is given and does not respond to changes in the world price. We then set the demand at the world price equal to the exports of the nine major producers. Finally, we abstract at this stage from the possibility of smuggling, say between Ghana and Côte d'Ivoire. This added constraint on the power of some countries to set export taxes will be taken up at a later stage.

Table 17–1 presents some of the basic data used in the analysis. The output levels (in thousands of metric tons)and the shares correspond to 1986, with the Côte d'Ivoire at 585 (36%), Brazil at 329 (20%), Ghana at 219 (13%), Malaysia at 125, (8%), Cameroon at 118 (7%), Nigeria at 110 (7%), Ecuador at 85 (5%), Indonesia at 32 (2%), and Oceania at 30 (2%). Table 1 also shows the 1986 world price, tax rate[8], and domestic prices, the long-run supply elasticities and the corresponding slopes and intercepts of the long-run supply curves.

The highest export tax rate (70%) corresponded to Ghana, followed by Nigeria (50%), Cameroon (40%), Côte d'Ivoire (25%), Brazil (20%), and a zero tax rate for Malaysia, Ecuador, Indonesia and Oceania. The more recent producers, Malaysia and Indonesia, have a high long-run supply elasticity (3), while those of traditional producers—like Ghana (0.71), Côte d'Ivoire (1.15), Nigeria (0.45) and Brazil (0.58) - are significantly lower.

Table 17–1. *Basic data*

	Output (000MT)	*Output share (%)*	*Export tax[a] (%)*	*Domestic price (US$/MT)*	*Elasticity[b]*	*Slope[c]*	*Intercept (000MT)*
Côte d'Ivoire	585	35.8	25.1	1550	1.15	0.434	-87.7
Ghana	219	13.4	70.0	621	0.71	0.250	63.5
Cameroon	118	7.2	40.0	1242	1.81	0.172	-95.6
Nigeria	110	6.7	50.0	1035	0.45	0.048	60.5
Malaysia	125	7.7	0.0	2070	3.00	0.181	-250.0
Indonesia	32	2.0	0.0	2070	3.00	0.046	-64.0
Oceania	30	1.8	0.0	2070	3.00	0.043	-60.0
Ecuador	85	5.2	0.0	2070	0.28	0.011	61.2
Brazil	329	20.1	20.0	1656	0.58	0.115	138.2

Note: [a] The non-zero export tax rates are from Imran and Duncan, Table 7, page 21, and refer to 1982 and 1983 (for Brazil). [b] The long-run elasticities for Brazil, Côte d'Ivoire, and Malaysia were obtained from Akuyama and Bowers, page 25. They apply to ten-year periods, using the highest production levels to obtain those values. We assume that the elasticities of Indonesia andOceania are equal to that of Malaysia. The other elasticities are from Behrman. [c] The slope is the change in metric tons for a one US dollar change in the domestic producer price.

World supply is 1633 and the world price is 2070. The elasticity of demand is 0.4, the intercept is 2286.3 and the slope is -315.6 M.T. per U.S. dollar.

Simulation Results

In this section, we perform five sets of simulations. This will allow us to examine the effects of some of the recent changes in the cocoa market under actual taxes and compare them with the effects in the case where countries choose Nash taxes. We also examine the effects under "myopic" taxes.

The simulations are as follows:

- First, we compare the initial results with actual taxes with those in the case of free trade, Nash taxes and myopic taxes. This is shown in tables 17–2 and 17–3.
- Second, we simulate an exogenous outward shift in Ghana's supply curve of 100,000 tons (MT), and examine the results under existing and Nash taxes (see table 17–4).
- Third, we simulate an exogenous shift in Malaysia's supply curve of 100,000 M.T. (see table 17–5).
- Fourth, we simulate the impact of Côte d'Ivoire reducing its tax rate to zero (see table 17–6).
- Finally, we simulate the impact of 2-4 above simultaneously (see table 17–7).

1. Table 17–2 shows the tax rates, outputs and total profits (or income) with actual taxes and under free trade. The results on output are presented in thousands of metric tons, and those on profits (and revenues in the other tables) are in millions of U.S. dollars.

The actual world price is 2,070, while under free trade it is 1,562 or 24.5 percent lower. Hence, output falls for countries with an initial tax rate below 24.5 percent (all five Non-African countries) and rises in the countries with an initial tax rate above 24.5 percent (all four African countries). The fall is largest in the countries with high elasticity (by 74% in Malaysia, 75% in Indonesia and 77% in Oceania). Output increases by over 100 percent in Ghana (the domestic producer price increases by over 150%, from 621 to 1,562, and the elasticity is .71). Africa's output increases by over 30 percent (from 1032 to 1,349), world output increases by about 10 percent (from 1,633 to 1,793), and Africa's share increases from 63 percent to 75 percent.

Profits, on the other hand, fall everywhere. They fall most for the small producers with elastic supply (by 93% in Malaysia, by 93.6% in Indonesia and by 94.5% in Oceania). The fall in profits is also significant in Côte d'Ivoire (42.6%), in Cameroon (37.7%), in Brazil (31.9%), in Ecuador (27.9%) and in Nigeria (24.8%). The only country where the fall in profits is negligible is Ghana. Both the actual tax rate of 70 percent and the zero tax rate are sub optimal. In fact, we show below that the optimal (myopic or Nash) tax rate lies between these two extremes.

Profits for Africa fall from 1,443 to 1,043 (or by 27.7%), and world profits fall from 2,182 to 1,512.3 (by 30.7%). Thus, even though Africa's output (and share in world output) would rise under free trade, it would come at the expense of a reduction in profits.

Table 17–2. ***Initial results with actual, Nash and myopic tax rates***

	Tax rates (%)			*Output (000MT)*		*Profit [1] (millions of US$)*			*Revenue (millions of US$)*	
Country	*Actual (1)*	*Nash (2)*	*Myopic (2)*	*Actual (3)*	*Nash (4)*	*Actual (5)*	*Nash (6)*	*Myopic (6)*	*Actual (7)*	*Nash (8)*
Côte d'Ivoire	25.1	25.2	29.9	585	490	698	496	705	304	220
Ghana	70.0	19.5	20.5	219	421	405	493	544	318	146
Cameroon	40.0	8.2	9.8	118	184	138	125	172	98	27
Nigeria	50.0	5.7	6.3	110	139	202	182	221	114	14
Africa				1032	1234	1443	1296		834	407
Malaysia	0.0	2.8	5.1	125	63	43	14	44	0	3.2
Indonesia	0.0	0.7	1.2	32	17	11	3.4	11.3	0	.2
Oceania	0.0	0.6	1.3	30	16	11	3.2	11.1	0	.2
Ecuador	0.0	3.2	3.3	85	80	151	126	152	0	4.6
Brazil	20.0	13.4	14.7	329	315	523	424	525	136	75
World				1633	1725	2182	1866.6		970	490.2

Note: World price (U.S.$/MT) Actual: 2,070 Nash: 1,779. [1] Profits are defined to include producers' surplus and government revenue. Actual profits are derived by assuming that the calibrated demand and supply curves are true demand and supply curves. These profits will be different in general from actual observed profits (inclusive of tax revenues).

Table 17–3 presents the initial results with actual, Nash and myopic taxes. Let us first consider myopic taxes. For a given country, this tax is derived by maximizing profits under the assumption that other rates are held fixed at their current (actual) levels. In terms of figure 17–2, if *B*'s tax is frozen at t^1_B, *A*'s myopic tax is given by t^1_A. In table 17–3, if tax rates of countries other than Côte d'Ivoire are kept at the levels shown in column (1), Côte d'Ivoire's (myopic) optimal tax is 29.9% (column 2). Similarly, if tax rates of countries other than Ghana are held at the levels shown in column (1), Ghana's myopic optimal tax is 20.5% (column 2).

In contrast with myopic taxes, Nash taxes allow other countries to adjust their taxes optimally. Thus, for each country, the tax rate in column (2) is optimal when other countries choose tax rates at levels shown in the same column. In contrast to myopic taxes, Nash taxes do not leave room for profit-increasing changes in tax rates for any country as long as the others keep their taxes at Nash levels.

Not surprisingly, as seen by comparing columns (5) and (6), each country taken one at a time can increase its profits under myopic taxes. But while this country maximizes profits, others do not do so under the myopic tax assumption. Reactions by others eventually lead to the Nash equilibrium.

The most interesting comparison is between actual and Nash taxes. A striking result here is that profits under actual taxes are higher than under Nash behavior for all countries except Ghana. With as many as nine participants in the market, Nash behavior leads to excessive tax competition and results in lower profits for all participants except one. Also, as expected, profits under Nash taxes are larger than under free trade for all countries (compare column (5) in table 17–2 and column (6) in table17–3).

The world price under Nash taxes is lower by U.S. dollars 291 per metric ton (MT) than under actual taxes, falling from 2,070 to 1,779 or by 14.1 percent. This is due to the significant fall in tax rates for several producers: from 70% to 19.5% for Ghana, from 50% to 5.7% for Nigeria, and from 40% to 8.2% for Cameroon. These lower Nash tax rates result in significantly higher output for Ghana (from 219 to 421), Nigeria (from 110 to 139), and Cameroon (from 118 to 184).

The Nash taxes are somewhat lower than actual taxes for Brazil (13.5% versus 20%), they are not significantly different from actual taxes for Côte d'Ivoire (25.2% versus 25.1%), and they rise from zero to 3.2% for Ecuador, to 2.8% for Malaysia, to .7% for Indonesia and to .6% for Oceania. Côte d'Ivoire's output falls by 16.3% (from 585 to 490) because of the 14.1% fall in the world price, combined with a long-run supply elasticity slightly larger than one. The outputs of Malaysia, Indonesia and Oceania fall by about 50 percent (a combination of the 14.1% fall in the world price, a small rise in the tax rate and an elasticity of 3).

World output rises from 1,633 to 1,725 (by 5.7%). Hence consumers in developed countries gain from the lower world price and higher output. Africa's output under existing taxes is 1,032 and its share is 63.2 percent. Under Nash taxes, Africa's output rises to 1,234 or by 19.6%, and its share rises to 71.5 percent, or by 8.3 percentage points. One of the concerns for Africa has been the fall in its share of world cocoa

Table 17–3. *Initial results with actual, Nash and myopic tax rates*

	Tax rates (%)			*Output (000MT)*		*Profit* [1] *(millions of US$)*			*Revenue (millions of US$)*	
Country	*Actual (1)*	*Nash (2)*	*Myopic (2)*	*Actual (3)*	*Nash (4)*	*Actual (5)*	*Nash (6)*	*Myopic (6)*	*Actual (7)*	*Nash (8)*
Côte d'Ivoire	25.1	25.2	29.9	585	490	698	496	705	304	220
Ghana	70.0	19.5	20.5	219	421	405	493	544	318	146
Cameroon	40.0	8.2	9.8	118	184	138	125	172	98	27
Nigeria	50.0	5.7	6.3	110	139	202	182	221	114	14
Africa				1032	1234	1443	1296		834	407
Malaysia	0.0	2.8	5.1	125	63	43	14	44	0	3.2
Indonesia	0.0	0.7	1.2	32	17	11	3.4	11.3	0	.2
Oceania	0.0	0.6	1.3	30	16	11	3.2	11.1	0	.2
Ecuador	0.0	3.2	3.3	85	80	151	126	152	0	4.6
Brazil	20.0	13.4	14.7	329	315	523	424	525	136	75
World				1633	1725	2182	1866.6		970	490.2

Note: World price (U.S.$/MT) Actual: 2,070 Nash: 1,779. [1] Profits are defined to include producers' surplus and government revenue. Actual profits are derived by assuming that the calibrated demand and supply curves are true demand and supply curves. These profits will be different in general from actual observed profits (inclusive of tax revenues).

exports, and the rise in the share of Malaysia and Indonesia. A Nash strategy would have resulted in a higher share for Africa, as well as in a fall in the share of Malaysia from 7.7% to 3.6%, and in a fall in the share of Indonesia from 2.0% to 1.0%. Such an expansion for Africa would have come at the cost of lower profits, however.

Profits under Nash taxes, in millions of US dollars, increase by 88 in Ghana (or by over 20%), and fall by 202 (29%) in Côte d'Ivoire, by 99 (19.2%) in Brazil, by 29 (67.4%) in Malaysia, by 25 (16.5%) in Ecuador, by 20 in Nigeria (10%), by 13 (9.4%) in Cameroon, and by about 8 (69.7%) in Indonesia and Oceania (70.7%). Overall profits fall by 315 or by 14.4 percent. Profits in Africa fall from 1443 to 1296 or by 10.2 percent. Africa's share in total profits rises by three percentage points (from 66.5% to 69.5).

The beneficiaries from Nash taxes rather than actual taxes are the cocoa consumers and Ghana, while the other producers lose. The main losers, in terms of the proportional fall in profits, are Oceania, Indonesia, Malaysia and Côte d'Ivoire.

Government revenues increase slightly in Ecuador, Malaysia, Indonesia and Oceania, and they fall everywhere else. They fall by 87 percent in Nigeria, 72 percent in Cameroon, 53 percent in Ghana, 46 percent in Brazil and 27 percent in Côte d'Ivoire. The proportional fall in revenues is directly related to the reduction in tax rates.

These results are preliminary. Assuming that they hold after careful empirical analysis is done, it would lead us to conclude that when providing policy advice to any one country in the case of commodities such as cocoa, the various bilateral and multilateral donors should take into account both the effect on other countries as well as their possible reaction to the recommended policy change. This would help avoid a potentially undesirable outcome such as the Nash solution where most countries end up worse off despite the expectation of a welfare improvement. Even if some countries gain—such as Ghana in this case—the actual gain would be substantially lower than what might have been expected based on the assumption that myopic taxes in various countries are mutually consistent.

2. The effects of a 100,000 M.T. outward shift in Ghana's supply curve under both sets of taxes are shown in table 17–4. The difference in the effects of actual and Nash taxes in this case are very similar to those in the original case (table 17–3). Hence, we first compare the results in the original case and in the present case when *actual* taxes are levied. The comparison of results when Nash taxes are used is similar and is briefly discussed later.

Clearly, Ghana gains from the exogenous productivity increase. Its profits rise from 405 to 579 (see column 5 in tables 17–3 and 17–4) under actual taxes. The increased productivity in Ghana leads to a fall in the world price, from 2,070 to 1,993 (or 3.7%).

The lower world price implies a loss in profits in all other countries. The loss is 56 in Côte d'Ivoire (8.%), 28 in Brazil (5.3%), 17 in Cameroon (12.3%), 11 in Nigeria (5.7%), 7 in Ecuador (4.6%), 9 in Malaysia (20.9%), 2.7 in Indonesia (24.5%) and 3.3 in Oceania (30%). Interestingly, the overall profits for the nine countries change very little. They rise only from 2,182 to 2,222, or by 1.8 percent. In other words, even though Ghana gains from the increase in its productivity, the losses by the other countries are such that overall industry profits remain practically un-

Table 17–4. *Increasing Ghana's intercept by 100,000MT*

	Tax rates (%)		*Output (000MT)*		*Profit* [1] *(millions of US$)*		*Revenue (millions of US$)*	
Country	*Actual (1)*	*Nash (2)*	*Actual (3)*	*Nash (4)*	*Actual (5)*	*Nash (6)*	*Actual (7)*	*Nash (8)*
Côte d'Ivoire	25.1	25.3	560	470	642	460	280	205
Ghana	70.0	23.5	313	492	579	631	437	200
Cameroon	40.0	8.1	109	175	121	114	87	24
Nigeria	50.0	5.8	107	137	191	173	107	14
Africa			1089	1274	1533	1378	911	443
Malaysia	0.0	2.5	111	53	34	10	0	2.3
Indonesia	0.0	0.6	28	15	8.3	2.5	0	.2
Oceania	0.0	0.6	26	14	7.7	2.3	0	.1
Ecuador	0.0	3.3	83	79	144	122	0	4.6
Brazil	20.0	13.7	322	309	495	405	128	73
World			1659	1744	2222	1919.8	1039	523.2

Note: World price (U.S.$/MT) Actual: 1,993 Nash: 1,722.

changed. Ghana gains 174, while the other countries lose 134, so that overall profits increase by 40. Hence, the principal gainers from Ghana's increase in productivity are Ghana and the consumers who benefit from a lower price.

Africa's profits increase from 1,443 to 1,533, and its share in total profits rises from 66.1 percent to 69 percent. Non-Africa's profits fall from 739 to 689, and its share falls from 33.9 percent to 31 percent.

The lower world price leads to a reduction in output in all countries except Ghana (given the unchanged tax rates), and world output increases by 26 (from 1,633 to 1,659, see column 3 in tables 17–3 and 17–4), i.e., by one quarter of the shift in Ghana's supply curve. Total revenues fall for all countries with positive tax rates other than Ghana because of a lower output and lower world price.

Interestingly, in the case of Nash taxes, producers as a whole are able to retain more profits from Ghana's productivity increase than in the case of existing taxes, even though world prices are lower in the former case. Under Nash taxes (see column 6, tables 17–3 and 17–4), overall profits rise by 53.2 (from 1,866.6 to 1,919.8), while under existing taxes (column 5, tables 17–3 and 17–4), they only rise by 40. Also, Nash tax rates in this case are similar to those in the original case for all countries except for Ghana, whose Nash tax rate increases from 19.5 percent to 23.5 percent.

In this case, under existing taxes, Africa as a whole gains about half (90) of what Ghana gains (174), while all producers gain about 23 percent (40) of what Ghana gains. These findings are again preliminary. If they hold under closer empirical scrutiny they would lead us to conclude that when considering the financing of investment projects in commodities such as cocoa, bilateral and multilateral donors should take into account the negative impact on the other producers. Moreover, simultaneous expansion of output in several countries would not generate the returns on the investment projects which would result from output expansion in one country only. These points were raised in an a somewhat different context in an early paper by Goreux (1972).[9] The World Bank policy of denying lending for output expansion in commodities such as cocoa (except in special circumstances) is based on the above considerations.

3. The effects of a 100,000 M.T. outward shift in Malaysia's supply curve are quite similar to those when the shift occurs in Ghana's supply curve, and are shown in table 17–5. In the case of existing taxes, the effects are identical for all countries other than Ghana and Malaysia. World supply in both cases is 1,659 and the world price is 1,993. In the present case, Ghana's profits fall from 405 to 380 (and its output falls form 219 to 213), while Malaysia's profits rise from 43 to 123 (and its output rises from 125 to 211).

World profits are lower (2,112) in this case than in the case where Ghana's supply shifts outward (222). The reason is that in the present case, Ghana's output is lower (than when the supply shift occurs in Ghana) by 100, while Malaysia's output is higher by 100 (see tables 17–4 and 17–5, column 3), and Ghana is the lower-cost producer. As is shown in table 17–1, at the actual outputs, Ghana's producer price (and marginal cost) is 621 while that of Malaysia is 2,070.

In fact, the outward shift in Malaysia's supply curve leads to overall immiserization for the exporting countries as a whole. Total profits fall to 2,112, compared to 2,182 in the actual case. Hence, even

Table 17–5. *Increasing Ghana's intercept by 100,000MT*

	Tax rates (%)		*Output (000MT)*		*Profit*[1] *(millions of US$)*		*Revenue (millions of US$)*	
Country	*Actual (1)*	*Nash (2)*	*Actual (3)*	*Nash (4)*	*Actual (5)*	*Nash (6)*	*Actual (7)*	*Nash (8)*
Côte d'Ivoire	25.1	25.2	560	470	642	457	280	203
Ghana	70.0	19.7	213	408	380	463	297	138
Cameroon	40.0	8.1	109	174	121	113	87	24
Nigeria	50.0	5.8	107	136	191	172	107	14
Africa			989	1188	1334	1205	771	373
Malaysia	0.0	6.6	211	140	123	70	0	16
Indonesia	0.0	0.6	28	14	8.3	2.4	0	.1
Oceania	0.0	0.6	26	13	7.7	2.2	0	.1
Ecuador	0.0	3.3	83	79	144	121	0	4.5
Brazil	20.0	13.7	322	309	495	404	128	73
World			1659	1743	2112	1804.6	899	467

Note: World price (U.S.$/MT) Actual: 1,993 Nash: 1,717.

Table 17–6. *Comparing the initial equilibrium and the effect of fixing Côte d'Ivoire's tax at zero*

	Tax rates (%)		*Output (000MT)*		*Profit*[1] *(millions of US$)*		*Revenue (millions of US$)*	
Country	*Original (1)*	*New (2)*	*Original (3)*	*New (4)*	*Original (5)*	*New (6)*	*Original (7)*	*New (8)*
Côte d'Ivoire	25.1	0.0	585	739	698	629	304	0
Ghana	70.0	70.0	219	206	405	352	318	275
Cameroon	40.0	40.0	118	100	138	105	98	76
Nigeria	50.0	50.0	110	105	202	179	114	100
Africa			1032	1150	1443	1265	834	451
Malaysia	0.0	0.0	125	95	43	70	0	0
Indonesia	0.0	0.0	32	24	11	2.3	0	0
Oceania	0.0	0.0	30	22	11	2.1	0	0
Ecuador	0.0	0.0	85	82	151	120	0	0
Brazil	20.0	20.0	329	313	523	396	136	119
World			1633	1686	2182	1855.4	970	570

Note: World price (U.S.$/MT) Actual: 2,070 Nash: 1,905.

though Malaysia's profits rise by almost 200 percent, total industry profits fall. Comparing the present case with that when Ghana's supply shifts outward (with world output and price being the same in both cases), the reason for which profits are only 2,112 rather than 2,222 is that a larger share of the given total output is produced by a higher-cost producer (Malaysia) rather than by a lower-cost producer (Ghana).

Africa's profits in this case are 1,334 out of a total of 2,112, so that its share is 63.2 percent. That compares with the higher profits (of 1,443) in the actual case (and a share of 66.1%), and profits of 1,533 (share of 69%) when Ghana's supply curve shifts outward.

In the case of Nash taxes, tax rates in the present case in all countries other than Ghana and Malaysia are extremely close to the tax rates when the supply shift occurs in Ghana. The tax rates for Ghana and Malaysia are quite different in each case. In the present case, Ghana's Nash tax rate is 19.7 percent and that of Malaysia is 6.6 percent. When Ghana's supply shifts outward, the Nash tax rates are 23.5 percent for Ghana (i.e., 3.8 percentage points higher) and 2.5 percent for Malaysia (4.1 percentage points lower).

With Nash taxes, profits for all countries other than Ghana and Malaysia are extremely close in both cases. Malaysia's profits rise from 10 (when Ghana's supply shifts outward) to 70 in the present case, while Ghana's profits fall from 631 to 463. Total industry profits in the present case are 1,804.6 rather than 1,866.6 in the original cae, i.e., they fall by 62. How-

Table 17–7. *Comparing the initial equilibrium and the effect of fixing Côte d'Ivoire's tax at zero and increasing the intercepts of Ghana and Malaysia by 100,000MT*

	Tax rates (%)		*Output (000MT)*		*Profit*[1] *(millions of US$)*		*Revenue (millions of US$)*	
Country	*Original (1)*	*New (2)*	*Original (3)*	*New (4)*	*Original (5)*	*New (6)*	*Original (7)*	*New (8)*
Côte d'Ivoire	25.1	0.0	585	673	698	522	304	0
Ghana	70.0	70.0	219	295	405	483	318	362
Cameroon	40.0	40.0	118	84	138	80	98	59
Nigeria	50.0	50.0	110	102	202	160	114	89
Africa			1032	1154	1443	1245	834	510
Malaysia	0.0	0.0	125	167	43	77	0	0
Indonesia	0.0	0.0	32	17	11	3.0	0	0
Oceania	0.0	0.0	30	15	11	2.8	0	0
Ecuador	0.0	0.0	85	80	151	124	0	0
Brazil	20.0	20.0	329	299	523	412	136	105
World			1633	1732	2182	1863.8	970	615

Note: World price (U.S.$/MT) Actual: 2,070 Nash: 1,754.

ever, with existing taxes, they fall from 2,182 to 2,112 or by 70. Hence, the use of Nash taxes enables the industry as a whole to suffer smaller losses from the supply shift than with existing taxes.

Thus, Africa as well as the industry as a whole loses both with actual and Nash taxes. This would seem to reinforce the point made earlier that donors should take into account the interdependencies among producing countries when assessing investment projects.

4. Up to mid-1989, the Caisse de Stabilisation in the Côte d'Ivoire did not pass on the significant reduction in world cocoa prices to its producers. The Caisse was still paying the producers more than the price it obtained (net of marketing costs). It was paying 400 CFA/MT, implying a cost for the Caisse of well over the FOB price of 500 CFA/MT. We simulate the reduction in Côte d'Ivoire's export tax rate by setting it equal to zero and keeping the tax rates of the other countries at their initial level. This is shown in table 17–6.

We notice a dramatic increase in Côte d'Ivoire's output, from 585 to 739, or by 26.3 percent. The world price falls from 2,070 to 1,905 or by 8 percent, so that all countries other than Côte d'Ivoire lose. However, Côte d'Ivoire also loses as its profits fall from 698 to 629, or by 10 percent. Hence, the strategy followed by Côte d'Ivoire was certainly not optimal from its own viewpoint (unless the Caisse aimed to stabilize the producer price and believed that the changes leading to a fall in the world price were transitory). With its large share in the world market, a zero tax rate does not maximize profits.

5. Finally, we examine the impact of the three previous simulations simultaneously, i.e., a 100,000 M.T. supply shift in Ghana and Malaysia and a zero export tax rate in Côte d'Ivoire. The results are shown in table 17–7. The output of Côte d'Ivoire rises from 585 to 673 because of the fall in its tax rate (but rises by less than under the previous simulation), and it rises for Ghana and Malaysia because of the supply shift. Output falls in all other countries. Total output rises from 1633 to 1732, and the world price falls from 2,070 to 1754 (by 15.3%). Profits fall in all countries except Ghana and Malaysia. Total profits and Africa's profits also fall.

Concluding Comments

The results presented in this paper are preliminary. The demand and supply functions were linearized based on point estimates of the relevant parameters. The exact form of these functions is crucial in determining the equilibrium tax rates, outputs, profits and revenues. In future work, we plan to closely examine the form of the relevant functions. We will also examine the properties of other Nash equilibria. These include revenue-maximizing Nash taxes, and using export quotas rather than the export tax as the policy variable. Our findings so far seem to suggest that in providing policy advice and support of investment projects in the case of commodities such as cocoa, the donor community should take into account the effects on and possible reactions of the other producing countries.

Notes

We would like to thank Shantayanan Devarajan for useful comments and Brendan Kennelly for excellent research assistance . The views expressed in this paper are the authors' and do not necessaily reflect those of the World Bank or its affiliated organizations.

1. This viewderives from the theory of the second best which emphasizes that distortions should be attacked at the source. See Bhawati (1971) for further details.

2. See Bhagwati (1988) on this issue.

3. For example, see the classic article by Johnson (1954) and the more modern treatment in Dixit and Norman (1980, Ch. 6).

4. If we feel that general equilibrium analysis may be useful, we will engage in the future. This may be the case if we wnat to include the implicit tax on exports due to tariffs on imported manufactures (Lerner Symmetry Theorem).

5. For a survey of recent game theorey models, see Dixit (1986) and the references therein.

6. We assume that the importer, C, does not levy import tariffs.

7. Power to affect the price in the market for a country's exports of a competitively produced commodity is a case where optimal export taxes will raise welfare for the exporting country.

8. The tax rates correspond to 1982 and 1983. reliable estimates of export tax rates for 1986 were not available.

9. He examined the returns on investment projects alternatively from the viewpoint of the firm of the investing country, of all producing countries, and of the world as a whole.

References

Akiyama, T. and A. Bowers (1984); "Supply Response of Cocoa in Major Producing Countries," Division working paper No. 1984-3, Commodities Studies and Projection Division, EPD, World Bank.

Balassa, B. (1978); "Exports and Economic Growth: Further Evidence" *Journal of Development Economics* (June): 181-89.

Behrman, J.R. (1968); "Monopolistic Cocoa Pricing". *American Journal of Agricultural Economics.* 50: 702-19.

Bhagwati, J.N. (1971); "The Generalized Theory of Distortions and Welfare," in J. Bhagwati, et al., *Trade, Balance of Payments and Growth,* Amsterdam: North-Holland.

Bhagwati, J.N. , (1978). *Anatomy and Consequences of Exchange Control Regimes.* Cambridge, Mass.: Ballinger

Bhagwati, J.N., (1988). "Export-Promoting Trade Strategy: Issues and Evidence". *World Bank Research Observer* 3 (1): 27-57.

Cairncross, A. (1962); *Factors in Economic Development.* London: Allen and Unwin.

Dixit, A. and V. Norman (1980); *Theory of International Trade,* London: Cambridge University Press.

Dixit, A., (1986), "Comparative Statics for Oligopoly," *International Economic Review,* 27(1): 107-22.

French, B.C. and J.L. Matthews (1971); "A Supply Response Model for Perennial Crops," *American Journal of Agricultural Economics* 53.

Goreux, L.M. (1972); "Private, National and International Returns; An Application to Commodity Lending". *Eurpean Economic Review.* 3(2): 131-80.

Humphreys C. and W. Jaeger (1989); "Africa's Adjustment and Growth," *Finance and Development,* June. Humphreys C. (1989); "Response to Singer," Finance and Development, December: p. 52.

Imran, M. and R. Duncan (1988); "Optimal Export Taxes for Exporters of Perenial Crops," Division Working Paper No. 10. International Commodity Markets Division, WPS 10, World Bank.

Johnson, H.G. (1954); "Optimum Tariffs and Retaliation". *Review of Economic Studies* 21, 142-53.

Keesing, D. (1967); "Outward-Looking Policies and Economic Development". *Economic Journal* (June):303-20.

Krueger, A.O. (1961); "Export Prospects and Economic Growth: India: A Comment" *Economic Journal* (June): 436-42.

Landell-Mills, P., Agarwala R. and S. Please (1989); "Sub-Saharan Africa: From Crisis to Sustainable Growth – a Long-term Perspective Study", World Bank.

Panagariya, A. and M. Schiff (1990); "Optimal and Revenue-Maximizing Export Taxes in a Multi-Country Model". World Bank, CECTP.

Singer, H.W., (1989); "African Adjustment," *Finance and Development,* December: p. 52.

World Bank (1989); "Strengthening Trade Policy Reform", SecM89-1454.

18

Response to Relative Price Changes in Côte d'Ivoire: Implications for Export Subsidies and Devaluations

Victor Lavy, John L. Newman, Raoul Salomon, and Philippe de Vreyer

The ability of firms to expand their output and exports in response to changes in relative prices has been central in the debate regarding the use of export subsidies as a trade policy instrument. Since the early 1980s, the introduction of export subsidies has been proposed as a way to counteract the adverse effects of an exchange rate overvaluation among member countries of the West African Monetary Union. Despite deteriorating terms of trade and mounting external debt, these countries and France have opted not to change the CFA exchange rate with the French Franc.

Faced with an exchange rate out of equilibrium, the economy must adjust—either through adjustments in quantities (output and employment) or in prices. As quantity adjustments typically involve higher foregone consumption during the adjustment period, adjustments generated by changing the relative price of traded to nontraded goods are generally preferred. In the absence of a nominal devaluation, one way to attempt a real devaluation would be to reduce nominal wages through tight monetary and fiscal policies.

An alternative means of altering the relative price of traded to nontraded goods is to attempt to mimic a devaluation by raising import tariffs and export subsidies by the same magnitude. This is, indeed, the policy that has been proposed for Côte d'Ivoire. There has been some disagreement over whether the plan has been implemented as proposed. While an export subsidy alone might be thought sufficient for expanding exports and improving the balance of payments, such a policy would not increase the price of imported goods and, thus, would not have as large effects on relative prices as that of a devaluation. In order to address the major concern of expanding output and employment in the tradeables sector, proponents of the tariff cum subsidy scheme argued that it was necessary to mimic the effects of a devaluation more completely and increase import prices as well.

Even with the increase in both export subsidies and import tariffs, this policy will still not mimic exactly a nominal devaluation. A devaluation would affect services and capital movements as well. Moreover, export subsidies will lead to higher export prices only if firms choose to participate and take advantage of the subsidy program. If, due to lack of credibility of the program or high transaction costs, firms do not participate, their export prices will not rise. Finally, the budgetary consequences of the subsidy cum tariff policy could be substantially different from that of a devaluation.

Opponents of the tariff cum subsidy policy claimed that it is doomed to fail because:

- producers in Africa are slow to adjust to a change in market signals
- the short run export supply elasticity is very small since African exporters cannot compete and capture international markets due to their

high labor cost compared to their competitors and to the protectionist policies of the developed countries
- the export subsidies will lead to a comparable increase in domestic prices, leaving unaltered the relative profitability of exporting
- the import surcharges will not raise enough revenue to finance the export subsidies, implying a heavy fiscal burden on other revenue sources. The expected long run nonsustainability of this deficit will undermine from the beginning the credibility of the program and will discourage firms from responding to it.

These views have not been based on extensive empirical evidence, as evidence on export and output supply response is even more scarce for African countries than it is for other developing countries. One of the few studies is by Balassa (1987) who presents results suggesting that the response of exports of goods and agricultural products to price changes is actually greater in Sub-Saharan African countries than in other developing countries. (for evidence on other developing countries see, for example, Balassa et al 1986; Nogues 1989; Milanovic 1986; Artus and Rosa 1978; Goldstein and Khan 1978; Bauman and Braga 1988).

The objective of this paper is to address the first three issues above and to provide empirical evidence on export and output supply responses that could inform policy discussions on the tariff cum subsidy scheme introduced in Côte d'Ivoire in 1986. We base the empirical work on data from two sources for the six years that preceded the implementation of the program (1980-85). Information on sector level output, exports, domestic sales, capital, and variable input use is obtained by aggregating information from individual firm-level panel data obtained from the Banque des Donne es Financie res. The individual data was aggregated because export, import, domestic output, and input price indices obtained from the Ministere du Plan et de l'Industrie are available only at the sectoral (3 digit) level.

The paper models the short-run response of firms to exogenous changes in export and import prices, taking into account the possibility that firms may sell to both domestic and foreign markets. In such a framework, firms may alter sales in response to the relative profitability of the two markets. The net output response depends on the response in both markets. In world markets, Ivorian industries are assumed to be price takers. In the domestic market, even if individual firms face exogenous prices, the industry demand curve will be downward sloping, making the domestic price endogenous from the viewpoint of the industry. Allowing for the endogeneity of the domestic price requires a model of the domestic demand for the output of the manufacturing sectors. Thus, we model jointly firm output supply and input demand functions and domestic demand for domestically produced goods.

The short-run focus of this study is motivated by methodological as well as policy considerations. First, analyzing the long run response implies treating the capital stock as endogenous and modeling investment behavior and credit markets. This task is further complicated by difficulties in measuring the price of capital services. Second, from a policy point of view, the uncertainty associated with this particular policy reform in Côte d'Ivoire suggests that a sensible approach is first to evaluate the immediate short-run effects.

On the supply side, our empirical model follows closely the production theory framework of Diewert and Morrison (1988) (see also Kohli, 1978). The major new element that we introduce to their framework is the modeling of the demand side. Recent studies such as Zilberfab (1980), Aspe and Giavazzi (1982), and Faini (1988) have discussed the role of domestic market conditions and relative prices in modeling export supply, but have not treated domestic prices as endogenous and identified empirically the influence of exogenous export and import prices on domestic prices. Our work also differs in that we use sector-level data aggregated from individual firm records, whereas Diewert and Morrison use aggregate data. We also allow for non-constant returns to scale.

We compare results from a model that estimates only the supply side with one that estimates jointly supply and demand. We simulate the model to yield domestic supply and export supply elasticities, as well as measures of the sensitivity of the domestic price and domestic demand to exogenous variation in export and import prices. The empirical results are very sensitive to the assumption regarding the endogeneity of the domestic price and to the inclusion (or exclusion) of the demand function in the model. Not allowing for domestic prices to change and affect the relative profitability of exporting versus selling domestically leads to a large bias in the estimate of the export elasticities.

The main and somewhat surprising result is that manufacturing producers in Côte d'Ivoire are able to expand their exports in the short run in response to an increase in export prices. However, most of this expansion comes at the expense of sales in the domestic market. The net short run output and employment responses are small. A second important and less surprising result is that the domestic supply curve is much more sensitive to price changes than the export supply function. This implies that any exogenous shocks that lead to an increase in domestic demand, such as an increase in import prices, will have a sizable effect on domestic sales and output, and a contractionary effect on exports. Finally, increases in export prices alone were estimated to have a much smaller effect on output and employment than would increases in both export and import prices, as would occur in a devaluation or in the tariff cum subsidy program.

Empirical Specification

As has been observed elsewhere, the evidence from Côte d'Ivoire indicates that firms tend not to specialize in exporting but to sell their products in both domestic and foreign markets. Between 1980 and 1985 about a third of all manufacturing firms exported to foreign markets. Of those who exported, roughly sixty percent of the value of their sales was to the domestic market. We therefore model firm behavior by allowing firm and industries to produce two different products in a joint production process. The products could be identical or differentiated from each other. Often export products will be similar to the product destined for the domestic market, but distinguished by being of a higher and more uniform quality.

Our empirical specification of joint production is based on the production theory approach employed by Diewert and Morrison (1988) in their study of aggregate export supply and import demand functions in the U.S. In their approach, one specifies a single profit function and derives supply functions to domestic and foreign markets by partially differentiating the profit function with respect to the output price in the respective market. Our work differs in two main respects. First, we work with sectoral rather than aggregate data. It is not our intention to model the complete trade balance, but rather to model short run responses to price changes among firms in the manufacturing sector. Second, we model the domestic demand for domestically produced goods and imports.

Firms are assumed to operate under perfect competition in factor markets. Thus, all input prices are treated as exogenous. Export prices are also assumed to be exogenously determined. We consider two alternative ways of accounting for the endogeneity of domestic prices. In the first approach, we simply instrument for the endogenous price of domestic goods and estimate how input demands and outputs supplied to domestic and foreign markets vary as functions of the instrumented domestic prices and exogenous export and input prices. In the second approach, we explicitly model the domestic demand for domestic output and for imports. Domestic demand and import demand functions are then estimated

Figure 18-1. Bias in Estimated Net Export and Output Responses

The export market

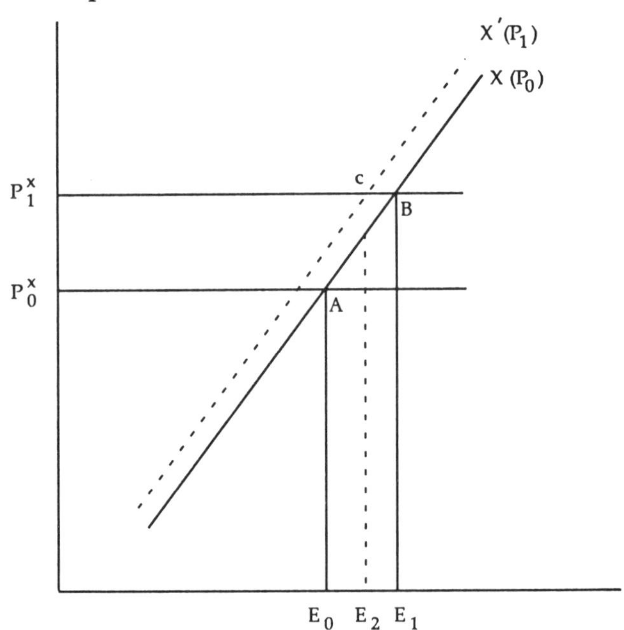

The domestic market

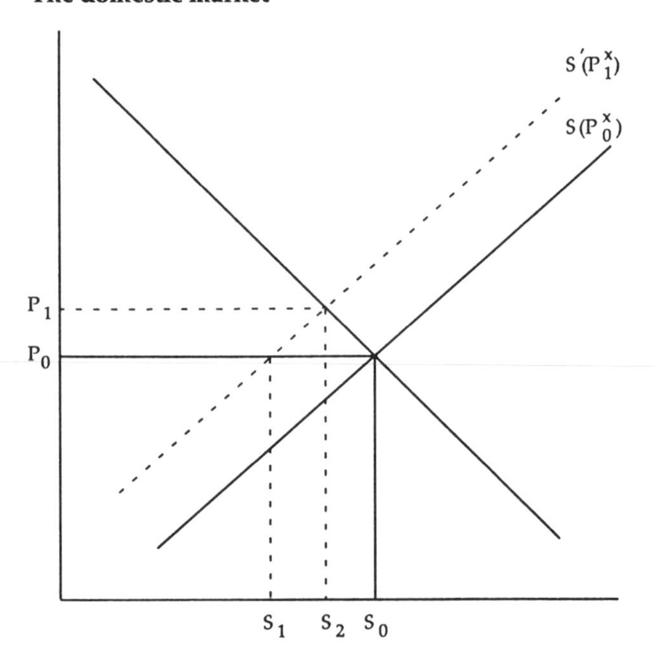

jointly with the domestic output supply, export supply, labor demand and intermediate goods demand functions. The domestic market equilibrium condition for domestically produced goods determines the domestic price level and the quantity sold. The advantage of the first approach is that the estimation procedure is simpler, being very similar to the procedure under the assumption of exogenous domestic prices. Using instrumented domestic prices purges the system of simultaneous equations bias and allows one to estimate the model as a system consisting only of output supply and input demand equations.[1]

The main disadvantage of this approach is that the simulations obtainable from this approach do not tell us exactly what we would like to know. Either under the assumption of exogenous domestic prices or with instrumented prices, it is only possible to obtain estimates of the change in domestic supply or export supply with a change in export prices, holding domestic prices constant. However, one would not expect domestic prices to remain constant.

Export supply and domestic supply functions are derived holding the output price in the other market constant. Thus, an increase in export prices that increases the relative profitability of exporting will shift the domestic supply function to the left. If domestic prices are assumed exogenous, the leftward shift of the domestic supply curve will not lead to any change in domestic prices. However, if as argued earlier, it is more reasonable to assume that the industry demand curve is downward sloping, then the shift in the domestic supply function will lead to an increase in domestic prices. As domestic prices change, this would call forth a leftward shift of the export supply function. The net result after the domestic price adjustment would be a smaller substitution away from domestic sales to exports and, consequently, a smaller increase in exports than if there had been no adjustment.

The bias in the estimat of the net ex[ort and output responses from neglecting to account for the adjustment of the domestic price is illustrated in figure 18–1. An increase in the export price from P_o to P_1 leads to firms to move along the export supply function and to reduce the amount they are willing to supply to the domestic market at a given price, a leftward shift in the domestic supply function from S to S'. Even if all firms take prices as given, the reduction in the industry supply curve will lead to an increase in the domestic price if the industry demand curve is downward sloping. The increase in the domestic price leads to a shift of the export function from X to X'. The new domestic priceand output will be at P_1 and S_2, while exports will be at E_2.

If the domestic price is forced to remain at P, then the export and domestic responses can be seen to be upwardly biased (cf. E_1 versus E_2 and S_1 versus S_2) and the increase in total output will be downwardly biased. The magnitude and the importance of these biases will be demonstrated later in the discussion of the empirical results.

Despite the bias, it is still useful to consider the simpler model as it provides an upper bound to the expected size of the export supply response to export price changes. If there is no observed response of export supply to changes in export prices when domestic prices are held constant, it is unlikely that there will be any response once the influence of export prices on domestic prices is taken into consideration. Thus, we first specify the model using only the production relations and then indicate how this approach is modified to take account of the demand side.

The Supply Side

We derive econometric specifications of producer supply and input demand functions, consistent with profit-maximizing behavior, from a generalized McFadden restricted profit function of the type employed by Diewert and Morrison (their equation 8.4). This form of the profit function is one of the class of flexible functional forms and can be specified with or without an assumption of constant returns to scale.[2,3] For a nonconstant returns to scale technology, the restricted profit function is:[4]

$$
\begin{aligned}
(18\text{–}1)\quad \Pi^t(P_t^d, P_t^e, w_t^l, K_t) = {} & b_{ss}P_t^d K_t + b_{ll}w_t^l K_t + b_{gg}w_t^g K_t \\
& + \alpha_s P_t^d t + \alpha_x P_t^e t + \alpha_l w_t^l t + \alpha_g w_t^g t \\
& + b_{st}P_t^d t K_t + b_{xt}P_t^e t K_t + b_{lt}w_t^l t K_t + b_{gt}w_t^g t K_t\} \\
& + b_s P_t^d + b_x P_t^e + b_l w_t^l + b_g w_t^g\} \\
& + \beta_s P_t^d K_t^2 + \beta_x P_t^e K_t^2 + \beta_l w_t^l K_t^2 + \beta_g w_t^g K_t^2 \\
& + \gamma_s P_t^d t^2 K_t + \gamma_x P_t^e t^2 K_t + \gamma_l w_t^l t^2 K_t + \gamma_g w_t^g t^2 K_t \\
& + \frac{1}{2} b_{sx} \frac{(P_t^d)^2}{P_t^e}) K_t + \frac{1}{2} b_{lx} \frac{(w_t^l)^2}{P_t^e} K_t + \frac{1}{2} b_{gx} \frac{(w_t^g)^2}{P_t^e} K_t \\
& + b_{sl} (\frac{P_t^d w_t^l}{P_t^e}) K_t + b_{sg} (\frac{P_t^d w_t^g}{P_t^e}) K_t + b_{lg} (\frac{w_t^l w_t^g}{P_t^e}) K_t
\end{aligned}
$$

where: S_t = output supplied to domestic markets; X_t = output supplied to foreign markets; L_t = real expenditure on labor; G_t = real expenditure on intermediate goods; w_t^l = the cost of labor; w_t^g = the cost of intermediate goods; P_d^t = the domestic price of domestically produced goods; P^f_t = the foreign price of domestically produced goods; K_t = the capital stock; t = time; The b's α's β's and γ's are parameters to be estimated.

π^τ is linearly homogeneous in prices. A constant returns to scale profit function would be obtained by imposing the restrictions that $b_i = 0$, $\alpha_i = 0$, $\beta_i = 0$, and $\gamma_i = 0$, $i = s,x,l$, and g.

As Diewert and Wales (1987) indicate, in order for the profit function to satisfy the theoretical curvature properties, namely, that the profit function be convex with respect to prices, the matrix of second derivatives of the restricted profit function with respect to prices must be positive semidefinite. This will be satisfied if and only if the matrix

$$(18\text{–}2) \qquad B = \begin{matrix} b_{s_x} & b_{s_l} & b_{s_g} \\ b_{s_l} & b_{l_x} & b_{l_g} \\ b_{s_g} & b_{l_g} & b_{g_x} \end{matrix}$$

is positive semidefinite. We did not impose positive semidefiniteness in the estimation, but did check whether the conditions were satisfied.

Partially differentiating the restricted profit function with respect to input and output prices yields: (1) a domestic supply function; (2) (minus) the labor demand function; (3) (minus) the intermediate input demand function; and (4) the export supply function. The resulting solutions are analogous to equations (8.5-8.9) in Diewert and Morrison (1988). As they suggest, we divide by K_t to make the assumption of homoskedasticity of the error terms more plausible.

$$(18\text{–}3) \; \frac{X_t}{K_t} = b_{xx} + b_{xt}t + \alpha_x t\,(\frac{1}{K_t}) + b_x(\frac{1}{K_t}) + \beta_x K_t + \gamma_x t^2$$

$$- \frac{1}{2} b_{sx}(\frac{P^d_t}{P^f_t})^2 - \frac{1}{2} b_{lx}(\frac{w^l_t}{P^f_t})^2 - \frac{1}{2} b_{gx}(\frac{w^g_t}{P^f_t})^2$$

$$- b_{sl}\frac{(P^d_t w^l_t)}{(P^f_t)^2} - b_{sg}\frac{(P^d_t w^g_t)}{(p^f_t)^2} - b_{lg}\frac{(w^l_t w^g_t)}{(P^f_t)^2} + \varepsilon_x$$

$$\frac{L_t}{K_t} = b_{ll} + b_{lt}t + \alpha_l t\,(\frac{1}{K_t}) + b_l(\frac{1}{K_t}) + \beta_l K_t + \gamma_l t^2$$

$$+ b_{sl}(\frac{P^d_t}{P^f_t}) + b_{lx}(\frac{w^l_t}{P^f_t}) + b_{lg}(\frac{w^g_t}{P^f_t}) + \varepsilon_l$$

$$\frac{G_t}{K_t} = b_{gg} + b_{gt}t + \alpha_g t\,(\frac{1}{K_t}) + b_g(\frac{1}{K_t}) + \beta_g K_t + \gamma_g t^2$$

$$+ b_{sg}(\frac{P^d_t}{P^f_t}) + b_{lg}(\frac{w^l_t}{P^f_t}) + b_{gx}(\frac{w^g_t}{P^f_t}) + \varepsilon_g$$

$$\frac{S_t}{K_t} = b_{ss} + b_{st}t + \alpha_{s}t\,(\frac{1}{K_2}) + b_s(\frac{1}{K_t}) + \beta_s K_t + \gamma_s t^2$$

$$+ b_{sx}(\frac{P^d_t}{P^f_t}) + b_{sl}(\frac{w^l_t}{P^f_t}) + b_{sg}(\frac{w^g_t}{P^f_t}) + \varepsilon_s$$

The exact forms of the output supply and input demand equations were determined by limitations of the available data. While we had information on production, domestic sales, exports, and input use of individual firms in the manufacturing sector, we did not have information on the output or export prices of the individual firms. Information on prices was available only in the form of sous-branche level price indices, corresponding to a three digit SITC classification.

In the absence of output prices of individual firms, we estimated the model at the sous-branche level to ensure a close correspondence between the price data and the output supply and input demand functions.[5] To arrive at sous-branche level values of domestic sales, exports, and expenditure on inputs we summed the values of individual firms in the sous-branche. We estimated our models for two different samples—one where the sous-branche level values were formed by summing the individual values over all firms in the sous-branche and the other formed by summing over only the firms that had exported in that year.

We considered two samples because it was not obvious what aggregation would lead to the closest connection between firm decisions and the relevant price data. The export price index, based on quantity weights of year t, does not capture changes in export prices of firms who happened not to export during year. However, these unobserved price changes might be expected to affect the observed sous-branche changes in output and input demands from year t-1 to year t. Limiting the sample to exporting firms and correcting for the selectivity bias in forming that sample might present a closer connection between the export price and the output of the firms. However, limiting the sample to exporting firms weakens the connection between the domestic price index and output decisions since we do not have access to a domestic price index specific to exporting firms.

A second feature of the data was that because the base used in calculating the published price index changes every year, it was impossible to use a constant base-period weighted price indices in the analysis. Because the value of the published price in- dex in year reflects the change in prices between year -1 and , maintaining a close connection between the theoretical model and the econometric specification implies that the estimating equations should relate changes in output supplies and input demands from year t-1 to t to the value of the price index in year t. Therefore, we estimated the model in differenced form. This also has the advantage of eliminating any fixed unobserved sous-branche specific effects.

The Demand Side

While maintaining the same specification of the supply side, we now add a simple model of the demand side. Domestic consumers may consume either domestically produced goods or imported goods, which are available at an exogenously determined world price. As was done on the supply side, we follow a dual approach and derive domestic consumers' demands for domestically produced goods and imported goods by partially differentiating their expenditure function with respect to domestic prices and imported prices respectively.

At a very high level of aggregation, one may posit only two goods—domestically produced goods and imported goods. However, as we are working with sous-branche level data, there are many more goods, each with its own price. In general, the expenditure function would be a function of the prices of all goods in the system. In order to simplify the model, we assume strong separability among goods in different sous-branches and across other non-manufacturing sectors. This implies that demand for domestically produced goods will be a function only of the prices of domestically produced goods and imports within that sous-branche. Our implicit assumption is that we are imposing a set of zero restrictions on parameters in the expenditure function. We do not test for these restrictions, but, in principle, they could be relaxed at the cost of reducing degrees of freedom. This is obviously an assumption one would want to relax at a later stage.

We adopt the same functional form for the expenditure function as we did for the restricted profit function, namely the generalized McFadden form. The expenditure function is defined as:

$$\text{(18–4)}\quad E(P_t^d, P_t^m, Y_t) = \frac{1}{2} d_{dm} \frac{(P_t^d)^2}{P_t^m} Y_t + b_{dd} P_t^d Y_t$$

$$+ b_{mm} P_t^m Y_t + b_d P_t^d + b_m P_t^m + b_d P_t^d t Y_t + \alpha_d P_t^d t$$

$$+ \alpha_m P_t^m t + \beta_d P_t^d Y_t^2 + \beta_m P_t^m Y_t^2 + \gamma_d P_t^d t^2 Y_t$$

$$+ \gamma_m P_t^m t^2 Y_t$$

Partially differentiating the expenditure function with respect to the domestic price of domestically produced goods and imported goods yields the domestic demand for domestically produced goods and the demand for imports. Again, given the nature of the price indices, we estimate the model using differences in demands. Since the equilibrium condition is that the domestic supply equal the domestic demand of domestically produced goods, the measures of supply and demand must be in the same units. Because domestic supply S_t is deflated by real capital, we also deflate domestic demand D_t by real capital. Thus, the demand side of the model is given by:

$$\text{(18–5)}\quad \frac{D_t}{K_t} = \frac{D_{t-1}}{K_t} \cdot \frac{Y_t}{Y_{t-1}} + d_{dm} \left(\frac{P_t^d/P_{t-1}^d}{P_t^m/P_{t-1}^m}\right) \frac{Y_t}{K_t}$$

$$+ b_d \frac{1}{K_t}\left(1 - \frac{y_t}{Y_{t-1}}\right) + b_{dt} \frac{Y_t}{K_t} + \alpha_d \frac{t}{K_t}\left(1 - \frac{y_t}{Y_{t-1}}\right)$$

$$+ \alpha_d \frac{1}{K_t} \cdot \frac{Y_t}{Y_{t-1}} + \beta_d \frac{Y_t}{K_t} + (Y_t - Y_{t-1})$$

$$+ \gamma_d \frac{Y_t}{K_t}(2t-1) + u_d$$

$$\frac{M_t}{K_t} = \frac{M_{t-1}}{K_t} \cdot \frac{Y_t}{Y_{t-1}} - \frac{1}{2} b_{dm} \left(\frac{P_d^t/P_{t-1}^d}{P_t^m/P_{t-1}^m}\right)^2 \frac{Y_t}{K_t}$$

$$+ b_m \frac{1}{K_t}\left(1 - \frac{Y_t}{Y_{t-1}}\right) + b_{mt} \frac{Y_t}{K_t} + \alpha_m \frac{t}{K_t}\left(1 - \frac{Y_t}{Y_{t-1}}\right)$$

$$+ \alpha_m \frac{1}{K_t} \cdot \frac{Y_t}{Y_{t-1}} + \beta_m \frac{Y_t}{K_t}(Y_t - Y_{t-1})$$

$$+ \gamma_m \frac{Y_t}{K_t}(2t-1) + u_m$$

where: D_t is the domestic demand for domestically produced goods; M_t is the demand for imported goods; Y_t is real GDP; $P^m{}_t$ is the price of imported goods. The b's, α's, β's, and γ's are parameters to be estimated, and all other variables are defined as before.

The Data

The data used in the estimation are drawn from two sources. The export, output, and import price indices were obtained from the Ministere du Plan et de l'Industrie. These series were only available from 1980. The intermediate goods index was calculated by applying the output price indices to the input-output matrix for production in Côte d'Ivoire. It was available only at the two digit level. To ensure conformability with the other indices, we calculate a labor cost index as the median nominal wage of all firms in the sous-branche in year relative to the median nominal wage in year t - 1. The median was chosen to reduce the effect of outliers.

Data on the value of imports within the sous-branche classification are also obtained from the the Ministe re du Plan et de l'Industrie. The imports refer to imports of final goods within this classification, not the imported inputs used by domestic firms within the sous-branche.

All other data on domestic sales, exports, expenditure on labor, and expenditure on intermediate inputs was calculated from information provided to the Banque des Donne es Financie res by individual firms in the manufacturing sectors. The value of domestic sales was measured by the value of gross sales minus the value of exports. Capital was measured as net cumulated investment, deflated by an aggregate capital goods price index. All differences are real differences, expressed in prices prevailing at time t - 1 and deflated by the appropriate price index.

This information was available going back to 1976, but we could not consider a longer sample period owing to the limitation on the price data. Details on the construction of the data and a brief description of the data set are available from the authors.

Estimation and Results

We estimated models with and without the demand side equations and with and without the restrictions of a constant returns to scale technology for two different samples.[6] The models with only the output supply and input demand equations were estimated using iterative seemingly unrelated regression. The models with the demand side were estimated using iterative three-stage least squares. The results are presented for the sample obtained from an aggregation over all firms in the sous-branche in table 18–1.

We also estimated the model with only exporting firms using a two-stage correction for possible selectivity bias, but found little difference in the results with the more complicated selectivity bias correction.[7] (The result of these regrssions are presented in Lavy, Newman, and de Vreyer 1990). The main points to make with respect to these estimates are that:

- the parameters are all of the expected signs;
- on the whole, the parameters on the price variables are estimated quite precisely;
- the results with exporting firms only are similar to those with all firms, but with a somewhat greater sensitivity to price variation;
- the restrictions implied by a constant returns to scale production process are rejected in at the one percent level in all cases; and
- the theoretical curvature conditions, that the profit function be convex with respect to prices, are satisfied in all but one of the models estimated. For all cases except for the estimation of all firms with the demand side and a constant returns to scale technology, all the eigenvalues of the matrix of price coefficients were positive, indicating a positive semidefinite matrix.

Because of the nonlinearities present in all the models, it is difficult to draw inferences from tables of the estimated coefficients of the structural model. Model simulations based on the estimated coefficients provide a clearer picture of the effect of price changes on output and exports.[8] By simulating the model one hundred times, one can calculate not only the mean elasticity but also a standard error of the simulation. For over 90 percent of the sous-branches and years, one can not reject the hypothesis that the distribution of the one hundred solutions of the endogenous variables is normal. Thus, these standard errors can be used to form confidence intervals for the simulated elasticities.

Table 18–2 presents arc elasticities based on simulations of the model with supply equations only and non-constant returns to scale for the sample aggregated over all firms. The reported elasticities capture the partial effects of a change in one exogenous variable, holding all other exogenous variables constant. In particular, it is important to note that the elasticities with respect to changes in export prices, labor costs, and intermediate goods costs were obtained holding domestic prices constant. These elasticities pertain to the short run, keeping capital fixed. Standard errors of the simulations are presented in parentheses. There is very little dispersion around the mean effect. The low standard errors reflect the fact that the price variables that are changed in the simulations were precisely estimated in table 18–1.

Table 18–2 indicates a positive export supply response to increases in export prices. Given the estimated elasticity of 0.34, a 40 percent increase in export prices (of the order of magnitude contemplated in the subsidy cum tariff program) would be expected to lead to a 12 percent increase in exports. This is very close to the value of the short-run own price export supply elasticity obtained by Diewert and Morrison (1988) using aggregate U.S. data. Using a very different methodology and aggregate quarterly data for Greece, Balassa et al (1986) find a short-run elasticity of around 0.6. Their estimated long run elasticities are considerably higher.

Table 18–1. *All firms*

Parameter	Supply side only		Joint supply and demand	
	CRTS	Non-CRTS	CRTS	Non-CRTS
b_{sx}	4.38*	3.26*	39.87*	39.85*
	(1.35)	(1.30)	(1.48)	(1.97)
b_{sl}	-0.60*	-0.47*	-3.62*	-3.12
	(0.17)	(0.15)	(0.20)	(0.23)
b_{sg}	-2.92*	-2.28*	-27.37*	-30.04
	(0.97)	(0.95)	(1.08)	(1.54)
b_{lx}	0.42*	0.44*	0.67*	0.64*
	(0.04)	(0.04)	(0.05)	(0.04)
b_{lg}	0.17	0.07	2.20*	2.08*
	(0.12)	(0.10)	(0.18)	(0.21)
b_{gx}	2.52*	2.19*	19.01*	23.22*
	(0.74)	(0.73)	(0.98)	(1.41)
b_{xt}	0.60*	0.64*	1.75*	1.15*
	(0.12)	(0.15)	(0.22)	(0.25)
b_x		-1.04		-27.3
		(11.6)		(19.8)
α_x		2.77		23.7*
		(5.19)		(9.4)
β_x		1.15E-5		5.96E-5
		(9.98E-5)		(1.43E-4)
γ_x		-0.03		0.01
		(0.02)		(0.03)
b_{lt}	-0.01	-0.07	0.91*	0.49*
	(0.06)	(0.05)	(0.09)	(0.10)
b_l		-2.23		13.2**
		(1.84)		(7.3)
α_l		-2.56*		-13.7
		(0.83)		(7.3)
β_l		1.59E-5		4.4E-5
		(1.57E-5)		(5.12E-5)
γ_l		0.0078		0.01
		(0.0034)		(0.01)
b_{gt}	0.13	0.06	7.28*	5.90*
	(0.33)	(0.33)	(0.64)	(0.87)
b_g		-16.3		1.38E+2*
		(12.3)		(64.1)
α_g		-2.79		-1.10E+2*
		(5.50)		(30.4)
β_g		8.69E-6		2.71E-4
		(1.06E-4)		(4.6E-4)
γ_g		-0.03		0.04
		(0.02)		(0.10)
b_{st}	1.95*	2.25*	-10.59	-8.39*
	(0.47)	(0.53)	(0.97)	(1.23)
b_s		-74.6		-2.12E+2*
		(30.5)		(93.8)
α_s		54.74		1.75E+2*
		(13.64)		(45.2)
β_s		3.62E-4		-4.91E-4
		(2.65E-4)		(6.3E-4)
γ_s		-0.06		-0.08
		(0.06)		(0.014)
b_{dm}			0.00016	-3.75E-4**
			(0.0002)	(0.0002)

Table 18–1. *(continued)*

b_{dt}			0.0016	0.0021*
			(0.0010)	(0.0009)
b_d			-2.92E+3	-2.29E+4*
			(1.44E+4)	(1.17E+4)
α_d			-2.57E+2	-2.21E+2*
			(99.91)	(84.8)
β_d			-6.01E-8	-4.47E-7*
			(2.76E-7)	(2.24E-7)
γ_d			0.0012	0.0010*
			(0.0004)	(0.004)
b_{mt}			-0.11*	-0.24*
			(0.06)	(0.06)
b_m			2.56E+6*	1.62E+6
			(1.08E+6)	(1.13E+6)
α_m			1.19E+4*	2.92E+4*
			(6.20E+3)	(6.89E+3)
β_m			4.93E-5*	3.12E-5
			(2.06E-5)	(2.16E-5)
γ_m			-0.05*	-0.14*
			(0.03)	(0.03)
Objective*N	546.00	530.00	239.34	211.95
N	139.00	139.00	139.00	139.00

Note: Standard errors are in parentheses. *Significant at the 5 percent level. **Significant at the 10 percent level.

The low output supply elasticity (0.03) and the negative domestic supply elasticity (-0.11) with respect to an increase in export prices, suggests that the increase in exports comes about primarily from a decrease in sales to the domestic market. Holding domestic prices constant, the increase in the relative profitability of exporting due to a 40 percent increase in export prices is estimated to lead to roughly a 4 percent decrease in domestic sales. Since the output effect is small, the effect of the increase in export prices on input demands for labor and intermediate goods is also estimated to be small.

Increases in input prices have the expected effects. An increase in labor costs reduces labor use, export supply, supply to the domestic market, and output. The own price elasticity for both labor and intermediate goods is close to minus one. Output appears to respond more in the short run to changes in the cost of intermediate goods than to changes in the cost of labor.

The third column is included to allow for a comparison of our results with other approaches that assume that domestic prices are exogenous and do not respond to changes in export or import prices. The simulations suggest that both the supply curve for domestic sales and for exports are responsive to changes in domestic prices. The own price elasticity is 1.12 for domestic sales and 0.51 for exports, based on a similar 10 percent increase in prices.

The supply curve for domestic sales is clearly more responsive to price that that of exports. In other words, it takes a higher increment in price to induce an equal increase in output for export goods than for domestic goods. Given our framework, we can only speculate why this is the case. The differences may reflect higher costs of transportation, higher costs of achieving a standard quality for export markets, higher transaction costs, or increased risk.

Consistent with the notion of a fairly flat supply curve for domestic sales, the simulations suggest relatively high elasticities of output, use of labor, and use of intermediate goods with respect to the domestic price. As output would have to increase substantially before domestic prices would rise, higher domestic prices would be associated with a large reduction in exports. We do not want to place undue emphasis on the results of a change in domestic prices, since we do not believe that they should be considered exogenous.

Table 18–3 presents quantity responses based on simulations of the model with joint estimation of supply and demand, with non-constant returns to scale and for the sample aggregated over all firms. Again, the simulations pertain to the short run, with

Table 18–2. *Exporting firms only*

Parameter	Supply side only		Joint supply and demand	
	CRTS	Non-CRTS	CRTS	Non-CRTS
b_{sx}	7.32*	6.64*	54.29*	35.41*
	(1.93)	(1.98)	(1.68)	(1.72)
b_{sl}	-1.75*	-1.40*	-2.39*	-3.41
	(0.35)	(0.36)	(0.47)	(0.25)
b_{sg}	-3.90*	-3.72*	-40.49*	-25.93
	(1.42)	(1.43)	(1.26)	(1.33)
b_{lx}	1.04*	0.84*	0.97*	0.66*
	(0.13)	(0.13)	(0.11)	(0.04)
b_{lg}	0.16	0.08	0.63*	2.20*
	(0.26)	(0.13)	(0.41)	(0.21)
b_{gx}	3.23*	3.27*	31.81*	19.35*
	(1.13)	(1.14)	(0.98)	(1.20)
b_{xt}	0.65*	0.73*	2.16*	1.10*
	(0.16)	(0.19)	(0.27)	(0.23)
b_x		-14.81*		-42.44
		(5.56)		(46.55)
α_x		2.07**		16.13*
		(1.17)		(10.16)
β_x		2.14E-6		5.80E-5
		(8.68E-5)		(1.31E-4)
γ_x		-0.01		0.014
		(0.02)		(0.029)
b_{lt}	-0.56*	-0.44	0.85*	0.65*
	(0.14)	(0.15)	(0.15)	(0.11)
b_l		-8.80*		15.95
		(2.91)		(18.74)
α_l		-0.68		-1.11E+1*
		(0.61)		(4.10)
β_l		1.70E-5		3.70E-5
		(4.54E-5)		(5.20E-5)
γ_l		0.01		0.013
		(0.01)		(0.012)
b_{gt}	0.40	0.31	9.04*	5.39*
	(0.48)	(0.50)	(0.93)	(0.75)
b_g		-7.99		1.76E+2
		(8.30)		(1.39E+2)
α_g		-0.77		-9.94E+1*
		(1.74)		(30.12)
β_g		3.06E-5		2.30E-4
		(1.30E-4)		(3.9E-4)
γ_g		-0.01		0.036
		(0.04)		(0.088)
b_{st}	1.66*	2.13*	-12.86	-7.73*
	(0.47)	(0.82)	(1.30)	(1.06)
b_s		-0.86		-2.33E+2
		(20.9)		(2.0E+2)
α_s		5.65		1.57E+2*
		(4.39)		(80.29)
β_s		2.62E-4		-9.80E-7
		(3.27E-4)		(3.81E-7)
γ_s		-0.09		-0.0014
		(0.09)		(0.036)
b_{dm}			0.00012*	-4.5E-4**
			(0.0003)	(0.0002)

Table 18–2. *(continued)*

b_{dt}			0.00185	0.0030*
			(0.0015)	(0.0008)
b_d			-6.83E+4	-5.12E+4*
			(4.36E+4)	(1.98E+4)
α_d			-5.50E+1	-3.02E+2*
			(1.38E+2)	(80.29)
β_d			-1.31E-6	-9.80E-7*
			(8.42E-7)	(3.81E-7)
γ_d			0.0014	0.0014*
			(0.00062)	(0.00036)
b_{mt}			-0.22*	-0.80*
			(0.04)	(0.046)
b_m			7.58E+6*	3.23E+6*
			(2.7E+6)	(1.49E+6)
α_m			1.68E+4*	6.61E+3*
			(4.07E+3)	(5.04E+3)
β_m			1.46E-4*	6.17E-5*
			(5.23E-5)	(2.86E-5)
γ_m			-0.063*	-0.27*
			(0.019)	(0.22)
Objective*N	530.00	510.00	276.42	202.60
N	134.00	134.00	134.00	134.00

Note: Standard errors are in parentheses. *Significant at the 5 percent level. **Significant at the 10 percent level.

capital fixed. In these simulations, domestic prices are permitted to respond to changes in one of the exogenous prices. In the previous simulations presented in table 18–2, the increase in exports due to an increase in export prices resulted from a movement along the export supply curve. As the domestic prices were kept constant, no shift in the supply curve for exports took place. Thus, the previous simulations provide evidence of elasticities of the supply curve. In contrast, the simulations presented in table 18–4 do not yield elasticities of supply and for that reason the term elasticities is in quotes. The simulations do present percentage changes in quantities with respect to percentage changes in prices, but the changes are calculated by comparing two different equilibrium points after allowing domestic prices to adjust. As a change in domestic price shifts the export supply curve, the two equilibrium points correspond to two different short-run export supply curves.

Table 18–3 indicates that domestic prices, in fact, do respond to changes in exogenous prices. Domestic prices are responsive to changes in export prices, labor costs, and costs of intermediate goods—all changes that would shift the domestic supply curve. With an elasticity of approximately 0.2, a 40 percent increase in export prices would be expected to increase the domestic price by 8 percent. Domestic prices are not very responsive to changes in import prices, which would shift the domestic demand curve. These results are consistent with the fairly flat domestic supply curve, suggested from the simulations of the previous model, and with a fairly inelastic domestic demand curve.

Allowing domestic prices to respond reduces the estimated supply responses as theory would predict and as discussed previously. An increase in export prices would lead to a movement along the firms' export supply function. This increase in relative profitability of exports shifts back the domestic supply function and because the domestic demand is downward sloping leads to an increase in the domestic price. With a 40 percent increase in export prices, domestic prices are estimated to increase by roughly 8 percent. This increase in domestic prices reduces the incentive for firms to substitute away from domestic sales and towards exports. While the sign pattern is the same as with the supply side only, the estimated change in exports and domestic output are one third of their previous values in table 18–1. The same 40 percent increase in export prices leads to a 4 percent increase in exports, compared to the 12 percent increase in the previous case when domestic prices were not allowed to vary.

The fact that the export and output responses in this simulation are smaller is not an indication of a lack of responsiveness to price change in the manufacturing sector. As the export supply and domestic supply functions have exactly the same form in the two models, one can directly compare the partial derivatives of the supply functions with respect to the output price. The results indicate that the estimated slopes of the supply function are actually larger in the model where demand and supply are estimated jointly. Moreover, the evidence has indicated that domestic prices do respond to changes in export prices and that firms react in a theoretically consistent manner.

As before, the net effects on output and the use of labor and intermediate goods are small. One would expect the estimated output response to be greater when both export and domestic prices increase than when only export prices are allowed to increase. However, it appears that the net output response is too small to measure this effect.

Accounting for a downward sloping domestic demand curve also leads to a much smaller drop in output and input use with an increase in input prices. However, the same pattern prevails of output being more responsive to changes in prices of intermediate goods than to changes in labor costs. In table 18–3, labor and intermediate goods are estimated to be substitutes. In table 18–3, the cross-price elasticities were negative, but imprecisely estimated.

Using the model with demand equations as the basis for the simulations allows us to consider the effects of an increase in import prices, alone, or in combination with an increase in export prices (as would occur under a tariff cum subsidy program). Given the apparently relatively flat domestic supply curve, increases in import prices that would shift out the domestic demand curve have a sizable effect on domestic sales and output. Holding all other prices constant, an increase in import prices of 40 percent is estimated to decrease imports by 1.2 percent and increase domestic sales by 5.4 percent. As the increase in domestic demand holding export prices constant makes sales to the domestic market relatively more attractive, exports would fall. The 40 percent increase in import prices is estimated to lead to a 2.4 percent drop in exports. The net effect on output is to raise total output by 3.6 percent.

As there is a significant effect on output with the increase in import prices, the use of labor and intermediate goods must go up to generate the additional production. Expenditure on labor is estimated to go up by 3.2 percent and expenditure on intermediate goods by 6.4 percent with the 40 percent increase in import prices.

The simulated effect of the tariff cum subsidy program (a simultaneous 40 percent increase in import and export prices) was estimated to increase output by 4 percent. The percentage increase in domestic sales is more than double that of exports (5.2 to 2 percent). Thus, more of the increased output comes about from an increase in domestic sales rather than exports. All firms benefit from the increase in the import price of their competing products. Exporting firms benefit directly from an increase in export

Table 18–3. *Arc elasticities based on estimates of supply side model*
(with non-constant returns to scale, sample of all firms)

	Change in price				
Change in real quantity	*Export (+10%)*[a]	*Export (+40%)*	*Domestic (+10%)*	*Labor (+10%)*	*International goods (+10%)*
Exports	0.51 (0.01)	0.34 (0.01)	-0.48 (0.04)	-0.06 (0.01)	-0.17 (0.03)
Domestic sales	-0.14 (0.01)	-0.11 (0.01)	1.12 (0.04)	-0.16 (0.01)	-0.80 (0.03)
Total output	0.06 (0.01)	0.03 (0.01)	0.62 (0.03)	-0.13 (0.01)	-0.60 (0.02)
Labor	0.08 (0.01)	0.07 (0.01)	0.87 (0.03)	-0.84 (0.01)	-0.11 (0.02)
International goods	0.04 (0.01)	0.03 (0.01)	1.09 (0.04)	-0.03 (0.01)	-1.07 (0.03)

a. The percentage changes in the prices that were used in calculating the arc elasticities are presented at the top of each column. Because the model is nonlinear, the point estimates of the elasticities calculated on the basis of a 10 percent change in price are different from those calculated on the basis of a 40 percent change.

b. Standard errors of the simulations are presented in parentheses.

prices, while nonexporting firms benefit only to the extent that higher export prices increase domestic prices.

In drawing inferences about a subsidy program from these results, it is important to note a critical distinction between an export subsidy and a nominal devaluation. A nominal devaluation will always lead to an increase in export prices, while export subsidies will lead to higher export prices only if firms choose to participate and take advantage of the subsidy program. If, due to lack of credibility of the program or high transaction costs, firms do not participate, their export prices will not rise. Only if enough firms participate in the subsidy program to affect the domestic price level, will firms that do not export benefit from the export subsidy program.

Previous evaluations of export subsidies have appraised their performance on the basis of the extent to which they lead to an increase in exports. They paid less attention to net effects on output and employment. Our simulation results suggest that in the short run a greater increase in exports (4 percent as opposed to 2 percent), would occur if the export subsidy was not accompanied by an import tariff. Indeed, we ran an additional simulation (not reported in table 18–3) that suggested that a still larger increase in exports (8 percent) would be obtained if a 40 percent increase in export prices was accompanied by an equal decrease in import prices. This is consistent with the finding of Nogues (1990) that export subsidies failed (in the sense that they did not lead to increased exports) in Argentina where there was high import protection, but succeeded in Brazil when they were accompanied by import liberalization. However, at least in the short run in Côte d'Ivoire, increases in exports do not compensate for the decreases in domestic sales that occur when domestic producers confront lower prices for competing imports.

The Program and its Stylized Facts

The behavioral relationship estimated and reported in this paper could be used to predict output and export supply under the set of export and import prices that prevailed with the Ivorian export subsidy cum import tariff program. These predictions should then be compared to, and evaluated against the actual changes of exports, imports, and output. An estimate of the net effect of the program could be obtained by comparing the predicted supply under the actual export and import prices with the predicted supply under the export and import prices that would have prevailed in the absence of the program. Unfortunately, neither complete price nor output information is available to us at the moment. We

Table 18–4. ***Arc "elasticities" based on joint estimation of supply and demand***
(with non-constant returns to scale, sample of all firms)

	Change in price						
Change in real quantity	*Export (+10%)*[a]	*Export (+40%)*	*Import (+10%)*	*Export & import (+10%)*	*Import (+40%)*	*Lab or (+10%)*	*International goods (+10%)*
Exports	0.15 (0.02)[b]	0.10 (0.01)	-0.07 (0.01)	-0.06 (0.01)	0.05 (0.01)	-0.03 (0.01)	-0.05 ((0.02)
Domestic sales	-0.03 (0.002)	-0.03 (0.01)	0.20 (0.01)	0.16 (0.01)	0.13 (0.01)	-0.02 (0.002)	-0.16 (0.01)
Total output	0.03 (0.01)	0.01 (0.004)	0.12 (0.004)	0.09 (0.004)	0.11 (0.01)	-0.02 (0.005)	-0.12 (0.01)
Imports	0.01 (0.001)	0.01 (0.001)	-0.04 (0.002)	-0.03 (0.002)	-0.03 (0.002)	0.01 (0.003)	0.04 (0.002)
Domestic price	0.19 (0.002)	0.19 (0.002)	0.02 (0.001)	0.01 (0.001	0.21 (0.002)	0.10 (0.004)	0.72 (0.002)
Labor	0.04 (0.008)	0.03 (0.006)	0.10 (0.006)	0.08 (0.005)	0.12 (0.007)	-0.64 (0.008)	0.48 (0.01)
International goods	0.01 (0.01)	0.003 (0.01)	0.20 (0.01)	0.16 (0.01)	0.17 (0.01)	0.12 (0.002)	-0.34 (0.01)

a. The percentage changes in the prices that were used in calculating the arc elasticities are presented at the top of each column. Because the model is nonlinear, the point estimates of the elasticities calculated on the basis of a 10 percent change in price are different from those calculated on the basis of a 40 percent change.

b. Standard errors of the simulations are presented in parenthesis.

therefore do not attempt to evaluate the effectivness of the program. Instead, we only sketch the main elements of the subsidy-tarrif program, present what is known about its implementation, and attempt to draw a picture of the main changes during its first two years. We also analyse the likely causes of the observed pattern, relating it to the elasticities and simulations reported above.

The Export Subsidy Cum Import Tax Scheme

The tariff cum subsidy program was announced at the end of 1985, with the first disbursements taking place in the middle of 1986. Initially, an export subsidy at a rate equal to the tariff rate on a similar product was instituted for three sectors—wood products, textiles, and agro-processing. The program was implemented selectively primarily because of the concern over its budgetary impacts. The gross subsidy rate varied from 10 to 40 percent even within the same subsector. In January, 1988 the export subsidy was expanded to the chemical and rubber industries. Other industries such as machinery, electronics, etc. were still excluded. The payments were to be made within three months, but, in practice, have been paid out with a longer lag which reduces the value of the subsidy. At the beginning of the program, import tariffs were adjusted and increased with the objective of achieving enough revenue to finance the export subsidy.

From 1986 to August 1988, 52 firms took advantage of the new export incentives. We have identified most of these firms in the sample, but only 42 had continuous data for the years since their creation. Most of these firms are "veteran" exporters. They are much larger than other exporters: in 1987 their mean size (1,020 workers) was more than twice the average size of all exporters (438). Their export level (in nominal CFA) is again almost twice that of other exporting firms, reflecting not only the size differential but also a much higher propensity to export. The premium receiving firms exported on average in 1987 almost 60 percent of their total output compared to only 28 percent for non premium receiving firms.

No thorough analysis of the impact, or even the potential impact of this scheme, has yet been conducted. (The study by Noel and Gilles (1984) studied the impact of uniform export subsidy and import tariff within a general equilibrium SAM model.) Preliminary evidence, in the form of firm surveys conducted by the Government, indicates that to date the scheme has not had its desired effect. Firms currently exporting state that the subsidy has helped them to maintain their market share, but has not affected their plans for investment. Producers of import substitutes assert that the scheme has had very little effect on the prices of imports and, therefore, has not enhanced their competitiveness.

In the next sub-section we present the evidence regarding the performance of the exporting sector and of the premium-receiving firms after the implementation of the program. Since the firm data from the Banque des Donne es Financieres is incomplete for 1987 and 1988, we base the inference of changes only on firms for which data is available for pairs of adjacent years (t and t-1). The annual files are matched and merged for every two preceding years, allowing a comparison using identical firms in both t and t+1. Using information on the date of creation, surviving firms are distinguished from "true" entry and also from "true" exiting firms. This distinction is required in order to detect any response of entry into more rewarded sectors, and exit from less rewarded sectors as a response to the change in the trade regime. All the relevant nominal values were deflated by the appropriate price deflators.

The Experience with the Program: 1986-1987

During 1986-1987 the nominal (CFA) value of exports declined by 11 percent in 1986 and by 1.0 percent in 1987. However when exports are deflated by export prices the results are just the opposite: the volume of exports increased sharply both in 1986 and 1987 (15 and 22 percent respectively). To verify this result, we compared it with the volume of exports data obtained from the customs figures. This alternative source provides an independent series on the quantity of exports and its confirms our earlier results, suggesting also a significant increase in the quantities exported. This growth took place while export prices (in domestic currency) in most sectors declined, mainly as a result of the large appreciation of the CFA versus the U.S. dollar. Thus the total volume and the dollar value of exports increased, though exports values in terms of CFA or French francs have decreased.

The output and export behaviour of the premium firms resemble closely that of the export sector. During the period 1981-1985, the export performance of the two groups of firms was similar, averaging a volume growth of 13 percent annually. During 1986, however, the premium firms experienced a lower growth of exports compared to other firms (6 versus 15 percent). In 1987 the export growth rate was almost equal to that of the non premuim firms (23 percent).

The output and export performance (expansion) in 1986-87 could be based on three possible effects. First, the export subsidy could have contributed to this growth by offsetting to some extent the decline in export prices. Second, the appreciated CFA would have led to lower imported input prices leading to a shift to the right of the supply curve. Thirdly, the decline in domestic demand as well as the shift to the right of the supply curve, could explain the export performance in 1986 and 1987. The model and the simulations presented in the previous section suggest that, in the past, all three of these effects have contributed to growth in real exports. The elasticities reported in tables 18–3 and 18–4 suggest a large sensitivity of exports to domestic prices and a more modest export response to export and input price changes. However, in order to evaluate the exact relative importance of the three factors more information is needed about the actual changes of export and input prices as well aggregate demand.

Conclusions

From a methodological standpoint, two results emerged from this paper. First, the exercise of estimating firms' output supply and input demand functions using flexible functional forms was successful. The estimates satisfied theoretical curvature properties and the price effects were precisely estimated. Second, joint estimation of supply and demand leads to considerably different estimates of export and output supply responses than estimates based on supply relations alone. Simply instrumenting for endogenous domestic prices may take care of simultaneous equations bias, but is not an adequate substitute for modeling of the demand side. Without modeling the demand side, it is not possible to obtain estimates of the net effect of changes in export prices on exports and output. Although our specification of the demand side was relatively simple, the results were consistent with theoretical expectations. Further work using less restrictive specifications of the demand side would be useful, as would estimations based on alternative flexible functional forms.

Contrary to prior expectations, our results suggest that firms in Côte d'Ivoire do sell more to the foreign market when it is more profitable to do so. Exports respond positively to increases in export prices and negatively to increases in import prices. However, the fact that exports would be lower if an export subsidy were combined with an import tariff is not an argument for introducing an export subsidy alone. Firms producing in the tradable goods sector suffer from an overvalued exchange rate not only because they would receive a lower price for their exports, but also because they must compete against lower priced imports. Our estimates indicate that the introduction of an export subsidy alone would be insufficient to increase output of the tradable goods sector. The combination of an export subsidy with an import tariff, which comes closer to mimicking the effects of a devaluation, would serve to counteract some of the short-run adverse output effects of an overvalued exchange rate. We cannot speak to the longer run effects of such a program, to the budgetary implications, or to implementation issues.

The results presented in this paper do not in any way constitute an evaluation of the actual tariff cum subsidy program that was adopted in Côte d'Ivoire in 1986. The subsidy program has not resulted in the type of uniform increase in export prices within a sous-branche that was analyzed in this paper. As of August 1988, only 52 firms had taken advantage of the new export incentives. These firms comprise a third of all eligible exporters, but a lower proportion of all manufacturing firms within the sous-branches eligible for the subsidy. Most of the firms which applied for the subsidy are large, veteran exporters, employing on average more than 1000 workers and exporting more than 60 percent of their output.

The program had started with long delays in subsidy payments, which may have discouraged firms from applying for the subsidy. Even though tariff rates on imports were raised, no increases in tariff revenues could be detected. By the fall of 1988, the program's budgetary burden led the government to stop all payments and the program was de facto abandoned. However, this program remains a policy option under discussion.

Notes

1. Properly speaking, errors in the first-stage estimation (the instrumenting equation) should be taken into account in the calculation of the second stage standard errors (see for example, Duncan, 1987; Pagan, 1986).

2. As defined by Diewert (1974) and as stated in Diewert and Wales (1987), a flexible functional form of cost function is one that would provide a second order differential approximation to an arbitrary twice continuously differentiable cost function that satisfies the linear homogenity in prices property at any point in the admissable domain. Flexibility of the functional form of a profit function is defined analogously. The expression "restricted" in the profit function refers to capital being fixed.

3. We also specified producer supply and input demand functions derived from the more restrictive Cobb-Douglas production function, but did not obtain reasonable results. For example, the effect of increases in export prices on exports was negative.

4. See Diewert and Wales (1987).

5. All efforts to estimate models using individual firms as the unit of observation and replacing the unobserved firm prices by sous-branche level price indices were unsuccesful. We suspect that measurement error in the price data was responsible for the poor resultas.

6. All models were estimated using SAS's PROC SYSNLIN.

7. In the first stage we predicted the probability of each firm in the sous-branche would export, constructed a Mills ratio for each exporting firm, and estimated the models using sous-branche level data with the sum of all individual firm's Mills ratios. This has the implicit assumption that the covariance matrix of the errors in the export probability equations and the second stage model equations is the same for all the firms.

8. All simulations were done using SAS's PROC SIMNLIN.

References

Aspe, P., and F. Giavazzi. (1982). "The Short-Run Behavior of Prices and Output in the Exportables Sector." Journal of International Economics 12:83-93.

Artus, J. and S. Rosa. (1978). "Relative Price Effects on Export Performance: The Case of Non-Electrical Machinery." IMF Staff Papers 25, pp. 25-47.

Balassa, Bela. (1987). "Economic Incentives and Agricultural Exports in Developing Countries." World Bank, DRD Discussion Paper DRD250, Washington, D.C.

Balassa, B., E. Voloudahis, E.P. Fylahtos and Sub Suhtai. (1986). "Export Incentives and Export Growth in Developing Countries: An Econometric Investigation." World Bank, DRD Discussion Paper DRD159, Washington, D.C. (February)

Baumann, Renato, and Braga, Helson C. (1988). "Export Financing in LDCs: the Role of Subsidies for Export Performance in Brazil." World Development (UK). 16:821-33.

Diewart, W. E., and C. Morrison. (1988). "Export Supply and Import Demand Functions: A Production Theory Approach." in Feenstra (ed.) Empirical Methods for International Trade, MIT Press, Cambridge, Mass.

Diewert, W. E. and T.J. Wales. (1987). "Flexible Functional Forms and Global Curvature Conditions", Econometrica 55: 43-68.

Duncan, G. (1987). "A Simplified Approach to M-Estimation with Applications to Two- Stage Estimators", Journal of Econometrics, 34: 373-389.

Faini, R., (1988). "Export Supply, Capacity, and Relative Prices", The World Bank, WPS Paper #123.

Goldstein, N., and M.S. Khan. (1978). "The Supply and Demand for Exports: A Simultaneous Approach." Review of Economics and Statistics, 60: 275-86.

Kohli, U. (1978). "A Gross National Product Function and the Derived Demand for Imports and Supply of Exports." The Canadian Journal of Economics 11:167-182.

Lavy, V., J.Newman and P. de Vreyer, 1990. "Export And Output Supply Functions With Endoggenous Domestic Prices", mimeo, The World Bank, October.

Lorch, Klaus. (1989). "Côte d'Ivoire: Industrial Competitiveness During Economic Crisis and Ajustment" Mimeo, The World Bank, Washington, D.C.

Milanovic, B. (1986). "Export Incentives and Turkish Manufactured Exports, 1980-1984." World Bank Staff Working Paper, No. 660.

Ministere de l'Industrie, (1988). "Schema Directeur du Developpement Industriel de la Côte d'Ivoire" Republique de Côte d'Ivoire, (in four volumes).

Nogues, Julio. (1990). "Experience of Latin America with Export Subsidies." Weltwirtschaftliches Archiv, Federal Republic of Germany, 126, No. 1:97-115.

Pagan, A. (1986). "Two Stage and Related Estimators and Their Applications", Review of Economic Studies, 53: 517-538.

Zilberfarb, B. (1980). "Domestic Demand Pressure, Relative Prices and the Export Supply Equation: More Empirical Evidence." Economica 47:443-450.

19

Do African Countries Pay More for Imports? Yes

Alexander J. Yeats

The debt crisis and declining living standards require careful husbanding of critically scarce foreign exchange in most African countries. But economic theory suggests that smaller countries, which import from only a few international suppliers and cannot support competitive markets and infrastructure, would be likely to pay more, rather than less for imports. Analysis of import unit values for 1962-87 shows that the 20 African former French colonies paid a price premium of 20 to 30 percent on average over other importers for iron and steel imports from France. The losses associated with these adverse prices total approximately $2 billion by 1987. The study also finds that similar price premia (of 20 to 30 percent) were paid by former Belgian, Portuguese and British colonies in Africa for imports of these products from their former rulers.

Development of optimal trade and commercial policies in developing countries depends crucially on factors such as whether transnational corporations extract excessive profits, or whether multiple sources of supply will produce lower import prices. This is important since the poorest countries must pay the lowest possible prices for imports of industrial equipment and production inputs required for economic growth. However, if market imperfections exist, or if competition is less vigorous than it might be, some developing countries may pay more than competitive prices for imports, or receive less than competitive prices for exports (for examples of studies that have found evidence of this, see Hewett 1974, UNCTAD 1975, and Yeats 1978 and, for a review of the recent industrial organization literature on these issues, see Bresnahan 1989).

Institutional factors may prevent developing countries from attaining the best terms for imports. Helleiner (1978) argues that restrictive trade practices, national and international cartels, or some countries' lack of countervailing power may work against the efficient functioning of international markets. In addition, antitrust laws are often weak, nonexistent, or unenforceable at the international level. Similarly, Edwards (1972) documents the adverse effects of restrictive inter-firm practices such as agreements for the allocation of territorial markets; pooling and allocation of patents, trademarks, and copyrights; fixing prices and discriminatory pricing; allocation of export business shares among firms; and establishment of reciprocal, exclusive, or preferential dealing. At the national level, inter-firm agreements on exports extend not only to the allocation of foreign markets, but even to individual foreign customers, allocation of specific goods to be exported, fixing of prices and levels of bidding on foreign contacts and the selection in advance of the firm that will submit the lowest bid. All of these factors can lead to higher prices than those that would prevail under more competitive conditions.

To determine if such "excess" prices are being charged to African importers, this article first examines the distribution of import prices paid by developing countries which have highly concentrated trade with a major exporting country (France) using extensive time series data on unit values for homogenous goods. These prices are compared with those paid to France by other countries whose imports come from more diversified sources and, where evidence of "excess" prices are found, the level of economic costs is quantified. In addition, the analysis employs correlation and regression tests to account for the influence on relative prices of other economic and institutional factors such as the degree of market concentration, or the size of the importing market. Next, the article examines the pricing policies of other European countries (Belgium, Portugal and the United Kingdom) with their former colonies. It closes with an overall assessment of the implications of these findings for developing countries' trade and commercial policies and suggest related research that appears to have a high priority in addressing these issues.

The Methodological Approach

By comparing various European countries' share in the trade of their former colonies with similar data for a control group of developing countries, Kleiman (1976) develops an index of relative trade concentration.The results suggest that former colonial countries' trade with the United Kingdom was three times the normal level for developing countries, for the French associates about eight times as high and for the Italian, Belgium and Portugese colonies the concentration was even higher. As this might suggest, while France was selected as the main focus for study due to its very high trade intensity with its former colonies, the analysis will also show that the findings can be generalized to Belgium, Portugal, the United Kingdom, and, quite probably, other countries.

The first step in the empirical analysis was to compile annual data on the quantity and value of French exports (free on board -- f.o.b.) on a joint product-by-country basis from United Nations Series D Commodity Trade Tapes and to compute unit values for every five-digit Standard International Trade Classification (SITC) iron and steel product exported by France from 1962 to 1987. Where more disaggregated data were not available, similar statistics were drawn for several higher level products (four-digit SITC). An effort was made to hold the four-digit items to a minimum, however, since their unit values can be affected by differences in their product mix. Several products had to be excluded from further analysis when tests showed they were exported to too small a number of countries, or when full 1962-87 value and quantity data were not available. This left 11 four- and five-digit SITC steel products that comprised 40 to 60 percent of French iron and steel exports over the 25 year period. The products and the 20 African French-associate countries are listed in table 19–1.

Next, the size of any overall price margins that French-associated countries may have paid relative to other importers was estimated. Unit values, *U*, of good *i* imported by French associates (*f*) (or the comparator group, *g*), were calculated as the total f.o.b. value (V_i) over the quantity (Q_i). "Price" margins for each good were compiled as the ratios of unit values-for the French associates to unit values for the comparators:

$$Mi = Vif/Qi \,/ Vig/Qig$$

To derive an aggregate across goods, $M_{f,g}$, the individual goods' margins were weighted by the share of the value of each good in the total value of the sampled iron and steel shipments imported by the associated French countries (V_{Tf}) and summed:

$$(19\text{–}1) \quad M_{f,g} = \sum (V_{if}/Q_{if} \times Q_{ig}/V_{ig}) \times V_{if}/ V_{Tf}$$

Equation(19– 1) computes the aggregate price differential that French associates pay (positive or negative) relative to other countries, weighted by the value of shipments to the associates. The results are presented for two year time periods in an attempt to smooth out the effects of any unrepresentative trade values that might influence annual figures.

A second measure of the costs (or benefits) of these price differentials (*Ef*) is derived:

$$(19\text{–}2) \quad Ef = \sum (Uif - Uig)\, ' Qf$$

where *Uif* and *Uig* are unit values for the French associates and other developing countries (i.e., the *V/Q* terms in equation (19–1) for the imported product. By multiplying the difference between the unit values of the two country groups times the quantity of imports, this equation computes how much more (or less) the associated countries pay for their total imports of the product. These calculations are then summed over all iron and steel imports, and are expressed in current and present value terms.

Several points about the export unit value statistics employed in this study should be noted. Analyses based on unit values must generally be treated with caution since differences among products grouped in the same SITC category, or differences in quality among otherwise homogenous products may be reflected as price differences. The statistics on some coated five-digit SITC steel products, for example, do not differentiate between zinc-coated, tinplate and electric-sheet or other similar products. Several previous studies involving five-digit iron and steel products, however, suggest that the overall effects of such variations may be minor. In fact, this homogeneity is such that studies by Stigler and Kindahl (1970), McAllister (1961), and others have used iron and steel unit values to assess the accuracy of wholesale price quotations employed by the United States Bureau of Labor Statistics.

An additional factor that could affect the quality of the trade statistics is the invoicing practices of importers and exporters. Exporters and importers may over- or underinvoice customs vouchers to reduce tariff liabilities, to evade restrictions on the use of foreign exchange, or to illegally obtain subsidies (Bhagwati 1967, Sheikh 1974, UNESCO 1974). In some situations it may be possible to uncover evidence of false invoicing by comparing reported exports with the partner country's reported imports. For several reasons this procedure cannot be used satisfactorily with African data: there are major gaps in many of the African countries' data, many of the records are United Nations "estimates", and some African countries only report trade data at the three-digit level - only four report at the five-digit level employed here. While these factors cause major discrepancies in the official import statistics of many African countries (Yeats 1990), the analysis in this study is based on French and other developed countries' offic ial export statistics which are not subject to the same magnitude of error as the African data. It is unlikely that major incentives exist to induce the iron and steel exporters studied here to falsify their export vouchers.

The Empirical Findings

Initial comparisons of the relative prices paid by the French associated and other developed and developing countries over the 1962-87 period are reported in table 19–2. These statistics show that the French-associated countries are paying more for their imports than other developed or developing

Table 19–1. ***The value and destination of French iron and steel exports: 1962 to 1987***

				Share by importing country groups (percentage)[a]							
	French Iron and steel exports			*Developed countries*				*Developing countries of which*			
Year	*All products ($ million)*	*Sampled products[b] ($ million)*	*Sampled products share (percent)*	*All countries*	*of which: EEC[c]*	*EFTA[d]*	*Total less French associates*	*Latin America*	*Asia*	*French associates[e]*	*Socialist countries*
1962	786.8	461.1	60.0	68.4	48.0	11.4	12.7	4.7	2.9	11.7	6.0
1965	966.4	556.6	57.6	74.7	46.4	11.2	13.3	4.0	3.5	7.8	3.6
1968	1,013.1	561.0	55.3	73.9	48.0	9.7	10.7	3.7	2.0	8.2	6.0
1971	1,532.1	814.0	53.1	77.6	48.8	9.0	10.5	3.3	2.2	7.0	4.4
1974	3,978.5	2,181.6	54.8	73.8	48.4	8.4	11.2	3.1	1.4	7.2	6.8
1977	4,279.3	1,938.3	45.2	68.8	46.4	5.7	12.3	3.4	1.5	8.8	9.3
1980	7,290.0	3,035.2	41.6	69.9	51.7	6.8	14.4	5.0	2.5	7.58	7.5
1983	4,854.1	1,33.9	39.8	69.0	46.4	6.3	15.2	3.3	4.7	6.4	7.4
1986	6,152.5	2,446.5	39.8	75.7	53.0	6.2	12.2	2.7	4.1	4.8	7.1
1987	6,642.7	2,619.0	39.4	76.9	53.8	6.3	11.7	2.4	3.9	3.6	6.6

Notes: a. The developed, developing and socialist country trade shares may not sum to 100 since the destination of some French exports is not shown in the database.

b. (SITC numbers in parentheses): iron and steel simple wire excluding rod (677.01); iron and steel plates and sheets of other than high carbon or alloy steel (674.81); bars and rods of other than high carbon or alloy steel (673.21); tubes and pipes of iron other than cast iron (678.3); plates and sheets, less than 3 mm thick, of other than high carbon or alloy steel (674.31); iron and steel simple big sections (673.41); tube and pipe fittings of iron and steel (678.5); wire rod of other than high carbon or alloy steel (673.11); heavy plates and sheets of iron and steel other than high carbon or alloy steel (674.11); medium plates and sheets, 3.5 mm to 4.75 mm thick, of other than tinned plates and sheets (674.21); angles, shapes and sections, less than 80 mm thick, of other than high carbon or alloy steel (673.51).

c. European Economic Community (10 member countries).

d. European Free Trade Association.

e. These countries consist of Algeria, Benin, Burkina Faso, Cameroon, Central African Republic, Chad, Congo, Cte d'Ivoire, Gabon, Guinea, Madagascar, Mali, Mauritania, Mauritius, Morocco, Niger, Reunion, Senegal, Togo and Tunisia. The declining importance of these countries as a destination for France's iron and steel exports is due primarily to major reductions in France's share of the associate's total iron and steel imports. An additional factor was that the growth in total import demand in these countries generally lagged well below that of other regions. See appendix table 1 for statistics on France's share of the associated countries' iron and steel imports.

Source: Author's calculations based on United Nations Series D Commodity Trade Tapes.

countries. For the full 26 year period, the unit values for French-associated countries always exceeded those of developed market economy countries (their average premium for this period was approximately 24 percent), and in only one two-year period (1976-77) did the associates' price fall below that for all other developing countries. The French associates paid an average premium of 23 percent above the unit value for other developing countries over the full 1962-87 period.

The total cost of the price differential of French-associates relative to all other developing countries, calculated using equation (19–2), was around $430 million, or, in present value terms, close to $900 million over the 1962-87 period. It should be remembered, however, that these are losses only from imports of the sampled steel products which constituted only 39 to 60 percent of French steel and iron exports. If the same pattern of price premia holds for all iron and steel shipments the present value of the associated losses on imports would be approximately $2 billion. To place this ($2 billion) figure in context, note that it is equivalent to 60 percent of the total gross international reserves of 18 of the 20 countries and exceeds the long-term debt of 12 of the associated countries in 1987.

The average premiums or discounts paid by individual associated countries for all imports of sampled iron and steel products are shown in table 19–3. As before, these price comparisons are based solely on French export unit values. For the full 1962-87 period the individual country premiums average close to 27 percent; some of the lowest values were recorded for 1974-77 and 1982-83. The most striking point to emerge from table 19–3 is the extreme variance of average premiums paid both among countries in any one period and by any one country over time. For example, over the 26 year period these premiums averaged 3.1 percent for Morocco, but for Algeria, Gabon and Mauritania, they were at least 15 times greater. The structure and size of markets in these countries are examined below to indicate if they can suggest possible reasons for these price margins.

An Analysis of Differences in Unit Values

Why are there such major differences between the f.o.b. export unit values for different countries of

Table 19–2. *Comparative unit values for France's exports of iron and steel products*

	French f.o.b. exports to associated countries		*Premium or discount paid by French-associated countries compared to:developing countries*				
	Value ($ thousands)	*Unit value ($)*	*Total* [a]	*All developed*	*All non-French*	*Latin America*	*Middle East*
1962-63	118,446	167.00	37.9	40.5	36.9	26.8	50.6
1964-65	98,593	151.50	27.5	29.8	21.8	20.4	23.5
1966-67	86,042	143.80	24.6	26.8	21.0	21.6	18.9
1968-69	101,180	150.00	28.5	31.3	23.9	32.7	14.2
1970-71	119,695	199.30	29.6	32.6	16.7	13.3	13.0
1972-73	187,362	234.80	23.0	26.9	18.6	22.0	16.6
1974-75	368,537	386.70	18.1	26.4	8.1	16.7	17.2
1976-77	341,378	375.80	13.1	20.4	-3.6	10.2	2.9
1978-79	465,702	496.60	19.5	19.8	26.1	24.1	12.1
1980-81	489,195	581.20	25.4	28.6	20.9	26.3	-11.2
1982-83	350,566	458.30	6.6	8.3	8.6	6.0	-13.8
1984-85	318,623	442.90	17.4	15.7	36.2	34.2	16.8
1986-87	269,537	668.00	40.1	37.0	66.5	54.7	10.9

Net revenue gains or losses ($ thousands)	
Actual dollar amount[b]	432,199
Present value[c]	876,183

a. Excludes the French-associated countries in Africa.

b. Calculated as: $A_{df} = (U_f - U_o) \times Q_f$ where Q_f is the quantity of French associated country imports and U_o is the average unit value paid by all other developing countries. These values are then summed over the 1962-87 period.

c. The present value in 1987 of all annual gains or losses computed as above (A_{df}) discounted at 8 percent.

Note: See Yeats (1989) for similar statistics for each of the four and five-digit SITC products included in these computations. The French-associated-country premium or discount P/D_f is calculated as: $P/D_f = [(U_f - U_g)/U_g] \times 100$, where U_f is the unit value for the French associates and U_g is the unit value for the comparator group of countries. The premiums (or discounts) are averaged over all sampled iron and steel products.

Source: Author's calculations based on United Nations Series D Commodity Trade Tapes.

destination? Since these items are generally homogenous in nature, differences in product characteristics should have a fairly limited influence on unit values and, because these are poorer countries, one would expect that they would purchase lower price, poorer quality goods if any product or quality differentiation does exist within the categories.

The relationships of these unit value premia with seven possible explanatory factors are analyzed through simple pairwise correlations. The concentration of import supply among a small number of firms, for instance, could lead to oligopolistic pricing practices. To determine the extent of this concentration the share of iron and steel supplies originating in the largest, and three largest exporting countries was measured. Because each country could have many exporting firms, these measures alone do not provide direct evidence on the potential for collusive practices and over pricing. In support of the country ratios, however, is the fact that iron and steel production is generally among the most concentrated of industries in developed countries so the potential number of exporting firms is limited. Firms headquartered in the same exporting nation may have a greater opportunity and tendency to participate in cartel arrangements or collusive decisions on foreign prices. During the 1962-87 period there appear to have been only 3 or 4 French firms producing the sampled steel products for export, and at various times the links between these companies were reinforced by nationalization.

The size of the export market or export shipments might also be expected to influence the pattern of relative prices. In a study of United States machinery exports, for example, Hufbauer and O'Neill (1973) find evidence "that a small importing country pays a much higher price for its machinery" (p. 272). Thus, the quantity of each country's iron and steel imports from France was computed to determine if larger shipments were associated with lower import prices.

Table 19-3. *Average unit value differentials paid by French-associated countries relative to other countries: Selected iron and steel imports from France, 1962-87*
(percentage)

Importing country	*1962-63*	*1964-65*	*1966-67*	*1968-69*	*1970-71*	*1972-73*	*1974-75*	*1976-77*	*1978-79*	*1980-81*	*1982-83*	*1984-85*	*1986-87*	*1962-87*
Morocco	14.1	6.9	-2.1	4.7	3.2	3.5	0.3	0.9	10.9	0.9	-14.6	-5.6	17.0	3.1
Togo	17.2	21.0	11.7	2.7	-1.0	-4.2	-5.9	-14.3	-3.0	18.5	11.0	25.8	69.9	11.5
Burkina Faso	29.6	29.6	37.1	27.8	12.0	21.6	5.6	6.7	-1.0	2.8	-2.8	-1.0	10.7	13.7
Senegal	21.5	22.8	23.2	17.2	9.8	7.2	5.6	-3.2	6.9	12.4	2.3	21.8	52.3	15.4
Reunion	18.2	20.9	20.6	24.2	13.2	9.4	-1.5	-1.5	10.6	14.4	8.1	26.4	55.5	16.8
Cote d'Ivoire	28.2	27.8	34.7	28.8	16.0	5.9	8.0	0.5	0.2	40.0	8.0	17.4	36.9	19.4
Madagascar	22.1	31.9	26.5	18.3	12.9	-0.3	8.5	4.2	15.3	35.5	18.4	19.5	40.2	19.5
Central African Republic	29.8	26.7	28.0	19.9	13.1	11.3	-0.8	1.6	26.8	4.4	5.7	29.0	60.4	19.7
Benin	36.1	22.4	33.4	11.7	1.9	2.8	3.2	-3.5	1.9	24.0	44.0	20.5	79.6	21.4
Mali	28.7	32.0	73.8	57.1	46.8	10.8	7.6	8.8	10.1	9.9	-2.5	-2.4	16.3	22.8
Mauritius[a]	n.a.	-8.2	-17.7	-2.7	21.1	42.2	60.4	66.1	n.a.	n.a.	n.a.	24.9	n.a.	23.3
Chad	19.7	36.0	30.2	23.8	10.4	26.6	6.7	9.8	18.9	15.9	10.1	34.1	75.5	24.4
Cameroon	30.8	46.2	44.0	34.2	22.7	8.8	18.0	9.2	23.3	38.2	19.4	20.8	78.0	30.3
Congo	27.3	50.4	48.6	20.4	25.9	32.7	10.4	-0.8	22.0	20.6	40.7	46.8	97.0	34.0
Tunisia	15.1	31.3	45.8	48.5	68.4	46.0	42.4	22.1	31.9	18.3	-1.6	15.0	66.0	35.0
Niger	17.4	41.4	14.6	29.0	34.6	15.1	9.6	20.3	41.9	47.7	12.2	73.1	100.5	35.2
Guinea	43.8	55.0	59.2	38.6	45.5	66.0	51.0	29.2	30.8	45.7	49.4	36.6	34.4	45.0
Mauritania	28.3	60.0	49.0	36.3	35.3	35.7	35.0	20.4	62.9	30.8	27.1	48.4	32.6	46.3
Gabon	51.2	49.4	63.8	60.5	58.6	47.5	55.2	5.5	55.9	33.7	28.4	22.4	81.3	47.2
Algeria	77.9	41.0	43.6	50.5	70.8	60.7	33.2	135.9	65.2	58.3	18.1	27.0	22.9	54.2
Weighted average French-associate	37.9	27.5	24.6	28.5	29.6	23.0	18.1	13.1	19.5	25.4	6.6	17.4	40.1	26.9

n.a. — not available

a. For some specific years low import volumes precluded computation of a relative unit value.

Table 19-4. ***Correlation between iron and steel relative import unit values and indicators of market sructure and effective size: 1968-69 and 1986-87***

	Market structure variables						*Market size variables*				*Dummy variabes*			
Size and Structure	*Relative price*		*No. of trade partners*		*Share of 3 largest country suppliers*		*Relative quantity of imports*		*Total imports*		*Associaed countries*		*Developed countries*	
variable	*1968-69*	*1986-87*	*1968-69*	*1986-87*	*1968-69*	*1986-87*	*1968-69*	*1986-87*	*1968-69*	*1986-87*	*1968-69*	*1986-87*	*1968-69*	*1986-87*
Number of trade partners in selected products	-0.448	-0.564*												
Share of 3 largest country suppliers of selected products	0.384*	0.472*	-0.762	-0.569										
Relativequantity of imports of selected productsa	-0.134	-0.708	0.355*	0.711*	-0.150	-0.510								
Total imports (value)	-0.17	-0.626*	0.474*	0.831*	-0.219*	-0.471	0.842*	0.817*						
Asociated country group	0.064*	0.447*	-0.778	-0.671*	0.680*	0.407*	-0.216*	0.507*	-0.377*	-0.753*				
Developed country group	-0.200*	-0.663	0.593*	0.5588	-0.441	-0.423*	0.454*	-0.717*	0.598*	0.688*				
GNP per capita	-0.287*	0.572*	0.560*	0.727*	-0.385	-0.466*	0.575*	0.755*	0.716*	0.820*	-0.517*	-0.745*	0.799*	0.748*

Note: The selected products are noted in table 1. An asterisk (*) indicates statistical significance at the 99 percent confidence level. a Defined as the ratio of iron and steel shipments(in tons) to an individual countru to the average tonnage shipped by France to all trading partners.

Source: Author's caluculations based on United Nations Series D Commodity Trade Tapes.

In addition, the absolute size of each nation's total imports from all sources was measured, to indicate whether there might be economies of scale associated with larger shipments, or wheth- er French pricing policies are different for large export markets where countervailing power may be influential. The correlation of the number of altern- ative (country) suppliers with unit value differenc- es is estimated to determine whether a large variety of contacts and potentially greater sources of information on competitive prices are related to import prices. Since quality differences in imports of machinery have been found to be positively associated with real income (Hufbauer and O'Neill 1972), the correlation of each country's GNP per capita with unit value differences was estimated. Finally, a dummy variable was used to designate transactions between France and another developed country, while a second dummy was used for shipments between France and a former colonial country.

The results of the correlation analysis are summarized for the 1968-69 and 1986-87 periods in table 19–4. Since 1968-69 France considerably broadened its trade contacts thus providing a larger base and range of country characteristics for the price comparisons than was available earlier. To provide the widest possible interval for the intertemporal comparisons of correlation results the period 1986-87 was also selected. Although many of the 1986-87 correlations appear stronger than those for the earlier period, the variables that were significantly correlated with prices and market structure in 1968-69 had a similar relationship in 1986-87.

As shown in the first column of the table, five variables had a significant relationship with relative French export prices in 1968-69 with all of the variables being significant in 1986-87. As is the case with industrial country market studies, variables relating to market structure are strongly correlated with relative prices. For example, a highly significant positive relation ($r = 0.384$) exists between relative prices and the share of imports controlled by the three largest supplying countries in 1968-69 and the relation was even stronger ($r = 0.472$) in 1986-87. A significant inverse association also exists between relative prices and the number of trading partner (country) contacts. Thus, those importing countries maintaining trade relations with a larger number of exporters, and theoretically benefiting from greater competition and information on comparative prices, pay less for their exports.Unfortunately, from the view of development policy, there is evidence that the smaller, poor countries may not be able to sustain a larger number of trading contacts since this variable was significantly and positively correlated with GNP per capita, market size, relative quantities purchased and the developed country dummy.

Somewhat surprisingly, although a strong, negative and significant association between relative prices and both the market size variables is evident in 1986-87, in 1968-69 these associations were lower, and only significant at the 95 but not the 99 percent confidence level. The indirect effects of size appear important in both periods, however, since table 19–4 shows that both these variables are strongly and significantly correlated with market structure which, in turn, influences market prices.

While the correlations between relative prices and these variables are the primary focus of this analy-sis, some of the intercorrelations between the independent variables are also of interest. For example, compared to the developed countries developing nations had significantly fewer trade contacts, smaller markets and higher concentration ratios, all of which undoubtedly contribute to higher import prices.

Additional Evidence From Other African Countries

The previous analyses raise the question of whether other industrial countries' exports show similar pricing patterns. For a test of this proposition, f.o.b. unit values were computed for the United Kingdom's exports of major iron and steel products to former African colonies (Gambia, Ghana, Kenya, Nigeria, Sierra Leone, Sudan, and Tanzania) as well as to all other developing countries. Next, similar computations were made for Belgium (for Burundi, Rwanda and Zaire) and Portugal (Angola and Mozambique). These data were then used to compute the average premia or discount that the Belgian, Portuguese or British former colonies paid over the 1962-87 interval reported in table 19–5.

Over the full 1962-87 period the average premia paid by the former Belgian and French colonies are remarkably close (23.7 and 23.2 percent, respectively) while the former British colonies paid a slightly lower premium of 20.0 percent. The same pricing pattern emerges during 1962-75 for Portugal's exports to its former colonies, but from 1976 on the premiums more than tripled and averaged over 120 percent. It appears that the hostilities in Angola were a major factor behind this dramatic rise as domestic firms may have employed excess pricing as a means of transferring resources out of the country. The statistics in table 19–5 are important, however, as they show that the payment of price premia for imports is widespread among African countries.

Alternative Hypotheses on the Causes of Price Differentials

The empirical approach employed in this paper has close parallels to previous structure-performance studies of industrial countries, which found that prices and profits were higher in markets where aggressive competition was absent. A series of other factors that could influence price margins include: (i) institutional arrangements at the national level between governments; (ii) transnational firm linkages or special commercial arrangements between enterprises; (iii) factors limiting access to international markets due to information, transport, service, or marketing and distribution constraints; and (iv) financial risk differences.

An example of institutional arrangements at the national level which may adversely influence African import prices is the practice of tying bilateral aid, so that recipients must use the funds to buy goods produced in the donor country. Because they are in a sense "captive importers" the African countries may not be offered prices that would prevail in international markets more generally. The bargaining power of African countries may also be limited by the rules of origin under the Lome' Convention. In order to qualify for preferential market access when they are exported to Europe, any assembled iron and steel or other fabricated products must use components produced in the European Community. Similarly, under "reverse preferences", imports from France or Britain were admitted into their former colonies at tariffs considerably below those paid by other exporters until the early 1970s. Such arrangements reduced competitive pressures on domestic European firms and allowed them to raise f.o.b. export prices above those of other (non-African) countries.

A second set of factors that can influence relative prices relate to transnational operations or inter-firm ties which may result in collusive practices. For example, subsidiaries of foreign firms may be required (formally or informally) to purchase from the parent company even when other international traders were offering goods of equal quality at lower prices (Kreinin 1988). This tie was found to be particularly strong for Japanese enterprises. Such overpricing could, of course, be used as a means of transferring profits and capital out of Africa.

Price premia may also reflect weaker African infrastructure for domestic transport, marketing, distribution, service and information systems. Smaller African markets and demand also may not be sufficient to support the required domestic distribution and service operations of more than one or a few foreign suppliers. And problems of language, finance or size or operations all could limit active competition.In addition, existing international transport lines may limit African countries' ability to trade with some low cost producers. Most African countries do not have direct access to North American or Far Eastern producers and imports from these geographic areas may require costly offloading and reloading in foreign ports in transit.

A final set of factors that could affect relative prices derive from the extent, variability and enforcement of government intervention in African markets. For example, taxes, government regulations and currency controls create incentives to falsify customs invoices to transfer foreign exchange abroad. A related question is whether African countries are somehow riskier than others, which is a function of the reliability of market support systems, demand, and government policies. Exports of durable capital goods are often fully financed by the exporter with the f.o.b. export prices reflecting these costs of finance and insurance. If the African countries were generally considered riskier than most alternative destinations their price premiums may reflect these higher finance charges.

Summary and Policy Implications

Using techniques which have been employed for analysis of domestic market performance in industrial countries, this study established that African countries paid higher import prices for iron and steel shipments than did other developing or developed countries. The magnitudes of these excess prices indicated that they have been an important drain on foreign exchange: the overall premia for 1962 to 1987 had a present value of close to $1 billion in 1987. If the same pattern of excess prices applied to all iron and steel imports (rather than just those products selected) the magnitude of the costs would approximately double. The differentials calculated here are relative to average unit values for other developing country importers of French products. If the equivalent price differentials were calculated relative to the lower-cost suppliers such as Japan, the total losses would be even greater. Whether collusive practices among importers, restrictive government polices or higher unit costs in the smaller African markets prevented them from importing steel and iron at these lower prices, however, is not known.

These figures calculated here relate solely to iron and steel shipments, and a key question is whether such excess price margins also apply to other capital goods imports. There is some tentative evidence in support of this proposition. Yeats (1978, p. 178) compared four-digit SITC product unit values for all French shipments to selected associated and non-associated African countries for 1962-69 and found that unit-values for the ex-colonials averaged between 13 and 18 percent higher. If this excess price margin applied to all manufactured imports this would mean that the associates could have been overcharged by approximately $25 billion. The fact that trade intensity ratios are lower for most iron and steel products imported by the French-associated countries than they are for other items (see appendix table 19–2) also suggests that the price margins found here may actually be less than those of other products for which supply is handled by fewer countries. Similar analysis is needed, however, to estimate such margins for other goods.

While this study established that Africa countries pay higher prices than other countries for iron and steel products, a question of key importance is why they do and have done so over such an extended period. As noted, the excess prices margins are fully consistent with both economic theory on the functioning of markets and results from investigations of markets where monopoly elements exist. However, it was not possible within the scope of the current investigation to precisely identify the factors that were adversely affecting the African countries. Among the possibile factors are: the relatively small size of their markets, which could be important given economies of scale in distribution, financing and insurance; the influence of factors that limit access to competitive suppliers such as tied aid, and established lines of international and domestic transport; a lack of information on competitive suppliers; the use of overpricing to facilitate graft and corruption; or transfer pricing by subsidiaries of foreign firms in the African countries. Definitive information on the relative importance of such factors will require a detailed analysis of the procurement practices and problems of African importers.

There are several lines that this related research might take. First, it would be useful to extend the procedures developed in this study to other types of homogenous products (i.e., glass, cement, nonferrous metals, etc.) to see if further evidence of discriminatory pricing exists for these items. Second, trade intensity and other structural variables

Table 19–5. ***Premium or discount charged by selected European countries on iron and steel exports to associated African countries***

	Average premium or discount charged to associated countries			
Year	*Belgium*[a]	*France*[b]	*Portugal*[c]	*United Kingdom*[d]
1962-63	20.7	36.9	12.7	4.0
1964-65	21.2	21.8	37.3	8.8
1966-67	25.7	21.0	25.6	14.4
1968-69	19.1	23.9	29.9	12.4
1970-71	15.2	16.7	43.7	13.0
1972-73	18.0	18.6	18.7	15.5
1974-75	26.4	8.1	42.9	9.9
1976-77	35.3	-3.6	n.a.	22.5
1978-79	37.0	26.1	n.a.	15.1
1980-81	17.1	20.9	n.a.	19.2
1982-83	25.5	8.6	n.a.	36.5
1984-85	16.0	36.2	n.a.	37.9
1986-87	31.5	66.5	n.a.	53.0

a. Burundi, Rwanda and Zaire.

b. See table 19–3 for French-associate countries.

c. Angola and Mozambique. From 1976-77 to 1986-87 the premiums on Portugal's exports rose dramatically and averaged over 120 percent. It appears likely that the hostilities in Angola were a major factor causing the large increase in premiums over those which prevailed during 1962-63 to 1974-75.

d. Gambia, Ghana, Kenya, Nigeria, Sierra Leone, Sudan, Tanzania and Uganda.

Note: Based on the four and five digit SITC products listed in table 19–1. The average premium or discount has been calculated relative to the average unit value for each product paid by other developing countries. premiums on Portugal's exports rose dramatically and averaged over 120 percent. It appears likely that the hostilities in Angola were a major factor causing the large increase in

Source: Author's calculations based on United Nations Series D Commodity TradeTapes.

(see appendix) could be computed for a large number of bilateral trade flows and the results used to distinguish outlier countries which may be subject to oligopoly or monopoly pricing. The procedures used in this study might then be applied to these specific countries to test for evidence of monopoly pricing. Third, the procedures should be applied to homogenous goods exported from developing countries to determine if they may be receiving less than competitive prices for this trade.

Appendix. Measures of Trade Concentration

This appendix presents summary statistics relating to market shares, trade intensity ratios and indices of import concentration in the French associated countries' markets. Appendix table 19–1 shows the share of France in the associates' total imports of iron and steel products (SITC 67) and all goods for selected years from 1962-85. Because some of the associated countries did not report their imports for specific years, France's share and the trade intensity ratios could not be computed for these years. The table also gives a trade "intensity" index (I_{ij}) defined as the share of country *i's* (France) exports to associate country *j* (X_{ij}/X_i) relative to the share of *j's* imports (M_j) in world imports net of i's imports ($M_w - M_i$). That is,

$$(19\text{–}3) \qquad I_{ij} = X_{ij}/X_i / M_j/M_w - M_i$$

The index can take values between zero and infinity with values above one indicating a greater intensity of trade between two countries than can be accounted for by the countries' importance in world trade. That is, a value of two would indicate that the intensity of trade between countries was twice as great as what would be expected on the basis of their share in world trade.

Appendix table19– 2 provides statistics on the concentration of associate countries' iron and steel imports from alternative major suppliers. A three-country import concentration ratio ($C3j$) was computed:

$$(19\text{–}4) \qquad C_{3j} = (M_{3j} / M_{Tj}) \times 100$$

where $M3j$ is the value of associate country *j's* iron and steel imports from the three largest supplying countries and MTj is the total value of imports. The Hirschmann concentration index (Hj) was also computed, This index may take values ranging from zero to one with the higher numbers indicating more concentrated markets. Similar statistics have been computed for the total imports of Brazil, the U.S., U.K., the Federal Republic of Germany, and all developed and developing countries.

Market structure indices like equations(19– 3) and (19–4) have been used extensively in structure-performance studies of industrial countries where they are based on data on individual firms. There is a potential problem in applying these measures to national trade data, because a high ratio at the national level may conceal a large number of (national) competing firms. In OECD countries, however, there are relatively few iron and steel firms (some of which are nationalized) so this should not be a major problem for the current study.

Two major points clearly emerge from these indices. First, France has maintained a dominant position in almost all the associated countries' markets (Mauritius is an exception) although many of the ratios declined over the period. The fact that 14 of the 20 countries have higher bilateral trade ratios for all imports than for iron and steel in 1985 suggests that "overpricing" may extend beyond this one sector to all goods. Yeats (1978, table 4, p. 178) shows that the average unit values for all four-digit SITC products imported by selected associate countries from France are consistently higher than those of other African countries.

Second appendix table19– 2 shows that the markets of the ex-colonial countries for iron and steel imports remain far more concentrated than those of developed or developing countries although the market structure indices also are falling. Still, by 1985 the three largest supplying countries controlled 70 percent or more (over 90 percent in the case of Chad and Reunion) of the associates' imports. In industrial market studies such very high levels of concentration have consistently been found to be associated with higher seller prices and profits.

Notes

The author is an economist in the International Economics Department of the World Bank. He would like to thank Azita Amjadi for assistance with much of the empirical

Table 19–A1. *The shares of French exporters in associated countries imports and their bilateral trade intensity indices:* 1962 to 1985

	Share of France in associates' imports (percentage)						French-associate bilateral trade intensity ratio[a]					
Country/product group	*1962*	*1965*	*1970*	*1975*	*1980*	*1985*[b]	*1962*	*1965*	*1970*	*1975*	*1980*	*1985*[b]
Algeria												
Iron & steel	n.a.	60.1	28.0	20.8	12.3	17.4	n.a.	4.35	2.36	2.06	1.04	1.64
All items	n.a.	70.4	42.4	33.5	23.2	26.0	n.a.	11.00	6.24	4.85	3.57	4.41
Benin												
Iron & steel	71.4	61.2	37.6	n.a.	33.0	25.7	4.43	4.43	3.17	n.a.	2.78	2.57
All items	59.3	54.8	42.2	n.a.	25.2	27.4	9.88	8.56	6.20	n.a.	3.87	4.13
Burkina Faso												
Iron & steel	83.1	89.1	49.2	64.0	72.5	50.5	5.16	6.46	4.15	6.33	6.12	5.05
All items	52.2	53.9	50.7	43.4	39.3	27.9	8.70	8.42	7.45	6.29	6.05	6.44
Cameroon												
Iron & steel	78.7	89.4	54.0	58.1	58.2	42.4	4.89	6.48	4.55	5.75	4.91	3.66
All items	54.5	58.1	50.5	46.3	44.7	42.1	9.08	9.08	7.43	6.71	6.88	6.19
Central African Republic												
Iron & steel	84.3	91.6	59.3	73.3	68.4	81.1	5.24	6.64	5.00	7.25	5.77	7.01
All items	60.5	60.9	58.4	57.0	60.7	52.7	10.08	9.52	8.59	8.26	9.34	9.95
Chad												
Iron & steel	91.9	97.1	47.3	52.3	72.5	86.7	5.71	7.04	3.99	5.17	6.12	8.67
All items	53.2	46.4	39.8	40.8	31.0	33.3	8.87	7.25	5.85	5.91	4.77	5.08
Congo												
Iron & steel	89.8	79.9	55.4	76.4	76.7	44.6	5.58	5.79	4.67	7.56	6.47	4.22
All items	67.7	61.2	55.1	49.7	47.8	45.5	11.28	9.56	8.10	7.20	7.35	7.72
Côte d'Ivoire												
Iron & steel	84.5	76.2	52.7	67.7	63.0	44.9	5.24	5.61	4.45	6.70	5.32	4.25
All items	66.7	62.4	46.2	39.1	40.8	32.1	11.12	9.75	6.79	5.67	6.27	5.44
Gabon												
Iron & steel	84.1	91.0	69.7	71.0	56.8	65.4	5.22	6.69	5.89	7.02	4.79	6.54
All items	61.9	58.5	56.6	66.9	58.4	54.2	10.32	9.14	8.32	9.70	8.98	9.57
Guinea												
Iron & steel	n.a.	n.a.	n.a.	n.a.	31.1	58.0	n.a.	n.a.	n.a.	n.a.	2.62	5.80
All items	n.a.	n.a.	n.a.	n.a.	32.6	32.3	n.a.	n.a.	n.a.	n.a.	5.01	5.34
Madagascar												
Iron & steel	93.1	88.9	59.1	67.1	45.5	78.2	5.78	6.44	4.98	6.63	3.83	7.40
All items	74.9	62.5	54.7	40.9	37.6	29.5	12.48	9.76	8.04	5.93	5.78	5.00
Mali												
Iron & steel	90.0	38.7	43.4	72.4	62.3	46.7	6.21	2.80	3.66	7.16	5.25	4.67
All items	39.2	24.1	38.4	34.1	36.3	25.3	6.53	3.77	5.65	4.94	5.58	4.15
Mauritania												
Iron & steel	97.2	90.5	57.6	78.0	81.1	41.4	6.05	6.56	4.86	7.72	6.84	4.14
All items	72.5	44.4	35.7	42.3	34.6	23.8	12.08	6.94	5.25	6.13	5.32	3.90
Mauritius												
Iron & steel	4.9	10.1	0.6	3.0	1.6	10.4	0.30	0.73	0.05	0.30	0.14	1.04
All items	4.8	5.7	6.9	8.6	10.7	11.8	0.80	0.89	1.01	1.25	1.65	1.93
Morocco												
Iron & steel	75.1	73.8	41.8	50.4	31.7	31.2	4.66	5.34	3.53	4.99	2.68	2.95
All items	42.7	38.0	31.0	30.4	24.8	22.8	7.12	5.94	4.56	4.41	3.82	3.86
Niger												
Iron & steel	95.0	84.6	73.4	73.5	64.6	30.1	5.90	6.13	6.19	7.27	5.45	3.01

Table 19–A1. *(continued)*

Reunion												
Iron & steel	92.7	67.7	67.9	80.0	68.7	66.0	5.75	4.90	5.73	7.92	5.80	6.24
All items	68.8	67.6	62.1	62.6	65.3	65.0	11.4	10.5	9.13	9.07	10.05	11.02
Senegal												
Iron & steel	90.6	90.5	71.5	52.8	71.7	74.1	5.63	6.56	6.03	5.22	6.05	7.41
All items	65.0	53.1	51.2	41.5	34.1	43.2	10.83	8.30	7.52	6.01	5.25	7.08
Togo												
Iron & steel	51.0	52.2	32.4	30.7	54.8	30.1	3.17	3.78	2.73	3.04	4.62	3.01
All items	33.5	31.2	29.5	35.1	25.0	19.6	5.5	4.88	4.38	5.09	3.85	3.21
Tunisia												
Iron & steel	70.4	37.3	43.5	59.9	33.5	22.4	4.37	2.70	3.67	5.92	2.83	2.12
All items	52.2	39.0	34.7	34.4	25.2	27.6	8.70	6.09	5.10	4.99	3.88	4.68

a. The index represents the share of France in all exports to the associated country divided by the share of France in world trade (see equation 3). A value greater than unity indicates a greater intensity of trade than would be expected based on France's share in world trade.

b. Because more recent information was not available for Benin, Burkina Faso, Central African Republic, Chad, Gabon, Guinea, Mali, Mauritania, Mauritius, Niger, Senegal and Togo the statistics shown in these columns are for 1983. Because 1985 data were not available for Cameroon the information shown relates to 1986 trade.

Source: Author's calculations based on United Nations Series D Commodity Trade Tapes.

Table 19–A2. *Supplier share and concentration indices for iron and steel imports: 1962 to 1985*

	Share of imports from three largest suppliers (percentage)						*Import supply concentration index*[a]					
Country	*1962*	*1965*	*1970*	*1975*	*1980*	*1985*	*1962*	*1965*	*1970*	*1975*	*1980*	*1985*
Algeria	99.3	92.1	57.5	62.3	60.0	57.8	0.98	0.86	0.40	0.49	0.39	0.39
Benin	99.9	98.4	93.0	84.2	81.0	72.3	0.82	0.71	0.55	0.51	0.50	0.47
Burkina Faso	99.3	95.0	87.8	88.2	89.2	71.5	0.92	0.83	0.73	0.73	0.73	0.48
Cameroon	95.8	88.9	80.5	84.4	73.6	75.7	0.82	0.77	0.52	0.62	0.64	0.65
Central African Republic	98.7	97.4	92.6	93.0	93.7	88.9	0.87	0.90	0.69	0.70	0.68	0.69
Chad	99.0	98.0	86.9	97.0	88.4	96.2	0.96	0.92	0.6	0.66	0.74	0.68
Congo	97.7	91.4	77.7	84.6	93.2	70.8	0.90	0.79	0.56	0.69	0.79	0.45
Côte d'Ivoire	96.9	98.2	81.1	88.2	79.5	85.2	0.87	0.79	0.58	0.74	0.66	0.61
Gabon	97.1	96.2	84.6	87.6	86.8	54.1	0.90	0.90	0.66	0.71	0.65	0.67
Guinea	45.7	91.9	96.9	90.8	83.1	70.5	0.75	0.81	0.74	0.60	0.53	0.49
Madagascar	98.5	95.0	89.7	95.0	87.0	81.9	0.94	0.86	0.66	0.76	0.63	0.61
Mali	99.9	99.7	98.2	94.5	93.8	74.6	0.97	0.71	0.58	0.69	0.72	0.51
Mauritania	99.9	98.1	82.3	94.7	87.8	86.1	0.97	0.85	0.66	0.86	0.81	0.57
Mauritius	76.7	68.4	64.6	72.1	87.7	87.5	0.88	0.88	0.79	0.78	0.84	0.84
Morocco	97.8	94.6	71.5	76.6	81.3	82.7	0.84	0.84	0.53	0.55	0.53	0.49
Niger	99.3	91.0	95.7	88.1	63.4	64.9	0.98	0.82	0.81	0.74	0.58	0.44
Reunion	98.0	97.2	92.1	98.6	96.4	97.6	0.59	0.43	0.44	0.52	0.55	0.70
Senegal	99.0	95.9	92.5	71.0	89.6	83.8	0.92	0.82	0.71	0.54	0.76	0.60
Togo	91.9	90.9	83.0	79.1	88.6	75.3	0.58	0.82	0.54	0.52	0.62	0.46
Tunisia	93.3	73.1	69.4	82.8	78.8	72.3	0.79	0.49	0.52	0.68	0.50	0.45
Brazil	67.4	65.4	67.6	69.7	65.3	64.7	0.41	0.42	0.43	0.46	0.39	0.40
Germany, ed. Rep.	78.2	67.7	64.2	58.7	53.7	48.7	0.51	0.45	0.43	0.39	0.37	0.33
United Kingdom	38.2	43.3	41.1	44.0	49.5	52.6	0.26	0.31	0.34	0.34	0.35	0.38
United States	52.9	63.8	66.8	67.5	63.6	55.9	0.34	0.45	0.48	0.49	0.44	0.38
All developed countries	58.6	49.7	46.6	50.2	46.4	40.2	0.37	0.34	0.33	0.34	0.31	0.30
All developing countries	52.6	53.7	60.7	64.1	59.9	57.4	0.37	0.37	0.43	0.47	0.46	0.47

a. Hirschmann concentration index: $H_j = \sum (\Xi_{ijp}/\Xi_{jp})^2$.

Source: Author's calculations based on United Nations Series D Commodity Trade Tapes.

analysis and Paul Meo for many helpful comments and suggestions.

References

Avramovic, Dragaslov (1978). "Common Fund, Why and What Kind," *Journal of World Trade Law*, 12 (October), pp. 370-43.

Bhagwati, Jagdish (1967). "Fiscal Policies, the Faking of Foreign Trade Declarations, and the Balance of Payments," *Oxford Bulletin of Economics and Statistics*, (February).

Bresnahan, Timothy F. 1989. "Empirical Studies of Industries with Market Power," in Richard Schmalensee and Robert D. Willig, eds., *Handbook of Industrial Organization*, Vol. II, Amsterdam: North-Holland, pp. 1011-55.

Edwards, Corwin (1972). "Barriers to International Competition: Interfirm Competitive Behavior," in R. Hawkins and I. Walter (eds), *The United States and International Markets*, Lexington: D.C. Heath.

Helleiner, G. (1978). *World Market Imperfections and Developing Countries*, Washington: Overseas Development Council.

Hewett, E.A. (1974). *Foreign Trade Prices in the Council for Mutual Economic Assistance*, London: Cambridge University Press.

Hufbauer, G.C. and J.P. O'Neill (1972). "Unit Values of U.S. Machinery Exports," *Journal of International Economics*, vol. 2, pp. 265-276.

Kleiman, Ephraim (1976). "Trade and the Decline of Colonialism" *Economic Journal*, vol. 86 (September), pp. 459-480.

Kreinin, Mordechai (1988). "How Closed is the Japanese Market? Additional Evidence," *The World Economy*, (December), pp. 529-542.

McAllister, Harry (1961). "Statistical Factors Affecting the Stability of the Wholesale and Consumer Price Indexes," in U.S. Congress, Joint Economic Committee, *Government Price Statistics Hearing*. Washington: U.S. Government Printing Office.

Sheikh, Munir (1974). "Underinvoicing of Imports in Pakistan," *Oxford Bulletin of Economics and Statistics*, (November).

Stigler and Kindahl (1970). *The Behavior of Industrial Prices*, (New York: National Bureau of Economic Research).

United Nations Economic and Social Council (UNESCO) (1974). *International Trade Reconciliation Study*, (E/CN.3/454), Geneva.

United Nations Conference on Trade and Development (UNCTAD) (1975). *The Control of Transfer Pricing in Greece, Geneva: United Nations.*

Yeats, Alexander (1978). "Monopoly Power, Barriers to Competition, and the Pattern of Price Differentials in International Trade, *Journal of Development Economics*, 5 (June), pp. 167-180.

Yeats, Alexander (1989). "Do African Countries Pay More for Imports? Yes," *Policy Planning and Research Working Paper No. 265*, Washington: World Bank, September.

Yeats, Alexander J. (1990). "On the Accuracy of Economic Observations: Do Sub-Saharan Trade Statistics Mean Anything?" *The World Bank Economic Review* 4, no. 2.

20

Comments on Trade Policy

Ademola Ariyo and C. D. Jebuni

Ademola Ariyo

These three papers address very interesting issues which throw light on the relationship between trade and some macroeconomic indices of less developed countries (LDCs) of Africa. Given the potential policy relevance of all the papers, my discussion is influenced by this policy orientation.

The Panagariya and Schiff Paper

The first paper, by Panagariya and Schiff, attempted to provide an insight into the ongoing debate regarding the potential effect of an expansion of the production of a primary product (i.e. cocoa) on the terms of trade, export revenues and real incomes of LDCs. In this regard, the simple and simultaneous effects of an alternative tax (Nash) arrangement, increased production efficiency and trade liberalization (devaluation) on real income, export and tax revenues were simulated. While the authors indicated that additional studies need to be carried out, the main tentative conclusion was that each simulation has a differential impact on the chosen macroeconomic aggregates. The results also indicate that there is a high degree of interdependency whereby the policy actions of a country have significant effects on the macroeconomic indices of other countries concerned. It was therefore the view of the authors that the donor communities should coordinate policy advice and support of investment projects in primary products, such as cocoa.

While the authors' efforts deserve commendation, a number of issues need to be addressed, especially as they may influence future studies on the subject matter. First, the paper attempts to throw light on a policy issue affecting a set of producers of a primary product. Hence, as much as possible, the institutional factors should be incorporated into the analysis. For example, even if a Nash tax framework were to prove beneficial, one is not sure if each (sovereign) nation would behave in a manner consistent with the model given. For example, the short-run impact of changes in trade tax on government revenue may affect a particular governments decisions on cocoa pricing.

Some major policy related developments have also been recorded in some of the countries concerned. For example, Côte d'Ivoire recently put an embargo on the release of cocoa output into the world market as a means of boosting prices. If this had succeeded, it probably would have encouraged others to do likewise. In addition, Nigeria has totally liberalized cocoa trade. Under this arrangement, in addition to the excise tax charged by the federal government, each cocoa producing state within the country is free to impose additional levies on graded cocoa. Unless this practice exists in all producing countries, it might not be possible to accurately determine the trade tax on cocoa in Nigeria to talk less of comparing same with other countries. This problem is further compounded by the fact that the state-level tax rates are not uniform.

Following closely is the paper's implicit assumption that these cocoa producing countries would always continue to export raw cocoa beans or that only an insignificant proportion of the total output could be processed locally. Again, recent developments in Nigeria suggest the need to modify this assumption. For example, the Federal Government of Nigeria has banned the exportation of cocoa beans effective from the 1991/92 cocoa season. It is recognised that representations have been made to the effect that existing local processing capacity could accommodate only about one-third of total output. However, they are mainly concerned with a phased implementation, rather than an abandonment, of the proposal. It is desirable to anticipate the demonstration effects of this development on other producers and its implications on the issues addressed in the paper.

There were some conclusions contained in the paper which may require further elaboration. For example, the authors indicated that a Nash strategy could enhance Africa's share of the cocoa market by reducing that of, say, Malaysia. One is tempted to argue that Malaysia's emergence and prominence in the cocoa trade reflects the positive impact of a religiously implemented long term strategic plan. Hence, other countries need similar plans rather than mere Nash strategy to alter Malaysia's position. Furthermore, the authors suggest that any country experiencing significant distortions and disequilibria in its economy should embark on a structural adjustment program and then adjust its tax rate to the optimal value. The possible non-feasibility of adjusting tax as suggested had been referred to earlier. With regard to the former, all the countries concerned have similar macroeconomic problems and virtually all have embarked on structural adjustment programs, including devaluation. Hence, the authors could have discussed the implication of similar action by all the countries rather than a partial analysis that was done for Ghana only.

As the authors rightly observed, trade policy interventions should be implemented with a view to improving long term efficiency of resource allocation. In addition, it is agreed that the price elasticity of demand for cocoa is very low. One then wonders if devaluation and a Nash type tax regime are the best methods for achieving this long term objective. To my mind, the long run interest of primary producing nations will be better served by accelerating their industrial development efforts through deliberate and articulated policies and programs. An initial focus on less capital intensive, agro-allied industries may ease the demands for the take-off.

Some additional observations may also be made as they may impact future studies on the paper's subject matter. First, the assumption of absence of smuggling is convenient for the theoretical analysis. However, smuggling does exist in practice, and suggestions aimed at eliminating or significantly reducing the phenomenon would be of policy interest. Second, there is a renewed emphasis on intraregional economic cooperation. The possible impact of this on trade flows, terms of trade and other macroeconomic aggregates identified by the authors should be kept in view. Third, tax rates for 1982 and 1983 were used for the analysis based on 1986 data. The authors need to check or confirm that there has not been a significant change in the tax profile between 1982 and 1983 or beyond, with a view to making appropriate adjustments.

The Newman, Lavy, Salomon and de Vreyer paper

The paper by Newman, Lavy, Salomon and de Vreyer attempts to evaluate the differential impact of two possible trade related policies: tariff-cum-subsidy versus devaluation. The findings suggest that the latter would be more beneficial for the economy. The reasons adduced for the plausibility of the observed results were indicated by the authors. Among these are the inability of (manufacturing) producers in Africa to promptly adjust to changes in market signals and the smallness of short run export supply elasticity. The authors also believe that export subsidies ultimately leave unaffected the profitability of exporting activities. They however admitted that the results are sensitive to the assumed endogeneity of the domestic price and to the inclusion or exclusion of the demand factor.

Although the authors stated that the essence of the analysis was not an evaluation of the reasonableness or otherwise of the tariff-cum-subsidy program in Côte d'Ivoire, it is difficult to appreciate the isolation of the findings from its potential policy relevance. This is important given that data based on Côte d'Ivoire were used. It is in this light that one would raise some observations regarding the assumptions underlying the model employed. In doing this, I must state upfront that I had a problem interpreting the word "price" in may instances in the paper regarding when the word refers to input or sales price of either exports or domestic sales. Hence, any misinterpretation of some statements may be attributable to this limitation.

First, the authors should not be surprised that manufacturers in Côte d'Ivoire are able to expand their exports in the short run in response to an increase in export prices. The behavior is expected of rational economic entities. Furthermore, the nature of production may be amenable to 'shock'. In addition to the diversion of output for domestic sales to the export market as the authors found, the possibility exists for some excess capacity which could be utilized without affecting the quantity of output available to the domestic market.

Second, the assumed endogeneity of domestic prices is in order except that the validity rests upon empirical facts. For example, the assumption could hold in an open economy only when the import content forms an insignificant proportion of total manufacturing inputs. It is probably because of the contextual irrelevance of this assumption that the authors found that domestic prices were sensitive to the prices of intermediate inputs, as expected in an LDC environment. In fact, some of the results indicate that domestic prices do respond to changes in exogenous prices.

Third, the authors had no information on the output or export prices of individual firms. Given the heterogeneity of the operations of the firms, the results based on aggregate data would have related to no firm in particular, nor to any specific industry.

In spite of the above observations, one needs to commend the efforts of the authors for the significant methodological improvements compared to previous studies. The findings provide useful empirical evidence bearing on the possible outcomes of alternative policy options. However, as noted earlier, additional work is required to enhance the policy relevance of the study's findings.

The Yeats Paper

The paper by Yeats provides another valuable revelation regarding the outcome of trade relations between the developed economies and their former colonies. Although the study focussed primarily on France and its former colonies, there is reason to believe that similar results could be expected with respect to other developed countries and their former colonies. The possible causes and the estimated costs of the observed scenario were well documented. In order to avoid a repetitive discussion of what the participants would have known from reading the paper, no attempt is made here to discuss the reasons adduced for the results reported. However, as indicated in the paper, the explanatory variables accounted for about 56 percent of the determinants of the observed phenomenon, suggesting that other equally important factors might have been left out.

To enhance the ability of African countries to meaningfully address this issue, the following observations are noteworthy. First, it should be noted that prior to and at independence, most of the prominent trading houses were overseas affiliates of the home based companies and hence, naturally patronized these companies abroad. Over time, this could have become a habit. An issue of interest therefore is an investigation of the mix of the states of origin of the trading houses in developing countries since then.

Second, developed countries usually have economic (trade) agreements or treaties with former colonies as a means of consolidating their market share. Hence, the nature of trade flows could persist thereafter, especially for public sector transactions. Hence, an assessment of the differential effect of treaties on trade between former colonies and non-colonies could provide some useful insight. In addition, the private sector in these former colonies is expected to develop over time and engage meaningfully in international trade. It would thereby be expected to make more contacts than the public sector, which is already constrained by the provisions of some treaties. Hence, a distinction between public and private sector activities is recommended.

Third, it may be useful to compare purchases funded through external loans or those which are contractor financed with those purchased with internally generated funds. Such analysis may provide some indication of the effect of tied aid or external loans, as well as the source of the trading activities, on the observed results. Finally, the results did not show a discernible trend, with 1982/83 appearing to be totally out of tune with other years. The author should endeavor to identify the causes of this phenomenon.

Assuming that the analysis in the paper validly accounted for the observed results, a major policy issue relates to an assessment of the prospects for reversing this undesirable trend. The following suggestions are offered. First, additional studies should be embarked upon as recommended above and also by the author in order to identify the major causes of the observed results. The findings should also be widely disseminated, especially by being brought to the attention of both the developed countries and their former colonies.

Second, with the current macroeconomic problems confronting many LDCs, there is a likelihood of greater reliance on foreign loans and/or aid for executing their recovery or development programs. Hence, the problems associated with tied aid or its equivalent may be intensified. The upward trend in the premium as reported on table 2 of the paper may be instructive in this regard. It is suggested that loans or aid through multilateral sources should be preferred to bilateral arrangements, since the latter will most likely confine the recipient to a specified source of purchase.

Third, there is need to address squarely the problem of transfer pricing and deliberate over-invoicing of purchases (especially in the public sector). This may be easier to do between long-established trading links than with new ones, especially in those countries that give little room for this practice.

Finally, many of these former colonies engage the services of inspection agencies. Hence, the author should evaluate the usefulness of the services of these inspection agencies. This could be done by analysing the differences in the premium paid by those former colonies which engaged the services of inspection agents relative to those that did not. A similar test could be carried out between the former colonies and other relevant countries.

C.D. Jebuni

Trade policy is one of the key elements of the structural adjustment policies adopted almost everywhere in Africa. The package usually pivots around an export-led growth strategy. The three papers presented, by concentrating on the visible side of the current account and the relevant policies therein, can contribute positively to the process of structural adjustment taking place in Africa.

Balassa's (1979) work on changing comparative advantage and some of the references cited in these papers have dispelled part of the pessimism concerning less developed countries' manufactured exports as a fallacy of composition. The more recent concern, however, is that a simultaneous expansion of primary commodity exports resulting from the adoption of structural adjustment programs may lead to declining export revenues and real incomes.

The papers by Panagariya and Schiff, and Newman, Lavy, Salomon and de Vreyer concentrated on policies and/or strategies for increasing export volumes or revenues and real income from exports. But equally important in terms of the current account of the balance of payments are imports. In the face of increasing current account deficits, the concern about imports relates to ways of reducing expenditure on imports without adversely affecting domestic production and reducing consumption below levels that may generate political and social unrest. The usual policy prescriptions have aimed at reducing imports expenditures through a cut in consumption of imports either as final goods or immediate imputs. No analyses, to my knowledge, have investigated the possibility that considerable foreign exchange savings could be made on imports through the prices we pay for our imports. In that context, Yeats' paper and investigations have potentially useful policy and balance of payments implications.

The Panagariya and Schiff Paper

The paper by Panagariya and Schiff addresses this issue. Their results show that uncoordinated donor policy advice and support of investments projects leading to simultaneous expansion of exports of commodities such as cocoa, could result in a fall in export and tax revenues.

Our experience, at least in Ghana between 1985 and 1989, is consistent with this finding. As a result of this experience, policymakers are shifting to a position where policy will aim at maximizing expo- rt revenues from cocoa. This may involve output changes or increased processing of cocoa for export. In this connection, the next phase of the authors' paper, in which they plan examining revenue, ma- ximizing strategies will be critical. This exercise would be more useful if it incorporated the possibility of processing part of the cocoa produced before exports.

It is strange that all the exercises performed by the authors related to output increasing strategies and policies. What will be the effect on export revenues and real incomes if output fell? In a meeting of the Cocoa Producer Alliance in Accra in March 1990, consideration was being given to the possibility of controlling supply with a view to increasing export earnings. An exercise involving a decrease in cocoa exports will therefore be useful.

In simulating the effect of a devaluation on cocoa exports, the authors failed to take into consideration the institutional arrangements for the marketing of the product. Cocoa, in Ghana, is purchased by a marketing board which also fixes the price. Cocoa production and exports respond to changes in cocoa prices, but their response to devaluation is doubtful (Jebuni, Sowa and Tutu 1990). This is because the marketing board may not pass on the currency devaluation to producers. Failure to take this into consideration would bias the results for a devaluation. Even in cases where exports are not purchased by a marketing board but by a middleman, this mechanism has to be taken into consideration.

A broader issue of increasing concern regarding agricultural exports and real incomes is the domestic economy repercussions of such export expansion. I will illustrate this using cocoa in Ghana. The stylised facts to take into consideration include:

1) cocoa is marketed by the Ghana Cocoa Marketing Board which may pass on part or all of policy changes to cocoa farmers through increased producer prices.

2) Prices in the other subsectors of agriculture are determined by supply and demand in weekly village markets and are not directly affected by government policies.

3) The connection between these village markets and the consuming urban markets is through middlemen or middlewomen, who buy from the village markets and sell in the urban markets and who may not behave as perfect competitors.

4) Cocoa producers' demand for products of the other agricultural subsectors is zero, and;

5) the Arthur Lewis unlimited supply of labor is no longer applicable.

In these circumstances an increase in cocoa prices as a result of a devaluation or tax cuts, increases cocoa production and exports. With limited labor supply, cocoa farmers bid up the price of labor in the rural areas. At the same time, because of the nature of price determination in other sectors in agriculture, producer prices may not increase even though urban prices are rising. The combination of increasing labor costs and probably declining terms of trade for the other subsectors could cause a decline in their production or the rate of growth of output. Loxley (1988) and IFAD (1988) have documented this process for the economic recovery period in Ghana. As a result of increasing cocoa prices, agricultural wages doubled while the terms of trade for the food subsector, the largest subsector of agriculture in Ghana, declined. Further complications could be added if we consider the prices of imported inputs in the other sectors and the effects of the polices on urban real incomes. Incorporating these considerations will require a macroeconomic approach to modelling the relations between commodity exports and real incomes.

The Newman, Lavy, Salomon and de Vreyer Paper

The paper by Newman, Lavy, Salomon and de Vreyer addresses the policy dilemma faced by the Francophone countries in using devaluation. The problem for these countries is that a devaluation of the CFA franc will require the cooperation and agreement of the countries in the Zone and France. For an individual country, the alternative it would seem is to mimic a devaluation. The authors consider the possibility of using export subsides and import tariffs as a policy option. On the basis of the short run responses of manufacturing firms in Côte d'Ivoire to relative price changes, the authors have sought to dispel part of the skepticism surrounding this policy option. Their results can make a major contribution to this debate.

However their concentration on short run responses may still raise questions among the skeptics. As their results show, in the short run, firms may respond to increased profitability in exporting by diverting domestic sales to exports. In the long run, there will be a need for capacity building. In the meantime, the diversion of domestic sales to exports could affect domestic prices and inflation and have repercussions for the rest of the economy. Thus, the addition of a long run perspective will be illuminating.

An important issue in the debate is the mode of financing the export subsidy and how this impacts on fiscal and monetary policy and in turn, influences the domestic price level and the ensuing effects. The export subsidy must by financed and it is not clear that the additional revenue from the increased import tariff can do this. This could affect the government budget and have wider implications for the economy depending on the size of the budgetary effect.

In examining policy options one might wish to incorporate the reactions of neighbors, particularly when they have similar economic structures. Currently, considerable amounts of Ivorian Dinor Oil can be found in the Ghanaian market. Meanwhile Ghanaian producers cannot sell their oil. Even though the Ghanaian government is committed to import liberalization, it is coming under increasing pressure to do something about the situation. The argument by Ghanaian manufacturers that these products are subsidized in Cote d'Ivoire is convincing.

The Yeats Paper

Using techniques of organization (market structure) on domestic prices, Yeats has demonstrated that French-associated African countries pay higher prices for their iron and steel imports. In the 26-year period from 1962 to 1987, French-associated African countries' unit value of iron and steel imports always exceeded those of developed market economy countries by an annual average of 24 percent. They paid an average premium of 23 percent above the unit value for other developing countries for the same period. The present value of the losses associated with these premiums could amount to $2 billion if all iron and steel imports by these countries from France are considered.

The paper shows that this phenomenon is not confined to the French-associated African countries, but may also apply to iron and steel imports by other African countries from their former colonial masters.

This fact, in combination with the association of these premia with size, number of contracts, etc. suggests that considerable foreign exchange savings could be made by bulk orders, diversification of the sources of purchase, etc. Another strategy for reducing these premia, not discussed by the author, is the use of international companies for vetting the quality and prices of imports.

The analysis, as the author noted, has to be extended in several directions in order to realize its full policy potential and implications for balance of payments arrangements. There is a need to extend the analysis to cover other imports and imports from other countries and also exports by African countries. Such an extension will tackle the questions of whether this is a general phenomenon for all Africa's imports from all countries or whether it is peculiar to their relations with their former colonial masters. It will also address the question of whether African export marketing strategy results in losses from receiving export unit values below what they could have obtained. This is important for maximizing export revenues.

On methodology, while it is enough for purposes of calculating the premium or discounts relative to other countries by using French export unit values to other countries, this may not be appropriate for estimating the costs of such pricing practice. The true cost of paying these premia must relate the cheapest source of these imports. The appropriate unit values for this purpose must be based on the cheapest source unit values of the relevant imports. If this were done, it would be seen that the present values of losses presented in the paper will be underestimates of the true losses.

Earlier analysis had shown that less developed countries pay more for "aid" imports. Thus, there might be two sets of forces operating to generate these premia those related to "aid" and those related to monopolistic practices. It should be noted that the analysis covers relations between some African countries and their former colonial masters. These former colonial masters, it can be expected, also extend the most bilateral aid to these countries. Overpricing associated with "aid" imports need not be related to market characteristics. It is essential to make this distinction if we are to identify the determinants of the prices paid by African countries for their imports. By ignoring this distinction, market characteristics may be conveying influences which are due to other forces.

References

Balassa, B. (1979), "The Changing Pattern of Comparative Advantage in Manufactured Goods", Review of Economics and Statistics, Vol. 61.

IFAD (1988), Report of the Special PRogramming Mission to Ghana", Report No. 0105-9H, Rome.

Jebuni, C.D., Tutu, K.A., Sowa, N.K. (1990), "Exchange Rate Policy and Macroeconomic Performance in Ghana". Paper presented at African Economic Research Consortium Workshop. May 27-31, Nairobi.

Loxley, John, (1988). "Ghana: Economic Crisis and the Long Road to Recovery". The North-South Institute, Ottawa, Canada.

21

Integration Efforts in Sub-Saharan Africa: Failures, Results and Prospects — A Suggested Strategy for Achieving Efficient Integration

Ali Mansoor and Andras Inotai

The recent economic performance of Sub-Saharan Africa has been unsatisfactory. Over the period 1980 to 1985, real GDP declined by almost 1 percent annually, while population increased by 3.3 percent a year on average. Since then, GDP has grown at an average annual rate of just under 4 percent, thus leaving per capita income substantially below the level at the start of the decade.

To restore significant per capita income growth, the World Bank has supported reforms aimed at making Sub-Saharan African economies more flexible. By the end of 1989, 33 countries in Sub-Saharan Africa had adopted some form of structural adjustment program with the support of the World Bank, and 17 countries had accepted the need to liberalize their trade regimes. However, despite these developments and some encouraging progress in countries that have implemented strong reform programs,[1] far more needs to be done to improve the outlook for Sub-Saharan Africa. Emphasis needs to be given to measures that will strengthen ongoing reform efforts to achieve more rapid and extensive improvements in efficiency. In practice this requires initiatives in many directions, including strengthening and deepening of the adjustment process through regional integration and cooperation.[2]

It is with these considerations in mind that this paper reviews the performance of various regional groupings formed to promote regional economic integration. This experience has been generally negative because of the inefficient resource allocation that has resulted from the emphasis on regional import substitution and attempts at regional industrial planning. However, the progress with outward oriented adjustment in Sub-Saharan Africa sets the stage for and will facilitate the implementation of a new approach to regional integration, that will emphasize the complementarity between external trade liberalization and regional liberalization of factor markets. In turn, such regional integration will facilitate and extend ongoing adjustment efforts by increasing internal competition and providing greater flexibility for factors to adjust to changing economic conditions. Selectivity and adaptability will be key concepts in implementing the new approach.

Section two of this paper looks at the performance of integration schemes in Sub-Saharan Africa and among other developing countries. Section three presents the rationale for regional integration efforts in Sub-Saharan Africa. Section four discusses the scope for change and the main elements of a new approach to regional integration in Sub-Saharan Africa. Section five contains concluding remarks.

Regional Integration in Sub-Saharan Africa and Among Other Developing Countries

The eight major economic integration arrangements in Sub-Saharan Africa aim at industrial cooperation and trade liberalization in one form or another.[3] Their objectives are to promote collective self-reliance and to establish a common market. These objectives are generally consistent with the Lagos Plan of Action, which was adopted in April 1980. Except for the West African Economic Community (CEAO), the regional organizations have not achieved any significant increase in intraregional economic activity.

In general the results of integration schemes in Sub-Saharan Africa have been more disappointing than elsewhere in terms of generating regional trade.[4] Table 21–1 summarizes some basic indicators for the Central American Common Market (CACM), the Andean Pact,the Association of South East Asian Nations (ASEAN), the Latin American Free Trade Area (LAFTA), the European Free Trade Area (EFTA), and the European Community.

The groupings in Sub-Saharan Africa have experienced the greatest degree of opening up of all unions. This is all the more remarkable given the rhetoric of self-reliance and de-linking from the world trading system that underlies much of the official justification for such groupings.[5] However, the share of regional trade and intraunion trade creation in Sub-Saharan Africa are generally lower than in other regions. It would appear that the opening up of the African groupings is more a reflection of the failure to achieve extensive and efficient import substituting industrialization than of explicit policy decisions. The CACM achieved large increases in regional trade. It has been the most successful union in the developing world, in terms of intraunion trade creation. However, its trade gains were based on an import substitution strategy in which consumer goods were produced at prices exceeding those on the world market. The inputs for this production were financed by buoyant commodity export receipts. When commodity prices collapsed in the late 1970s and foreign financing became unavailable in the early 1980s, the earlier gains were reversed (see World Bank 1989c).

Outside Sub-Saharan Africa, all organizations except for ASEAN and the Andean pact generated more internal than external trade. In Sub-Saharan Africa the reverse is true. In the case of ASEAN this result is not surprising given the negligible emphasis placed on intraunion trade liberalization. Despite this, the sh-are of regional trade within ASEAN is relatively high, while ASEAN also displays the greatest degree of openness. Similarly, the CEAO is both the most open group in Sub-Saharan Africa and the one with the largest share of intraregional trade. This suggests that regional exchange may be favored by openness.

The comparisons in table 21–2 suggest that unions in Africa have been less successful than those elsewhere, except for the Andean Pact, which failed through overemphasis on industrial planning. Surprisingly, African unions have witnessed more external than internal trade generation and have moved from being relatively closed in the 1960s to about the same degree of openness as unions in other parts of the world.[6]

Regional Integration Among Developing Countries

Economic integration among developing countries was a major policy issue in the 1960s and the early 1970s. Subsequently it appeared to have has lost its relevance. However, despite adverse economic and political conditions, regional integration schemes did not dissolve. Their policy relevance is indicated by the continuing membership of most developing countries (except in Asia) to at least one regional integration scheme. More importantly, in the last few years new initiatives have appeared. Thus, while some countries are considering closer ties to neighboring economies for strategic rather than economic reasons (Maghreb and other Arab regional groupings), many established integration schemes are redefining policy priorities and instruments. This activity has been partly spurred by regionalization trends in the developed market economies (European Community 1992, US-Canada free trade pact, Pacific Rim initiatives).

Original motivations. At the end of the 1950s, economic integration among developing countries was a response to the limits of national import substitution and the emergence of the European Economic Community (EEC). It aimed for import substitution on the regional level through the use of trade policy instruments such as those being applied by the EEC. However, the attempt to emulate the EEC did not sufficiently take into account the characteristics of the members of the EEC. These included the size and level of development of the market, the high share of intraregional trade before integration, historically developed production and financial linkages, and the high income and industrialization levels which provided considerable financial resources for compensation payments. More importantly, the EEC aimed at achieving greater regional integration while pursuing, in parallel, external trade liberalization. The commitment of all the members of the EEC to an increasingly liberal world trading regime, under the auspices of GATT, ensured that the initial movement

Table 21–1. *Basic indicators for selected economic and trade groups*

(percentage)

Group	Share of regional trade 1/ (Trade=Exp+ Imp)	Degree of openness (Trade/ GNP)	Net trade gains 1/ (Change in degree of opennes) 2/	Crude trade creation 3/	Trade diversion (-) or gains (+) with non-partners 3/	GDP: US$ bils 1986	GDP: Avg. anual 1965-80	GDP: Growth rates 1980-86	POP (mils) mid-1986	GDP p.c. (US$)	Distribution of GDP in %: Agr 1986	Distribution of GDP in %: Ind 1986	Distribution of GDP in %: Manufa 1986	Area (thousands of square kms)
Outside SSA														
CACM	2	43	8	8	—	22	4.6	0.0	24	914	23	27	19	423
EC9	53	48	13	7	6	3067	3.7	1.5	264	11629	5	37	20	1527
EC6	n.a.	31	3	5	-2	2508	3.8	1.2	198	12645	3	36	24	1169
EFTA	24	36	-1	2	-3	475	3.5	2.3	32	15047	5	38	23	1236
Asean	23	58	13	1	12	192	7.8	3.5	295	652	18	33	22	3064
AndeanPact	4	41	9	1	8	120	5.6	0.0	83	1458	16	33	19	4719
Lafta	11	15	-6	—	-6	550	5.3	0.1	351	1566	15	35	22	19311
In SSA														
CEAO	13	57	31	3	28	18	4.1	0.7	47	379	37	22	13	4444
ECOWAS	4	33	30	2	28	78	6.3	-1.8	179	436	40	26	9	6093
PTA	7	41	18	1	17	33	3.9	2.1	146	227	40	22	14	4973
UDEAC	2	42	35	1	34	18	5.9	6.2	23	805	19	36	5	1708
MRU	1	4	35	1	34	4	3.3	0.2	12	346	41	23	3	429
CEPGL	—	47	11	—	11	9	2.4	1.3	43	208	35	31	14	2399
IOC	3	42	n.a	n.a.	n.a.	7	n.a.	2.5	13	517	n.a.	n.a.	n.a.	562

Note: 1/Data is for 1983. 2/For SSA, calculations are based on difference between 1965 and 1983. For other groupings, various dates are used corresponding to a period before and after the union became effective. 3/Here trade creation/diversion is measured as the cahnge in rade relative to GDP. A more precise measure should take account of production and demand in a well specified model (see Corado and De Melo (1986). Note that Net Trade Gains (column 4) = Crude Trade Creation (column 5) + Trade Diversion (-) or Gains with non-partners (column 6). Crude Trade Creation is the change in the share of regional trade in GDP following the creation of the Union.

Source: Inotai 1986 (Table 1, p. 44), Robson 1987, OECD Various years, and International Monetary Fund 1988.

Table 21–2. ***Characteristics of trade among members of selected economic integration and cooperation schemes***

Group	*Major trade-related characteristics*[a]	*Export among members in 1987 (US$ millions)*	*Intrascheme exports as a percentage of total exports*					
			1970	*1975*	*1980*	*1983*	*1985*	*1987*
Central American Common Market	1,2,7	492	27	23	22	22	15	12
Andean Group	1,2,7	683	3	5	3[b]	4	3	3
Caribbean Community	1	323	7	7	6	9	8	6
UDEAC in Central Africa	1,2,4	38	3	4	4	2	8	6
W. African Economic Comm. (CEAO)	2,3,4	383	9	7	7	13	9	688
East African Common Market	5	142	17	13	8[b]	7	7	7
Economic Community of West Afrcian States (ECOWAS)		885	2	3	4	4	4	6
RCD (Iran, Pakistan, Turkey)		1,305	1	1	5[b]	9	10	5
Latin American Integration Association	6	8,103	10	14	14	11	9	11
Association of Southeast Asian Nations	6,7	14,529	15	16	18	23	17	218
Memorandum item: European Economic Community		555,616	49	49	53	53	55	59

Note: [a]1=Free trade among members; 1=common external tariff; 3=redistribution of proceeds from tariff to settle payments imbalances among members; 4=common currency; 5=now defunct; 6=some preferential trade treatment among members; 7=joint positions international trade negotiations. [b]1981. [c]This total and the shares do not include Singapore's very large exports to Indonesia, which are not reported by mutual agreement.

Source: Inotai 1986 (Table 1, p. 44), Robson 1987, OECD Various years, and International Monetary Fund 1988.

to a common external tariff involved, in general, a lowering of tariffs.

Performance. As a consequence of the "fallacy of transposition" (see Laghammer and Hiemenz 1989), overambitious goals were formulated. Regional integration was expected to become an engine of growth which would increase international competitiveness by making use of the advantages of specialization, larger economies of scale, enhanced regional competition, and a regional training ground. It appeared to offer collective self-reliance and decreased dependency on the developed world.

Instead, the generally low level and modest share of intraregional trade, accompanied by rather modest export-to-GDP ratios in various countries, failed to alter the determining role of extraregional trade in growth (see tables 21–1 and 21–2). The regional market reamined too small to secure significant economies of scale. This was compounded by the tendency to consider integration of markets by product groups. Progress was further handicapped by weak (if any) industrial structures, the lack of intraindustry linkages, and usually nonexistent or underdeveloped infrastructure.

Additional constraints emerged from the implementation of trade policy instruments. Originally scheduled tariff reductions did not take place or required time-consuming procedures. As liberalization approached the highly protected domestically produced goods, the liberalization process was stopped and in some cases reversed. The integration schemes failed to generate the benefits that could have been expected from exposing the highly inefficient national import substitution industries to regional competition.

Larger and more developed countries became the main beneficiaries of the regional market while smaller and less developed economies registered growing trade deficits. Payments arrangements and preferred treatment status granted to deficit economies provided at best a temporary remedy. Particularly in regional integration schemes involving countries with inconvertible national currencies, this problem often proved disruptive and contributed to the reduction of intraregional trade volumes. Trade liberalization was not accompanied by factor market liberalization, and thus impeded the flow of capital from surplus countries to deficit countries to offset trade imbalances.

Table 21–3. *Gross domestic product, 1987*
(US$ billions)

Africa region		*Asia region*		*EMENA region*		*LAC region*	
Lesotho	0.3	Bhutan	0.3	Yemen PDR	0.8	Haiti	2.3
Mauritania	0.8	Laos, PDR	0.7	Yemen Arab Republic	4.3	Jamacia	2.9
Sierra Leone	0.9	Nepal	12.6	Jordan	4.3	Nicaragua	3.2
Chad	1.0	Papau New Guinea	3.0	Oman	8.2	Honduras	3.5
Liberia	1.0	Sri Lanka	6.0	Tunisia	8.5	Trinidad and Tobago	4.3
Central African Rep.	1.0	Bangledesh	17.6	Morocco	16.8	Costa Rica	4.3
Malawi	1.1	Malaysia	31.2	Kuwait	17.9	Bolivia	4.5
Burundi	1.2	Philippines	34.6	United Arab Emirates	23.7	Paraguay	4.6
Togo	1.2	Thailand	48.2	Syrian Arab Republic	24.0	El Salvador	4.8
Mauritius	1.5	Indonesia	69.7	Hungary	26.1	Dominican Republic	4.9
Mozambique	1.5	Korea, Republic	121.3	Pakistan	31.7	Panama	5.5
Botswana	1.5	India	220.8	Portugal	34.3	Uruguay	6.4
Benin	1.6	China	293.4	Egypt, Arab Republic	34.5	Guatemala	7.0
Burkina Faso	1.7	Yugoslavia	60.0	Ecuador	10.6		
Somalia	1.9	Turkey	60.8	Chile	19.0		
Mali	2.0	Algeria	64.6	Columbia	31.9		
Zambia	2.0	Saudi Arabia	71.5	Peru	45.2		
Madagascar	2.1	Venezuela	49.6				
Rwanda	2.1	Argentina	71.5				
Congo, People's Rep.	2.2	Mexico	141.9				
Niger	2.2	Brazil	299.2				
Tanzania	3.1						
Gabon	3.5						
Uganda	3.6						
Senegal	4.7						
Ethiopia	4.8						
Ghana	5.1						
Zimbabwe	5.2						
Zaire	5.8						
Kenya	6.9						
Côte d'Ivoire	7.7						
Sudan	8.2						
Cameroon	12.7						
Nigeria	24.4						
Subtotal	126.1		849.4		491.7		727.0
Percentage to grand total	6.0		39.0		22.0		33.0
Grand total	2,194.2						

Source: World Bank 1989b, Table 3.

Some groups made an attempt at joint industrialization. However, those sectors selected as integration industries were capital and technology intensive, which further distorted the production pattern of the member countries and resulted in inefficient allocation of scarce imports and investment resources. This policy could not prevent redistributional disputes; it merely transferred the problem to another, politically highly sensitive area. The distribution of industries was based on political viability, thus contributing to the inefficient location of investments.

In the EEC, the high share of intraregional trade required the introduction of a common external tariff to prevent differences in competitive position arising from different national tariffs towards third countries. In economic integration among developing countries, given the modest intraregional share of total trade, the common external tariff was mainly required to support the common industrialization policy by providing protection to the new integration industries. Thus, a generalized decrease in tariffs was precluded and instead the common external tariff often resulted in a substantial tariff increase for the relatively most open and developed economies of the group. In this way, it increased the level of protection in the member countries that were relatively most integrated into the world economy. Consequently, the common external tariff became an instrument against later trade liberalization (e.g. the CACM). An additional shortcoming of a high common external tariff is that, within the integration area, it protects multinational companies that settled down in the period of national import substitution. These become the main beneficiaries of enhanced protection and can be expected to oppose later most favored nation tariff reductions.

Evaluation of trade performance. According to customs union theory, the success of regional integration schemes can be measured by the share of intraregional trade in total trade. However, a growing intraregional trade share can only be seen as a positive development if it is accompanied by increasing relative weight of the region in world trade. The Latin American Integration Association (LAIA) and ASEAN are the only regions with intraregional trade flows of some international importance—above 8 billion dollars (see table 21-2). Intraregional trade reached more than 10 per cent of total trade only in ASEAN,[7] the Central American Common Market, and LAIA. However, intraregional trade shares show a stagnating or decreasing trend, between 1975 and 1987. If one takes account of unrecorded trade, which in some cases surpasses the officially registered figures (especially in Sub-Saharan Africa), it is even harder to find evidence of gains.

A less homogeneous picture is provided by the share of member countries in intraregional trade. Smaller, landlocked economies generally show figures above average, with intraregional trade shares of between 30 and 60 percent of total trade. There is a marked difference between intraregional trade and extraregional trade, with manufactured goods and nontraditional products accounting for a larger share in intraregional trade. Different manufactured goods are sold in the regional and international markets. In most countries, goods based on domestic resources and labor intensive consumer products are oriented to the international market, while capital intensive goods and products based on imported inputs have a relatively higher intraregional share in total exports.

Response to crisis. By the mid-1970s, fundamental changes in the international economic environment had exerted a decisive impact on economic integration among developing countries. First, the commodity price boom and subsequent terms of trade changes had a differential impact on the member countries of the same integration area according to their export and import pattern. Second, adjustment to the new realities required an export oriented development strategy as opposed to regional integration based on regional import substitution.

Large trade deficits and growing debt burdens pressured countries to increase exports to earn convertible currencies, and curtail domestic investment and the ability to import inputs for processing into regional exports. The result was a relative decline in the share of regional trade. Exports were diverted from intraregional to extraregional markets;[8] reduced investment had an adverse impact on production capacities and infrastructural projects; and reduced import capacity led to concentration on "essential items" generally only available from outside the region. Large and competing devaluations added to the shift of exports to external markets. The previously achieved level of regional trade policy coordination was diluted by "temporary" reintroduction of previously lifted quantitative restrictions, slowdown or stopping of originally agreed trade liberalization schemes, implementation of "administered trade", and increased tariffs in intraregional trade.

Member countries with substantial intraregional trade surpluses became unwilling (or unable) to finance the deficits of other members. At the same time, regional imbalances were considered as secondary issues compared to the servicing of international debts. Economic integration among developing countries has not provided the capital, technology, and human skill that has become increasingly important in the international division of labor. It has not taken account of the necessity of connecting trade liberalization with factor market liberalization. And it has not provided the rapid and flexible response required due to institutional bottlenecks, protracted decision-making processes and constraints imposed by national interests in the search for a harmonized regional response.

Countries following an export oriented path of development were able to carry out a more rapid adjustment with fewer losses than were inward looking economies. In regional integration schemes with less emphasis on regional import substitution (ASEAN), member countries could adjust more easily to changing world economic conditions than in integration schemes based on the philosophy and practice of import substitution (Latin America).

New orientations. As a result of the past failures and the newly emerging consensus on the importance of an outward orientation, the pattern of regional integration has begun to change. Regional integration schemes are becoming more flexible and selective. They allow increasingly for optional measures to be implemented by a subgroup with greater complementarity of interests.[9] Cautious member countries, which preiously blocked development, may be bypassed. Member countries that are more prepared to further integrate into the international economy may undertake more substantial steps than others. In this spirit, the green light has been given to establishing individual relations with nonmember countries. Similarly, some countries have lowered their tariff level below the common external tariff imposed by the integration scheme (Costa Rica in CACM).

The earlier trade policy approach of regional integration has given way to other approaches[10] and the trade policy instruments which are still applied have experienced fundamental modifications. In overall trade strategy, regional protection and "training ground" arguments favoring gradual global competitiveness through the regional market have been replaced by the viewpoint that regional competitiveness is the outcome and consequence of global competitiveness. Liberalization of national trade policies has enabled industries (enterprises) exporting to the extraregional market to increase their intraregional exports as well (ASEAN). One reason for this is the comparative advantage provided by global specialization. World market orientation offers large room for making use of intraindustry linkages which can be supported by the increasing participation of international direct investment (Japanese investment in ASEAN). Another reason is the generally higher economic growth characteristic of world market oriented economies, which results in higher demand for domestic and regional goods. Additionally, export oriented economies generally have a lower debt ratio and more liquidity. Therefore trading with them may be easier and will not require the use of different compensation mechanisms, strict bilateralism or other trade minimizing instruments. Finally, adjustment to the international economy usually fosters policy coordination among participating countries. Once their national currency becomes convertible, as a result of the adjustment policy, a major barrier to intraregional trade can be eliminated.

Results From Ongoing Integration Efforts in Sub-Saharan Africa

The performance and results of each of the main ongoing integration efforts currently underway in Sub-Saharan Africa are reviewed in the World Bank report, "Intra Regional Trade in Sub-Saharan Africa" (World Bank 1989d). The major themes which emerge from this review are similar to those for economic integration among developing countries in other regions. The only relatively successful union, the CEAO, is characterized by currency convertibility and capital mobility. These features are shared only by UDEAC, which has been one of the least successful unions. The CEAO is also the only union with significant labor mobility. Thus it would appear that currency convertibility and capital mobility may be necessary, but not sufficient, for success. And it would appear that it may be necessary to implement extensive labor mobility to achieve successful integration and enhanced intraunion trade.

The erection of a common external tariff around any of the major unions is likely to lead to some diversion of recorded African trade, in addition to diverting trade from the rest of the world. Except for the CEAO, none of the Unions has been particularly successful in creating intraunion trade or achieving a large overall share of intraunion trade. However, recorded intra-African trade is not necessarily low.[11] Of the top 10 largest economies in Sub-Saharan Africa, only Côte d'Ivoire and Kenya have shares of intra-Sub-Saharan Africa trade above 6½ percent. But 17 out of 39 Sub-Saharan African countries for which data was available had more than 10 percent of their recorded trade with Sub-Saharan African partners. Seven of these countries achieved a share of at least 20 percent. The smallest unions, the MRU and CEPGL, have little relevance for the promotion of intra-African trade. Unless a compelling case can be made for other functions, it is unlikely that their continued existence could be justified.

There may be a positive relationship between degree of openness and the share of intraregional trade. This would fit in with the observations in the Caribbean and ASEAN. However, while this appears to be true in comparing the performance of the different unions in Sub-Saharan Africa, the opposite is observed at the country level. This suggests an inverse relationship between the degree of openness and the share of Sub-Saharan African trade.

Most of the countries with substantial shares of intra-Sub-Saharan African trade are located in West Africa. None of the Southern African countries has a share of intra-African trade exceeding 10 percent. In East and Central Africa, except for Rwanda and Uganda, all the countries have less than 15 percent of their trade accounted for by regional partners. It is possible that currency convertibility explains part of these variations, since seven of the ten West African nations with more than a ten percent share of intra-African trade are members of the Franc Zone.

The landlocked countries (which tend to be more closed) are those with the largest shares of intraregional trade. Of the 11 landlocked countries for which data are available, all but three have at least a 12 percent regional trade share. It is possible that re-exports by neighbors might be inappropriately classified as exports from these countries. This would artificially inflate the recorded share of regional trade for these countries. Zimbabwe, Zambia, and the Central African Republic have unusually low shares of regional trade for landlocked countries, which may suggest the presence of large unrecorded trade flows. This is especially true for Zambia, which has a very open economy.

The Rationale for Integration Efforts in Sub-Saharan Africa

Does Regional Integration in Sub-Saharan Africa Make Economic Sense?

The above review suggests that the failure of past models of economic integration was based on import substitution behind regional barriers. This is consistent with Lachler's (1989) evaluation of growth performance, which determined that for a given level of external protection, achieving a greater degree of intraregional openness will yield significant economic benefits. Lachler's findings can be explained by the efficiency gains that arise from intraregional factor mobility and competition.

Benefits from integration would, therefore, arise from increasing competition within Sub-Saharan Africa by liberalizing factor and goods and services flows across African boundaries, rather than activating customs unions that would grant high external protection against third parties. Without economic integration, Sub-Saharan Africa may improve performance by moving to an outward orientation, but it will fail to seize the benefits of horizontal and vertical integration available to larger economies. Thus, for example, by pursuing outward oriented adjustment during the 1980s, Mauritius achieved high rates of growth and eliminated unemployment. However, future growth is handicapped by lack of access to the labor and raw materials readily available in neighboring countries.

The proposed approach to regional integration assumes, as argued by Balassa (1979), that intraregional trade is not in itself more valuable than other international trade. It requires an outward oriented strategy that "provides incentives which are neutral between production for the domestic market and exports" (World Bank 1987, page 8). Experience shows that "the important lesson is that the strongly inward oriented economies did badly" (World Bank 1987, page 8); the validity of this lesson is not affected by the size of the economy (Lachler 1989). Table 21—3 presents data on gross domestic product by country and by region for developing countries in Sub-Saharan Africa; Asia; Europe, the Middle East and North Africa; and Latin America. This could be considered a crude indicator of domestic and regional market size.

The emergence of a single European market in 1992, the US-Canada Free Trade Pact, and suggestions for a Free Trade Agreement between the US and Pacific rim countries reflect concern that generalized concessions within the GATT may be harder to achieve in the future compared to the past. The emergence of large blocs may help mitigate the free rider problem and thus facilitate the movement to ever freer international trade. Sub-Saharan Africa as a whole has the economic size of a small industrialized country (such as the Netherlands) and so the relevance of a Sub-Saharan Africa bloc should not be overemphasized. However, to the extent that internal liberalization within Sub-Saharan Africa facilitates a lowering of external barriers, this reduction in external protection could be turned to credit from the GATT.

Regional Integration and Trade Liberalization

After a decade of structural adjustment efforts in Sub-Saharan Africa, progress with trade liberalization has been slow. In general, trade regimes in Sub-Saharan Africa remain restrictive. Less than one third of the countries have eliminated all quantitative restrictions (import licensing and/or foreign exchange allocation systems).[12] African governments seem to be aiming at average protection of 30 to 40 percent compared with less than 10 percent in industrial countries (World Bank 1987, page 136).

A regional approach may offer a new dimension to supplement the unilateral and uncoordinated national efforts of the sort currently being engaged in with World Bank and IMF support. Binding reforms through an international treaty will make it harder for opponents to challenge than national legislation.

According to Nellis (1986) and Steel and Evans (1984), African countries have attempted to industrialize by relying on the public sector to set up operations geared at small national markets. As a result, many public enterprises in Africa are inefficient, badly managed, and overdimensioned. "The persistence of capacity underutilization... suggests that a temporary shortage of foreign exchange is not the fundamental problem. The underlying problems are the dependence of production on imported rather than domestically produced inputs, and excessive growth of production relative to the growth of import capacity—and in many cases relative to the size of the market" (Steel and Evans 1984, page 55).

Steel and Evans (1984) also note that high effective protection encourages new investment, even with unused capacity. "High cost operation is an inevitable consequence of these problems of capacity underutilization, inadequate infrastructure, declining productivity, and excessive capital intensity. At the same time, incentives to reduce costs have been blunted by high effective protection, import prohibitions, restricted competition, and administered pricing systems ... As a result, a large share of industrial production in Africa takes place at costs that are not competitive in terms of world market prices, and it is not uncommon to find some firms that actually use more foreign exchange than they save" (Steel and Evans 1984, pages 59-60). This supports Nellis' (1986) view that making firms competitive requires competition. However, he notes that in "most African countries internal markets are so small that at least large manufacturing firms frequently acquire automatically a monopolistic or oligopolistic position." (Nellis 1986, page 44). A regional approach may be politically more acceptable, not only by building on the pan-African sentiment of the continent, but also by reducing the cost of adjustment.

A program of rapid liberalization with regional partners may diminish the costs of adjustment by forcing competition first with firms that are of comparable levels of (in)efficiency. This would allow a reduction of costs, through mergers, acquisitions and takeovers, that may be significant enough to facilitate survival in the world market. Such an opening of factor and product markets within a program of general liberalization may induce investment that would not be forthcoming in a purely national context. Under efficient management, possibly from a neighboring country, the restructured firm will have a large enough internal demand to eliminate "wasteful overcapacity".[13]

A similar approach has been advocated for the Central American Common Market (CACM). Until 1980 the CACM was the most successful regional grouping, in terms of generating intraunion trade, in the developing world (see table 21-2). Thus, it is argued that "the removal of intraregional trade barriers in this context would enable efficiency gains ... Moreover, the reduction in intraregional trade barriers, as recommended here, would be accompanied by lower extra-regional trade barriers. This would also reduce the danger of trade diversion... These arguments present a case in favor of achieving regional reintegration in the context of an overall free trade environment, but not as an alternative to external trade liberalization since Central America, even if fully integrated, would still be better off by liberalizing trade with the rest of the world." (World Bank 1989c, pages iv & v).

Thus, the proposed strategy emphasizes extension of the size of all product and factor markets as a means of increasing competition and efficiency, thereby facilitating a lowering of external barriers. This contrasts with traditional approaches that emphasize the extension of the market for selected products to exploit the economies of scale perceived to be necessary for import substitution behind high barriers. Economies of scale tend to be more relevant for heavy industry. They are irrelevant for agricultural goods and for services such as marketing and transport where significant increases in economic activity could be expected in Sub-Saharan Africa. Freeing agricultural trade is particularly important, both to support national reforms to shift the terms of trade in favor of the agricultural sector and to enhance food security. As noted by the *World Development Report 1987* (World Bank 1987), "even large economies, if cut off from international trade, would lack stimuli for efficient industrial development." (page 3). This explains why, despite occasional temptation to rely on scale economies for achieving import substituting industrialization, Brazil has tended to have an outward orientation (World Bank 1987, Figure 5.1, page 83).

None of the existing subregional groupings in Sub-Saharan Africa are large enough to rely on internal trade. The case of Nigeria is relevant in emphasizing that significant economic benefits cannot arise simply from extensive economic integration. None of the economic groupings (including ECOWAS without Nigeria) has as large and unified a market, total factor mobility, harmonized investment and tax codes and common external tariff, to the same extent as Nigeria (viewed as an economic unit). The poor economic performance of Nigeria reinforces earlier arguments concerning the importance of generalized liberalization to secure major economic benefits. Integration is only a useful means to that end.

Sub-Saharan African countries depend for their growth "upon their ability to trade relatively freely with the rest of the world." (World Bank 1987, page 3). Notwithstanding any dynamic or static gains from regional trade, the most important benefits reside in expanding external trade. Regional integration should be supported as a means of achieving greater eventual outward orientation. This is why the focus should be on creating competition in regional factor and product markets as a step to external liberalization.

The Way Ahead

Current efforts at regional integration have not proceeded as rapidly as hoped for in Sub-Saharan Africa. Thus the Organization of African Unity states that, "Five years after the adoption of the Lagos Plan of Action and the Final Act of Lagos, very little progress has been achieved in the implementation of the Plan and the Act" (Organization of African Unity 1985). This is why the regional organizations are increasingly turning to the World Bank and other donors for support.[14] There is thus a need for African governments, regional organizations, and donors to develop a framework for responding to such requests in a manner supportive of the objectives of efficient integration and of national adjustment efforts.

Specific details and problems have to be resolved in a more concrete context. Nevertheless a general framework is required to ensure that integration contributes to improved growth prospects in Sub-Saharan Africa by strengthening and deepening ongoing national adjustment efforts. The following suggestions should be seen as a first step towards achieving a consensus on the critical elements. To reconcile these elements, it is proposed that the starting point should be the existing undertakings and agreements by the member states. To date these have generally amounted to little more than declarations of intent to reduce tariffs. Existing organizations should allow and indeed encourage subgroups of states to implement more rapid and extensive elimination of trade barriers and obstacles to labor and capital mobility between them, while recognizing that some members would not want to proceed further than already agreed.

Main Elements of a New Approach

Most favored nation trade liberalization. Any new approach must start with the recognition that the regional market can be increased through most favored nation trade liberalization pursued by each member country. At present nontariff barriers and tariffs in intraregional trade tend to be higher than barriers imposed by extraregional partners. In other words, intraregional trade is in a dispreferred situation. A number of products now directed to the external market would find their way to regional markets if trade conditions were identical. Particular importance bears on the rapid lifting of nontariff barriers. Concerning tariff reduction, a compulsory time schedule for reduction is required, including unilateral measures.

Consistency of national and regional liberalization. Regional integration should provide an additional means of moving groups of African countries towards overall economic liberalization by improving conditions for a more active role by private agents across the frontiers of Africa. The approach needs to be consistent with and promote liberalization efforts at the national level and not set back programs that any individual country may want to undertake on its own. At the same time, the prospects for efficient economic integration, based on a regional liberalization of factor flows combined with external trade liberalization, are enhanced by the increasing recognition by African governments of the need to implement outward oriented adjustment strategies.

Existing subregional organizations. Closer economic links should be based on existing and potential complementarities and trade flows. To the extent possible, new initiatives should work with and through the existing subregional organizations, in particular ECOWAS, PTA, UDEAC and SADCC. New initiatives should promote factor mobility and the free movement of goods and services within the group as well as generalized trade liberalization with the rest of the world; they should not aim at reversing the failure to activate customs unions with high barriers against nonmembers including those on the continent.. Specific provision will have to be made for interunion liberalization and to avoid excluding nonmember African trading partners that may have more liberal trading arrangements and/or have extensive trade relations with some group members, such as Botswana and Zimbabwe in the context of the PTA and Zaire and Congo in the case of UDEAC.

Donors' role. To support these efforts, donors could assist those individual member States that are prepared to relax controls on cross-border flows of factors of production and goods and services. The main justification for donors to support African requests would be to provide national authorities with the necessary leverage (arising from technical expertise and financial resources) to overcome political obstacles that would not otherwise be removed and which hinder economic progress.

Consistency With the Aims of the Existing Regional Groupings

In 1980 African Heads of State adopted the Lagos Plan of Action that advocated "a far-reaching regional approach based primarily on collective self-reliance" (Organization of African Unity 1980, Preamble, Article 1). In the current climate of adjustment, the objectives of the Lagos Plan need to be interpreted even more broadly. "To achieve the goals of rapid self-reliance and self-sustaining development and economic growth" (Organization of African Unity 1980, Preamble, Article 3) requires the emergence of strong African firms able to compete in the world market. Thus, the spirit of the Lagos Plan is fully consistent with the promotion of extensive regional liberalization as the first step to generalized liberalization.

Economic integration may offer a powerful means for achieving an outward orientation[15] in some Sub-Saharan African countries. Regional cooperation should result in increased openness (and thus a reduction in protection vis-à-vis the rest of the world) on the part of the member countries of the regional union. This should lead to a reduction in barriers to trade in services and factors of production in addition to trade in goods.

The groupings usually acknowledge, as in the case of the PTA, that "unless simultaneously with the reduction and eventual elimination of tariffs, nontariff barriers are also eliminated, the effort of eliminating tariffs can be easily nullified" (PTA 88-017, page 4). Therefore, it would be consistent with the agreed objectives to concentrate initial efforts on eliminating non-tariff obstacles to trade.

Similarly, the proposed emphasis on factor mobility is fully consistent with the broad objectives of the major economic groupings in Sub-Saharan Africa. They see the current agreements "as a first step towards the establishment of a Common Market and eventually of an Economic Community ... to promote co-operation and development in all fields of economic activity particularly in the fields of trade, customs, industry, transport, communications, agriculture, natural resources and monetary affairs with the aim of raising the standard of living of its peoples, of fostering closer relations among its Member States, and to contribute to the progress and development of the African continent." (PTA Treaty, Chapter 2).

The major area of divergence concerns the granting of regional preferences. The objective of existing unions is to pursue import substituting industrialization behind high barriers against third parties, while the World Bank would argue for a lowering of external barriers. In practice regional preferences to date have been few and generally insignificant. Further, there is an increasing commitment of African governments to outward oriented adjustment. Thus, the stage is set for the Common External Protection to be interpreted as an objective of sustaining lower external barriers than would be feasible in individual countries, thanks to the efficiency gains from factor mobility and increased internal (regional) competition. Such an approach has been informally accepted by the Secretariats of the PTA and ECOWAS.

Consistency With Ongoing World Bank Operations

The World Bank is supporting the efforts of Sub-Saharan African countries geared to the elimination of economic distortions and greater integration into the World Trading System. This involves greater reliance on market clearing prices for the allocation of resources, including foreign exchange; efforts geared at rationalizing and reducing effective protection through tariff reform coupled with the elimination of quantitative restrictions; and measures designed to increase private sector economic activity. Bank supported regional operations will need to reinforce these actions.

Greater reliance on market clearing prices. Securing trade and investment flows that contribute to lasting growth will involve establishing some market clearing price mechanism for regional activity. It will be an important prerequisite to allow free pricing of all transactions by the private sector, including trade of regional currencies. This may lead to distortions arising from multiple exchange rates. To minimize these, mechanisms will be required to ensure (1) that there is rapid movement towards exchange rate unification at a level consistent with an open external current account, and (2) that arbitrage through existing parallel markets is fully exploited to limit deviations of regional cross-rates from those on the parallel market for each regional currency against hard currencies.

Tariff reform. Customs unions, with their potential for trade diversion behind a high common external tariff, should not be the primary objective of integration efforts. Inevitably, a regional strategy will give some preference to regional partners, but avoiding a common external tariff will limit potential trade diversion and emphasize trade creation.[16] A common external tariff requires joint agreement by several countries for changes and this will needlessly complicate trade liberalization.

The short term aim would be to complete the dismantling of intraregional nontariff barriers and implement most favored nation trade liberalization according to an accelerated time-table. Over the medium term, external tariffs should be reduced towards the levels in the more open developing countries. Once tariffs are relatively low, it would probably be welfare enhancing to dismantle intraregional tariffs (to secure the full benefits of regional competition) provided this was accompanied by a further reduction of external tariffs. In no case should intraregional tariff preferences be excessive (say more than 20 percent) to avoid significant trade diversion and inefficient investment. In the long run, there must be an opening up to the discipline of the international market and therefore any regional trade preferences should not be excessive and should be reduced after a clearly specified adaptation period (say 5 to 10 years).

Increased private sector economic activity. Ideally, a regional initiative would involve complete nontariff liberalization of trade and capital flows, and full labor mobility among regional partners. There would be free flow of inputs and dividends across national boundaries within the group. Freer market access and greater factor mobility is needed to attract the foreign private investment that is sorely lacking in the region. Discrimination according to ownership of firms and product must end, at least for those countries to benefit from World Bank financial assistance. Uncompetitive behavior must be dealt with by appropriate regulation without excluding potentially dynamic actors from contributing to increased regional economic activity.

Compensation

The purpose of integration should be to improve resource allocation, which will raise absolute incomes. The relative incomes of some partners may diverge, and the emergence of a few poles of industrialization should be expected.

The primary objective of compensation should be to equalize benefits to private economic agents rather than to national governments. This can best be achieved by extending employment and investment opportunities and opening up goods and services markets in the more advanced countries to those from the economically weaker ones. Nevertheless, political sensitivity will have to be taken into account in addition to purely economic arguments. Thus, there will have to be workable means of transferring compensation between governments. In implementing any such scheme, it will be essential to avoid negating the benefits that should accrue to the private sector.

Compensation schemes should move away from the existing pattern in Sub-Saharan Africa, which emphasizes the financing of supposedly regional projects in the poorer countries. The danger is that funds specifically set aside for regional projects could be spent for political reasons on schemes with a low or even negative rate of return. There is also a danger that such funds would finance plant and equipment that would be incompatible with a rational deployment of resources from a regional perspective.

A more attractive option would be a direct transfer to the budget of the weaker economies, on the model of the Southern Africa Customs Union (SACU). SACU shares trade taxes with deliberate overcompensation to the weaker members. To avoid conflicts between external trade liberalization and fiscal revenue objectives, it may be preferable to share total revenue or sales taxes (on the model of the EC, which shares value added tax receipts). This is especially relevant in the sort of framework advocated here which emphasizes increased competition and factor mobility instead of tariff concessions to partners, while maintaining high barriers against third parties.

African governments should avoid reliance on donors to finance compensation schemes. This would discourage countries from participating mainly to benefit from foreign aid and to ensure that the mechanism is sustainable. This should not preclude donors from financing regional projects and infrastructure, but such financing should be provided on the merit of the projects and not as a means of providing compensation. With budgets remaining severely constrained for the foreseeable future, the temptation to renege on regional responsibilities may be difficult to resist. This suggests some urgency in devising a mechanism with some automacity in the transfers to weaker states.

Summary Recommendations and Conclusion

The major risks of this approach are lack of interest by Sub-Saharan African governments and regional organizations, and an implementation that results in diverting trade from efficient suppliers because the expected generalized liberalization does not materialize. The failure of existing groups largely reflects the unrealistic expectation that it is feasible for many countries to simultaneously agree on major and rapid liberalization. It is essential to accept that different countries are unlikely to be able and willing to liberalize at the same rate. Provided a few countries might be willing to accelerate adjustment through a regional apProach, it may be worthwhile to explore this option.

The integration effort must be placed within the framework of a general process of liberalization to avoid diverting efficient trade away from industrialized countries and other developing countries (including those in Sub-Saharan Africa not in the arrangement). However, because of policy and market failures in Sub-Saharan Africa, there may be potential for efficient trade switching to less expensive African exporters from costlier traditional suppliers. This potential could be as high as between four and five billion dollars (see World Bank 1989d). There is also an unrealized trade potential in agricultural commodities that is hindered by national regulations on both the export and import side. Harmonization of policies to eliminate unofficial trade flows will also have to be an important element of any integration strategy.

While economists have always agreed on the benefits of free trade, in practice no economic development experiment, including that of the Newly Industrialized Countries, has ever involved systematic and extensive free trade. The issue of the rate and scope of external trade liberalization should be seen in a practical context of movement towards a system that is more open and less biased against exports.

Efforts should be redirected from the activation of customs unions or the granting of tariff preferences. Instead, emphasis should be put on the removal of restrictions on cross-border factor movements and on the dismantling of nontariff barriers to the free regional movement of goods and services. It is difficult to envisage how the World Bank could support African governments in implementing tariff preferences.

From a purely theoretical perspective that ignores social and political costs, rapid and far-reaching unilateral most favored nation trade liberalization with the commensurate exchange rate action should accompany the regional integration of factor markets. This would lead to rapid and extensive restructuring driven by the competition from imports that would channel factors of production to newly profitable activity.

Where possible, African governments should adopt such an approach. In practice, however, we are far from a first best situation and a wide gap persists between the objectives and achievements of Bank supported trade liberalization efforts in Sub-Saharan Africa (see World Bank 1989d). It is noteworthy that even the Bank's current objectives do not aim at anything close to the level of protection in industrial countries, let alone full trade liberalization. This is why the regional approach may have practical relevance for advancing trade liberalization, quite apart from its emphasis on regional factor mobility.

Nevertheless, the recent study of liberalization experiences summarized by Papageorgiou, Choksi and Michaely (1987) and Michaely (1988) argues that one-shot liberalization may also be preferable for practical reasons. The study finds that expanding sectors grow fast enough to offset unemployment in the adjusting sectors when liberalization is immediate, while phasing the process provides opportunities for contracting sectors to overturn the reform by providing opportunities to generate political resistance.

The Papageorgiou review includes no Sub-Saharan African countries, and Sub-Saharan Africa may not be comparable to other regions concerning the impact of trade liberalization. This is because in Sub-Saharan Africa the economic base is so narrow, the distortions so extensive, and the supply of investment so constrained that there may be few sectors or firms that can provide an offset to the general contraction induced by the liberalization. In this regard, even in Nigeria, which is one of the most diversified economies and the largest economy in Sub-Saharan Africa, the experience with liberalization has been that unemployment rises in the short run. This is because the expanding sectors grow from a low base, while the contracting sectors involve larger firms (Zanini 1987, van Eeghen 1988).

Thus, the Papageorgiou findings notwithstanding, it is unrealistic to expect most Sub-Saharan Africa governments to adopt immediate and extensive trade liberalization. It is this view together with the substantial progress still possible that leads to a pragmatic view that regional liberalization may yield more lasting and worthwhile results in some cases where extensive unilateral liberalization is resisted.

Appendix

Active Regional Integration Organizations in Africa

CEAO — Communaute Economique de l'Afrique de l'Ouest (7)
Founded: 1972. Established: 1974
Benin, Burkina, Côte d'Ivoire, Mali, Mauritania, Niger and Senegal.

CEPGL — Communaute Economique des Pays des Grands Lacs (3).
Founded: 1976
Burundi, Rwanda and Zaire

ECOWAS — Economic Community of West African States (16 members — 15 initial but Cape Verde split from Guinea-Bissau). (Also known as CEDEAO — Communaute Economique des Etats de l'Afrique de l'Ouest).
Founded: 1975
Benin, Burkina, Cape Verde, Côte d'Ivoire, Gambia, Ghana, Guinea, Guinea-Bissau, Liberia, Mali, Mauritania, Niger, Nigeria, Senegal, Sierra-Leone and Togo. (Integrates the CEAO members and MRU members and Cape Verde, Gambia, Ghana, Guinea-Bissau, Nigeria, Togo).

IOC — Indian Ocean Commission (5)
Founded: 1982. Established: 1989
Comoros, France (Reunion), Madagascar, Mauritius and Seychelles.

MRU — Mano River Union (also known as Union du Fleuve Mano) (3).
Founded: 1973. Established 1974
Guinea, Liberia and Sierra-Leone
Guinea only joined in 1980

PTA — Preferential Trade ARea (17).
Formed: 1982. Started 1984 (July 1)
Angola, Burundi, Comoros, Djibouti, Ethiopia, Kenya, Lesotho, Malawi, Mauritius, Mozambique, Rwanda, Somalia, Swaziland, Tanzania, Uganda, Zambia and Zimbabwe. Sudan and Zaire are actively considering membership.

SADCC — Southern African Development Coordination Conference (9)
Founded: 1980
Angola, Botswana, Lesotho, Malawi, Mozambique, Swaziland, Tanzania, Zambia and Zimbabwe.

UDEAC — Union Douaniere et Economique de l'Afrique Centrale (4)
Founded: 1964. Established: 1966
Cameroon, Central African Republic, Congo, Gabon, Chad and Equatorial Guinea

Notes

This paper was prepared for the African Economic Issues Conference to be held in Nairobi, Kenya June mports.5-7, 1990. We are grateful to Michael Sarris, Paul Meo, Ajay Chhibber, Charles Humphreys and Jorge Culagovski for comments and help with this paper. The paper does not necessarily reflect the views of the World Bank or its affiliated organizations.

1. See *World Bank and United Nations Development Programme* 1989.
2. See *World Bank* 1989a for a detailed discussion.
3. See appendix for a list of organizations, membership and date of formation.
4. Note that in this section a positive evaluation is made in terms of the objectives of the unions to increase intraregional activity. No normative judgement should be inferred from references to success or failure in achieving increased intraregional trade.
5. It should be noted that the heavy dependence on export commodity production, the increases of import prices, especially of fuels, and the increased reliance on food imports have importantly contributed to this result. It is also true that in some cases import substitution requires imported inputs to produce output with negative value added at world prices. Nevertheless, regardless of such explanations this increased dependence on the world economy serves to emphasize the folly of autarchic policies for Africa.
6. This is because of the failure of the unions to achieve their stated objectives of regional import substitution.
7. It is significant that ASEAN has achieved the highest sustained level of intraregional trade while being the group with the most outward orientation and the least emphasis on preferential trade arrangements.

8. In the PTA, for example, recorded intraregional trade was almost halved from about $1 billion in 1984 to about $500,000 in 1988.

9. For example, bilateral agreements between Argentina and Brazil give special preferences to the capital goods sector, create a special compensation mechanism in case of large bilateral trade imbalances by including capital flows into the deficit country, and consider bilateral economic relations as a component of planning and development programs.

10. The modified role of regional integration schemes is perhaps best reflected in the increasing attention paid to traditional cooperation areas, without economic and trade policy coordination. Joint activities in developing physical and human infrastructure may substantially increase the "integration capabilities" of a region. In Latin America, the ASEAN and the Arab region, there are ambitious plans to interconnect national electrical systems and develop the region's energy sector. Also the development of information technologies requires close cooperation. Additional possibilities may be opened in the exploitation of geological resources, joint prote ction of the environment and the utilization of hydrological resources. In this paper, however, we are more concerned with the benefits of factor and goods market integration. See Berg 1988 for a discussion of these issues in the Sub-Saharan African context.

11. This is not to deny that in general intra-African trade is lower than in Asia or Latin America. However, notwithstanding the possible distortions from re-exports to landlocked countries, one third of Sub-Saharan African countries achieve recorded trade with their neighbors that is not out of line with experience in other parts of the developing world.

12. See World Bank 1989d, table 2.

13. This is not an argument in favor of import substitution in a large internal market. Given existing widespread underutilization of capacity, a policy of regional liberalization will lead to fewer plants at greater levels of capacity utilization, but this should provide the basis for external trade liberalization rather than be used as a justification for continued third party protection.

14. The World Bank has received specific requests from PTA, ECOWAS, CEAO, UDEAC, and SADCC.

15. Outward orientation simply means not creating anti-export bias. Many countries have achieved this by offsetting some of the anti-export bias of import barriers rather than by totally dismantling such barriers (see World Bank 1987, page 81).

16. Trade diversion occurs when more expensive union partners substitute for less expensive external trade partners as a result of preferences. Trade creation is the substitution of expensive domestic production by less expensive partner imports.

References

Balassa, Bela. (1979). "Intra-Industry Trade and the Integration of Developing Countries in the World Economy", In Herbert Giersch, ed., *On the Economics of Intra-Industry Trade.*

Berg, Elliot. (1988). "Regionalism and Economic Development in Sub-Saharan Africa." A study prepared for the United States Agency for International Development. October.

Inotai, Andras. (1986). *Regional Integrations in the New World Economic Environment.* Budapest: Akademia Kiado.

International Monetary Fund. (1988). *Direction of Trade Yearbook.* Washington, D.C.

Lachler, Ulrich. (1989). "Regional Integration and Economic Development." Industry Series Working Paper Number 14. Washington, D.C.: World Bank. Processed.

Laghammer, Rolf J. and U. Hiemenz. 1989. *Regional Integration among Developing Countries — Survey of Past Performance and Agenda for Future Policy Action.* Kiel: Institute for World Economics.

Michaely, M. (1988). "Trade Liberalization Policies: Lessons of Experience." World Bank. Paper presented at the conference in Sao Paulo for a New Policy Towards Foreign Trade in Brazil. Washington, D.C.: World Bank. April.

Nellis, John R. (1986). "Public Enterprises in Sub-Saharan Africa." World Bank Discussion Paper 1. Washington, D.C.: World Bank. Processed.

Organization of African Unity. (1985). *Africa's Priority Programme for Economic Recovery 1986-1990.*

Organization of African Unity. (1980). *Lagos Plan of Action for the Implementation of the Monrovia Strategy for the Economic Development of Africa.* Addis Ababa.

Organization for Economic Co-operation and Development. Various years. *Foreign Trade Statistics. Paris.*

Papageorgiou, D., M. Michaely and A.M. Choksi. 1987. "The Phasing of a Trade Liberalization Policy: Preliminary Evidence." World Bank Discussion Paper. Washington, D.C.: World Bank. Processed.

Preferential Trade Area. Report of the Twelfth Meeting of the PTA Council of Ministers.

Robson, P. (1987). *The Economics of International Integration.* London: Allen and Unwin.

Steel, William F. and Jonathan W. Evans. (1984). "Industrialization in Sub-Saharan Africa - Strategies and Performance." Washington, D.C.: World Bank.

van Eeghen, Willem. 1988. "Nigeria: Survey of 33 Manufacturing Enterprises." Washington, D.C.: World Bank.

Wonnacott, Paul and Mark Lutz. 1988. "More Free Trade Associations?" Unpublished paper prepared for a conference of the Institute for International Economics.

World Bank. (1987). *World Development Report 1987*. New York: Oxford University Press.

(1989a). *Sub-Saharan Africa: From Crisis to Sustainable Growth. A Long-Term Perspective Study*. Washington, D.C.: World Bank.

(1989b). *World Development Report 1989*. New York: Oxford University Press.

(1989c). "Trade Liberalization and Economic Integration in Central America." Report Number 7625-CAM. Washington, D.C.: World Bank. Processed.

(1989d). "Intra-Regional Trade in Sub-Saharan Africa." Washington, D.C.: World Bank. (December).

World Bank and United Nations Development Programme. (1989). *Africa's Adjustment and Growth in the 1980s*. Washington, D.C.: World Bank.

Zanini, G. (1987). Note on SFEM Impact on 32 Industrial Companies in Nigeria Surveyed in February and May 1987, unpublished report prepared by the West Africa Projects Department (Industry and Finance Division). Washington, D.C.: World Bank.

22

The Record and Prospects of the Preferential Trade Area for Eastern and Southern African States

Nguyuru H.I. Lipumba and Louis Kasekende

The quest for the political and economic integration of Africa preceded the political independence of African states in the late 1950s and early 1960s. The Pan Africanist movement was largely founded by African Americans and Caribbeans. It was joined by leading African nationalists, including Osagefyo Kwame Nkrumah and Mzee Jomo Kenyatta, after the Second World War. It called not only for the formal independence of African states, but also for political and economic unity. After Ghana's independence in 1957, Nkrumah envisaged that all independent African states would immediately form a unitary government and a military high command to liberate those countries that were still under colonial rule. The political realities of African states after independence have prevented an all African political and economic union. However, despite the poor performance of most of the African regional economic arrangements, interest in the formation of regional trade areas and common markets has persisted.[1]

Most African states have a GDP that is less than the value of output of goods and services produced in a medium sized town in Europe and North America. In order to address the problem of economic balkanization and to promote African economic integration, the African heads of state adopted the Lagos Plan of Action. The Lagos Plan envisaged the formation of an African Common Market by the year 2000. For purposes of implementing the Lagos Plan, Sub-Saharan Africa was divided into three regions: West Africa, Central Africa, and Eastern and Southern Africa. Each region was supposed to pass through three stages before the formation of the African Common Market. First, each region would gradually establish a free trade area in which tariff and nontariff barriers would be removed. Second, each region would become a customs union by introducing a common external tariff. Third, a full fledged economic community with a monetary union, preferably using a common currency, would be formed. It would allow freedom of movement of goods and services and labor, capital, and entrepreneurs. Under the economic union, fiscal and monetary policies and investment codes would be harmonized. After the attainment of an economic union in each region, interregional barriers would be removed and an African Common Market would be formed.

The Preferential Trade Area for Eastern and Southern Africa (PTA) is the brain child of the Economic Commission for Africa and the Lagos Plan of Action. It was formed to incorporate 20 countries of Eastern and Southern Africa.[2] The PTA was formally established in 1981 when 12 member states ratified the Treaty. Operations were begun in July 1982. At the time of writing (May 1990) only Botswana, Madagascar, and Seychelles have not formally ratified the Treaty and joined the PTA. Sudan, which according to the Lagos Plan of Action was to be part of the Central Africa region, has joined the PTA.

The objective of this paper is to analyze the performance and prospects of the PTA in promoting intraregional trade and economic integration. After this introduction, the second section briefly analyzes the advantages and problems of economic integration and sets out a conceptual framework for analyzing the performance of the PTA and its future prospects. The third section briefly analyzes the economic characteristics of member states and discusses the objectives of the PTA and policy instruments to be used to attain these objectives as provided in the Treaty. The fourth section looks at the PTA's performance in trade promotion, cooperation in monetary affairs, and cooperation in transport and communications, industry, and agriculture. Section five discusses problems and prospects facing the PTA.

Advantages and Problems of Economic Integration

According to international trade theory, the first best choice is not regional economic grouping,[3] but free world trade. Free trade areas and customs unions are considered as second best choices. Preferential trade arrangements would be welfare increasing if trade creation (i.e. the replacement of expensive domestic production by cheaper imports from other members) would be larger than trade diversion (i.e. the replacement of cheaper imports from the rest of the world by more expensive imports from partner states).[4] Article 24 of the GATT permits the formation of customs unions. This runs against the nondiscriminatory principle of the most favored nation status, but it is believed that trade creation will dominate trade diversion in a customs union. This section looks at some of the advantages and problems of economic integration, and presents a conceptual framework for a developmental approach to regional economic integration.

Advantages of Economic Integration

The perceived advantages of economic integration are related to increased productivity and efficiency. Increased market size would facilitate the expans-ion of production based on economies of scale, specialization, and comparative advantage. It would enable the setting up of core industries for industrialization and development. And it would increase the viability of projects, which would improve countries' abilities to attract capital for project funding. Regional payment and clearing arrangements and increased intraregional trade would enable countries to save on hard currencies. Greater factor mobility within the customs union would foster increased productivity. And successful economic integration would require and at the same time enhance political stability.

Problems of Economic Integration

One of the problems of economic integration is that tariff reduction may not necessarily confer trade preferences to partner states. The reduction of tariffs among members of a regional economic arrangement while maintaining higher tariffs on third country imports will produce trade preferences if tariff inclusive import prices are the effective determinant of import sourcing. However, in many countries domestic currencies are not convertible and sourcing of imports depends more on securing import licenses and foreign exchange allocation.

Another problem is that the formation of a customs union with a high common external tariff may be used to support import substituting industrialization at the regional level. The Central American Common Market (CACM) succeeded in promoting industrialization in some of its member countries;[6] however, the increased level of trade within the CACM was largely trade diversion. Establishing internationally competitive industries is a necessary condition for sustained growth of manufactured output even under a regional common market.

There are problems related to the unequal distribution of costs and benefits among member countries. The customs union will confer more benefits to the more developed member countries, which will attract increased investment in manufacturing. The less developed member countries may suffer from a low level of industrial development because most of the manufactured consumer goods will be supplied by the advanced member countries. Also, in many small less developed countries, particularly in Africa, customs duties account for a large percentage of total government revenues. Large trade diversion under a preferential trade arrangement may drastically reduce government revenues which may not be easily recouped by other tax measures.

There are problems related to regional industries. More advanced member countries, such as Kenya within the East African Community, oppose planned distribution of industries. They see the allocation system as an unnecessary constraint to the expansion of their manufacturing sector. Planned regional location and market guarantees for regional industries may conflict with the objective of fostering competition to improve efficiency. And it may be difficult to plan and run regional industries that involve many governments. Regional executives are usually not given adequate powers to effectively complement regional projects, and some member countries may not give the necessary priority to regional projects.

A Developmental Approach to Regional Economic Integration

A developmental approach to regional economic integration would protect the regional market from outside competition. For the less developed member states, a policy of unilateral tariff reduction is superior to joining a custums union.[7] A regional economic arrangement should emphasize the broadening of the production base by planning and implementing joint projects in infrastructure and manufacturing industries. Particularly important are the large-scale heavy industries such as iron and steel, fertilizer and other chemicals, transport equipment, etc.

A developmental approach to regional integration demands more political commitment than does market oriented integration. It requires the delegation of powers to a regional institution. Governments would be closely involved in planning, implementing, and mobilizing finance for development projects. In the planning of the location of industries, governments would be expected to address the problem of equitable distribution of benefits and costs as well as the efficiency criteria.

The Economic Characteristics of the PTA Countries and the Objectives of the PTA

The establishment of the PTA was intended to address the region's economic development problems. This section provides data on the economic characteristics of the PTA countries and discusses the main objectives of the PTA.

Economic Characteristics of the PTA Countries

When the PTA started operations in 1982, only 12 states ratified the Treaty. By 1990 the organization had 17 member states. The total population of the PTA countries in 1987 was estimated to be 200 million (see table 22–1). In 1987 the total GDP of the PTA region was about 50 billion dollars; average GDP per capita was $257. Only in Mauritius was per capita GNP greater than $1,000. Fourteen of the countries had a per capita GNP of less than $400.

The overall economic performance of the PTA countries during the 1980s was dismal. Per capita GNP in all the PTA countries except Malawi, Mauritius, and Swaziland decreased during the period (see

Table 22–1. *Basic economic indicators of the PTA member states*

Country	Population (millions) mid-1987	Area (thousands of square kilometers)	GDP (millions of dollars) 1987	GNP (per capita dollars) 1987	Average annual growth rate (percentage) 1980-1987	Average annual growth rate (percentage) 1982-1987
Angola	9.2	1,247	7,740	—	—	—
Burundi	5.0	28	1,150	250	-2.0	2.6
Comoros	0.4	2	199	370	—	—
Djibouti	0.4	22	206*	572*	—	—
Ethiopia	44.8	1,222	4,800	130	-1.6	0.9
Kenya	22.1	583	6,930	330	-0.9	3.8
Lesotho	1.6	30	270	370	-0.9	2.3
Malawi	7.9	118	1,110	160	0.0	2.6
Mauritius	1.0	2	1,480	1,490	4.4	5.5
Mozambique	14.6	802	1,490	170	-8.2	-2.6
Rwanda	6.4	26	2,100	300	-1.0	2.4
Somalia	5.7	638	995	290	-2.5	2.2
Sudan	23.1	2,506	8,210	330	-4.3	-0.1
Swaziland	0.7	17	369	700	1.2	3.3
Tanzania	23.9	945	3,080	180	-1.7	1.7
Uganda	15.7	236	3,560	260	-2.4	0.4
Zambia	7.2	753	2,030	250	-5.6	-0.1
Zimbabwe	9.0	391	5,240	580	-1.3	2.4
Total PTA	198.7	9,568	50,959	256	—	—

Note: *1985.

table 22–1). Foreign exchange shortages and unsustainable balance of payments deficits characterized most of the PTA economies. Inadequate foreign exchange led to the underutilization of the import dependent manufacturing industries. The manufacturing sector accounted for at least 20 percent of GDP in only Mauritius, Swaziland, Zambia, and Zimbabwe (see table 22–2).

The main objectives of the PTA. The PTA has two main objectives. The first is to promote growth in intraregional trade by removing tariff and nontariff barriers, gradually moving towards a common market, and eventually forming an Economic Community for Eastern and Southern Africa. The PTA's second main objective is to promote regional economic development through planning and mobilizing funds for financing regional projects.

Promoting Intraregional Trade. According to Article 12 of the Treaty, the objective of promoting intraregional trade was to be achieved in two ways. First customs duties and nontariff trade barriers among member states would be gradually reduced and eventually eliminated. Second would be the gradual evolution of a common external tariff with respect of all goods imported from third countries, with a view to the eventual establishment of a common market among member states.

The Treaty allowed for a period of 10 years for the elimination of all tariffs and nontariff barriers. Tariffs of commonly agreed upon commodities were to be reduced at different rates, depending on the classification of the goods. After the initial tariff reduction, member states were to reduce tariffs by 10 percent every two years and remove the remaining tariffs in 1992. The initial common list of products whose tariffs had to be reduced contained 212 items; by the end of 1989 it contained over 700 commodities. Member states also agreed not to increase tariffs on any imports from the region above the levels prevailing in September 1982.

Lesotho, Swaziland, and Botswana (a nonmember) are members of the South Africa dominated Southern African Customs Union (SACU). They have a common external tariff and cannot reduce tariffs to PTA member states without violating the SACU Treaty. Swaziland and Lesotho were granted temporary exemption from implementing the tariff reduction measures of the PTA Treaty. Comoros and Djibouti argued that they were very small and dependent on customs duties for government revenue. The Treaty recognized the special economic condi-

Table 22–2. *Other economic indicators of the PTA member states, 1987*

Country	Share in Agriculture	GDP (percentage) Manufacturing	Current account balance (millions of dollars)	Total debt	Debt service as a percentage of exports	Annual rate of inflation (percentage)	Current account balance as a percentage of GDP
Angola	46	3	—	—	—	—	—
Burundi	59	9	-185	755	38.9	7.3	-16.1
Comoros	37	4	-61	203	31.2	—	-30.7
Djibouti	—	—	—	181	—	—	—
Ethiopia	42	12	-475	2,590	39.9	-2.4	-9.9
Kenya	31	11	-639	5,950	30.2	5.2	-9.2
Lesotho	21	.15	-16	241	4.01	5.2	-5.9
Malawi	37	—	-53	1,363	32.0	25.2	-4.8
Mauritius	15	24	47	775	6.1	0.5	3.2
Mozambique	50	—	-676	—	—	—	-45.4
Rwanda	37	16	-250	583	13.9	4.1	-11.9
Somalia	65	5	-59	1,965	—	-28.3	-5.9
Sudan	31	8	-702	11,388	93.4	19.5	-8.6
Swaziland	24	20	-2	293	5.9	12.5	-0.5
Tanzania	61	5	-605	4,335	65.7	29.9	-19.6
Uganda	76	5	-200	1,405	31.22	38.0	-15.6
Zambia	12	23	-12	6,400	58.7	43.0	-0.6
Zimbabwe	11	31	-22	2,512	25.41	12.5	-0.4

Source: World Bank, *Sub-Saharan Africa: From Crisis to Sustainable Growth, 1989.*

tion of these countries and granted temporary exemptions from the full application of certain provisions of the Treaty. They were required to reduce tariffs for commodities in the common list by only 25 percent of the prescribed percentages.

To be granted PTA preferences by other member states, goods had to pass stringent rules of origin. The products must have been produced in the member states by enterprises which are subject to management by a majority of nationals and at least 51 percent equity holding by nationals of the member states or a Government or Governments of the member states or institutions, agencies, enterprises or Corporations of such Government or Governments, annex III Rule 2(a).

For goods to qualify for PTA tariff preferences, the value added by PTA countries could not be less than 45 percent. Certain goods that are designated by the PTA council of ministers as goods of particular importance to the economic development of member states could qualify for tariff preference as long as they contain not less than 25 percent of value added from the member states.

The ownership rules of origin were included in order to prevent foreign companies in one country from taking advantage of the larger market. This rule is illegal under GATT principles and has led to heated discussion because foreign ownership in the manufacturing industry is predominant among the more industrialized countries including Zimbabwe, Mauritius, Kenya, and Swaziland. Some countries, particularly Zimbabwe, requested temporary exemptions from the ownership rules of origin. This rule was temporarily relaxed by adopting a three tier sliding scale preferential formula. The level of tariff reduction would be based on the ownership shares of PTA nationals in the producing enterprises. This complicates the administration of tariff preferences within the PTA. The classification of potential importables and exportables within the PTA region into 10 commodity groups with different initial tariff reduction levels adds to the administrative complexity. The value added requirement that is not strictly uniform also complicates the administration of tariff preferences.

The Treaty recognizes that nontariff barriers are the major impediments to the promotion of intra-PTA trade. These include quantitative restrictions, export and import licensing, foreign exchange licensing, stipulation of import sources, advance import, deposits, conditional permission for imports, and special charges for acquiring foreign exchange licenses. The Treaty calls for preferential treatment of the PTA member states in the allocation of quotas and in the issuing of import and foreign exchange licenses during the 10 year interim period before the complete removal of the nontariff barriers. However, the Treaty does not explicitly recognize that the nonconvertible currencies of most member states make tariff preferences ineffective instruments for promoting intraregional trade.

To promote and facilitate trade in goods and services, Article 3(4)(iii) of the Treaty aims to establish appropriate payments and clearing arrangements among themselves that would facilitate trade in goods and service. The main objectives of the PTA Clearing and Payments Arrangements as stipulated in Annex VI of the Treaty include:

- promoting the use of national currencies expressed in the Unit of Account for the Preferential Trade Area (UAPTA) in the settlement of eligible transactions among member states,
- providing machinery for the multilateral settlement of payments among the member states, and
- undertaking regular consultations among themselves on
- monetary and financial matters.

The PTA Clearing House was established and started operations in February 1984. Member states agreed that the Clearing House should be housed in the Reserve Bank of Zimbabwe until an independent Clearing House could be established. The PTA secretariat has urged the member states to direct all their intraregional trade through the Clearing House using PTA currencies. It was agreed that multilateral clearing of payments would take place every 75 days. Net debtor countries would pay the net creditor countries in convertible currencies. The maximum limits of net debit and net credit for each member state would depend on the level of its intra-PTA trade. A country that exceeds its net debit position would have to pay the excess amount in hard currency within three business days.

To facilitate travel within the region so as to promote trade in goods and services, the PTA, through its Eastern and Southern Africa Trade and Development Bank, introduced its own UAPTA travellers cheques. The PTA travellers cheques are cleared in the PTA Clearing House. Net creditors are paid in dollars at the end of the transaction period.

Promoting regional economic development. The second main role of the PTA is to promote overall regional economic development through cooperation by member states. The Treaty recognized the importance of efficient transport and communication links...for the development of the Preferential Trade Area (Article 23). Transport and communication links established during the colonial period mainly linked the export enclaves to the colonial power. The lack of adequate intra-African transport and communication links has continued even after three decades of independence.

The PTA Treaty also calls for member states to cooperate in the field of industrial development to promote collective self reliance, complementary industrial development, the expansion of trade in industrial products and provisions of related training facilities within the Preferential Trade Area (Article 24). Priority areas for cooperation include basic and heavy industries (metallurgical, chemical, and petrochemical industries, and mechanical, engineering, electrical, and electronics industries), manufacturing and processing industries for consumer goods production, and facilities in relation to raw material and related infrastructure such as programs for the development of power and energy (Annex VII Article 3 (c)).

The Treaty recognizes that individual member states' markets are too small for the establishment of some of the large-scale industries. Member states are encouraged to establish multinational industrial enterprises where combined markets of more than one member state are required to make an enterprise profitable and require huge amounts of capital investment. Guidelines on the formation, location, and regulation regarding ownership and management of multinational industrial enterprises were to be developed by the committee on industrial cooperation.

In the field of agricultural development, the Treaty calls for cooperation in the following:

- development research, extension, and the exchange of technical information and experience;
- production and supply of foodstuffs that will ensure surpluses of food, especially grains, and the establishment of adequate storage facilities and strategic grain reserves;
- coordination of the export of agricultural commodities;
- harmonization of programs in agricultural and livestock production;
- development of land and water resources;
- sharing of agricultural service and technology;
- and the marketing and stabilizing prices of agricultural commodities (Annex IX article 2).

The objective of cooperation in the industrial and agricultural sectors makes PTA not only a Preferential Trade Area, but also a development institution. The Treaty called for the establishment of the Eastern and Southern African Trade and Development Bank, with the objective of providing financial and technical assistance to promote economic and social development in the member states. It would promote the development of trade among member states by financing trading activities and projects designed to make the economies of member states more complementary to each other. It would also supplement the activities of the member states' national development agencies by jointly financing operations and using such agencies as channels for financing specific projects. The PTA Trade and Development Bank was established in 1985 with its headquarters in Bujumbura, Burundi.

Performance of the PTA Since Launching its Operational Phase

The PTA integration program is aimed at promoting cooperation among member states in all sectors of the economy. A program of implementation is currently in place to eventually establish a common market. Many programs have been initiated by the PTA Secretariat. There has been increased contact between policy makers, business people, and politicians of the different member states of the PTA region. However, trade creation and diversion are yet to respond significantly to the signals in place. Performance of the PTA should be judged in terms of trade promotion, cooperation in monetary affairs, and cooperation in transport and communications, industry, and agriculture.

Trade Promotion

The PTA program for the promotion of intra-PTA trade includes:

- The establishment of a common list of selected commodities of import and export interest to member states (Article 3, annex 1, pg. 44),
- The relaxation and eventual elimination of tariff and nontariff barriers, and
- The setting up of a PTA clearing house to save on foreign currency (Annex VI, pg. 115).

The PTA common list of commodities and the tariff reduction program are in place. However, delays in implementing treaty provisions have resulted in postponement of the deadline for trade liberalization. The tariff and nontariff barriers to intra-PTA trade were originally to be eliminated by 1992; this has since been revised to the year 2000. The program is currently running behind the revised schedule.

Statistics on trade promotion are presented in table 22–3. The table does not show evidence of an increase in intra-PTA trade arising out of the implementation of the PTA program. The share of intra-PTA trade in total trade is less than 10 percent. A review of the trade patterns at a disaggregated five digit commodity level reveals that some 25 percent of total trade is in commodities in which PTA members are both exporters and importers.[8]

The binding constraints to trade are nontariff barriers, particularly the use of foreign exchange controls and import licensing. Even if a country reduced tariffs, there would be no prior grounds for expecting increased volumes of imports from the PTA or otherwise since imports would still be subject to import controls and foreign exchange allocations. It is not surprising that PTA trade has not yet responded significantly to the tariff reduction program. Few member states accord preferences to importers from the region when issuing import licenses and allocating foreign exchange. Only Zimbabwe is on record as having earmarked a proportion of foreign exchange reserves for financing intra-PTA trade. However, the amount set aside in 1989 was less than the value of imports from other PTA states in 1988.

Information gathering and dissemination. There is a lack of readily available trade information on export and import sources in the PTA region. The PTA secretariat, with the assistance of ITC, UNCTAD, and GATT, has implemented a project in information gathering and dissemination with the following objectives:

- to establish a regional trade and production information network with a central data base at the PTA secretariat inter-linked with designated national trade promotion organizations within each member state,
- to conduct systematic supply and demand market surveys on products identified by member states as being of priority import and export interest,
- to convene buyer and seller meetings as a complementary activity of the supply and demand surveys.

By November 1988, the PTA Trade Development Centre had created a computer data base with information on over 2,000 enterprises in the region and trade data on 1,000 products for the period from 1980 to 1985. The PTA has organized a series of buyer and seller meetings and trade fairs. These have increased contacts among business people and increased knowledge of potential sources of imports. It is estimated that at the second trade fair, held in Zambia in 1988, close to 100 million dollars in business was initiated. Nontariff trade barriers, in most cases, hinder further negotiations and the conclusion of such contracts.

Use of the clearing house. To encourage the use of national currencies in transactions between member states and the use of the Clearing House, the PTA authority decided that all transactions in goods and services between member states would be settled through the Clearing House. This is in line with the broad objectives of promoting and facilitating intra-PTA trade.

The Clearing House was established in February 1984. In the period up to end-December 1984 (see table 22–4) only six member states participated in the Clearing House. By its third year of operation, all member states (with the exception of Comoros and Djibouti) had participated in the Clearing House. Between 1984 and 1988, Clearing House payments and the ratio of payments to total intra-PTA imports

Table 22–3. *Trends in total and intra-PTA trade: 1982 to 1988*
(in millions of dollars and percentages)

	1986	*1983*	*1984*	*1985*	*1986*	*1987*	*1988*
Total exports	6,505	5,489	6,060	5,968	6,560	7,630	8,392
Total imports	5,344	5,015	5,349	5,203	6,035	6,511	7,105
Total trade	11,849	10,504	11,409	11,171	2,595	14,141	15,497
Intra-PTA exports	402	377	374	339	359	538	575
Intra-PTA imports	542	430	434	355	391	584	643
Total intra-PTA trade	945	807	809	695	750	1,122	1,218
Intra-PTA exports share	6.2	6.9	6.2	5.7	5.5	7.1	6.9
Intra-PTA imports share	10.2	8.6	8.1	6.8	6.5	9.0	9.1
Intra-PTA trade share	8.0	7.7	7.1	6.2	6.0	7.9	7.9

Note: These figures exclude Swaziland and Lesotho. Total exports are less than total imports largely because of underreporting of imports to evade customs duties.

increased significantly, and the net settlement in foreign exchange as a percentage of gross trade financed decreased. According to a report given by the Secretary General to the 14th meeting of the Clearing and Payments committee (November 1989), improved performance of the Clearing House continued in 1989; its volume of business amounted to UAPTA 334.8 million and only 41.3 percent of payments through the Clearing House were settled in foreign exchange currency. These figures indicate success in the use of the Clearing House as a multilateral institution for settlement of payments.

Cooperation in Monetary Affairs

Included in the PTA's cooperation in monetary affairs are the establishment of the PTA Trade and Development Bank, promotion of UAPTA travellers cheques, and establishment of a monetary union.

The PTA trade and development bank. The PTA Trade and Development Bank was established in November 1985 and became operational in January 1986. In May 1989 the Bank introduced financing schemes in respect of letters of credit confirming preshipment finance and providing credit to export oriented industrial projects. By October 1989, the Bank had confirmed letters of credit for import of critical inputs from third countries worth 4.2 million dollars. The Bank is currently concentrating on providing short term trade financing to facilitate intra-PTA trade. Also under consideration is the setting up of an Export Development Finance Department (EDF) in the Bank.

UAPTA travellers cheques. UAPTA travellers cheques were launched in August 1988 with the main objective of facilitating intra-PTA travel. By April 1989, a total face value of UAPTA 37.6 million had been distributed to 14 member states by the PTA Bank (refer to document PTA/TC/CP/XIII/10). During August to December of 1988, total sales of the cheques by member states amounted to UAPTA 1.5 million. In the first four months of 1989, total sales amounted to UAPTA 1.1 million.

Establishment of a monetary union. In a further endeavour to attain the objectives of trade liberalization and economic integration, the PTA is moving in the direction of establishing a monetary union. At the 13th meeting of the Council of Ministers held in Arusha (November 1988), it was agreed to undertake a study to examine the feasibility of setting up a monetary union and how the PTA could attain this goal on a gradual basis. The study is currently in progress.

Cooperation in Transport and Communications, Industry, and Agriculture

The medium and long term objective of the PTA is to restructure the economies of the PTA member states through cooperation to establish industrial, agricultural, transport, and other projects. This is intended to increase capacity utilization in existing industries and create new capacities. A charter on multinational industrial enterprises between two or more member states is already in place. Table 22–5 presents a list of projects and probable locations which have been identified by the ECA and PTA Secretariats. The list is not exhaustive.

Problems and Prospects of the PTA

The PTA is a group of poor countries with weak economies. At the formal poitical level there is appearance of a firm commitment to cooperation. However, many countries in Africa are too small to be viable economic units. The landlocked countries are unlikely to sustain economic development if their

Table 22–4. *Gross transactions financed through the Clearing House*

Period	*Payments (millions of UAPTA)*	*Payments Annual percentage rate*	*Intra-PTA imports (millions of UAPTA)*	*Payments as a percentage of intra-PTA imports*	*Net settlement in foreign exchange as a percentage of gross trade financed*
1984	37.2	—	-423.8	8.8	70.4
1985	48.9	31.5	350.0	14.0	86.9
1986	59.4	22.7	333.0	17.8	51.7
1987	87.9	47.9	451.9	19.5	54.8
1988	142.0	61.5	478.8	29.7	50.1

Source PTA Clearing House Executive Secretary's Report, June 1989.

neighboring countries are in economic turmoil. For these countries, economic cooperation with neighboring states is necessary, though not sufficient, for sustained economic development.

A major problem facing the PTA is the diversity of its member countries. Five of the member states have populations of less than one million. Lesotho and Swaziland cannot effectively participate because of their membership in the Southern African Customs Union. Comoros and Djibouti consider themselves too small and dependent on customs duties as a source of government revenue to be able to effectively implement the PTA tariff reduction program. Angola and Mozambique are facing civil wars that make tariff reduction and monetary union look like academic exercises. A few countries other than those that have specific derogations have published the reduced tariff rates applicable to goods imported from the PTA. However, the complex commodity classification and the three tier system of the rules of origin make it difficult for business people to understand clearly which goods have what preferences under the PTA program.

Another major problem the PTA faces is the elimination of nontariff barriers. These are imposed in different forms by virtually all member states. Most currencies are inconvertible. Foreign exchange allocation and import licenses are required before goods can be imported. Unless goods in the common list are put in an open general license by member states, tariff preferences will not be effective. Many member states, however, have expressed the fear that opening up in this manner will result in large foreign exchange outflows in settlement of the net indebted position arising from intra-PTA trade. They consider foreign exchange allocation, for example, as a necessary tool for management of scarce foreign exchange resources.

One of the key objectives of the PTA is the use of national currencies in intra-PTA trade and the multilateral clearing facility. To date there has been little use of the clearing house and many exporters have continued to invoice in hard currencies. This is explained largely by the fact that many of the PTA countries maintain overvalued and rapidly appreciating currencies. The export retention schemes in operation in Uganda and Tanzania also discourage invoicing in local currencies as exporters stand to benefit if they earn hard currencies. A large proportion of trade among some member states goes through unofficial channels to avoid official foreign exchange and import control and to evade customs and sales taxes.

Most of the member states have gone through a period of drastic economic decline and persistent crisis. Many are currently implementing World Bank and International Monetary Fund designed structural adjustment programs that require, among other things, trade liberalization. Overall trade liberalization has the effect of undermining tariff preferences to member states. At present, this is an academic point because tariff reduction does not confer effective preferences. Effective liberalization of import licensing and payment regimes may actually benefit intra-PTA trade by removing the foreign exchange control impediments to intraregional trade.

The fostering of economic cooperation will require coordination of macroeconomic and trade policies. The PTA is currently looking at the modalities of forming a monetary union which will require a common currency or fixed exchange rate regime and free movement of capital. It would be a pleasant surprise if the PTA countries were to adopt one currency. The monetary union which the PTA central bankers probably have in mind is coordination of monetary and exchange rate management to attain exchange rate stability among member countries. Similar monetary and fiscal policies will be necessary, though not sufficient, for achieving rates of inflation that do not differ excessively among member states.

The PTA has not confined itself to promoting and facilitating trade. It is also a development institution that will identify and promote development projects, in particular in industry and agriculture. The selection, planning, allocation, and implementation of multinational projects to serve the whole region is bound with political and technical problems. The use of the efficiency criteria is likely to lead to concentration of industries in countries with better infrastructure and existing industrial bases. Equitable distribution may lead to establishment of high cost industries. There is an inherent conflict between efficiency and growth of the region as a whole, and equity. The smaller and more coherent East African Community's experience in regional industrial allocation does not offer optimism that the larger and more diversified PTA region is likely to succeed in this area.

The PTA has a small secretariat and an ambitious program. It attempts to initiate cooperation on many fronts at the same time. It so far has not consolidated achievements in the specific area of promoting and liberalizing trade. There is a danger of its efforts being dissipated in too many areas without showing concrete results in any particular area. Moreover, Angola, Botswana (non-member), Lesotho, Malawi, Mozambique, Swaziland, Tanzania, Zambia, and Zimbabwe are members of the Southern African Development Coordination Conference (SADCC).

Table 22–5. *Sub-Regional projects identified for establishment in the PTA*

Projects and project ideas	*Location*	*Other participants*
1. Steel, steel products, and related activities	Zimbabwe Uganda Kenya Mauritius Tanzania Lesotho Malawi Swaziland	
2. Manufacture of engines for farm equipment and road transport components	Zimbabwe Kenya	Must identify other states to supply parts and components
3. Manufacture of diesel engine-mounted chassis	Ethiopia Tanzania	As above
4. Low cost multi-purpose vehicles	Uganda Zambia	As above
5. Agricultural machinery	Zimbabwe	As above
6. Irrigation equipment	Zambia	
7. Sections and bars for high tension electricity wires	Zimbabwe	
8. Energy equipment	Zambia	
9. Manufacture of transformers	Zambia	Participation encouraged
10. Fertilizers	Kenya Lesotho Malawi Ethiopia Burundi Tanzania Uganda Zimbabwe Mauritius Swaziland Zambia	Other states invited to participate
11. Caustic soda production	Kenya	
12. Building materials	Mauritius	

Table 22-5. ***(continued)***

Projects and project ideas	*Location*	*Other participants*
13. Processing of fish and other seafood (exploratory study proposed)	Djibouti Comoros Somalia	
14. Improvement of cement industry	(Idea being promoted in SADCC)	
15. Utilization of steel plant waste	Zimbabwe	
16. Production of pharmaceuticals	(Action plan formulated)	
17. Production of pesticides	(Action plan formulated)	
18. Manufacture, maintenance, and repair of railway rolling stock	(Being promoted in SADCC and involving Tanzania, Zambia, and Zimbabwe)	
19. Production of hospital equipment and spare parts	(Being promoted in SADCC)	
20. Aircraft maintenance	Ethiopia Tanzania Kenya Zambia Zimbabwe	
21. Aircraft training	Ethiopia Kenya Uganda Zambia	
22. Development of inland ports and strengthening national shipping companies	Ethiopia Mauritius Kenya Malawi Tanzania Uganda	

SADCC has promoted trade among its member countries through exchange of trade information and possibly counter trade. It has, however, rejected the use of tariff reduction as a means of trade promotion. Thus, within the PTA region, there is duplicity of institutions with somewhat similar objectives. There is a need for coordination to avoid conflict and duplication of activities.

Small countries such as the Comoros, Djibouti, Burundi, and Rwanda feel that they are least developed and should be granted special and differential treatment in the implementation of the Treaty Provision and, if possible, be assisted financially by other member states. Using the standard UN criterion of classifying the least developed countries based on income per capita, share of manufactured output in total GDP, and life expectancy, all PTA countries are least developed with the exception of Kenya, Mauritius, Swaziland, Zambia, and Zimbabwe. Mauritius and Swaziland are small countries with populations of less than one million each. Zambia's economy has collapsed. Among the large countries, only Kenya and Zimbabwe can be considered as more developed. It is not feasible for the two countries to offer economic assistance to the rest of the member states.

Discussion of the problems facing the PTA may make the reader pessimistic about successful economic integration in Eastern and Southern Africa. However, the potential for expanding intraregional trade does exist and should be exploited.

The PTA has succeeded in bringing business people and traders from the region together and they have formed a PTA Chamber of Commerce. Information on products produced by each member country is increasingly available. What is needed is a more liberal payment regime at least for intra-PTA trade and easy movement of goods and people. The PTA should and can concentrate and solidify measures of liberalization and facilitation of trade before attempting to move forward in the difficult area of selecting, planning, locating, and implementing multinational projects.

Notes

1. The success of the West African Community, CEAO, and the West AfricanMonetary Union in promoting intraregional trade is documented by Atsain (1990).

2. The twenty countries are Angola, Botswana, Burundi, Comoros, Ethipoia, Kenya, Lesotho, Madagascar, Malawi, Mauritius, Mozambique, Rwanda, Seychelles, Somalia, Swaziland, Tanzania, Uganda, Zimbabwe, and Zambia.

3. For a discussion of advantages and experiences of economic integration in different parts of the world, see El-Agraa (1988).

4. The conepts of trade creation and trade division were introduced by Viner (1950).

5. For a critical review of the failure of regional economic integration among developing countries, see Vaitsos (1978).

6. For a detailed analysis of the benefits and costs of economic integration in Central America, see Cline and Delgado (1978).

7. See Cooper and Masell (1965).

8. PTA document presented to the Inter-government Committee of Experts in Nairobi, November 1989.

References

Atsain, A. (1990): Intra African Trade of the West African Monetary Union, mimeo.

Cline, W.R., E. Delgado (eds) (1978): *Economic Integration in Latin America* (Washington, D.C.: Brookings Institute).

Cooper, C.A. and B. F. Massel (1965): Towards a general theory of customs unions in developing countries, *Journal of Political Economy*, Vol. 73.

El-Agraa, E.M. (1988): *International Economic Integration* (London: MacMillan).

Hazlewood (1975): *Economic Integration: the East African Experience* (London: Heinemann)

Hazlewood (1979): The End of the East African Community: What are the Lessons for Regional Integration Schemes?, *Journal of Common Market Studies.*

Kiggundu, S.I.: *A Planned Approach to a Common Market in Developing Countries* (Nairobi: Coign Publications).

PTA: A Brief to the Sixth Meeting of the PTA Authority on Problem-Areas that are Delaying the Implementation of Activities of the PTA, PTA/AUTH/VI/3.

PTA/CM/X/5: Report of the Eleventh Meeting of the Council of Ministers. November 1987.

PTA/CM/XII/5: Report of the Thirteenth Meeting of the Council of Ministers. November 1988.

PTA/TC/CP/XIV/10: Report of the Fourteenth Meeting of the Clearing and Payments Committee. November 1989.

PTA/TC/CT/VIII/3: Study on the Feasibility of Eliminating Customs Duties and Non-Tariff Barriers to Intra-PTA Trade by September 1992. April 1987.

PTA Report of the Study Team on the Equitable Distribution of Costs and Benefits in the PTA. November 1989.

PTA Treaty for the Establishment of the Preferential Trade Area for Eastern and Southern African States. June 1982.

PTA/TC/CP/XIII/10: Report on UAPTA Travellers Cheque Operations. June 1989.

Vaitsos, C.V. (1978): Crisis in Regional Economic Cooperation (integration) among Developing Countries: A Survey, *World Development*, Vol. 6.

Viner, J. (1950): *The Customs Union Issue* (New York; Carnegie Endowment for International Peace).

23

Enhancing Trade Flows Within the ECOWAS Sub-Region: an Appraisal and Some Recommendations

Ademola Ariyo and Mufutau I. Raheem

The Economic Community of West African States (ECOWAS) was formed in 1975. It was seen as one of the surest means of enhancing the accelerated development of the subregion. Although several appraisals of the performance of this organization have been done before, another look at this issue was necessary for the following reasons.

First, the consolidation period for the operation of ECOWAS was expected to last 15 years. Hence, 1990 marks an important milestone for evaluating its performance. Second, most of the previous studies were undertaken before the introduction of stabilization and structural adjustment programs by some member countries of the ECOWAS subregion. Hence, the possible incremental impact of this development as reflected in the structure and volume of intra-ECOWAS trade is yet to be documented. Third, little attention has been paid to the trade matrix amongst members of the community. Finally, attention had earlier been more focussed on distribution, and less on the production of tradeable goods within the region.

The aim of this paper is to address some of these issues. The major findings of the study are as follows. First, there has been a slight improvement in the volume of trade within the region. This may be due to the benefits of stabilization and structural adjustment programs which encourage intraregional economic cooperation. Second, the structure of trade flows has broken the language and geopolitical barriers, especially along the Anglophone-Francophone dimension. Third, export performance is better than import performance, mainly because the required imports are not being produced within the region. This finding forms the basis for the suggestion that the success of ECOWAS requires focus not only on distribution, but also on production. It requires a deliberate production arrangement to facilitate the availability of some of the required goods hitherto not available within the subregion.

The remainder of the paper is organized as follows. The second section provides a brief discussion of the rationale for the establishment of ECOWAS. The third section provides an in-depth analysis of the nature and growth of intra-ECOWAS trade. Some of the major causes of the unimpressive level of trade flows within the subregion are highlighted in the fourth section. Some suggestions for enhancing trade flows within the subregion are offered in the fifth section. Concluding remarks are presented in the last section.

Formation of ECOWAS: The Rationale

Regional economic integration is an issue that has been extensively discussed in the economic literature with respect to both developed and the less developed countries (LDCs). Recently emphasis has been on the need for LDCs to form economic groupings as a means of enhancing their growth and development

Table 23-1. *ECOWAS merchandise trade: degree of openness for 1987*

Country	Total exports	Total imports	Total trade	GDP	Degree of openness (percentage)
	Millions of dollars				
Benin	114	607	721	1,570	46
Burkina Faso	127	416	543	1,650	33
Cape Verde	9	124	133	158	84
Côte d'Ivoire	3,166	2,153	5,319	7,650	70
Gambia	100	182	282	172	164
Ghana	1,017	1,019	2,036	5,080	40
Guinea	453	336	789	2,166	36
Guinea-Bissau	4	70	74	135	55
Liberia	912	1,482	2,394	990	242
Mali	109	437	546	1,960	28
Mauritius	370	364	734	840	87
Niger	413	355	768	2,160	36
Nigeria	8,300	5,390	13,690	24,390	56
Senegal	645	1,174	1,819	4,720	39
Sierra Leone	212	189	401	900	45
Togo	233	561	794	1,230	65

Source: World Bank 1989; IMF International Trade Statistics

prospects. This suggestion was based on the potential benefits of regional economic interaction.

Several attempts have been made in the past three decades to initiate economic grouping schemes in Latin America, Asia, and Africa. Informing this effort has been the usual general argument for trade expansion among developing countries through trade liberalization. It is believed that, by opening up a large and growing market among developing countries, economies of scale and more broad based specialization can be efficiently achieved than would otherwise be the case (UNCTAD 1979). Consequently, some incentives will emerge that will stimulate production in the tradable sectors and ultimately lead to an expansion of trade and economic activities.

Economic integration among LDCs may occur in response to the present world economic conditions, which are characterized by slow growth in industrial countries, declining terms of trade in LDCs, mounting protectionism within the industrial countries against developing countries' exports, and huge external debt repayment which constitutes a serious burden on the LDCs. Cicin-Sain and Marshall (1983) have emphasized the need to conceive and establish preferential trading and payment arrangements among the developing countries to allow, inter alia, for a better achievement of individual countries' balance of payments objectives. They conclude that this can be facilitated through a faster expansion of intraregional trade.

It was against this background that ECOWAS was formed. The Agreement (or the Treaty) establishing the ECOWAS was signed in Lagos on May 28, 1975, and became effective in June 1975 when the required

Table 23-2. *Total and intra-ECOWAS trade*
(millions of dollars)

Year	Total exports (1)	Total imports (2)	Total trade (3)	Intra-ECOWAS exports (4)	Intra-ECOWAS imports (5)	Intra-ECOWAS trade (6)	(4) as a percentage of (1)	(5) as a percentage of (2)	(6) as a percentage of (3)
1975	11,747	9,713	21,460	502	392	894	4.3	4.0	4.2
1980	34,289	24,977	59,266	1,472	1,441	2,913	4.3	5.8	4.9
1984	19,031	13,150	32,181	1,083	1,207	2,290	5.7	9.2	7.1
1987	18,285	14,859	33,144	1,098	1,291	2,389	6.0	8.7	7.2

Source: IMF International Trade Statistics (various issues)

Table 23-3. *ECOWAS members' trade statistics: 1975*
(millions of dollars)

Country	Total exports	Total imports	Intra-regional exports	Intra-regional imports	Exports to ECOWAS countries as a percentage of total exports	Imports from ECOWAS countries as a percentage of total imports	Exports as a percentage of total ECOWAS exports	Imports as a percentage of total ECOWAS imports
Benin	57	197	4	16	7.0	8.1	0.5	1.9
Burkina Faso	44	151	24	35	54.5	23.2	0.4	1.5
Cape Verde	—	—	—	—	—	—	—	—
Côte d'Ivoire	1,188	1,126	181	57	15.2	5.1	10.0	10.8
Gambia	149	59	2	4	1.3	6.8	1.3	0.6
Ghana	807	791	11	19	1.4	2.4	6.8	7.6
Guinea-	143	164	5	19	3.5	11.6	1.2	1.6
Guinea-Bissau	6	38	1	1	16.7	2.6	0.1	0.4
Liberia	395	331	5	4	1.3	1.2	3.3	3.2
Mali	50	241	9	77	18.0	32.0	0.4	2.3
Mauritius	199	227	3	32	1.5	14.1	1.7	2.2
Niger	91	102	6	21	6.6	20.6	0.8	1.0
Nigeria	7,995	6,032	163	55	2.0	0.9	67.5	57.9
Senegal	462	581	78	38	16.9	6.5	3.9	5.6
Sierra Leone	136	196	4	7	2.9	3.	1.1	1.9
Togo	126	174	6	8	4.8	4.6	1.1	1.7
Total	11,848	10,410	502	393	4.2	3.8	100.0	100.0

Source: World Bank 1989; IMF International Trade Statistics

number of seven countries ratified it. The Community encompasses 16 countries with the membership comprising nine French, five English and two Portuguese speaking countries. The broad objectives of ECOWAS as contained in Article Two of the Treaty envisage the promotion of:

"cooperation and development in virtually all fields of economic activity, particularly in the fields of industry, transport, telecommunications, energy, agriculture, natural resources, commerce, monetary and financial questions, and on social and cultural matters, for the purpose of raising the standard of living of its people, of increasing and maintaining economic stability, of fostering closer relations among its members and contributing to the progress and development of the African continent."

Promotion and expansion of intra-Community trade is therefore the principal motive for the establishment of ECOWAS. To enhance the realisation of this objective, many provisions were inserted into the Treaty. A timetable for the implementation of these provisions was agreed upon as follows. It was stipulated that two years after the coming into effect of the Treaty, member states would be required not to impose new duties on intra-Community trade. They were not obliged to eliminate existing duties. From the third year, however, member states were to commence a reduction in the level of tariffs and to progressively eliminate quotas and other quantitative restrictions on trade. Harmonization of external tariffs was expected to have been fully implemented by the tenth year. Finally, the Treaty specifically indicates that a customs union would be established within the first 15 years of operation of the scheme.

The establishment of the ECOWAS was the culmination of several past efforts and initiatives to bring together these countries of diverse political and socioeconomic backgrounds. This historical antecedent has been extensively discussed by Hazelwood (1967) Sohn (1972), Onwuka (1982), Ezenwe (1984) and Diejomaoh (1985), among others. Prior to the formation of ECOWAS, economic cooperation was organized along two separate blocks—Anglophone and Francophone groupings. The speed with which the ECOWAS provisions could be implemented was probably affected by the existence of these subgroups.

ECOWAS is now about 15 years old. Many of its decisions, protocols, and Treaties remain unimplemented. As the Community embarks on the second phase of its existence, it is desirable to provide evi-

Table 23-4. ***ECOWAS members' trade statistics: 1980***
(millions of dollars)

Country	*Total exports*	*Total imports*	*Intra-regional exports*	*Intra-regional imports*	*Exports to ECOWAS countries as a percentage of total exports*	*Imports from ECOWAS countries as a percentage of total imports*	*Exports as a percentage of total ECOWAS exports*	*Imports as a percentage of total ECOWAS imports*
Benin	63	331	9	24	14.3	7.3	0.2	1.3
Burkina Faso	90	358	39	84	43.3	23.5	0.3	1.4
Cape Verde	4	68	1	6	25.0	8.8	0.0	0.3
Côte d'Ivoire	3,142	2,187	482	185	15.3	8.5	9.2	8.8
Gambia	30	167	9	14	30.0	8.4	0.1	0.7
Ghana	1,206	1,130	9	261	0	23.1	3.5	4.5
Guinea-	358	324	19	14	5.3	4.3	1.0	1.3
Guinea-Bissau	11	55	2	4	18.2	7.3	0.0	0.2
Liberia	600	535	11	10	1.8	1.9	1.7	2.1
Mali	148	458	15	168	10.1	36.7	0.4	1.8
Mauritius	235	316	1	38	0.4	12.0	0.7	1.3
Niger	566	594	71	111	12.5	18.7	1.7	2.4
Nigeria	26,802	16,517	549	227	2.0	1.4	78.2	66.1
Senegal	477	11,052	147	153	30.8	14.5	1.4	4.2
Sierra Leone	219	336	22	35	10.0	10.4	0.6	1.3
Togo	337	551	89	108	26.4	19.6	1.0	2.2
Total	34,288	24,979	1,475	1,442	4.3	5.8	100.0	100.0

Source: World Bank 1989; IMF International Trade Statistics

dence on the extent of its achievements relating to the growth, structure, and direction of trade flows, and to examine some of the major solutions for overcoming existing constraints. These will hopefully be useful, especially for the political leaders and policymakers in redesigning the strategy for enhancing the achievement of the laudable objectives of ECOWAS.

The Nature, Growth, and Relative Importance of Intra-ECOWAS Trade

Foreign trade is a very important economic activity for all the member states of ECOWAS. They depend on trade to meet their growing needs for foreign exchange, essential imports, and government revenue through import and export duties. The extent of this dependence, normally referred to as the degree of openness, is measured by the ratio of foreign trade to the gross domestic product. The higher the value of this indicator, the higher the degree of dependency on trade. Table 23-1 provides data bearing on this issue for member countries for 1987.

The figures suggest that most countries are open to external economies. This suggests that the economic performance of these countries is largely determined by their external sector. They are highly susceptible to the vagaries of the world economy.

Table 23-2 presents data on the total volume of foreign trade by ECOWAS member countries for the years 1975, 1980, 1984, and 1987. The proportion of intra-ECOWAS trade in total trade is also given. The data indicate that the region's total trade increased from $21 billion in 1975 to $59 billion in 1980. This represents a 176 percent increase over the period. Between 1980 and 1984, the value decreased by 46 percent to $32 billion. The downward trend is attributable to the global recession which adversely affected the performance of the member countries' economies. No significant improvement was achieved since then as the trade volume increased marginally to $33 billion in 1987.

The trend of intra-ECOWAS trade follows a similar pattern. The volume of intraregional trade increased by 226 percent, from $894 million in 1975 to $2,913 million in 1980. By 1984, it had decreased to $2,291 million, after which it experienced a marginal increase to $2,390 million in 1987.

In spite of its increase in absolute terms, the proportion of intra-ECOWAS trade to the total foreign trade remains unimpressive. However, there are con-

Table 23–5. *ECOWAS members' trade statistics: 1984*
(millions of dollars)

Country	*Total exports*	*Total imports*	*Intra-regional exports*	*Intra-regional imports*	*Exports to ECOWAS countries relative to total exports*	*Imports from ECOWAS countries relative to total imports*	*Member country's exports relative to total ECOWAS exports*	*Member country's imports relative to total ECOWAS imports*
Benin	114	396	9	43	7.9	10.9	0.6	3.0
Burkina Faso	80	255	16	76	20.0	29.8	0.4	1.9
Cape Verde	3	71	1	5	33.3	7.0	0.0	0.5
Côte d'Ivoire	2,700	1,507	368	240	13.6	15.9	14.2	11.5
Gambia	76	112	38	7	50.0	6.3	0.4	0.9
Ghana	530	621	5	118	0.9	19.0	2.8	4.7
Guinea	409	251	33	13	8.1	5.2	2.1	1.9
Guinea-Bissau	18	51	0	5	0.0	9.8	0.1	0.4
Liberia	451	363	12	49	2.7	13.5	2.4	2.8
Mali	136	341	42	102	30.9	29.9	0.7	2.6
Mauritius	270	331	17	19	6.3	5.7	1.4	2.5
Niger	228	298	39	88	17.1	29.5	1.2	2.3
Nigeria	13,142	7,116	359	89	2.7	1.3	69.1	54.1
Senegal	535	1,000	118	259	22.1	25.9	2.8	7.6
Sierra Leone	148	166	3	63	2.0	38.0	0.8	1.3
Togo	191	271	24	31	12.6	11.4	1.0	2.1
Total	19,031	13,150	1,084	1,207	5.7	9.2	—	—

Source: World Bank 1989; IMF International Trade Statistics

siderable variations in the value and share of intra-ECOWAS trade of individual countries compared to their global trade, as Tables 23–3 to 23–6 indicate for 1975, 1980, 1984, and 1987. The picture that emerges from these tables is that the largest exporters to the subregion include Nigeria, Côte d'Ivoire, and Senegal. The largest importers from the subregion are Côte d'Ivoire, Ghana, Mali, Mauritania, Senegal, Burkina Faso, and Nigeria.

With respect to exports, Nigeria's position is understandable because of increased petroleum exports to most ECOWAS member countries. However, Nigeria's performance is not impressive when related to its total volume of foreign trade. In relative terms, the landlocked countries—Burkina Faso, Mali, and Niger—conduct a substantial proportion of their foreign trade with other members of the subregion. They collectively account for the highest proportion of intra-ECOWAS trade for the years considered in this paper.

Tables 23–7 and 23–8 show the flow of each country's exports and imports to and from other members. Senegal exports to all other member countries, while Togo and Côte d'Ivoire trade with 10 or more countries in the group. Following closely are eight other countries (Benin, Burkina Faso, Gambia, Guinea, Mali, Mauritania, Niger, and Nigeria) which maintain dealings with between seven and nine other member nations. Ghana, Liberia, Sierra Leone, and Guinea Bissau deal with between three and six, while Cape Verdi appears to export to none of the countries within the subregion. Senegal and Togo have the most diversified import sources. Cape Verde is the least diversified, as it imports from only two member countries.

There is a wider spread in the countries' imports than in exports. Trade concentration in the subregion is influenced by monetary zone and language groupings. Countries in the same (table 23-7) monetary zone or language group tend to trade more with themselves than with other countries. For instance, there is a stronger and more intensive trade link among the Francophone countries than among the Anglophone countries. This may be due to the common historical and economic background within each group.

The commodity composition of the member states' exports and imports is also important. Tables 9 and 10 present data on the structure of merchandise exports and imports for ECOWAS member countries.

Table 23–6. ***ECOWAS members' trade statistics: 1987***
(millions of dollars)

Country	Total exports	Total imports	Intra-regional exports	Intra-regional imports	Exports to ECOWAS countries relative to total exports	Imports from ECOWAS countries relative to total imports	Member country's exports relative to total ECOWAS exports	Member country's imports relative to total ECOWAS imports
Benin	114	607	10	46	8.8	7.6	0.7	4.2
Burkina Faso	127	416	11	129	8.7	31.0	0.8	2.9
Cape Verde	9	124	3	5	33.3	4.0	0.1	0.9
Côte d'Ivoire	3,166	2,153	467	335	14.8	15.6	20.1	14.8
Gambia	100	182	44	4	44.0	2.2	0.6	1.3
Ghana	1,017	1,020	5	120	0.5	11.8	6.5	7.0
Guinea	—	—	—	—	—	—	—	—
Guinea-Bissau	4	70	0	3	0.0	4.3	0.0	0.5
Liberia	912	1,482	9	34	1.0	2.3	5.8	10.2
Mali	109	437	31	137	28.4	31.4	0.7	3.0
Mauritius	369	364	38	73	10.3	20.1	2.3	2.5
Niger	413	355	30	79	7.3	22.3	2.6	2.4
Nigeria	8,300	5,390	289	53	3.5	1.0	52.8	37.1
Senegal	645	1,174	111	212	17.2	18.1	4.1	8.1
Sierra Leone	212	189	2	23	0.9	12.2	1.3	1.3
Togo	233	561	9	20	3.9	3.6	1.5	3.9
Total	15,730	14,524	1,059	1,273	6.7	8.8	—	—

Source: World Bank 1989; IMF International Trade Statistics

Similar tables cannot be computed for the intraregional trade flows due to nonavailability of data. However, it is assumed that the tables presented here are representative of the intragroup trade. The export profile is dominated by fuels, minerals and metals, and other primary commodities, which includes food items and agricultural raw materials. The export of manufactures is very negligible with the exception of Mali, Senegal, and Sierra Leone, where manufactures accounted for more than 25 percent of total exports in 1987.

Food, fuels, and manufactured goods have dominated the region's import profile. The share of food in merchandise imports is greater than 15 percent in about half of the countries in 1987. The share of machinery and transport equipment in total imports is also relatively high. It accounts for more than 30 percent in Nigeria, Senegal, Ghana, Guinea Bissau, Burkina Faso, Mali, and Mauritania. It appears that ecological differences,

The level of intraregional trade in West Africa is still very low. All the ECOWAS member countries carry out a large proportion of their trade transactions with countries in Europe and America. However, there seems to be room for optimism. For example, previous studies asserted that the level of intraregional trade was between three and four percent during the 1970s and early 1980s. This study has shown that it had increased to about seven percent in the late 1980s. This may be a result of several efforts to induce intraregional trade. Most countries are now trading with a relatively larger number of other member countries than in the early 1980s, thus cutting across the monetary and language barriers witnessed in the past. Even so, the present level of about seven percent is very low. Some of the major factors militating against a significant improvement in intraregional trade flows, after 15 years of the existence of ECOWAS, are highlighted in the next section.

However, a substantial volume of intraregional trade is unrecorded. This may be due to the activities of smugglers and petty traders along the international boundaries where unofficial markets are located. Even though there is no certainty as to the magnitude and intensity of unrecorded trade, there is ample evidence which tends to suggest that the size may not be so small.

Table 23-7. ***Intra-ECOWAS trade matrix-exports***
(US$ millions)

Country	*Benin*	*Burkina Faso*	*Cape Verde*	*Côte d'Ivoire*	*Gambia*	*Ghana*	*Guinea*	*Guinea-Bissau*	*Liberia*	*Mali*	*Maurit-ania*	*Niger*	*Nigeria*	*Senegal*	*Sierra Leone*	*Togo*
Benin	—	0.12	—	0.45	—	1.77	—	—	0.01	0.25	—	3.24	1.46	0.06	—	0.60
Burkina Faso	0.26	—	—	1.74	—	0.67	—	—	—	0.49	—	0.75	0.22	—	—	0.34
Cape Verde	—	—	—	—	—	—	—	—	—	—	—	—	—	—	—	—
Côte d'Ivoire	16.6	98.0	—	—	—	28.4	—	—	14.0	47.6	6.10	27.04	12.4	36.7	—	27.7
Gambia	—	—	—	—	—	39.13	3.88	0.07	0.03	—	0.05	—	0.47	0.03	—	—
Ghana	—	2.80	—	—	0.10	—	—	—	1.10	—	—	0.10	0.20	—	—	0.10
Guinea	—	—	—	—	0.23	0.05	—	—	0.48	1.04	—	—	—	0.02	0.08	0.23
Guinea-Bissau	—	—	—	—	0.014	—	0.019	—	—	—	—	—	—	0.15	—	—
Liberia	—	—	—	0.10	—	0.30	0.70	—	—	—	—	—	5.90	—	1.20	—
Mali	—	1.32	—	3.87	—	—	—	—	1.55	—	0.85	0.29	4.45	0.16	0.10	—
Mauritania	0.53	—	—	12.92	—	0.70	—	—	1.66	—	—	—	3.53	2.29	—	0.70
Niger	2.84	2.16	—	1.61	—	0.02	0.08	—	—	0.76	0.01	—	21.58	—	—	0.08
Nigeria	4.00	0.10	—	1.40	—	36.0	—	—	6.00	—	—	6.00	—	74.0	1.90	1.40
Senegal	4.30	4.77	1.48	27.23	3.20	0.15	6.84	2.94	0.15	18.74	7.49	1.91	1.98	—	0.44	4.00
Sierra Leone	—	—	—	—	0.18	—	0.04	—	1.19	—	—	—	0.01	0.02	—	—
Togo	0.43	0.98	—	0.82	—	0.35	0.08	—	0.03	0.42	—	1.65	0.18	0.02	—	—

Source: Computed from IMF: International Trade Statistics.

Table 23–8. ***Intra-ECOWAS trade matrix-imports***
(US$ millions)

Country	*Benin*	*Burkina Faso*	*Cape Verde*	*Côte d'Ivoire*	*Gambia*	*Ghana*	*Guinea*	*Guinea-Bissau*	*Liberia*	*Mali*	*Maurit-ania*	*Niger*	*Nigeria*	*Senegal*	*Sierra Leone*	*Togo*
Benin	—	0.28	—	18.27	—	0.04	—	—	0.05	—	0.58	0.87	6.44	4.77	—	0.70
Burkina Faso	0.24	—	—	107.81	—	3.08	—	—	—	1.45	—	2.38	1.25	5.24	—	5.00
Cape Verde	—	—	—	0.90	—	—	—	—	—	—	—	—	—	1.60	—	—
Côte d'Ivoire	0.50	1.90	—	—	—	0.10	—	—	—	4.30	14.2	1.80	154.4	30.0	—	0.07
Gambia	—	—	—	—	—	0.11	0.25	0.02	0.02	—	—	—	—	3.52	0.20	—
Ghana	1.90	0.70	—	31.2	43.0	—	0.10	—	0.30	—	0.80	—	39.0	0.2	—	0.40
Guinea	—	—	—	—	4.27	—	—	0.02	0.72	—	—	0.09	—	7.53	0.04	0.06
Guinea-Bissau	—	—	—	—	0.078	—	0.038	—	—	—	—	—	0.02	3.23	—	—
Liberia	—	—	—	15.40	—	1.20	0.50	—	—	1.70	1.80	—	6.00	0.20	—	—
Mali	0.34	0.54	—	107.39	—	—	—	—	—	—	—	0.98	0.64	20.6	—	0.47
Mauritania	—	—	—	6.68	0.06	—	—	—	—	0.93	—	0.01	0.06	8.24	—	—
Niger	2.38	0.82	—	29.75	—	0.14	—	—	—	1.60	—	—	38.73	2.10	—	1.82
Nigeria	1.00	—	—	14.00	1.00	—	—	—	7.00	—	4.00	1.00	—	2.00	—	—
Senegal	0.06	—	0.01	40.32	0.04	0.44	0.02	0.02	0.01	0.17	2.52	—	81.53	—	0.30	0.02
Sierra Leone	—	—	—	0.23	—	—	0.09	—	0.36	—	—	—	21.26	0.49	—	—
Togo	0.29	0.58	—	8.17	—	0.58	0.13	—	0.18	0.11	0.72	0.45	1.0	4.89	—	—

Source: Computed from IMF: International Trade Statistics.

Table 23–9. *Structure of ECOWAS merchandise exports*
(percentage of share)

Country	Fuels, minerals and metals		Other primary commodities		Machinery and transport equipment		Other manufacturers	
	1980	1987	1980	1987	1980	1987	1980	1987
Benin	51	42	45	38	1	6	3	15
Burkina Faso	0	0	89	98	2	1	8	1
Cape Verde	—	—	—	—	—	—	—	—
Côte d'Ivoire	6	4	84	86	2	2	7	7
Gambia	1	1	96	92	0	0	4	7
Ghana	81	37	67	60	0	0	2	2
Guinea	—	—	—	—	—	—	—	—
Guinea-Bissau	—	—	—	—	—	—	—	—
Liberia	59	57	38	41	1	0	2	1
Mali	0	0	83	71	0	1	16	28
Mauritius	76	31	23	66	0	0	2	2
Niger	86	86	12	13	1	0	2	1
Nigeria	96	91	3	8	0	0	0	1
Senegal	22	17	49	43	2	3	26	37
Sierra Leone	13	22	31	19	0	1	56	58
Togo	76	66	16	26	1	1	6	7

Source: World Bank data

Table 23–10. *Structure of ECOWAS merchandise imports*
(percentage of share)

Country	Food		Fuels		Other primary commodities		Machinery and equipment		Other manufacturers	
	1980	1987	1980	1987	1980	1987	1980	1987	1980	1987
Benin	19	11	4	34	8	2	21	49	49	37
Burkina Faso	19	16	13	3	4	5	29	34	34	42
Cape Verde	—	—	—	—	—	—	—	—	—	—
Côte d'Ivoire	17	19	17	15	3	4	28	35	35	35
Gambia	11	43	7	4	3	14	19	60	60	30
Ghana	9	6	27	17	4	3	30	31	31	37
Guinea	—	—	—	—	—	—	—	—	—	—
Guinea-Bissau	7	5	15	9	3	3	34	41	41	49
Liberia	18	19	28	21	3	3	28	23	23	29
Mali	16	12	17	16	2	2	39	26	26	27
Mauritius	29	26	9	10	3	2	36	25	25	27
Niger	13	18	26	6	4	11	27	29	29	20
Nigeria	17	8	2	3	3	3	33	45	45	50
Senegal	15	10	17	8	3	3	30	35	35	43
Sierra Leone	19	17	14	9	5	4	18	44	44	49
Togo	19	20	20	6	3	6	20	28	38	40

Source: World Bank (1989)

Factors Accounting for Low Volume of Intraregional Trade in West Africa

Several factors have been identified as the causes of the low level of trade among ECOWAS member countries. One of these is the noncomplementary nature of the production structure arising from similarities in the characteristics of the national economies of member states. Virtually all the countries in the region rely to a large extent on the exportation of primary (agricultural and mineral) products. The import structure favors manufactured goods which are not produced domestically. Furthermore, some industrial outfits produce light manufactures such as consumables, textiles, and soft drinks, which are destined only for the local market. Hence, it may be difficult to satisfy the import requirements of a country by the exports of other member countries.

Another major factor is that some of the member countries have policies which are not conducive to the growth and development of intraregional trade. An important aspect of this is the nonliberalization of trade (especially with respect to unprocessed goods, traditional handicrafts, and industrial goods), which was considered necessary for the creation of a free trade zone within the region. The achievement of this goal has continued to be stalled by the failure of member nations to comply with the ECOWAS Resolution on Liberalization of Trade. While the tariff is still being retained in several instances, it has become fashionable among the countries to impose different kinds of nontariff barriers on trade flows. In particular, the retention of the existing tariff and the introduction of new tariffs are encouraged by the fact that indirect taxes (mostly in the form of import and export duties) account for a large proportion of government revenue in all the countries in the region. Tariff removal, therefore, does not appear attractive to these governments without an alternative arrangement that would provide adequate and appropriate compensation for those member countries whose import duties were to be lowered or removed through trade liberalization. The operation of the ECOWAS Fund for Compensation, Cooperation and Development (based in Lome, Togo) was meant to alleviate the problem of loss of revenue. Unfortunately, it is bedeviled with several operational problems. Furthermore, different kinds of nontariff barriers to trade flows have also been introduced.

Another factor contributing to the low level of intraregional trade is the traditional attachment of member countries to former colonial master countries, especially the United Kingdom, France, and Portugal. Even after attaining political independence, member countries continue to have strong economic links with the former metropolitan countries in Europe. This linkage is manifested in trade treaties and monetary arrangements between the two sets of countries. As Pobbi-Asaman (1980) pointed out, most of the former colonial powers have given preferential access to their former colonies' produce. These preferences and arrangements in effect have discriminated against countries which are not part of such a grouping.

Political factors, especially political instability in member countries, have also been a hindrance to the growth of trade within the subregion. There have been cases where political leaders of member countries have accused one another of direct or indirect interference in their domestic affairs. This usually provides an excuse for the closure of international borders with the attendant negative effect on trade flows within the ECOWAS subregion. Related to this is the problem of nonimplementation of the legions of protocols and decisions reached by the ECOWAS member countries. The ECOWAS Secretariat (1985) has lamented the low level of compliance with Community policies and the nonimplementation of Community decisions and programs which in effect have been the major constraint to the implementation of trade liberalization arrangements within the subregion. Available information suggests that this high default rate arises not necessarily because these decisions have been found unacceptable by the authorities of the member states, but rather due to the lack of political will to implement them.

The poor state of infrastructural facilities within the region is another factor militating against enhanced trade flows within the ECOWAS subregion. This relates particularly to transportation (air, rail, and road) and communication networks. For example, national railway systems were built according to different technical specifications. The implied gaps in the routes and differences in gauges, rolling stock, and facilities have made the cost of transport prohibitive. This is especially true in the case of the landlocked countries (UNCTAD 1978). The absence of regular shipping facilities and lack of suitable air connections aggravate an already bad situation. Perhaps with the exception of one or two member countries, there are no modern harbors conducive to efficient handling of import and export activities. Intracommunity communications are very expensive, irregular, and unreliable. In many instances, telephone, telegraphic, and telex networks often have to be channelled through European outposts.

The special problem of landlocked countries (Mali, Nigeria, and Burkina Faso), which have transit trade arrangements with the coastal states, deserves mention. These arrangements are usually bedeviled with problems relating to transportation mode, summary declaration of goods, harmonization and simplification of customs documents, and imposition of customs duties and other charges on transit traffic by coastal countries. As Alabere (1987) rightly pointed out, the inherent transportation difficulties and particularly high transportation costs make it difficult for exports of these inland countries to compete favorably with those of coastal countries. This problem has two main negative effects. First, these landlocked countries would have to lower the prices on exports to stay competitive, thereby lowering the income of rural producers. On the other hand, the higher prices which these countries must pay for their imports compared with coastal prices tend to fuel inflationary pressures thereby raising the real cost of living (IMF 1970).

Several studies (Olofin 1977, McLanaghan et al 1982, Cicin-Sain and Marshall 1983, Raheem 1983, ECA 1982 and 1984) have identified the widespread controls and restrictions on exchange transactions as a major factor impeding the development of intra-West African trade. These exchange controls and restrictions contribute to a large extent in making inconvertible most of the currencies of the member nations. The region is characterized by differences in payments restrictions applied by member countries. This is partly in response to the balance of payments problems they face, and partly influenced by divergent policy approaches. The proliferation of these mostly nonconvertible currencies constitutes a serious problem for intragroup payments on commercial transactions.

The dominance of cash transactions in intra-ECOWAS trade (Oke 1989) reduces the pace of development of intraregional banking and financial cooperation. For example, the extent of use of some conventional instruments like letters of credit, bills for collection, documentary credits, sight drafts, and bank financing of trade would be hindered by reliance on cash transactions. The negative effect of the problem has also been highlighted by the World Bank (1989), which emphasized that even if African suppliers could be competitive in price and quality, inadequate financial arrangements tend to place them at a disadvantage with non-African competitors. The establishment of the West African Clearing House (WACH) has facilitated the removal of some of the identified obstacles that emerge as a result of the lack of direct convertibility of the ECOWAS currencies. However, the operations of the WACH are being constrained by structural problems, regulatory difficulties, and some other weaknesses that have been pointed out in other studies (Cicin-Sain 1980; Frimpong-Ansah 1987). The cash dominated trading activities also encourage smuggling and unrecorded trading activities which thereby distort the true picture of trade flows within the ECOWAS subregion.

The existence of other economic subgroupings that run parallel with ECOWAS may also jeopardise the growth of intraregional trade within ECOWAS. Article 59 of the ECOWAS Treaty is considered to be very flexible and tolerant with respect to the establishment of other micro-integrative subsystems within the Community. The Article allows any member country to be a member of other regional or subregional associations with other member and nonmember countries. This has given rise to conflicts of interest between ECOWAS and other subgroups, and divided loyalty on the part of countries belonging to more than one group. Examples of such arrangements include the Economic Community of West Africa (CEAO) under the Treaty of Abidjan (1973), the Mano River Union, and the Free Trade Area between Cape Verde and Guinea Bissau. The Anglophone countries have no subregional grouping arrangements. However, there are a number of other bilateral trade arrangements between Community members.

Contradictions in the treaties of these different groups are bound to arise, and may retard the growth of trade within ECOWAS. For instance, there are glaring divergencies in terms of the 'rule of origin' issue (World Bank 1989). Also, the customs union proposed by ECOWAS, which involves the abolition of all tariffs, is incompatible with the CEAO's preferential system. The complications arising from the conflicting demands of the various subgroup arrangements have probably been a major cause of the nonimplementation of the phased program of trade liberalization envisaged in the ECOWAS Treaty.

Other factors militating against trade growth within ECOWAS include inadequacy of relevant information and lack of uniform and standard measurement and grading systems. There is a dearth of functional centers to provide information on the availability of products and related opportunities among member countries. There is no efficient means of contact between prospective business partners. The existence of the West African Chamber of Commerce has not helped much regarding information dissemination. The banking sector (international and local) in these countries has not been very helpful in this regard. The absence of uniform grading and measurement standards has made impossible meaningful price and quality comparisons across countries. The absence or inadequacy of efficient modern storage facilities for tradeable products such as meat, fresh fish, fruits, and vegetables constitutes another serious problem.

Enhancing Intra-ECOWAS Trade Flows: Some Suggestions

The preceding sections have highlighted the rationale for the establishment of ECOWAS, the structure of trade relationships among member countries, and the problems militating against the achievement of the objectives underlying the establishment of ECOWAS. The findings indicate that, in spite of the current low level of intraregional trade, ECOWAS member countries are striving to ensure the growth of trade flows within the region. The relative volume of intraregional trade flows has increased from about 4 percent in 1984 to over 7 percent in 1987.

Some of the previous studies commissioned by ECOWAS have emphasized the scope for increasing intraregional trade in goods being produced by existing industries in the region. Attention was expected to be focussed on semiprocessed and light industrial products. These include beverages, textiles, confectioneries, nails, timber and wood products, cement, and candles, among others. Our analysis indicates that, given the high level of import demand for manufactured goods in each of the member states, manufacturing holds the key to a meaningful and appreciable growth in intra-ECOWAS trade.

The potential effects of the industrial policies of member countries should not be overlooked. The countries have similar industrialization programs based on the import substitution strategy. Hence, for most industries the domestic market is the target. This leads to excess capacity problems, and massive importation of spare parts, which affect member states' balance of payments positions.

Recommendations

The fairly substantial level of smuggling within the region would have led to distortions and most probably an understatement of the actual level and direction of trade flows within the region. Hence, the first recommendation offered here is that efforts should be made to minimize this practice. This observation leads to the second suggestion that there should be a renewed determination to eliminate tariffs and other hindrances to trade within the region. Existence of and differences in tariff rates are major causes of smuggling within the ECOWAS subregion. Review and reconciliation of tariff structures should be given priority. However, this should be complemented with a frank assessment of the viability of the ECOWAS Fund to enable each member state to realistically assess the net effects of tariff review on its revenue.

Third, all impediments to the free flow of trade should be identified and the extent of amendments attainable should be quantified. Prominent among these is the existence of competing groupings whose terms of reference conflict with those of ECOWAS. Attempts should therefore be made to identify and reconcile the conflicts and appropriate adjustments should be made. The activities of these subgroups should be consistent with the objectives of ECOWAS.

Fourth, ECOWAS member states should prioritize trade profile targets. For example, our analysis showed that food items account for a sizeable proportion of total imports within the region. Hence, the desire for self-sufficiency in, say, food items within a specified number of years may be a priority. It was also noted earlier that there exists surplus exportable goods within the region. The hindrances to their exportation should be identified and appropriate incentives and proposals should be provided. Finally, there must be improved information flows and infrastructural facilities to encourage interaction among member states and their citizens.

New Areas of Focus

The suggestions presented here so far have related to eliminating some lapses in the existing decisions. It is also appropriate to suggest some new areas of focus. The ECOWAS Treaty is essentially concerned with the movement of tradeable goods (i.e. distribution) only. The Treaty has not paid particular attention to production. However, the analysis indicates that the volume of imports from outside the region still dominates intra-ECOWAS imports. Efforts should therefore be made to identify those items still being imported from outside the region and a joint production and investment program should be established. It should be possible to identify and establish some production units specifically meant to serve the ECOWAS market. The prioritization of the implementation of such projects may be guided by the multiplier effects, potential market size, technical feasibility, comparative advantage, geographical spread, and cost of implementation (UNCTAD 1986).

The existing ECOWAS Treaty does not seem to emphasize trade in raw materials and intermediate inputs. Goods in these categories could be purchased from outside only when not available from within the region. In addition, an overhaul of the financial system should be given priority to facilitate account settlements within the region. In this light we commend the existence of the West African Clearing House, although the scope and efficiency of its operations need to be strengthened. Finally, each member state should endeavor to formally incorporate the growth of intra-ECOWAS trade into its economic policy decisions. A good example is the construction policy recently launched in Nigeria, which made the ECOWAS region its primary point of call for required imports.

While the implementation of these suggestions would go a long way in enhancing intra-ECOWAS trade, it is pertinent to register two additional observations. The first relates to the fact that the absence of the political will to implement decisions reached by member countries constitutes a serious problem. Political leaders need to be reminded of the enormous advantages of the success of the ECOWAS arrangement. Issues of concern to these leaders should be made clear and appropriate remedial actions should be taken.

The second issue relates to the expected role of the private sector. In the expansion of intra-ECOWAS trade, efforts should be made to mobilize the private sector. This might allow the public sector to focus on the improvement of infrastructural facilities. The involvement of the private sector may also augur well for the free flow of resources. Entrepreneurs should be free to locate anywhere within the region, with appropriate incentives and protective measures, especially against imports from outside the ECOWAS subregion. The establishment of a limited convertibility arrangement and subsequently a monetary zone in the region should be pursued vigorously.

Concluding Remarks

Fifteen years after the formation of ECOWAS, most of the objectives of its founding fathers have yet to be achieved. Paramount among these is the inability to achieve an appreciable volume of intra-ECOWAS trade. For various reasons, member states still actively engage in trade with the developed economies of Europe and America, even in some commodities that could be supplied from within the subregion. Thus, the proportion of trade among ECOWAS countries is still less than eight percent of the total.

The study has identified some of the major causes for this, and has proffered some solutions. On one hand, promises and decisions are yet to be implemented. These relate to tariff elimination and trade liberalization proposals, and to the will of political leaders to ensure meaningful implementation of these proposals.

The study also identified new areas of focus. These relate to an expansion of the scope of activities covered by the existing ECOWAS Treaty, to include raw materials and intermediate inputs as tradeable items. The harmonization of production and investment proposals should be part of the package, with a view to producing items currently being imported from outside the region. The public sector may focus on the provision of infrastructural and communication facilities, as part of the incentives to mobilize and enhance the meaningful participation of the private sector.

Finally, two recent developments within and outside the subregion will significantly impact on future trade flows within ECOWAS. These are the adoption of stabilization programs by ECOWAS member countries and the formation of Europe 1992. The possible opportunities and threats to intra-ECOWAS trade arising from these developments should be studied in depth, with a view to designing appropriate strategies for capitalizing on the opportunities and averting any threats to enhanced intra-ECOWAS trade.

References

Albere, B.R. (1987), "Economic Integration in Less Developed Countries: The Case of ECOWAS". Unpublished M.Sc. Economics Project Report, Department of Economics, University of Ibadan, Nigeria.

Allen, R.L. (1961), "Integration in Less Developed Area". *Kyklos*, pp. 315-335.

Aitken, N.D. (1973), "The Effects of the EEC and EFTA on European Trade: A Temporal Cross Section Analysis". *American Economic Review*, Vol. 63, pp. 881-892.

Anusionwu, E.C. (1984), "West African Lessons From East African Failure in a Common Market" in Second Colloqium on the Economic Integration of West Africa.

Austin, D. (1967), "Economic Cooperation and Integration in Africa". Report of the Seminar Programme of Congress for Cultural Freedom, No. 11.

Balassa, B. and A. Stoutjesdijk (1975), "Economic Integration Among Developing Countries". *Journal of Common Market Studies*, XIV 1, September.

Cicin-Sain, A. (1980), "Strengthening the West African Clearing House and Adopting Limited Convertibility Arrangements or Steps Towards an ECOWAS Monetary Zone", Project RAF/77/032, December ECOWAS Secretariat.

Cicin-Sain, A. and J. Marshall (1983), "Study on the Limited Currency Convertibility Among ECOWAS Countries", UNCTRAD/ECDG/142 GE 83-55469.

Diejomaoh, V.P. (1985), "Economic Integration in the ECOWAS States: Problems, Progress and Prospects" in Second Colloqium on the Economic Integration of West Africa.

Diejomaoh, V.P. and M.A. Oyoha (eds.) (1980), *Industrialisation in ECOWAS* (Lagos: Heineman Educational Books Nigerian Ltd.)

Economic Commission of Africa, United Nations (1982), "Analysis of Bottlenecks to the Development and Expansion of Intra-Sub-Regional Trade and Remedial Measures in West Africa". Report submitted to the 5th Meeting of the Committee of Officials of the Niamey MULPOC in Banjul, The Gambia, February.

Economic Commission of Africa, United Nations (1984), "Monetary and Payments Implications of Border Trade in Africa". E/ECA/TRADE/15 August 1984.

Economic Community of West African States (1985), *Ten Years of ECOWAS: 1975-1985*. Lagos: ECOWAS.

Economic Community of West African States (1987), "*ECOWAS Economic Recovery*". Lagos: ECOWAS.

Edozien, E.C. (1975), "Intra-African Trade as a Step Towards an Inward-Outward Oriented Looking Developing Policy". *The Nigerian Journal for Economics and Social Studies*, Vol. 17, 65-76.

Ezenwe, U. (1984), *ECOWAS and the Economic Integration of West Africa* (Ibadan: West Books Publisher).

Frimpong-Ansah, J.H. (1987), Analysis of Suggested Measures to Supervise the Functionary of the West African Clearing House UNCTAD/ECD/TA/20, April.

Hazlewood, A. (1967), *African Integration and Disintegration* (London: Oxford University Press).

International Monetary Fund (1970), *Surveys of African Economies*, Vol. 3, Washington, D.C.: IMF p. 43.

McLenaghan, J.B. et. al. (1982), "Currency Convertibility the ECOWAS", IMF Occasional Paper No. 13.

Oke, B.A. (1989), "Legislative and Regulatory Obstacles to the Promotion of Intra-Regional Cooperation in Trade, Banking and Finance in Africa". *Economic and Financial Review*, Vol. 26, No. 4, December.

Olofin, S. (1977), "ECOWAS and the Lome Convention: An Experiment in Complimentary or Conflicting Customs Union Arrangements", *Journal of Common Market Studies*, Vol. XVI, September.

Onwuka, R.I. (1982), *Development and Integration in West Africa: The Case of ECOWAS* (Ile-Ife: University of Ife Press).

Pazos, F. (1973), "Regional Integration of Trade Among Less Developed Countries" *World Development* 1 (7), July. pp.1-12.

Pobbi-Azaman, K.O. (1980), "Prospects for Increased Economic Cooperation and Integration in Africa" in Second Colloquium on the Economic Integration of West Africa.

Pournakoris, M. (1969), "Economic Integration and Developing Economies with Similar and Different Economic Systems", *Economic Internationale*, Vol. 32 (1) pp. 112-125.

Raheem, M.I. (1983), "The Impact of Economic Integration on Intra-Regional Trade Flows: The ECOWAS Experience". Unpublished M.Sc. Economics Project, University of Ibadan, Nigeria.

Robson, P. (1980), *The Economies of International Integration*. (London: George Allen & Unwin Ltd.).

Robson, P. (1968), *Economic Integration in West Africa*. (London: George Allen & Unwin Ltd.).

Sohn (1972), *Basic Documents of African Regional Organization*. (New York: Oceana Publishing Inc.) Vol. 3.

Truman, E.M. (1969), "The European Economic Community: Trade Creation and Trade Diversion". *Yale Economic Essays*, vol. 9, pp. 201-257.

United Nations Conference on Trade and Development (1978), "Study of Recorded Trade Flows". Report prepared for ECOWAS Trade, Customs and Monetary Study Project, March.

United Nations Conference on Trade and Development (1979), "Preliminary Reports on Trade Liberalization Options and Issues for the ECOWAS". Geneva, June.

United Nations Conference on Trade and Development (1986), "Current Problems of Economic Integration: The Problems of Promoting and Financing Integration Projects". TD/B/6/.7/79.

World Bank (1989), *Sub-Saharan Africa: From Crisis to Sustainable Growth* (Washington, D.C.: World Bank).

24

Comments on Regional Integration

Achi Atsain and Sylviane Guillaumont Jeanneney

Achi Atsain

The papers presented by Mansoor and Inotai, Lipumba and Kasekende, and Ariyo and Raheem have raised major issues confronting the economic groupings of the developing countries. They present additional evidence of the overall poor performance of the economic groupings of the developing regions. Although Mansoor and Inotai tend to imply that ASEAN and CACM seemed to have performed relatively better than similar economic groupings in Sub-Saharan Africa, my own calculations show that the intraregional trade of these two groups has constantly declined between 1960 and 1987 as a percentage of their total exports.

This comment raises a methodological problem as data provided also by Ariyo and Raheem do not support my own findings. They find that the intraregional trade of ECOWAS has steadily increased between 1975 and 1987. My own data show declining values of exports from ECOWAS and that its share of intraregional exports with respect to its total exports was estimated at roughly 3.1 percent between 1975 and 1987.

Besides the problem of data source, the papers appear to have identified, although incompletely, the causes of the poor performance. Mansoor and Inotai tend to be more provocative by overemphasizing the industrial development strategy pursued by the Sub-Saharan African countries which favored, in their words, inefficient import substitution policy as the major cause of the poor performance of the Sub-Saharan African economic groupings. They argue that customs unions that have granted high internal protection against third parties have not succeeded in increasing intraregional trade. If protection means high tariffs against third parties, Lipumba and Kasekende have shown in their study that PTA has reduced tariffs over the years, but intraregional trade has not increased within this economic grouping. The countries of the WAMU-CEAO, which have had quite similar practices, have, on the contrary, experienced increases in their intraregional exports and appear to be the most relatively successful integration scheme in Sub-Saharan Africa. This is to say that trade liberalization would not be a necessary and sufficient condition to increase intraregional trade. Trade liberalization must be accompanied by complementary policies such as factors mobility, which has been correctly pointed out by the three groups of authors.

While Mansoor and Inotai have paid little attention to monetary factors, Lipumba and Kasekende, on the one hand, and Ariyo and Raheem, on the other, have linked the promotion of intraregional trade to the use of a convertible currency. Again one can argue that the use of a common currency may but partly explain the greater gains in intraregional trade for UDEAC, which uses CFA francs, compared to the CEAO, which has not experienced increases in intraregional trade over a comparable period. Increased intraregional trade, as argued by Landell-Mills and others, depends also on the availability of finance and financial instruments such as exports, credits, export insurance and other sources to traders and firms. It depends on the solidity of the financial institutions in Sub-Saharan Africa. The liquidity crisis that is experienced by many African commercial banks represents an impediment against increased intraregional trade. Unfortunately, sufficient attention was not paid to this factor in any of the three papers.

A further point that I would like to make is to stress that trade liberalization, financial instruments, availability of finance and labor and capital mobility will no longer be sufficient to expand intraregional trade without reforms of the general administration and the specialized institutions that have been established to promote intraregional trade. As argued by Landell-Mills and others, economic agents—firms and traders—need a trading system which is free from any regulatory and procedural barriers to intraregional trade. Although I do not firmly believe that unrecorded trade far exceeds officially recorded trade, it could be that the occurrence of the former must be the result of too many administrative regulations.

A further point that I would like to make and which has been discussed in the three papers relates to the direct relationship between improved infrastructure and increased intraregional trade. Infrastructure here includes information, as well as communication and telecommunications. Contrary to Mansoor and Inotai, who tend to make the World Bank financial assistance conditional to the removal of all forms of discrimination against third parties, I would argue that donors would greatly improve the prospects of increased intraregional trade by assisting economic groupings in Sub-Saharan Africa in removing these constraints. After all, no country in the world has removed all forms of barriers to trade with the rest of the world. Issues raised by free trade associations are complex and slippery and the most valid generalization is that it is very difficult to generalize. World free trade is an ultimate goal to achieve and the world is far away from free trade.

The last point of my comments has also received attention in the papers, although it is not sufficiently articulated. Political tempers have done a great deal in slowing progress towards economic cooperation in Sub-Saharan Africa. Political confidence and trust which nourishes political will must be sought through all forms of interaction if integration schemes in Sub-Saharan Africa are to bear more fruit for the African people. Political trust is a prerequisite for any form of development policy harmonization and for safeguarding income growth and the distribution of benefits from integration. If political trust is secured, viable regional projects could be found and funded by donors such as the World Bank or the African Development Bank. If not, I tend to agree with Mansoor and Inotai that no viable regional project could be found and those who continue to advocate integration by production rather than by market would find little empirical evidence to support their case.

Sylviane Guillaumont Jeanneney

The three papers I have to discuss are very interesting. The Mansoor and Inotai paper gives us an exhaustive survey of the benefits that developing countries can derive from economic integration and of the difficulties they meet. The Lipumba and Kasekende and Ariyo and Raheem papers enlighten us about two significant African experiences of integration: the Preferential Trade Area for Eastern and Southern Africa (PTA) and the Economic Community of West African States (ECOWAS).

The three papers reveal how difficult it is to evaluate the effectiveness of integration. So I first want to make some comments of method. Then I shall underline that two issues would have called for some greater attention because of their significance for a policy aimed at improving integration.

A Methodological Issue Raised by Smuggling

The overall importance of unrecorded trade, as well as its unequal weight from one country to another, is emphasized by the three papers. Unrecorded trade not only leads to an underestimation of the effective integration, but it also makes it difficult to evaluate the progress of intraregional trade. Indeed, an increase in recorded trade can be due to the fact that some informal trade has become an official flow, and the converse is also true.

However, one of the purposes of regional integration is perhaps just to reduce the part of informal trade to the benefit of the formal one, in order to widen the fiscal basis and so to increase public receipts as it is suggested in Ariyo and Raheem's paper.

Indeed it can be considered that a lowering of customs duties and nontariff barriers within a region aims at reducing the informal part of trade since it decreases the benefits of smuggling. Therefore, even if an increase in recorded trade results from reduced smuggling or from a lower underinvoicing of commercial flows, and so does not reflect a real growth of intraregional trade, it can be considered as a success of the customs union.

From the same point of view, we may discuss Mansoor and Inotai's conclusion that "to ensure that Most Favored Nation trade liberalization proceeds in parallel with the intraregional liberalization of labor and capital flows, a Common External Tariff should be avoided until tariffs are generally low." If it is true that there is a risk that a Common External Tariff would be excessively protective, it is also true that without a Common External Tariff intraregional smuggling is stimulated.

The PTA and ECOWAS papers have collected statistical data on official intraregional trade. The study of their evolution is a little more optimistic for ECOWAS than for PTA since in the first area "the findings suggest a slight improvement in volume of trade within the region" whereas in the second, there is "no evidence of an increase in intra PTA trade arising out of the implementation for PTA programs." However, we must notice that the share of official intraregional trade in total trade was by 1987 slightly higher in PTA than in ECOWAS, i.e. 7.93 percent vis-à-vis 7.21 percent, which both remain unimpressive. So we may consider that the performances of these two regional arrangements are rather similar.

These statistics on official trade are built from separate data on imports and exports. These records show a difference between total intraregional imports and exports, two flows which should be equal by definition. The gap is obviously due to a discrepancy in customs registration. It can not be interpreted as a specific performance of intraregional imports relative to intraregional exports (or conversely, according to the direction of the difference), as it is suggested in the ECOWAS study. Another explanation is necessary.

In many African countries, exchange restrictions have been made more severe since 1980. We can perhaps find here a reason why during the 1980s recorded intraregional imports are greater than recorded intraregional exports, whereas the opposite was the case during the 1970s. When the currency is not convertible, import registration is needed to obtain foreign exchange on the official market whereas unrecognized exports can be used to obtain foreign exchange outside the official market. But when there is no foreign exchange shortage, traders have an incentive to underinvoice imports more than exports because customs and fiscal duties are generally higher on the former than on the latter.

Two New Significant Ways for Improving Integration

There are two possible policies to improve African regional integration, that have been overlooked until recently. The three papers have presented a very interesting study of the impediments to African integration. All three have stressed the role of inconvertibility of African currencies. CEAO, which is mostly between UMOA countries, is presented as a counter example.

But I think that another major impediment is the instability of real exchange rates between African countries. This factor is all the more important since, in view of integration, the three papers say that production has to be coordinated and private investment has to be stimulated. So the first question is how to stabilize real exchange rates between members of ECOWAS, PTA or other regional arrangements.

It is not realistic to foresee a monetary union between so many and diverse countries as the members of ECOWAS or PTA. As written by Lipumba and Kasekende, "it will be a pleasant surprise of the century if PTA countries adopt one currency".

Let us consider the case of West Africa: the main problem is to stabilize the real rate of exchange between the naira and the other currencies of the region. Is the solution to peg the CFA franc, the CEDI, the Guinean franc, etc. to the naira? Is the solution to promote, through adjustment programs, a more stable real effective exchange rate of the different countries, including firstly Nigeria?

Until now, structural adjustment programs have led to very different real depreciations of African currencies. As the same international institutions (IMF, World Bank and now EEC) assist African governments in the implementation of their macroeconomic policies, should they not be more concerned about the implications that an adjustment program in a specific country has for its neighbors?

Moreover, part of the international financial assistance could be directed toward the stabilization of real exchange rates, while simultaneously aiming at greater convertibility. The means could be monetary cooperation between African governments and one of these international institutions. The experience of the Franc Zone provides a good example of what can be done in this field. Certainly the maximum degree of assistance, corresponding in the Franc Zone system to the possibility of unlimited negative balance in the operations account, is extremely unlikely to meet with much support in the international framework. But a range of alternative systems, implying a smaller degree of exposure, is possible. It authorizes a negative balance only for a specific limited term and/or places a fixed limit on the size of the allowable deficit. On the other hand, African monetary authorities should cooperate with international institutions to define their monetary policy; but this cooperation could be less tight than it is in the case of Franc Zone countries with French authorities. Surely this way by which international cooperation can promote African integration calls for further investigation.

The second question I want to ask is how to contribute to equity among the members of regional organizations. The main target of regional integration in Africa is generally to promote faster industrialization. But it is likely that the poorest countries will benefit less from industrialization. What kind of compensation has to be given to them?

The three papers seem to agree that a high common tariff may lead to an inefficient import-substitution industrialization, and that African integration has to be seen as a step toward multilateral trade liberalization, which is a more or less transitory step according to the authors. However, a means of compensation for the poorest countries can be a special Common External Tariff for basic agricultural production such as cattle or grain.

There is no evidence that trade liberalization must go at the same pace in agricultural and industrial fields. Indeed, the world conditions of marketing and price determination of agricultural and manufactured products are not the same, since the protection in industrialized countries is higher and cases of international dumping prices more frequent for the former than for the latter.

Another connected problem which requires further attention is whether financial compensation in favor of the poorest countries is better suited than commercial protection. And how can international institutions contribute to assess and finance these compensations?

Human Resources, Technology and Industrial Development in Sub-Saharan Africa

Sanjaya Lall

Sub-Saharan Africa remains one of the least industrialized regions in the world. Industrial production has stagnated or declined in many Sub-Saharan African countries over the past decade. Inefficiencies and external shocks, exacerbated by poor policies, have caused many industries to become a drag on their economies. This is the more worrying because Africa is still predominantly specialized in relatively simple, low technology industries. Its long-term development would entail entry into more complex activities, where technologies are changing rapidly. However, the prospects for growth in these highly demanding areas are not encouraging. The industrial lag of Africa is set to increase rather than diminish in the foreseeable future.

Two broad sets of solutions have been suggested for the problems of African industrialization, based on differing analyses of the causes of its poor performance. One has focused on adverse external circumstances, the other on the management of macroeconomic policies, trade regimes or internal industrial strategies.

Both approaches clearly have merit. Exogenous shocks have hit many African countries particularly hard, and economic management, both in policy formulation and its implementation, has suffered from many faults. The basic physical conditions for industrialization have not been very propitious in many (but not all) African countries: small, fragmented markets, poor infrastructure, remoteness and often a scarcity of relevant natural resources. Even when all these factors are taken into account, however, there remains a large unexplained "residual". Industry has fared worse in Africa than in many other developing regions with similar handicaps and policies. We have to look, therefore, to some structural features of African industry that account for its past record and impose limitations on its future development. These features (described in Lall 1989 and 1990a) have more to do with the human resource base available for industrialization than with physical conditions or trade and other policies; however, the latter influence the development of appropriate skills and capabilities and so affect industrialization directly as well as through human resources. This framework, broader and more comprehensive than the received approaches to industrialization, leads to the consideration of much longer term policy requirements than is currently fashionable.

The paper is laid out as follows. The second section describes the recent African experience of industrial development, evaluates its success more broadly than only in terms of output growth, and comments on some of its most striking features. The third section sets out a simple but general framework for analyzing the determinants of industrial development. The fourth section applies the framework to Africa. The fifth section concludes the paper with some recommendations on the promotion of healthy industrialization.

Background to Industrial Development in Sub-Saharan Africa

In 1965, manufacturing contributed nine percent of GDP in Sub-Saharan Africa, compared to 20 percent in middle-and low-income countries, 14 percent in South Asia, 26 percent in East Asia, and 23 percent in Latin America (World Bank 1989a, Table 3, and World Bank 1989b). By 1987, this figure had risen to 10 percent in Africa, compared to 18 percent in South Asia and 28 percent in Brazil, 30 percent in Korea, 25 percent in Mexico, 24 percent in Thailand, and some 40 percent in Taiwan. A few Sub-Saharan African countries had much higher shares for manufacturing; of the 45 countries in the region, in four (Zambia, Zimbabwe, Swaziland and Mauritius) manufacturing contributed 20 percent or more to GDP, while in another 10 it contributed 10 percent or more (World Bank 1989b, Table 3). At the same time, at least 10 African countries suffered declining or stagnant shares of manufacturing in GDP over the 22 years from 1965 to 1987 (World Bank 1989b), and some registered a larger share for manufacturing only because GDP itself was declining rapidly (as in Zambia). With the exceptions noted, therefore, the level of industrialization by this measure remained fairly low and stagnant in much of Africa.

The total value of manufacturing value added (MVA) in Africa came to $16.3 billion in 1986, which was 46 percent of that of India, 66 percent of that of Korea, and 28 percent of that of Brazil (World Bank 1989b, Table 8, and World Bank 1988a). Only four countries—Nigeria, Côte d'Ivoire, Cameroon and Zimbabwe—had an MVA of over $1 billion each, while 15 had an MVA of under $100 million each. Of the group of low-income countries, Nigeria had the largest MVA; of the middle-income countries, Zimbabwe, Côte d'Ivoire and Cameroon had the largest.

In terms of growth rates (table 25–1), African industry did fairly well in the 1965 to 1980 period (8.8 percent average annually). Though rates of growth appear overstated due to the small initial base, it is clear that the first flush of import substitution, built on aid and revenues from generally booming primary product exports, was vigorous. During 1980 to 1987, the African growth rate for manufacturing fell to an annual average of 0.6 percent, while the South Asian rate rose to 8.0 percent. East Asia kept up a healthy 10.4 percent while Latin America, beset by debt problems, fell to the same rate as Africa, 0.6 percent.[1]

The figures in table 25–1 suggest that the slowdown observed for Africa as a whole was in fact confined to low-income economies (which also did poorly in the 1970s), and was sharply affected by the Nigerian performance. Middle-income countries accelerated their manufacturing growth in the 1980s, and the performance of Côte d'Ivoire (8.2 percent per annum during 1980 to 1987), Congo (9.7 percent), Cameroon (8.5 percent) and Mauritius (10.9 percent) was fairly impressive by any standards. In the group of low-income countries, Burundi, Benin and Lesotho turned out average annual growth rates of 5 percent or more in this period.

This is not to deny that a large number of other African countries suffered from low or negative growth of manufacturing in the recession following the second oil crisis (some had stagnated for much longer). The worst affected were Zaire, Tanzania, Zambia, Uganda, Somalia, Sierra Leone, Central African Republic, Nigeria, Ghana and Liberia . The recession and its aftermath led to severe underutilization of capacity as foreign exchange for imported inputs and equipment was more severely rationed and domestic demand fell. The squeeze on modern industry led to considerable unemployment, and some of the unemployed entered into informal sector activity.

In Ghana and Nigeria, for instance, the expansion of informal sector activity was particularly noticeable in this adjustment period. Much of such activity was in low productivity, low technology activities that needed few imported inputs. It is important, however, not to "glorify" the resilience and capabilities of the informal sector. It fulfilled a valuable function of permitting survival, but it remained on the margins of subsistence. It may have provided the seedbed for some entrepreneurship and skills, but as such the existing informal sector does not have the

Table 25–1. ***Sub-Saharan Africa: Growth rates of GDP and manufacturing***

(average annual percentage)

	GDP			*Manufacturing*		
Africa	*1965-1973*	*1973-1980*	*1980-1987*	*1965-1973*	*1973-1980*	*1980-1987*
All Africa	5.9	2.5	0.5	10.1	8.2	0.6
Nigeria	4.0	1.7	2.3	n.a.	1.9	3.4
Low income economies	6.0	2.8	-0.4	10.7	10.2	-1.0
Nigeria	3.3	1.9	1.4	n.a.	1.5	1.4
Mid-income economies	5.2	1.4	3.8	n.a.	2.5	6.1
Six most populous economies	7.0	3.0	-0.8	12.0	12.9	-1.3
Sahelian economies	1.0	3.5	2.5	n.a.	n.a.	n.a.
Oil exporters	7.5	2.8	-0.5	13.5	15.0	-1.0
All low-income economies	6.0	4.6	6.1	9.1	8.1	10.3
China, India	5.9	4.3	1.7	8.3	10.7	3.9

Source: World Bank

dynamism to provide the basis of sustained growth. Once it is "modernized", with competitive industrial technologies and skills (in, say, the Italian mold), it may well be a source of growth, but this involves problems of capability and efficiency which also affect the formal sector.

African industrial growth (see Steel and Evans 1984 and Gulhati and Sekhar 1982) has been highly protected and overwhelmingly inward oriented. Launched primarily by foreign companies (or resident non-Africans) to serve local markets or process raw materials for export, it was led later by state enterprises. Yet the weakness in indigenous industrial entrepreneurship, which shows up most clearly in the paucity of our modern African small-scale industrial (as opposed to informal) activity, could not be remedied by setting up parastatals. Some activities, especially those which had simple technologies, had been in existence for a long time and had good (usually foreign) management, did achieve efficiency. However, a very large proportion of industry did not, especially in countries without ready access to a plentiful supply of foreign managerial, entrepreneurial and technical skills. The countries that have the best industrialization records (Zimbabwe, Côte d'Ivoire, Kenya, Gabon, and Mauritius) are the ones which could draw on expatriates or resident non-Africans,[2] or could continue to attract sufficient foreign investments.

The degree of inefficiency in African industry seems to rise with the degree of capital and skill intensity of the facilities set up. Many traditional industries also display considerable inefficiency in comparison to standards of other developing regions. Parastatal industries have tended to be among the most inefficient (Nellis 1986). Most African industry is highly import dependent and has remained so over time. Local linkages have been mainly confined to primary inputs, while manufactured components or intermediates, technical and consultancy services, and technological inputs have continued to be imported. The degree of import dependence has thus been much higher than in most other developing countries.

Import dependence per se would not matter if the transformation of imported inputs took place with sufficient efficiency to permit growth, diversification, and most importantly, penetration of foreign markets. It is in the sphere of manufactured exports, however, that the weakness of African industry shows itself most clearly. As data collected by the World Bank (1989b, Appendix Table 17), show, the total manufactured exports of Africa came to $3.5 billion in 1987. This was less than a tenth of manufactured exports of Hong Kong, Taiwan or Korea in that year (which were $44.6 billion, $47.3 billion and $43.9 billion respectively), and only 55 percent of that of Thailand. It contributed less than 1 percent of the $371.5 billion of the total manufactured exports of the Third World as a whole; in 1973 this share had been nearly four times higher.

Part of the poor export performance of African manufacturing is explained by the inward orientation of its trade regime. With the sole exception of Mauritius, there is no African economy that is strongly export oriented in the East Asian mode. However, the Mauritian manufactured export boom, which was almost entirely in knitwear and other garments, is based on factors which differentiate it from the rest of Sub-Saharan Africa. It has a strong indigenous entrepreneurial class, a well-educated (if not technically advanced) labor force, and a large influx (stimulated by the trade regime and the quality of labor) of direct investment from Hong Kong. The specific interaction of incentives and skills sparked off the Mauritian success. The only other African country that is "moderately" export oriented is Côte d'Ivoire (World Bank 1987). Its export success has been limited and directed mainly to neighboring Francophone countries. Moreover, its own skills have been strongly supplemented by expatriates at all higher levels of management, but this has not been sufficient to launch it on the course to becoming a new Newly Industrialized Country (NIC).

The export record of African industry thus reflects its inherent competitive weakness as well as the incentive structures provided by the trade regime. Detailed micro level studies of technical efficiency show very low levels of capability to operate or improve on imported technologies (Mlawa 1983 andWangwe forthcoming on Tanzania, Page 1980 on Ghana, and Mytelka forthcoming on Côte d'Ivoire), though where foreign skills are brought in to provide a minimum base and a "teaching" role the situation is much better (Pack 1987 on Kenyan textiles).

A Framework for Analyzing Industrial Success

Neither of the two conventional explanations of the problems of African industry comes to grips properly with its structural weaknesses. The conventional explanation based on exogenous shocks explains the fall in production because of lack of imported inputs and domestic recession, but does not deal with the reasons for its lack of competitiveness, dynamism or linkages, and its extreme dualism. The conventional explanation based on incentives (trade and industrial policies) explains part of the reasons for poor export performance and lack of competitiveness arising from protected inward orientation. However, it places the whole burden of the explanation on incentives rather than structural factors, and thus tends to be partial and incomplete, but not entirely wrong.

It is necessary to adopt a more comprehensive approach to the determinants of industrial performance. Recent research on the acquisition of technological capabilities at the firm level in developing countries provides a convenient starting point (Dahlman et al 1987, Lall 1987, Katz 1987, Fransman 1986). The antecedents of such work go back to economic historians like Rosenberg, originators of the "evolutionary theories" of growth like Nelson and Winter, analysts of technical change like Freeman, and theoreticians dealing with problems of innovation and information like Arrow. The application of these diverse approaches to problems of developing country enterprises has yielded many interesting new insights. While we cannot review these insights here at any length, a few relevant points may be usefully highlighted.

First, the process of becoming efficient even in a static sense (i.e. acquiring mastery of a given technology), can be slow and difficult. It goes well beyond simply setting up a physical facility, becoming familiar with its working in some passive sense, and realizing technical economies of scale. It requires firms to develop new skills, acquire new information, develop new supplier networks, and set up new organizational structures. In developing countries, where "ready made" skills, information, suppliers and institutions do not exist in the external environment, firms necessarily have to invest heavily in building up their capabilities. Thus, there is thus no predictable learning curve which all firms traverse over time. The process of becoming efficient is usually uncertain, risky and often incomplete. This is true of developed and developing countries, but the latter suffer greater lags, and larger dispersions round the norm, because of the weaknesses of institutions and support systems.

Second, efficiency cannot remain static when inputs, products and technological conditions are always changing. Dynamic efficiency requires more developed capabilities than simple mastery, to adapt, improve and innovate on existing technologies. No firm can be self-sufficient, since specialization is essential for competitive success, but in certain critical areas of its operation it must deepen and broaden its capabilities if it is to remain fully competitive.

Third, the development of capabilities is extremely sensitive to market incentives, especially those arising from competition. Competition provides the basic spur to investment in capability development, but is a double-edged weapon. Too little competition can lead to inadequate or misguided capability acquisition; too much can wipe out firms that cannot finance the costs of capability acquisition. The ideal set of incentives thus combines some competition (of the right sort, ideally from world markets) with protection for the period of learning when costs are high and quality low. Entry into more complex activities necessarily requires greater protection, but as the experience of Korea and Taiwan shows, can be accomplished if it is combined with incentives to enter export markets or face domestic competition in the near future (Lall forthcoming).

Fourth, incentives are only part of the story. The ability of firms to respond to incentives depends on their initial base of capabilities and their access to skills within the economy. It also depends on the development of infrastructure, a supplier network and an institutional structure that allows information and technology to flow, adequate standards to be set, research to be conducted, finance to be provided and labor to be trained. In-house capability acquisition has to be complemented by the external education and training system, the development of suppliers and service firms, and an institutional framework that allows markets to function efficiently. In developed countries these requirements are taken for granted; in developing countries they cannot be.

These firm level factors can be protected at the national level. Taking for granted that macroeconomic conditions and physical infrastructure are appropriate to industrial development, its progress depends on the complex interplay of three sets of factors: incentives, capabilities and institutions (Lall 1990 and forthcoming). Incentives comprise both the neoclassical prescriptions of export orientation (strictly speaking, neutral incentives to sell in foreign and domestic markets) and the internal competition espoused by Balassa, Bhagwati and Krueger, and sufficient selective protection to allow diversification and deepening to take place. The standard neoclassical recipe makes some allowance for infant industry protection to coexist with export orientation. However it recommends only low, uniform and short-lived protection across industries to minimize the "distorting" effects of selective protection. There is neither theoretical nor empirical support for such a recipe. Once the possibility of market failures due to dynamic and unpredictable learning, externalities or complementarities are admitted, protection has to vary by activity according to its complexity and linkages to other activities. If judiciously administered, selective protection is a necessary element of industrial deepening (Pack and Westphal 1986, Lall forthcoming).

Capabilities include three things: the launching of physical investment, the provision of human capital, and the undertaking of technological effort. The ability to set up physical facilities is such an obvious determinant of industrial growth that it needs little comment, except to point out that investment skills differ greatly between countries. Human capital for industrialization arises from the formal education system and employee training undertaken by industry (King 1984). The relevance of education and training for industrial development is obvious. However, the need for specialized technical training tends to be overlooked in many discussions of industrialization. This is fairly low at initial levels of industrial development (though some high level skills and training are always needed), when general education and flexibility are more important.

Skills and training become productive only when combined with technological effort in industry to absorb new knowledge, adapt and improve on it. While advanced innovation is accepted as the lifeblood of industrial success in developed countries, the need for more mundane technological effort there and (especially) in developing countries tends to be overlooked. Yet it is this which determines how successful newcomers to the industrial scene are in producing efficiently and building up their dynamic comparative advantage.

Finally, institutions here refer to those entities set up to facilitate the workings of markets. In the industrial context, these can provide finance, information, services, standards, export assistance, and the like, which comprise the whole network of external linkages that allow individual firms to operate efficiently. This network may not be thrown up automatically by the free market.

The experience of the NICs illustrates the relevance of this framework nicely. As far as incentives are concerned, three of the four East Asian NICs (the exception being Hong Kong) intervened both selectively and functionally to promote a chosen strategy of industrial deepening. Of the three, Singapore chose to rely most heavily on foreign investors for entrepreneurial leadership and technology. At the other extreme, Korea chose to rely primarily on national enterprises and to push massively into heavy industry. This entailed high degrees of intervention to promote selected activities and to build up the skills and technological capabilities needed in those particular activities. The capabilities developed followed the industrialization path set by interventions on incentives.

Table 25–2 (taken from Lall 1990) sets out some of the relevant data on performance and capabilities in the NICs and near-NICs. It shows, among other things, how the two medium-sized NICs invested heavily in education and training to back up their drive into advanced industry, and mounted substantial research and development efforts. Korea led the technological effort because of its unique combination (in the developing world) of nationalism, heavy industry push, and aggressive export orientation. The capabilities drive also involved interventions, both selective and functional in nature, since the education and technology "markets" could not produce the requisite output of skills or information needed without direction and assistance (Lall forthcoming).

Incentives, Capabilities and Institutions in Africa

We now apply this general framework to Africa and identify its comparable weaknesses.

Incentives

The framework sketched out here does not dismiss the role of correct incentives in the industrialization process. On the contrary, an export oriented trade regime is taken to be superior to an import substituting one. However, "correct" incentives are defined to include selective interventions to provide protection to infant industries. This kind of intervention, with its emphasis on selectivity and rapid gains in competitiveness, should be sharply distinguished from the by-now "classic" type of indiscriminate and permanent protection given in most developing countries. The pattern of interventions in the incentive regime in Africa have, sadly, been overwhelmingly of this sort (Steel and Evans 1984). The results have been worse than in India (Lall, 1987) because of smaller, more fragmented markets; lower levels of technical skill and technological effort; and a greater shortage of entrepreneurial abilities (exacerbated by the leading role assigned to parastatals). However, in African countries with more developed capabilities or access to foreign capabilities, the results have been distinctly better than in others. Thus, Zimbabwe set up an impressive and reasonably functional structure of capital and intermediate goods production in the years of isolation under UDI. Its integrated iron and steel mill (ZISCO) was the only one in Sub-Saharan Africa with acceptable levels of efficiency, and it was state owned and highly protected. Similarly,

Table 25–2. *Indicators of national technological capability in selected NICs*

	S. Korea	Taiwan	Hong Kong	Singapore	India	Brazil	Mexico	Thailand
Structure and Performance								
Value added $b. (1985)	24.5	22.2	6.7	4.3	35.6	58.1	43.6	7.7
Growth 1965-80/1980-86	18.7/9.8	16.4/12.9	17.0/7.0	13.3/2.2	4.3/8.2	9.6/8.2	7.4/0.0	10.9/5.2
Exports (1986) $b (1986)	31.9	35.9	32.6	14.7	7.2	9.1	4.9	3.9
Growth of merchandise-exports 1965-80/1980-86	27.3/13.1	19.0/12.7	9.5/10.7	4.7/6.1	3.7/3.8	9.4/4.3	7.7/7.7	8.5/9.2
Gross Domestic Invt. as % GDP (1986)	29.0	19.0	23.0	40.0	23.0	21.0	21.0	21.0
Capitol Goods Prod. as % of Total Manufacturing (1985)	23.0	24.0	21.0	49.0	26.0	24.0	14.0	13.0
Capitol Goods Imports $b (1985)	10.6	5.6	7.1	8.1	3.7	2.2	6.1	2.7
(as % MVA)	(43.3)	(25.2)	(106.0)	(188.4)	(10.4)	(3.8)	(14.0)	(35.1)
Stock of Foreign Direct Investment $b (1984-6)	2.8	8.5	6.0-8.0	9.4	1.5	28.8	19.3	4.0/5.0
Stock as % GDP	2.8	8.1	20-26	53.8	0.7	9.6	13.6	10.5-13.1
Education								
A) Education Expenditure as % household consumption (1980-5)	6.0	—	5.0	12.0	4.0	5.0	5.0	6.0
B) Public Expenditure %GNP	4.9	5.1	2.7	2.9	3.7	2.9	2.6	3.9
Year)	(1985)	(1986)	(1978)	(1980)	(1985)	(1984)	(1985)	(1984)
Central Government Expenditure on Total Government Expendure (1986)	18.1	20.4	—	21.6	2.1	3.0	11.5	19.5
Age Group Enrolled (1985)								
Primary	96	100	105	115	92	104	115	97
Secondary	94	91	69	71	35	35	55	30
Tertiary Education	32	13	13	12	9	11	16	20
Vocational Ed. Enrol. (1984) ('000)	815	405	32	9	398	1,481	854	288.0
as population working age	3.06	3.24	0.86	0.5	0.07	1.83	2.0	0.96
No. of tertiary level students in S/E fields (000)	585	207	36	22	1,443	535	563	360
% population	1.39							
(Year)	(1987)	(1984)	(1984)	(1984)	(1980)	(1983)	(1986)	(1985)
in engineering (000)	228	129	21	15	397	165	282	n.a.
% population	0.54	0.68	0.41	0.61	0.06	0.13	0.35	—
Science And Technology								
Patents Granted: Total (1986)	3,741	10,615	n.a.	598	2,500	3,843	2,005	n.a.
of which % local	69	56	n.a.	8	20	9	9	n.a.
RD % GP	2.3	1.1	n.a.	0.5	0.9	0.7	.06	0.3
(Year)	(1987)	(1986)		(1984)	(1984)	(1984)	(1984)	(1985)
RD in Productive Sector % GNP	1.5	0.7	n.a.	0.2	0.2	0.2	0.2	n.a.
Scientists/Engineers in RD Per million population	1,283	1,426	n.a.	960	132	256	217	150
All scientists/engineers								
(a) Total nos. (1000)	361.3	n.a.	145.5	38.3	1000-2000	1,362.25	565.6	20.3
(b) Per million population	8,706	n.a.	26,459	15,304	1282-2564	11,475	10,720	472
(Year)	(1986)	—	(1986)	(1980)	(1985)	(1980)	(1970)	(1975)

Sources: Asion Development Bank, *Foreign Direct Investment in Asia and the Pacific*, 1988; World Bank, *World Development Report*, various; World Bank, various country reports; U.N. , *Statistical Yearbook for Asia and the Pacific 1986-1987*, Bangkok; UNESCO, *Statistacal Yearbook 1988*, Paris, 1989; Republic of China, *Statistical Yearbook of the Republic of China 1988*, Taiwan; Republic of China, *Education Statistics of Republic of China 1984*, Taiwan; Republic of China, *Science and Technoly Data Book*, Taiwan, 1987; Republic of Korea, *Introductionof Science and Technolty*, Seoul, 1988; Evenson, R.E. (1989), "Intellectual Property Rights, R & D, Inventions, Technology Purchase and Piracy in Exonomic Development", (mimeo).

Mauritian enterprises, supplemented by investors from East Asia, successfully set up garment exporting facilities to take advantage of its export orientation. The technology involved was simple to master, but the organization of production and export marketing required skills and enterprise lacking in many other African countries. The textile industry in Kenya has also achieved a respectable degree of technological mastery, with substantial and prolonged infusions of skills from experienced technicians from India (Pack 1987). In general, however, the incentive structure in Africa has not been conducive to the healthy development of industrial capabilities (see Meier and Steel 1989, Lall 1989, World Bank 1989b).

But what would the response of Africa be to more liberalized incentives? In the sphere of manufacturing industry, apart from areas of obvious comparative advantage based on processing local mineral and agricultural resources, do sufficient capabilities exist to permit a spurt of export oriented growth? And, in the longer term, what would determine its dynamic comparative advantage? The answers to these questions depend on African capabilities and institutions.

Capabilities

Physical investments. The efficient implementation of industrial investments requires a broad spread of technical, design, organizational and construction skills. Many such skills can be imported from specialized engineering or consulting companies or capital goods manufacturers in developed countries. However, the cost of relying heavily on foreign contractors can be heavy, and certain critical investment functions (initial project preparation; negotiating technology design and transfer; participation in engineering, monitoring, and equipment selection) should be handled at least partly by the project sponsor. Otherwise the country risks biases in project design, technology, and location, and much higher investment costs than are necessary. The problems of inappropriate technology choice (Stewart 1977) and "white elephants" in Africa are well known, as are the very high costs of setting up projects (a steel mill in Nigeria costs three to four times as much as the same investment in other developing countries). What is perhaps less widely appreciated is that the lack of local participation in design leads to subsequent failure to master, adapt and improve upon imported technologies, and to the absence of linkages with potential local suppliers of investment goods (Lall 1990a). Thus, valuable opportunities for technological learning and spillovers are lost in comparison with the NICs, where investment costs are much lower and many externalities are captured.

Human capital. A significant part of the entrepreneurial, technical and managerial skills required for industrialization arises from the previous experience of commerce and industry. Africa has been, in comparison with other developing regions, particularly unfortunate in this respect. With the exception of a few trading communities, the indigenous populations of most of Africa have little traditional experience in modern commerce or manufacturing. However, there is no shortage of entrepreneurial drive: the informal sector in Africa is as active and vibrant there as in other developing regions (Page 1979), and the recent recession has swelled the ranks of informal entrepreneurs. However, the drive to profit from opportunity is not the same per se as the entrepreneurial capability required to organize, set up and run modern industry (Kilby 1971).

There tends to be a progression in most developing societies from commerce and informal industry to formal small- and medium-scale manufacturing, and within manufacturing from smaller or simpler to larger, more complex activities. There is, in other words, a process of learning entrepreneurial capabilities just as there is a process of learning technological or managerial capabilities. The inherited structures of African economies, shaped by colonial rule, have placed them far down on the learning curves for all three sorts of capabilities. Attempts to force the pace by "Africanization" of industry via public enterprises or small-scale industry promotion schemes have not been able to bypass the learning process. However, some learning has certainly occurred. West Africa in general has a better developed entrepreneurial class than Eastern or Southern Africa because of a smaller direct colonial presence in the past, and some communities have more advanced trading skills than others. But by and large, the inheritance of industrial capabilities is too small to permit a dynamic African industrial class to have emerged so far. The small-scale formal sector remains very weak and underdeveloped (Page 1979).

Past experience is only one source of capability acquisition; education and training are more important. The relationships between education and industrialization are important and binding, but the precise links between particular types of education and specific levels or forms of industrialization are not always easy to trace (King 1984). The operation of easy, low technology, activities with which industrialization generally starts, requires literacy and schooling, a range of basic technical skills and some high level technological and managerial skills. How do the structure and achievements of African education compare with other countries?

Table 25–3 below sets out aggregate data on gross educational enrollments in 1965 and 1985 for Africa and other developing regions. (A more detailed breakdown for a large sample of African countries and others for comparison is available from the author.)

The data show that Africa started with education levels far behind those of other regions and, while it made considerable progress in enrolling pupils at all levels, still lags far behind other developing regions (World Bank 1989b, Zymelman 1990). The most critical input for industrial development—secondary education—is particularly backward, while at the tertiary level the lag is even greater. In relation to enrollments in the "model NIC", Korea, Africa as a whole had 45 percent basic literacy, 17 percent secondary enrollment, and 6 percent tertiary enrollment in the 1980s. Nearly half of the African labor force had no schooling whatsoever, compared to 20 percent for Korea. University enrollments per 100,000 population in Korea came to 3,606 in 1986, compared to 330 for Zimbabwe (the highest in Africa), or only 10 in Mozambique. In terms of literacy and secondary level enrollments, the best education by far was provided by Mauritius, which provided the skill base for its export success.

Figures for gross enrollments are misleading because they do not show the proportions of students that stay on to complete the course. Dropout rates in Africa tend to be particularly high (World Bank 1988b), especially in comparison with East Asia. The figures also do not show the quality of training given, nor the technical orientation of the courses. There are reasons to believe that educational quality in Africa has been declining recently (World Bank 1988b), and

Table 25–3. *Gross enrollment ratios in education.*
(percentage of group)

Region	Primary		Secondary		Tertiary	
	1965	1986	1965	1986	1965	1986
Sub-Saharan Africa	41	66	4	16	0	2
East Asia	88	123	23	45	1	5
South Asia	68	84	24	32	4	5
Latin America	98	108	19	48	4	20
All developing countries	88	106	21	52	5	18

Source: WDR, 1989, Table 29

that the quality of vocational training is lower than in other developing countries (Middleton and Demsky 1989). Secondary education has a low technical content and poor quality of instruction and equipment (Zymelman 1990).

The available data on technical training at tertiary levels shows that the total number of tertiary students enrolled in scientific and technical fields in Africa came to 175,000 in 1983, which was below half that of Thailand (in 1985), below 30 percent that of Korea (in 1987), and only 5 times that of the tiny island economy of Hong Kong. As a percentage of the population, Africa had 0.04 percent in technical education, compared to 1.5 percent for Korea, 1.1 percent for Hong Kong and 0.7 percent for Thailand. No African country had a proportion higher than 0.17 percent (Madagascar). Of the more industrialized countries, Nigeria and Zimbabwe had 0.02 percent, Kenya and Côte d'Ivoire 0.06 percent; and Cameroon, Mauritius and Zambia 0.03 percent. It is interesting that Mauritius was fairly low on this scale, in contrast to its attainments in literacy and secondary schooling. The expansion of garments manufacturing does not require specialized technical skills. However, if Mauritius were to diversify and upgrade its export base in emulation of, say, Hong Kong, a massive push would be needed in its technical skills base.

Data on narrower technical fields are even more directly relevant to industrial skills. Total enrollments in engineering for Africa come to 48,000, which was slightly more than double that of Hong Kong, and only 21 percent of that of Korea. As a proportion of the population, Africa has 0.01 percent, Hong Kong 0.41 percent and Korea 0.54 percent. The highest for any African country, again Madagascar, is 0.02 percent, followed by Kenya and Gabon, with 0.016 percent each.[4]

UNESCO collects data on the total number of "potential scientists and engineers" in each region. Rough as the data are, they again indicate the small size of Africa's human capital base for industrialization. In 1985, Africa had 1,376 potential scientists and engineers per million population, compared to 11,730 for Asia (including Japan), 11,759 for Latin America, and 8,263 for all developing countries.

Total enrollment in vocational training for the region was 667.1 thousand in 1983 (World Bank 1988b). This was 80 percent of enrollments (in 1984) in Korea. As a proportion of the population, Africa has 0.16 percent and Korea 2.07 percent, Hong Kong and Thailand about 0.6 percent and Brazil 1.1 percent. The highest enrollments in Africa are registered by Zaire, Cameroon, Malawi and Gabon, with most of the large industrializers lagging in training their workers in basic technical skills.

While some of these data may not be accurate or strictly comparable, their broad implications are clear. "The educational structure of Sub-Saharan Africa is unsuitable for industrialization" (McMahon 1987, p. 19). If the data were adjusted for quality, and if firm-level training were taken into account, it is likely that Africa would fall further behind the NICs of East Asia and all the indications are that the gap is growing larger over time.

The low level of human capital relevant to industry suggests why the region presents a general picture of poor technological mastery and dynamism in industry. The exceptions that exist are explained precisely by efforts made at the firm-level to create skills and train workers. However, such efforts have not been widespread or intensive enough to make up for the general scarcity of skills. Firms are always reluctant to invest heavily in training when there is a risk of the benefits of their investment "leaking out" to other firms if workers leave. This is a classic case of market failure for which economists recommend subsidies for training or government sponsored training. In Africa such interventions do exist, but they are not sufficient to ensure firm level training comparable to that undertaken, say, in Korea (where 5 to 6 percent of firm turnover is required to be spent on training, McMahon 1987).[5]

Firm level training is not, of course, a substitute for the education system but a complement to it. If the system does not produce the base of literacy or formal training needed for industrial capabilities, firms cannot create such a base. African industry must, in consequence, operate with a small layer of relevant capabilities which have to be spread very thinly. It is something of a paradox, then, that this situation (of much of industry substituting with inadequately trained employees) coexists in many countries with unemployed engineers and technically qualified personnel in nontechnical occupations (Bennell 1984). This inability to exploit the potential capabilities that exist is itself a reflection of the low level of managerial competence, and of the lack of competitive pressures in most African economies and some others. To some extent it is also a reflection of the scarcity of technical personnel—they tend to give up "dirty" production jobs and move into easy administrative ones (Bennell 1984).

Technological effort. No enterprise can achieve efficiency, even if it is well endowed with skilled employees, if it does not undertake conscious, directed effort to collect and assimilate new technical knowledge (Lall 1987 1990b). It is very difficult to measure this kind of effort applied to routine production or investment activity. The best proxy is the incidence of engineers and technicians in the workforce, but it is a crude proxy. The evidence adduced earlier suggests that relatively little of such effort is undertaken in most African countries. The largest repositories of technological skills are probably long established firms with strong technical links abroad, e.g. affiliates of multinationals or local firms with good foreign management and technicians.

The component of technological effort which is most easily measurable is formal research and development. This is likely to be a small part of the total effort needed in most developing countries to master imported technologies. However, it is increasingly a critical input. As more complex technologies are imported, and as older technologies get fully mastered, local research and development becomes essential for assimilating, adapting and improving on these technologies (see Cohen and Levinthal 1989).

Data on research and development in Africa are patchy. The data available here cover research in all its forms (in agriculture and other nonindustrial sectors), and it is likely that in the region much of it is not devoted to manufacturing objectives.[6] Moreover, the data do not distinguish between research and development conducted by manufacturing enterprises as opposed to government laboratories separated from industrial activity.

The data show that the levels of total research and development in Africa are generally low (some of the figures, as for Togo, are dubious). The numbers of scientists and engineers involved in research and development are also low, though Ghana, Mauritius and Sudan stand out as exceptions when these are deflated by total population (they exceed Thailand and approach the levels reached by Brazil). The average for the region, 49 scientists and engineers in research and development per million population, is well below the norm for developing countries as a whole (127) and is below 4 percent of the figures achieved by Korea and Taiwan. However, as far as manufacturing is concerned, these greatly overstate the technological effort in Africa (see note 4).

The low intensity of formal technological effort in Africa is entirely expected and given its early stage of industrial development, is not entirely inappropriate. A booming export oriented economy like Thailand has managed well until now by relying passively on imported technology for most of its industrial needs. It is in the longer term, with growing deepening of the industrial structure, that research and development in industry becomes a significant determinant of competitiveness. At the stage reached by most African industry, what is more relevant is production related technological effort and capability development. This is where countries like Thailand and the East Asian NICs have established a distinct lead.

Zymelman (1990) surveys the African record of scientific effort as revealed by scientific publications. He finds that Africa's share of scientific publications in the world's total (0.4 percent) is much lower than its share of population (8.5 percent), and has lower-than-average citations per article. Publications are highly concentrated by country (Nigeria, Kenya and Sudan account for 70 percent), and by field, with only 3 percent of publications in engineering and technology (medicine and biology account for 82 percent). In this field, the most relevant for industry, Africa accounts for only 0.15 percent of relevant world publications in 1986.

Institutions

The development of industry related institutions (defined narrowly here to include those which facilitate the functioning of markets and the development of capabilities) is central to a broad based industrialization process. Firms cannot function efficiently as isolated units; they have to establish a variety of strong linkages with the rest of the economy. The economy, in turn, has to provide a variety of inputs, services, information and infrastructure, and standards and rules, to enable firms to produce, invest and grow. The primitive market structures with which developing countries start on industrialization suffer widespread structural deficiencies in furnishing all these linkages and services. Some deficiencies are remedied by private agents in response to market signals. Others require direct government intervention. Still others call for the setting up of permanent institutions to play facilitating roles, with autonomous status, specialized skills, and market accountability.

Some relevant institutions exist in several African countries, but the general scarcity of trained labor seems to have held back proper institutional development. Limited support is provided to manufacturers in terms of technical services, training, information or standards. Interlinkages between large and small firms are minimal, and are not facilitated by appropriate institutional assistance to potential small suppliers or subcontractors (World Bank 1989b). Where institutions exist, they are often poorly staffed and managed, given conflicting objectives, and starved of funds. The inward orientation of, and other interventions in, the economies reduce the incentive for private agents to find institutional solutions of their own.

Conclusions and Policy Implications

Successful industrial development requires a conducive macroeconomic setting, an adequate supply of foreign exchange, and adequate infrastructure. Policy makers must address the entire spectrum of determinants of industrial performance if they wish to broaden, deepen and improve their industrial base. The strategies open to different countries are bound to differ, but each country can act on the three main factors described above in some fashion.

It is generally agreed that trade and industrial policies in Africa have been too inward looking, interventionist and oblivious to needs of efficiency. They have permitted too many "white elephants" to come into existence and have held back potentially competitive activities from realizing their potential. Ownership patterns have been skewed to promote Africanization faster than is economically efficient. Loss making enterprises have been kept in operation for too long. Domestic impediments placed on competition and growth have added to the constraints imposed by high levels of indiscriminate protection.

While the correct policy response in most African countries would be to liberalize on domestic and foreign competition, allow freer entry to potential investors and ensure that incentives become more neutral between domestic and foreign markets, this may not amount to a case for free trade or even for low, uniform rates of infant industry protection. There is a good theoretical and empirical case for intervening in market driven incentives to build up competitiveness in progressively more difficult industries. Without such intervention, comparative advantage may well stay quite static.

The arguments against such selective intervention are well known. In essence they come down to three: (a) governments may be no better, and in practice can be much worse, than markets in "picking winners"; (b) the costs of entering new activities should be borne by capital markets (which should be tackled directly if they function imperfectly), while other arguments for protection dynamic and unpredictable learning, externalities, complementarities) are trivial; and (c) if a selective intervention goes wrong, it is difficult to rectify because vested interests build up in preserving "losers". There is clearly some merit in these arguments.

The arguments against selectivity can, nevertheless, be carried too far. The problem with past intervention has been non-selectivity and lack of strategy rather than selection and strategy based on experience, analysis and economic evaluation. Wholesale import substitution did not pay much attention to the process of reaching international competitiveness. Selective intervention in Korea, by contrast, concentrated on a few activities at a time, enforced a degree of market discipline by forcing export orientation in combination with domestic protection, and monitored the process of maturation closely (closing down or restructuring emerging losers) (Pack and Westphal 1986). More important, since protection itself is not a remedy for high costs or poor quality arising from poor training, institutional weakness, deficient infrastructure or lack of external support, the process of selective maturation requires a broader strategy which includes capability building and other supply-side measures as well as protection. Where the supply side elements are missing, protection only generates social cost. It sustains and encourages inefficiency, since firms cannot overcome handicaps arising in the external environment, and slowly slip into slothful routines.

The preconditions for the efficient deployment of selective intervention are quite demanding. It requires a strong and competent government which is driven by economic objectives. This government should be able to analyze technological information and select only activities which, given resource, skill and technological resources (and the "state of nature") can become efficient in a reasonable period. It should be able to monitor progress, ensure entry to foreign markets and a degree of foreign competition at home, act to remedy errors or close down losers. Most importantly, it should be able to progressively boost the capabilities and institutions that determine industrial efficiency.

Many of these conditions may not be satisfied in the majority of African countries. Governments may not be strong or objective enough to manage such strategies, and the existing structure of political economy may "hijack" sensible economic strategies (Biggs and Levy 1990). They may lack the human resources to mount the analytical, surveillance and promotion efforts needed. The given endowments with which they have to work may be so small that much of modern industry may be out of reach for the foreseeable future. Under these conditions of high risk of "government failure" or insuperable structural constraints, it may be best to "leave it to the market". The costs of past government failures are so high that many analysts veer to this view, certainly in the African context. But perhaps this is too pessimistic. There are countries with considerable industrial potential and past policy achievements. Instead of urging them to withdraw from selectivity altogether (which may be unrealistic in most cases), it may be most productive to guide them to a level of selectivity which is least prone to "hijacking" and creates the least long term distortions.

It would appear that the area of capabilities is the area with the greatest need for government action in Africa. It is beyond the scope of this paper to analyze the means by which general educational quantity, quality and relevance should be improved in Africa (see World Bank 1988b, Zymelman 1990, Middleton and Demsky 1989). It is relevant, however, to mention some measures which apply more narrowly to industrial skills and technology (Freeman 1989 has a brief and succinct review of the problems of latecomers to industrialization which has relevance to these issues).

It is essential that capability development in Africa be based on the present endowment of skills and social structures (Hyden 1983), and that it aim at a healthy and competitive industrial structure. Formal education and training are clearly needed for all levels of African industry. Though in a lot of informal and small-scale industry there is not much formal need for technical training, literacy and some familiarity with modern technologies are clearly important. Modern small-scale industry can, on the other hand, involve highly sophisticated operations, and sustained expansion and upgrading of this sector will inevitably call for the expansion of vocational (or similar) training and high level technical education. Formal large-scale industry needs far higher inputs of well trained workers, technicians, managers and engineers. It is imperative for governments to provide such training at the right level and cost, and of the right quality (Bennell 1984, Dougherty 1989, World Bank 1988b and 1989b).

For instance, firms need help, at least in the initial stages, in acquiring information on technology, equipment, materials, consultants, and so on. Certain technological functions cannot be undertaken inhouse because they are too "lumpy" or have too many externalities (public goods). Thus, only the government can set up standards or testing institutions, provide technical extension or conduct basic research. Only the government can establish a science and technology infrastructure linking industry with laboratories, consultants, universities and foreign entities. In many African countries, where even basic quality control or preventive maintenance skills are lacking, functions that are normally performed in-house may be "externalized" and provided by common units. This would conserve skills and provide "tutors" to in-house operators.

One important way to stimulate local technological activity and diffusion is the promotion of consultancy organizations. Consultants are the most important means of transferring technology in several industries (mainly with continuous processes). They are also repositories of technical knowledge gathered from experience in a variety of enterprises. Fostering the growth of consultants in more advanced fields often requires protection in the form of restricting the entry of foreign consultants or ensuring a twinning arrangement between foreign and local consultants. Like all protection, this is a tool which needs to be used with great care and discretion. In Africa, a comparative advantage in consultancy services can only be developed in the simple end of the spectrum of activities, concentrating on technologies which are in common use (construction, food processing, textiles, cement). While most consultants should be sponsored in the private sector (a common source of consultancy organizations is the specialized project divisions of large manufacturers), some may have to be set up in the public sector to provide technical assistance to small enterprises on a subsidized basis. Again, local consultants may benefit from "tutoring" by experienced foreign experts recruited for limited periods.

A promising path to promote institutional development in the technology and training area is to help industry associations set up facilities for their members. In Korea, industry associations provided many valuable technical functions in this way, and also acted as interlocutors for the implementation of government policy, export targeting and feedback of information. Well staffed and funded associations can be a substitute for several types of institutions that would take much longer to evolve.

Two areas of industry call for special strategies. The first is the engineering industry, the second is subcontracting between large firms and small suppliers. Basic metalworking skills are generic to a whole host of other industries, and the ability to build, copy, repair or improve capital goods (even of a simple sort) is widely regarded as the seedbed and but of technological progress (Fransman 1986). Such skills are scarce in most African countries, which adds to the countries' inability to utilize their existing capital stock efficiently or to raise its productivity over time. The inability to make spares locally or carry out troubleshooting services adds greatly to operating costs and downtime (when equipment lies idle). At the same time, the informal sector shows considerable ingenuity and dynamism in some metalworking jobs (utensils, tools) and repair (automobiles). If the level of skills and know-how can be raised to that required for more modern (small-scale) industrial engineering activity, this would provide a dynamic growth sector as well as facilitate the growth of other sectors.

Subcontracting and interfirm linkages in general have been slow to develop in African industry. This reflects the lack of capabilities on the part of both large firms (which have to expend considerable effort and transfer a lot of skills and technology in setting up subcontractors) and potential suppliers (which lack the entrepreneurial or basic technical skills to appear viable as reliable sources). It also reflects poor infrastructural facilities for small firms, biases in policy and credit markets and the lack of a technical extension network. Governments should try to remedy these deficiencies and correct existing biases, while encouraging large firms to establish local subcontracting systems. Needless to say, this potent tool for diffusing technology and skills and promoting small-scale industry should not be used too bluntly. Forcing the growth of inefficient local suppliers is not conducive to healthy development.

The broader economic setting should not be forgotten (Gulhati and Sekhar 1982). "The process of developing industrial capabilities is self-reinforcing, with different elements interacting to support each other. The general industrial environment affects its content and direction: a competitive, outward-looking regime is likely to call forth an appropriate set of technological responses. If macro policies and incentive structures are improved, and sufficient foreign resources provided to enable Africa to resume a strategy of long term growth, the ultimate determinants of how successfully it industrializes will be its education and training systems and increasingly its supply of technical labor and firm level investments in technological effort. Africa as a whole lags well behind the rest of the developing world in these areas. Even the "best" countries in Africa are very far from the best elsewhere, i.e. the NICs of East Asia.

The wide dispersion of industrial development within Africa will mean that future paths may diverge even further. The cumulative nature of capability development dictates that the better-off countries will, ceteris parabus, continue to industrialize more efficiently than others. If they plough back part of their growth in revenues into developing human and technological resources, their lead will increase further. Some countries with greater access to foreign exchange will be able to compress the learning curve by attracting foreign "tutors" and better technology.

Those with significant local supplies of non-African capabilities may similarly improve their position if they give full rein to those capabilities. Other countries, with neither foreign exchange nor large repositories of non-indigenous skills, may continue to lag industrially, whatever incentive policies they adopt. They will go up the learning curve only at the speed their own human capital development permits.

Notes

1. All these data are from World Bank 1989b.

2. For instance, some 70 percent of top managerial and technical positions in industry in Cote d'Ivoire and Zimbabwe are held by Europeans (World Bank country reports).

3. This subsection draws heavily on Lall (1990b and forthcoming).

4. Zymelman (1990) Notes that Africa graduates only 1.2 people per 100,000 population in science and engineering per annum, compared to 65 in industrialized countries. Engineering by itself is even weaker: "developed countries graduate 166 times more engineers per capital than do Sub-Saharan African countries" (p. 25).

5. As BAS (1989) Notes, on-the-job training is very weak in much of modern African industry, while traditional forms of apprenticeship training are inadequate for imparting modern technical skills.

6. UNESCO (1989) data on four African countries (Congo, Malawi, Mauritius and Zambia) show that of R&D performed in the "productive sector", agriculture, mining, utilities or construction, accounted for *all* spending or employment of scientists and engineers. *Manufacturing did not register any R &D at all in these countries.* No other African countries were included in these tabulations.

References

Bas, D. (1989), "On-the-Job Training in Africa." *International Labour Review* 128:4. P. 485-496.

Bennell, P. (1984), "The Utilization of Professional Engineering Skills in Kenya", in M. Fransman and K. King (eds.) *Technological Capability in the Third World.* London: Macmillan, p. 335-54.

Biggs, T. and Levy, B. (1990), "Strategic Interventions and the Political Economy of Industrial Policy in Developing Countries." in D. Perkins and M. Roemer (eds.), *Economic Systems Reform in Developing Countries.*Boston: Harvard University Press.

Chenery, H.B., Robinson, S. and Syrquin, M. (1986), *Industrialization and Growth: A Comparative Study.* New York: Oxford University Press, for the World Bank.

Cohen, W.M. and Levinthal, D.A. (1989), "Innovation and Learning: The Two Faces of R & D." *Economic Journal* 99, pp. 569-96.

Dahlman, C.J., Ross-Larsen, B. and Westphal, L.E. (1987), "Managing Technological Development: Lessons from Newly Industrializing Countries." *World Development* 15:6, p. 759-75.

Dougherty, C. (1989), "The Cost Effectiveness of National Training Systems in Developing Countries." World Bank, PPR Working Papers, WPS 171.

Fransman, M. (ed) (1982), *Industry and Accumulation in Africa.* London: Heinemann.

Fransman, M. (1986), *Technology and Economic Development.* Brighton: Wheatsheaf Press.

Freeman, C. (1989), "New Technology and Catching Up" in C. Cooper and R. Kaplinsky (eds.), *Technology and Development in the Third Industrial Revolution.* London: Cass, p. 85-99.

Gulhati, R. and Sekhar, U. (1982), "Industrial Strategy for Late Starters: The Experience of Kenya, Tanzania, and Zambia." *World Development.* 10:11, p. 949-72.

Hayden, G. (1983), *No Shortcuts to Progress: African Development Management in Perspective.* London: Heinemann.

Katz, J. (ed.) (1987), *Technology Generation in Latin American Manufacturing.* London: Macmillan.

Kilby, P. (ed.) (1971), *Entrepreneurship and Economic Development.* New York: Free Press.

King, K. (1984), "Science, Technology and Education in the Development of Indigenous Technological Capability", in M. Fransman and K. King (eds.), *Technology Capability in the Third World.* London: Macmillan, p.31-64.

Lall, S. (1987), *Learning to Industrialize: The Acquisition of Technological Capability by India.* London: Macmillan.

Lall, S. (1989), "Human Resource Development and Industrialization, with Special Reference to Sub-Saharan Africa." *Journal of Development Planning* 19, p.129-58.

Lall, S. (1990a), "Structural Problems of Industry in Sub-Saharan Africa." in *Background Papers: The Long-Term Perspective Study of Sub-Saharan Africa.* (Vol. 2), World Bank.

Lall, S. (1990b), *Building Industrial Competitiveness: New Technologies and Capabilities in Developing Countries* Paris: OECD.

Lall, S. (forthcoming), "Explaining Industrial Success in the Developing World." in V.N. Balasubramanyam and S. Lall (eds.) *Current Issues in Development* London: Macmillan.

McMahon, W.W. (1987), "Education and Industrialization," World Bank, Background Paper for *WDR 1987.*

Meier, G.M. and Steel, W.F. (1989), *Industrial Adjustment in Sub-Saharan Africa,* New York: Oxford University Press, EDI Series in Economic Development.

Middleton, J. and Demsky, T. (1989), "Vocational Education and Training." World Bank Discussion Papers, No. 51.

Mlawa, H.M. (1983), "The Acquisition of Technology, Technological Capability and Technical Change: A Study of the Textile Industry in Tanzania." University of Sussex, Dissertation.

Mytelka, L. (forthcoming), "Ivorian Industry at the Crossroads." in Stewart et al.

Navaretti, G.B. (forthcoming), "Joint Ventures and Autonomous Industrial Development: The Magic Medicine?" in Stewart et al.

Nellis, J. (1986), "Public Enterprise in Sub-Saharan Africa", World Bank Discussion Paper, No. 1.

Pack, H. (1987),*Productivity, Technology and Industrial Development*. Oxford University Press for the World Bank.

Pack, H. and Westphal, L.E. (1986), "Industrial Strategy and Technological Change: Theory versus Reality." *Journal of Development Economics*21, p.87-128.

Page, J.M. (1979), "Small Enterprise Development: Economic Issues from African Experience." World Bank Staff Working Paper, No. 363.

Page, J.M. (1980), "Technical Efficiency and Economic Performance: Some Evidence from Ghana," *Oxford Economic Papers* 23:2, p. 319-39.

Steel, W.F. and Evans, J.W. (1984), "Industrialization in Sub-Saharan Africa: Strategies and Performance", World Bank, Technical Paper No. 25.

Stewart, F. (1977), *Technology and Underdevelopment* London: Macmillan.

Stewart, F., Lall, S. and Wangwe, S. (eds.) (forthcoming), *Alternative Development Strategies in Africa*. London: Macmillan.

UNESCO (1989), *Statistical Yearbook* 1988, Paris.

Wangwe, S. (forthcoming), "Building Indigenous Technological Capacity." in Stewart *et al.*

World Bank (various years), *World Development Reports*. New York: Oxford University Press.

World Bank (1981), *Accelerated Development in Sub-Saharan Africa*. Washington, D.C.

World Bank. 1987. *World Development Report 1987*. New York: Oxford University Press.

World Bank. 1988a. *World Development Report 1988*. New York: Oxford University Press.

World Bank. 1988b. *Education in Sub-Saharan Africa: Policies for Adjustment, Revitalization and Expansion*. Wash., D.C.

World Bank. 1989a. *World Development Report 1989*. New York: Oxford University Press.

World Bank. 1989b. *Sub-Saharan Africa: From Cisis to Sustainable Growth*. Washington, D.C.

Zymelman, M. (1990), *Science, Education and Development in Sub-Saharan Africa*. World Bank Technical Paper No. 124, Africa Technical Department Series.

26

Entrepreneurship and Growth in Sub-Saharan Africa: Evidence and Policy Implications

T. Ademola Oyejide

Against the background of poor economic performance during the 1970s and 1980s, many African countries have implemented ongoing structural adjustment programs. In most parts of Africa there is increasing emphasis on private initiative and market forces as a means of reviving and restoring sustainable economic growth. It is presumed that this new orientation in economic management will improve the flexibility with which African economies can respond to rapid and unexpected changes in the international economic environment. The new orientation also implies an expanded role for the private sector in the development process.

There have been some fundamental changes in both the macroeconomic environment and the regulatory framework in many African countries in recent times. However, there is increasing concern that the supply response of private sector enterprises appears to be largely muted and uninspiring. This raises the question of whether there are enough entrepreneurs of the appropriate types to be able to respond effectively to the newly created opportunities. It also calls for study of the determinants of African entrepreneurship and its capability for response, and analysis of the factors which constrain the development of entrepreneurship and the private sector in Africa. Such study and analysis would provide improved knowledge of the origins, characteristics, constra- ints, and growth potentials of the private enterprise sector.

This paper is aimed at making a modest contribution to this broad research agenda. Its primary purpose is to survey the available literature as a means of identifying major elements of the dynamics of private enterprise development in Africa. In the second section the main characteristics of African entrepreneurship are analyzed. The third section examines the patterns of upgrading, graduation, and growth in African entrepreneurship. Section four contains a discussion of the constraints impeding enterprise development and growth. Emerging policy issues are identified and concluding remarks are offered in the fifth section.

Characteristics of African Entrepreneurship

Typically, the private sector in an African economy contains a wide variety of enterprises including informal enterprises; microenterprises; and small-, medium-, and large-scale modern firms. This section is concerned with the ownership and size distribution of these firms, their capabilities for development, and their overall economic growth enhancing capabilities.

Ownership and Size Distribution

The size distribution of private sector enterprises in Africa features several distinct characteristics. There is an overwhelming predominance of informal enterprises and microenterprises, which generally produce simple consumer goods. These enterprises are not integrated with the formal sector of modern small-, medium-, and large-scale firms. These enterprises account for the largest share of employment in the private sector. Surveys indicate that firms with less than 10 employees provide 59 percent of total private sector employment in Kenya, 75 percent in Nigeria, 83 percent in Zambia, and 90 percent in Sierra Leone (Page 1979, Liedholm and Mead 1986, Kilby 1988). The share of total employment accounted for by firms with between 10 and 50 workers averages less than 10 percent in many African countries.

Indigenous capital accounts for a small share of total private sector capital. Local capital is heavily concentrated in the informal and small enterprises. Foreigners, resident aliens, and the state control the medium- and large-scale firms (Kilby 1988, Levy 1990). These studies also show that African capital accounts for 20 percent of manufacturing capital in Nigeria, 80 percent of output and 20 percent of capital in Ghana, and 9 percent of value-added in Kenya.

Various attempts have been made in the past to modify this ownership pattern. An important example is the Nigerian case in which an indigenization program carried out in the 1970s transferred between 40 and 100 percent of the equity capital of many of the medium- and large-scale modern enterprises to local shareholders. The subsequent privatization program is in the process of sharply reducing the state's share in the transferred equity. The proportion of the total capital of Nigerian enterprises owned by private citizens and associations should increase substantially. A recent study laments the lack of adequate research on the growth of larger indigenous enterprises in Nigeria. It argues "that in terms of the scale of individual enterprise, the degree of corporate organization and the size and diversity of investment, Nigerian private capital has advanced well beyond African enterprise in Kenya, the Ivory Coast, Zimbabwe and other Sub-Saharan African countries" (Forrest 1989, p.1.).

Managerial Capability

Various studies of African private sector enterprises have identified a low level of managerial skills as one of the major deficiencies in the private sector. Owners of informal, micro and small-scale enterprises tend to be craftsmen-entrepreneurs who have probably acquired a level of technical proficiency in their production processes. However, they are largely without any extensive or formal training in business organization, financial management, and marketing. As a result, evidence abounds that many of the important functions associated with the efficient management of modern business enterprises are either not performed at all or badly carried out by the typical African craftsman-entrepreneur who owns and runs an informal or small-scale enterprise.

Decision making is often excessively concentrated in the hands of the owner-manager, whose authority is rarely delegated. In addition, there is limited propensity to form partnerships or take advantage of the limited liability public corporate form. Rather, the overwhelming majority of small African enterprises is organized as sole proprietorships (Marris and Somerset 1971, Akeredolu-Ale 1975, Kennedy 1980). This organizational form of ownership and control generates growth retarding enterprise succession problems.

Surveys show that the majority of African enterprises do not keep adequate financial records and that financial controls do not feature prominently in the management of the enterprises (Kilby 1962). There is a lack of experience, formal education, and training in record keeping, compilation, and use of numerical data. Even when rudimentary accounting records are maintained, they are often deficient and not frequently used as a management tool (Kilby 1962). This, in turn, leads to overcapitalization, maintenance of excessive stocks of raw materials, and incomplete separation of business from household and personal accounts. The managerial functions relating to marketing and promotional activities receive very little attention; market coverage tends to be limited, thus hampering enterprise expansion and growth.

Technological Capability

Owners of African informal, micro, and small-scale enterprises tend to be entrepreneurs who have emerged from the ranks of craftsmen. Their skills have usually been acquired through a process of apprenticeship, the length and formality of which reflect the extent of technical barriers to entry into the particular product group or industry. Surveys reveal that between 60 and 90 percent of such entrepreneurs had received training as apprentices before setting up their own enterprises (Gerry 1974, Aryee 1977). The system of apprenticeship is reported to be more formalized and older in West Africa than in the Eastern, Central, and Southern parts of Africa.

The heavy reliance on this traditional system of apprenticeship for the acquisition and accumulation of technical skills implies that these skills are not easily transferable to subsequent activities in the formal medium- and large-scale enterprises. Similarly, skills learned in the formal sector do not appear to form a basis on which small-scale enterprises are built. Thus, a kind of technological dualism seems to exist in which entrepreneurs with training and experience in the modern sector are concentrated in small formal sector firms, while less than 15 percent of informal and microenterprise owners claim prior experience in the modern medium- and large-scale industry.

The level of formal educational attainment among informal and microenterprise owners and managers is astonishingly low. Surveys indicate that estimates of the proportion of African entrepreneurs lacking formal education range from a low of 13 percent in Nigeria to a high of 77 percent in Sierra Leone. Furthermore, it is estimated that an average of about 50 percent of the entrepreneurs surveyed had not achieved functional literacy. Less than 15 percent of these entrepreneurs had attended technical or vocational schools prior to the establishment of their enterprises. The low level of functional literacy and technical training probably accounts, to a significant extent, for the rudimentary nature of the technology employed in the production processes of many African informal sector and microenterprises.

Enterprise Upgrading, Graduation, and Growth

From the point of view of enterprise development, the primary importance of the informal microenterprises and small-scale enterprises derives from their actual and potential growth dynamics. These enterprises contribute to the overall growth of the economy through increased efficiency and productivity, income generation, and employment creation. The number of enterprise start-ups and closures and the growth of enterprises are indicators of the development of entrepreneurship.

Enterprise growth has several dimensions. Firms may grow in number, in size, and by diversifying into different product lines. Each of these changes may occur with or without increased linkages between firms. Probably growth in firm size would be most conducive to increased efficiency, productivity, and income generation. Therefore an analysis of growth oriented entrepreneurship development should focus largely on this aspect of enterprise growth. But even when the growth of enterprises is so narrowly conceived, it is important to recognize that this growth can be accomplished through two interrelated but separate routes. Firms can experience a transition from an informal to a formal status in the process of expansion and growth. Subsequently or simultaneously, enterprises may go through a graduation process as they are transformed from microenterprises and small-scale enterprises into medium- and large-scale firms. Both of these constitute important features of entrepreneurship development in Africa.

It has often been suggested in the literature that the informal sector and microenterprises provide a valuable training ground for advanced entrepreneurial and managerial capabilities upon which more modern medium- and large-scale firms are established and sustained (Page 1979). To the extent that informal and small-scale enterprises play a significant role in the building of human capital, they would form an important bridge for the development of entrepreneurship in Africa.

In practice, however, the role of the informal sector in augmenting the supply of entrepreneurial and managerial skills for the overall economy is probably limited by several considerations. In so far as informal sector enterprises and microenterprises are predominantly owned and managed by craftsmen-entrepreneurs, the types of skills they embody may not be readily transferable to the management of modern medium- and large-scale firms. This limitation could arise from the fact that enterprise growth in the modern sector requires skills that are quite different from those possessed by craftsmen-entrepreneurs whose primary domain is the informal sector. In particular, their lack of formal education and failure to acquire adequate training in financial management and other organizational skills would limit the size and complexity of firms which they could efficiently run.

In any case, there apparently exists a strong tendency on the part of informal sector entrepreneurs to diversify their holdings of micro and small-scale enterprises rather than to concentrate resources towards expanding a single enterprise in the direction of a larger and more efficient size (Kilby 1962, Marris and Somerset 1971). Such growth by diversification could represent a rational response on the part of small enterprises to a structure of incentives which is hostile to growth in firm size. However, diversification tends to detract the entrepreneur's attention from efficiency concerns. It might be hypothesized, therefore, that modern small-scale firms owned by individuals who are more familiar with market opportunities and who have acquired greater educational, commercial, and managerial experience are likely to constitute a more effective bridge for enterprise development than the craftsmen-entrepreneurs in the informal sector.

The proportion of modern small- and medium-scale enterprises in Africa which came from the expansion of microenterprises in the informal sector is not entirely clear. Aggregate data on the size distribution of enterprises are not sufficiently revealing on the issue of the pattern of upgrading and graduation. But a broad interpretation of the evidence from four African countries indicates that in Nigeria, Rwanda, Botswana, and Sierra Leone, modern small- and medium-scale firms did not graduate from micro beginnings, but rather started life autonomously. Apparently some regional differences can be discerned from the available data. In Western Africa over 30 percent of the modern small- and medium-scale firms trace their origins to microenterprises. Twenty percent or fewer of such firms graduated from micro ranks in Eastern and Central Africa. Compared with Asia and Latin America, the rate of upgrading of smaller firms (those with between one and ten employees) to larger firms is much lower in Africa.

Enterprise Development Constraints

Constraints against enterprise development prevent informal enterprises from acquiring formal status and hamper the ability of microenterprises and small-scale firms to graduate into medium- and large-scale units. The identification and classification of these constraints has been approached in a number of different ways in the literature. For instance, Lall (1989) offers a comprehensive approach to the determinants of enterprise growth and performance primarily in the form of a complex interplay of three sets of factors, i.e. incentives, capabilities, and institutions. In this approach, incentives cover macroeconomic conditions, physical infrastructure, and internal competition. Capabilities include the ability of entrepreneurs to launch physical investments, the provision of human capital (in terms of skill, training, and ability to absorb and apply new knowledge), and the undertaking of technological effort. Institutions include entities established to facilitate the working of markets and the network of external linkages which allow individual firms to operate efficiently.

Another set of identification and classification criteria emanates from the functional approach to enterprise development which focuses on the bundle of activities that must be performed in a successful enterprise (Levy 1990). These include:

- initiative to identify and respond to opportunities for profit,
- discipline and skill to manage an enterprise so as to maximize revenues and minimize costs,
- ability to gain access to finance and other inputs (e.g. labor, raw materials, equipment, etc.) needed for production,
- ability to market the goods and services produced by the enterprise,
- ability to progressively improve the technical capabilities of the enterprise, and
- ability to manage relations.

Within this framework, three sets of obstacles to enterprise development can be identified. These are:

- shortfalls in the supply of individuals with the attributes of entrepreneurs,
- weaknesses in the external economic environment within which enterprises operate and associated high costs of gaining access to resources and markets, and
- regulatory environments which make enormously burdensome demands on enterprises in their relations with government bureaucracy.

The factors which relate to the internal capabilities of the entrepreneur (i.e. entrepreneurial, managerial, and technological) can be separated from the factors which relate to external constraints (i.e. the macroeconomic environment and regulatory framework). Many studies have concluded (e.g. Levy 1990) that there does not appear to be a serious shortfall in the supply of indigenous entrepreneurs. Hence, the factors responsible for the underdevelopment of Africa's private sector will probably be found in the macroeconomic and regulatory environments.

These factors form a barrier which hinders the transition of enterprises from informal to formal status, hampers the graduation of microenterprises and small-scale firms into medium- and large-scale units, and slows down the rate of start-ups of medium-scale enterprises. It would be useful to identify the policy induced distortions and institutional factors which constitute critical elements of this barrier. Analytically, the barrier can be represented as a hump in the typical firm's unit cost curve in relation to its size. The transition (from informal to formal) and graduation (from small- to medium-scale) issues are captured, in varying degrees, by the need to overcome the difficulties created by the hump. The existence of a hump in the firm's unit cost curve may also explain the reaction of entrepreneurs whose enterprises grow primarily by diversification into many product lines. This may be due to the perceived difficulties of going over the hump either by becoming formal or by expanding the scale of operations in the same product line.

An analysis of the firm's enabling environment (i.e. macroeconomic, institutional, and regulatory framework) should provide some pointers which would endow the hump hypothesis with more (and testable) specificity. Several elements of the enabling environment may be presumed to contribute to the location and magnitude of the hump in relation to the size and type of the enterprise. For instance, financial market distortions may play a role. Credit subsidization and non-price rationing policies generally favor large-scale enterprises and discriminate against (thus crowding out) smaller firms. Gaining access to institutionalized and increasingly cheaper credit may require making the transition from informal to formal status. This may, in turn, involve a relatively high up-front cost that, at the beginning, may outweigh the eventual benefits of improved access to and reduced costs of credit.

Similarly, labor market distortions emanating from minimum wage legislations or standards may play a role to the extent that the firm's unit labor cost increases sharply as it moves from an informal status to a formal status and/or graduates from a small- to medium-scale operation. This involves a movement from relatively heavy reliance on family cum apprentice labor to greater reliance on wage labor. What is involved is graduation into a firm size (usually 10 or more employees) for which minimum wage regulations apply and are enforced.

Other regulatory constraints often constitute entry barriers which may contribute to the sharp rise in the firm's unit cost. These include the need to obtain various licenses and satisfy other registration formalities that become mandatory as an enterprise attempts to make the transition from informal to formal status and/or to graduate from small- to medium-scale. Such formalities are often a requirement for becoming eligible for certain cost-reducing benefits enjoyed by larger firms. The costs of satisfying the requirements are normally incurred before the corresponding benefits accrue; and on a per unit basis, such costs constitute a greater burden for small than for large enterprises.

Surveys and studies of African entrepreneurship development have found deficiencies with respect to several aspects of the internal capabilities of enterprises, and key elements of the macroeconomic environment and institutional and regulatory framework. While entrepreneurs appear to exist in sufficient quantity in many African countries, their managerial and technological capabilities are either inadequate or inappropriate. Institutions which would enable indigenous enterprises to perform efficiently are in short supply (Lall, 1989). In terms of ranking the three major sets of constraints, Levy (1990) finds that in Tanzania regulatory controls are extremely restrictive. However, they constitute less of a binding constraint to dynamism on the part of indigenous small and medium enterprises than do weaknesses in the external economic environment (especially in the financial sector). Other studies (e.g. Schatz 1965) regard capital shortage as an illusionary constraint to enterprise development.

An important component of the economic environment is the degree and kind of internal competition in the field of entrepreneurship in African economies. Lall (1989) makes a persuasive case for some protection in this area. His plea finds strong support in previous studies, particularly in Nigeria, in which survey results have consistently cited the retarding effect of the activities of expatriate entrepreneurs on the development of indigenous enterprises through crowding out effects (Akeredolu-Ale 1975). It seems that the infant industry argument for protection may also find an application in the more general area of enterprise development.

Policy Issues

The accumulated evidence on Africa's informal sector and microenterprises suggests that few of these firms grow into modern firms of any size. It remains less clear, however, which constraints hinder the growth of these enterprises and the extent to which specific constraints are binding. Deeper understanding of the inhibiting factors and their incidence would constitute an important prerequisite for the formulation and implementation of policies aimed at relaxing the binding constraints.

This significant knowledge gap has not prevented African governments and the donor community from mounting various types of special assistance programs for eliminating perceived deficiencies in the development of African entrepreneurship. Policy actions can be taken to remove particular policy induced distortions or build up desired institutions to facilitate enterprise development. But special assistance programs have usually been targeted at comensating for the negative impact of given constraints. Specific assistance programs can be justified as compensatory measures in cases where the removal of existing distortions is, for some important reason, not immediately feasible. Assistance programs can be used as a means of off-setting externalities emanating from various types of market failures and as a means of strengthening desirable multiplier effects. These programs can be used to promote the development of entrepreneurship in the face of major constraining elements in the enabling environment.

However, the results of many previous attempts at using special assistance programs to promote entrepreneurship development in Africa have been less than desirable. Special credit assistance programs have not been effective in many countries for a variety of reasons, including inappropriate design and implementation, and inadequate proportion of enterprises covered. Special training programs have also mostly fallen short of their targets.

Aspects of the ongoing structural adjustment programs in many African countries generally address various deficiencies in the macroeconomic environment as well as the institutional and regulatory framework. But putting the right incentives in place will not elicit an appropriate response until the ability of indigenous entrepreneurs to respond to the new opportunities is enhanced. There can be a br - oad general approach for reforming the macroeconomic environment and regulatory framework, which may apply across most African countries. However, policies to enhance the internal capabilities of enterprises (i.e. entrepreneurial, managerial, and technological capabilities) are more likely to be successful if they are specifically designed to fit the particular circumstances of each country. A similar approach is probably necessary for building up special institutions designed to facilitate the working of markets.

The formulation, design, and implementation of such country specific policy interventions will require an improved knowledge base about the characteristics of African entrepreneurs and the environment in which they operate. It will also require an approach which does not insist that owners and managers of modern medium- and large-scale enterprises should evolve from the traditional operators whose domain has been the informal sector and microenterprises. It should not be surprising if these are eventually supplanted by a new breed of entrepreneurs who are more literate and knowledgeable in relation to both managerial and technological capabilities, and much less apologetic about the contributions which the private sector can and should make to the economic development of Africa.

References

Akeredolu-Ale, E.O. (1975). The Underdevelopment of Indigenous Entrepreneurship in Nigeria. Ibadan University Press.

Aryee, G. (1977). Small Scale Manufacturing Activities: A Study of the Interrelationships between Formal and Informal Sectors in Kumasi, Ghana. WEP Working Paper. Geneva, ILO.

Gerry, C. (1974). Petty Producers and the Urban Economy. WEP Research Paper 8, Geneva, ILO.

Kennedy, Paul. 1980. Ghanaian Businessmen: From Artisan to Capitalist Entrepreneurs in a Dependent Economy. Munich: Weltfoum Verlar.

Kilby, Peter. (1962)_. The Development of Small Industries in East Nigeria. Enugu: Ministry of Commerce and U.S. Agency for International Development. p. 6. Table 1. 1988. "Breaking the Entrepreneurial Bottleneck in Late Developing Countries: Is there a Useful Role for Government?" *Journal of Development Planning*, No. 18, Special issue on Entrepreneurship and Economic Development.

Lall, Sanjaya. (1989). "Structural Problems of African Industry." Presented at the workshop on Alternative Development Strategies in Africa. Oxford: Queen Elizabeth House. (11-13 December).

Levy, Brian. (1990). "Private Sector Development in Tanzania: Obstacles and Opportunities." Washington, D.C.: World Bank. Mimeo.

Liedholm, Carl and Donald C. Mead. (1986). "Small Scale Industry in Sub-Saharan Africa: Empirical Evidence and Strategic Implications." In Bos Bero and Jennifer Whitaker, eds., *Strategies for African Development*. Berkeley: University of California Press. Abstract in Meier and Steel, eds. 1989.

Marris, P. and A. Somerset. (1971). "African Businessman: A Study of Entrepreneurship and Development in Kenya." Nairobi.

Page, John M. (1979). "Small Enterprise in African Development, a Survey." World Bank Staff Working Paper No. 363. Washington, D.C.: World Bank. (October) Processed.

Schatz, S.P. (1965). "The Capital Shortage Illusion: Government Lending in Nigeria." *Oxford Economic Papers*, Oxford, XVII, No. 2. (July).

27

Comments on Human Capital and Entrepreneurship

William F. Steel and Samuel M. Wangwe

William F. Steel

I would like to review the papers on human resources from the point of view of the prospects for industrial development in Africa. Although I am normally reasonably optimistic about the way industrial structure is adjusting to policy reforms in Africa, I must confess that these papers are rather discouraging about the prospects for sustained growth of productivity in industry.

These papers suggest three important conclusions for industry about human resource development:

- substantial investment in human capital is a necessary condition for industrial development to be indigenous rather than just a transplantation of foreign machinery and managers;
- Africa's resource gap with the rest of the world may be increasing rather than closing; and
- other competing claims on scarce public resources may drain away the investments that are needed to raise productivity.

Before discussing these further, I should first note that there are some positive signs of progress. I agree with Oyejide's conclusion that entrepreneurs exist in substantial numbers. There certainly is a long entrepreneurial tradition in the commercial sector, especially in West Africa.[1] East Africa also has come a long way in entrepreneurial development since the colonial days when indigenous commercial activity was explicitly discouraged. Although few existing entrepreneurs may have the skills and capital needed to lead industrial transformation, there is growing evidence of what Oyejide referred to as a "new breed of entrepreneurs" who are better educated and trained than their predecessors and who are applying their skills to exploit new opportunities, not simply following their parents' trades.[2]

The question, however, is not whether entrepreneurial and labor force abilities have been improving, but whether they are expanding fast enough. Several responses in Lall's paper give cause for concern. First, he points out that "the drive to profit from opportunity (in small and informal activities) is not *per se* the same as the entrepreneurial capability required to organize, set up and run modern industry." Second, he shows that Africa still lags far behind the rest of the world in education and training, which countries such as Korea have used successfully to lay the basis for technological and industrial growth. This does not mean that no progress has been made: enrollment and literacy have improved. But they have continued to improve just as fast in other developing regions as well, so the gap remains. (It would be useful if the paper could also include statistics to show how Africa today compares to Korea, Hong Kong, Brazil and Thailand before they began industrializing rapidly.) Third, he argues that the technological effort is particularly weak in Africa, making it difficult to follow the adaptive approach that was in large part responsible for the success of the East Asian newly industrializing countries.

Even more discouraging are the indications that the situation may be getting worse. Educational expenditures have been squeezed by budgetary problems and the quality of education has probably fallen in many countries. Technical education continues to be underemphasized and lack scientific equipment.

Perhaps it should count as a positive sign that we are much more aware today of the importance of early development of labor and entrepreneurial skills than policymakers were at the time of independence. In fact, many leaders promoted an import-substitution industrialization strategy on the assumptions that this would generate missing labor skills and that the state could substitute for the apparent lack of entrepreneurs.[3] But in the 1980s, we woke up to the fact that many state-owned industries were being designed and run by bureaucrats, not enterprizing managers, and that labor productivity was falling instead of rising.

The emphasis shifted to liberalizing markets so that increased competition would force firms to be more efficient. In many countries, however, while inefficient firms are indeed feeling the pinch, investment in more efficient firms seems slow in coming. There are entrepreneurs who would like to adapt, but most are lacking the experience, finance, marketing information, or technological know-how to do so.

Today, industrialization is seen as a continuous process that requires substantial input of human and technological capabilities and that cannot be sustained simply through protection that stimulates capacity creation regardless of cost, on the one hand, or unsupported exposure to international competition, on the other. However, this more sophisticated view makes it difficult to design a practical strategy. For instance, the guidelines for the Second Industrial Development Decade for Africa mention an incredible range of things that need to be developed as a basis for industrialization. These include the following for human and technological development: technical schools and universities, research and development bodies, consultancy services, industry associations and entrepreneurship training, among others. For physical infrastructure they include roads, bridges, ports, railways, telecommunications, energy, market sites, storage, etc. All these are important, but the tremendous cost and effort involved are beyond the means of low income, less developed countries.

With limited resources, a strategic approach is needed, that is, some first principles must be emphasized. These papers do not give a clear sense of where to begin, although Lall argues for selective intervention to promote certain strategic industries. He quite rightly notes that this approach can impose high costs if the supply conditions are not right. I am concerned that the selective protection he argues for would foster those high costs in precisely the wrong industries: those with high linkages to the rest of the economy. I would rather suggest that efforts to stimulate such industries concentrate on making conditions more conducive for them through direct incentives, reduction of regulatory and entry barriers, training programs and a high level commitment to solving their supply problems.

I would draw two strategic conclusions from the papers:

- Investment in education, and especially technical training, is critical if the process of industrialization is to be a flexible one of adapting to change and applying new techniques. Raising the skills and access to information of the general labor force is the only effective way to achieve the objective of growth through rising productivity while pursuing the goal of equity through greater participation in the economy.
- It may be a strategic error to focus too much on large-scale industry as the primary source of industrial development in Africa over the next decade. The conditions for large industries to be competitive in world markets have not been adequately established and African consumers can ill afford the heavy costs of maintaining inefficient large industries through high protection or subsidies to state enterprises. To develop a self-reliant, sustainable model of industrialization, as put forward in the Lagos Plan of Action and the Industrial Development Decade for Africa, much more attention is needed to building up the needed experience, skills and institutions. Much of the capability-building is likely to occur in small- and medium-scale activities.

S.M. Wangwe

The Lall Paper

The paper by Sanjaya Lall on Human Resources, Technology and Industrial Development in Sub-Saharan Africa examines the background to industrial development in Sub-Saharan Africa and argues that the two conventional explanations of the problems of African industry (i.e. that based on exogenous shocks and that based on incentives—trade and industrial policies) do not come to grips with the structural factors of industrial performance. The author develops an alternative framework based on an interaction of three sets of factors, i.e incentives, capabilities and institutions.

The analysis of the interaction of the three sets of factors permits recognition of the important role of the process of learning and capability development as a determinant of industrial success. In this sense the framework adopted in this paper departs from the static approaches to industrial development. The process of learning and capability development has been a central characteristic of the current industrial revolution (since 1945) and is likely to continue to be important well into the 21st century.

The framework of analysis adopted in this paper has a merit of permiting a balance between market and administrative failures and permits a very important but changing role of the state in bringing about industrial development in Africa. In addition, a balanced analysis of protection based on the principles of objectivity and selectivity is permissible.

This paper brings out the important role of human capital in industrial development. This is a significant departure from the more common emphasis on physical capital based on the Harrod-Domar type of models. While a shift towards empasis on human capital is being suggested one lesson ought to be learned from the experience of the past two to three decades in Africa. The adoption of the development model which emphasized the shortage of capital and the need to mobilize investment capital also tended to take utilization of capital for granted. By the 1970s the problem of capacity underutilization began to emerge and has worsened in the 1980s. It is now recognized that investment in new capacity has to be balanced with adequate attention to the determinants of capacity utilization. This lesson is important in the proposed shift towards investment in human capital so that utilization of human resources is accorded at least as much attention as investing in the creation of new stocks of human capital. This would ensure that the existing stock of human resources translates into capability development. On this point the data which the paper presents already suggests that in Africa there seems to be a weak relationship between indicators of the stock of scientific personnel and technological development. For instance, the average share of engineers in the population of Sub-Saharan Africa is only 0.01 percent, but the corresponding figure for some of the African countries which are expected to have acquired above average technological capabilities are below this average. The share for Zimbabwe is only 0.005 percent, that for Côte d'Ivoire is 0.004 percent, and that for Nigeria is 0.005 percent. It is unlikely that these countries have below average levels of technological development in spite of their below average shares of engineers in total population. In terms of training in science and in vocational training, Nigeria and Zimbabwe are well below the average for Sub-Saharan Africa. While this unexpected relation may partly be a reflection of data problems, levels and patterns of utilization of scientific and technological personnel could be more important.

While effort is made to raise the low level of the stock of scientific and technological personnel in Sub-Saharan Africa, the question of effective utilization deserves attention. More specifically, the following are examples of determinants of utilization which should be accorded attention at the same time.

- making adequate supporting staff available at all levels. For example, engineers require an adequate number of technicians and artisans to be effective as their roles are complementary. Appropriate ratios of different levels and types of technological skills will have to be made available.
- scientific and technological personnel cannot work effectively if supporting tools and facilities are not put in place. Technological institutions and scientific laboratories, for instance, will need to be equiped with the necessary facilities for personnel to perform their work effectively.
- motivation and reward for achievement.
- upgrading, retraining and systematic on-the-job training programs will need to be put in place.
- foreign personnel will continue to be employed but attention should be paid to their utilization in training and in complementing not substituting local personnel.

As regards institutions, the role of research and development will be central to the process of learning and acquisition of technological capabilities. At least three aspects of research and development deserve attention. First, during the economic crisis of the 1980s many research and development activities have been starved of funds and facilities following an attempt to reduce government expenditure. This needs to be reversed to ensure that research and development activities are adequately funded and equipped. Second, the link between research and development institutions and industry needs to be established and/or strengthened. Third, the priority setting and policymaking government ministries should work closely with research and development institutions.

The framework adopted in this paper raises the question of the relatioship between local technology and foreign technology and the role of the latter. The relationship has often been more competitive than complementary in Africa. The role of imported technology in augmenting local technological efforts in a complementary way will receive high priority in the suggested framework.

The Oyejide Paper

The paper by Oyejide on Entrepreneurship and Growth in Sub-Saharan Africa surveys available literature on the determinants of entrepreneurship with a view to identifying major elements of the dynamics of private enterprise development in Africa. The paper examines various aspects of enterprise upgrading, graduation and growth. The issues raised in this paper draw attention to the barrier that can be represented as a hump in the typical firm's unit cost curve in its relation to size. This hump inhibits transition from informal to formal activities and inhibits graduation from small- to medium-scale operations.

An important policy issue which arises from this survey is how best to disaggregate the contents and structure of this hump. There are at least three relevant but interrelated factors here. First, there are capital requirements to facilitate jumping over the hump. Second, there are technology requirements for the transition and/or graduation to take place, which often call for replacement of indigenous technology by imported technology. Graduation is inhibited largely because two sides of the same coin are usually not given adequate attention—i.e upgrading local technology or combining it with relevant aspects of foreign technology in a coplementary way and adapting imported technology to suit the scale which is affordable by local entrepreneurs in terms of capital and skill requirements. Third, the pattern and structure of organization of production may or may not be supportive of transition and/or graduation. These factors will have to be tackled in a framework which captures these interrelated factors.

Notes

1. The paper on rural-urban linkages which was removed from the program at the last minute contains the historical example of the state-owned monopoly for coconut oil processing, which proved no match for "enterprising urban women [who] descended on Nzimaland to participate in a restructuring of commodity production and attendant social relations." By connecting rural coconut producers with urban consumers and by introducing consumer goods in the rural areas, they stimulated a pattern of growth with strong spillover effects into transport, housing and marketing. The state-owned processing factory had few such linkages because it tended to meet all its own needs from within. Despite its supposed technical superiority, it simply could not compete in economic terms with informal producers.

2. Recent studies that provide corroborating evidence include Jonathan Dawson, "Small-Scale Industry Development in Ghana: A Case Study of Kumasi", (London:" Overseas Development Administration, ESCOR, 1988, processed) and William F. Steel and Leila M. Webster, "Ghana's Small Enterprise Sector: Survey of Adjustment Response and Constraints", (World Bank, Industry Development Division, draft working paper, 1990, processed).

3. A state-centered approach suited some leaders' desire to avoid building up an indigenous business class that might serve as an opposition power base. The notion that indigenous entrepreneurs didn't exist conveniently justified this approach. In no case did leaders adopt strategies to build up the entrepreneurial class.

28

Growth and Adjustment in Sub-Saharan Africa

Benno J. Ndulu

Deceleration of growth in Sub-Saharan Africa set in during the second half of the 1970s. It was sustained during the 1980s mainly as a result of exogenous shocks and policy misdirection. Key among the exogenous shocks were a collapse of primary commodity prices, an increase in interest rates on foreign debt, and a steep cut in net foreign resource inflows during the first half of the 1980s. Policy misdirection was of two types. In the first, which has been referred to as 'capital habit' (Wheeler 1984), investment was driven beyond sustainability. Foreign resources played a significant role in easing constraints for financing capacity expansion; however, utilization of this capacity relied dominantly on own resources whose rate of expansion fell short of what was required. In the second, the anti-export policy bias of the 1970s contributed to the decline in export growth. Structural maladaptation was also related to a wrong expectation that the exogenous shocks of the late 1970s were short term and could be ridden out like the first oil crisis of 1973.

Most countries, even before adopting International Monetary Fund (IMF) programs, undertook stabilization measures. Programs adopted with pressure from the IMF, the World Bank, and bilateral donors continued with this perspective of adjustment. The perspective subsequently shifted to adjustment with growth. This change was premised on broad acceptance of the contention that sustained economic growth is central to an adjustment strategy aimed at achieving sustainable internal and external macroeconomic balance.

This paper emphasizes a medium term perspective. It does not take up distributional issues in the process of adjustment. Longer term issues related to structural transformation, institutional development, and the development of human capacity are definitely very fundamental to the development process in general and in African economies' capacity to respond (adapt) to changing conditions in particular. These sets of issues and the political economy of the adjustment process merit complete treatment on their own. Killick (1990) and Helleiner (1990) have excellent reviews on issues concerning the relationship between the structure of the economy and its capacity to adapt to changing conditions, and the relationship between structural adjustment and long term development.

There are definite links between medium term objectives and long term development. Sustained growth is fundamental to sustainable redistributive policy and poverty alleviation. Stable and healthy export growth is as important for medium term growth as it is for long term growth. The same study concluded that for low-income countries investment in human capital is a very significant determinant of long term growth. The last decade has shown that economic stagnation in the medium term constrains such investment to the detriment of longer term growth. Many of the features of growth with adjustment are thus implicitly and in some cases explicitly consistent with longer term growth and development.

This paper attempts to highlight the key trade-offs between growth and macroeconomic balance through discussion of resource gaps for growth, various channels for closure of such gaps, and implications for macroeconomic balance. In the second section, the peculiar features of the growth process in Sub-Saharan Africa are presented. These require analysis beyond conventional growth models. In the third section, an attempt at modelling growth incorporating these peculiar features is undertaken. Empirical testing of the major conclusions for the model is provided and adjustment experience is discussed in light of the growth model. The fourth section provides concluding remarks.

The Peculiar Features of the Growth Process in Sub-Saharan Africa

The process of growth entails not only the trade-off between current consumption and capacity growth (via investment), but also between capacity growth and capacity utilization in an import compressed situation. Thus actual growth targets and estimates of resource gaps for growth should take into account the possibility of much higher actual growth potential than that implied by investment projections alone. While net foreign resources are currently critical for the recovery of export performance, policy measures and resource allocation in favor of a diversified export sector remain crucial for sustained growth performance in the longer run.

Import Compression, Capacity Utilization, and Growth

A large part of the literature on growth has focussed on the relationship between capacity growth and investment. Growth of capital stock has been the focus of a majority of the models that deal with labor-surplus economies, such as those in Sub-Saharan Africa. These models have ranged from fixed-proportions in production emphasizing incremental capital-output relationship (Chenery and Strout 1966) to perfect factor substitutability production (Solow 1956, Robinson 1971, Fischer 1987).

Over the last fifteen years, there has been a deceleration of real growth in spite of reasonably high rates of investment in Sub-Saharan Africa (see tables 28–1 and 28–2). Therefore it has become necessary to emphasize the distinction between capacity growth and actual growth.

Real growth in a situation where capacity underutilization obtains is the sum of capacity growth g_p, which is driven by the investment rate, and growth in the rate of capacity utilization g_u:[1]

$$g = g_p + g_u \qquad (28\text{–}1)$$

Table 28–1. *Growth of production, investment and external trade: Sub-Saharan Africa*
(average annual percentage)

Indicator	*1965-73*	*1973-80*	*1980-87*
Gross domestic product			
Sub-Saharan Africa	5.9	2.5	0.5
Low-income economies	6.0	2.8	-0.4
Agriculture			
Sub-Saharan Africa	2.2	-0.3	1.3
Low-income economies	2.2	0.0	1.0
Manufacturing			
Sub-Saharan Africa	10.1	8.2	0.6
Low-income economies	10.7	10.2	-1.0
Gross domestic investment			
Sub-Saharan	9.8	4.0	-8.2
Low-income economies	10.3	4.5	-9.9
Export volume			
Sub-Saharan Africa	15.1	0.2	-1.3
Low-income economies	16.9	-0.6	-3.6
Import volume			
Sub-Saharan Africa	3.7	7.6	-5.8
Low-income economies	3.4	8.1	-7.8
Implicit GDP deflator			
Sub-Saharan Africa	7.5	6.8	15.2
Low-income economies	8.2	17.4	17.0

Source: World Bank 1989a *Sub-Saharan Africa: From Crisis to Sustainable Growth.*

High investment rates could coexist with declining actual growth as long as the decline in capacity utilization, g_u, exceeds the increase in capacity growth, g_p. For a given resource constraint, there may be a trade-off between capacity growth and capacity utilization. A number of African economies experienced this during 1975 to 1981, as capacity underutilization accelerated while investment rates increased. This was reflected in the study by Gulhati and Datta (1983), which observed rapid increase in ICORs and decline in investment productivity from 83.8 percent during 1961 to 1973, to 6.2 percent during 1980 to 1987 (World Bank 1989, p. 26). The process of growth does not only depend on factors that influence capacity growth but also those affecting the rate of capacity utilization. This contention is noteworthy given the fact that savings and hence investment ultimately depend on actual income growth. Resource allocation between capacity expansion and capacity utilization becomes as important as resource mobilization for investment to enhance capacity growth.

The critical role of import capacity in the realization of real growth can be looked at from the perspective of these two components of actual growth. Investment requires capital goods that are not domestically produced and therefore domestic savings cannot be transformed into investment goods unless foreign exchange is available (Ram 1985). On the other hand, it has been shown that intermediate imports are the single most important constraint to capacity utilization, especially in the "presumed" dynamic sector, industry (Gulhati and Datta 1983, Wangwe 1983, Ndulu 1986, Rattso 1987). Import instability hence translates into growth instability (Helleiner 1986).

Import compression occurs under two conditions. First, if substitution between imported and home goods is imperfect, a cut in capital and intermediate goods imports will result in a cut in real activity growth (Khan 1974, Khan and Knight 1988, Balassa et al 1986). Unlike in the extreme case of the fixed coefficients relationship, the reduction in real activity growth will be proportionately less than the rate of decline in imports, since some substitution away from imports may take place.

The remarkable stability of the ratio between real imports of capital goods and real investment over the last decade suggests a close to fixed proportional relationship. For Sub-Saharan Africa as a whole, the ratio of real capital goods imports (machinery and transport equipment) to real investment[2] increased very slightly from 35. 4 percent in 1980 to 36. 8 percent in 1987. In spite of a steep deceleration in import volume growth, the share of capital goods in total imports increased slightly between 1980 and 1987, from 31 percent to 32 percent (World Bank 1989a). However, the economywide ratios of real intermediate imports to value added seem to have shown a compression during the last decade of foreign exchange supply bottlenecks. While the annual growth rate of import volumes decelerated from 7. 6

Table 28–2. *Basic indicators of resource gaps for growth: Sub-Saharan Africa, selected years*

Indicator	*1965*	*1980*	*1985*	*1987*
Percentage od GDP				
Gross domestic investment				
Sub-Saharan Africa	14.0	20.2	12.1	16.0
Excluding Nigeria	15.0	20.0	15.7	16.0
Gross domestic savings				
Sub-Saharan Africa	14.0	22.0	12.6	13.0
Excluding Nigeria	15.0	14.3	13.8	11.0
Resource balance				
Sub-Saharan Africa	1.0	1.4	0.4	-3.3
Excluding Nigeria	-5.0	-5.7	-2.0	-4.9
Fiscal defect (excluding grants)				
Sub-Saharan Africa	—	-9.6	-10.6	-11.4
Excluding Nigeria	—	-9.9	-10.9	-11.8
Ratio				
Import surplus*				
Sub-Saharan Africa	1.06	0.89	0.78	1.11
Excluding Nigeria	1.06	1.15	0.90	1.12

*Imports/exports.

percent during 1973 to 1980 to 5. 8 percent during 1980 to 1987, the rate of growth of production declined from 2. 5 percent to 0. 5 percent. This suggests imperfect substitution. The capacity utilization rate declined proportionately less than the reduction in real intermediate imports. This was achieved through change in the relative sectoral contribution to GDP in favor of the less import-dependent sectors in production, particularly agriculture. However, substitution in the individual production processes is more rigid, as evidenced by the steeper decline in growth observed in the more import-dependent industrial sector (for agriculture and manufacturing see table 28–1).

Import compression can also occur when import volumes in low-income countries are predominantly constrained by import capacity rather than determined by relative prices and real income as traditional models postulate (Hemphill 1974, Moran 1990). In a model that takes into account import capacity directly, Moran (1990, p. 288) estimates short run income elasticity for import demand of 0.2 and price elasticity of -0.1. The argument here is that when foreign exchange constraints are explicitly considered, import restrictions dampen the transmission mechanism of relative prices.

As long as imports are imperfect substitutes for domestic goods, a squeeze on the capacity to import will result in a decline in investment and capacity utilization. Import compression thus ceteris paribus leads to reduced capacity growth and rate of capacity utilization. The severity of import compression effects varies from country to country depending on the extent of rigid dependence of investment and production on imports.

The two main sources of import capacity are export earnings and foreign savings. Export performance and the rate of net foreign resource flows are central to alleviation of import compression and growth. However there are other measures conducive to enhanced efficiency in the utilization of import capacity, especially in the longer run.

The above suggests that the problem of import compression in relation to growth is not merely that of the absolute availabilityof real import capacity, but also of its allocation between capacity epansion and utilization. However, in some cases selective complementary investment—particularly in infrastructure to remove bottlenecks for utilization of previously installed idle capacity—may have a higher pay off. Export performance and the rate of net foreign resource inflows are central to the alleviation of import compression and growth. However, net foreign inflows have often been tied to investment projects. Where foreign savings constitute a dominant proportion of import capacity, such rigidity has led to a steep decline in capacity utilization, and growth has largely been exogenous.

Exogeneity of Growth in Sub-Saharan Africa

The bulk of savings, and hence investment, is dependent on income growth itself. Where growth is constrained by foreign exchange, the level of exports, which is an important part of income, plays a crucial role in enabling the savings-investment process. Increased exogeneity of growth reduces the efficacy of policy efforts and has strong implications for medium term options. There are a number of exogenous factors that influence growth. Key among these are the external terms of trade, net foreign resource inflows, weather vagaries in economies dominated by rain-fed agriculture, and discovery and exhaustion of mineral resources. The extent of exogeneity of the growth process depends on the relative dominance of these factors over endogenous factors affecting growth.

Two main sets of factors explain the performance of real exports (income terms of trade):export volume changes, which are policy related, and changes in the barter terms of trade, which are exogenously determined. In a study covering 33 countries in Sub-Saharan Africa during 1970 to 1985, the explanatory contributions of these two factors were decomposed (Svedberg, 1990). The changes in export volume factor dominated in 19 countries, the barter terms of trade factor dominated in 11 countries, and three countries had a mixture of the two. Among those most affected by barter terms of trade were oil exporters and monocultural economies. A similar decomposition analysis for the period from 1980 to 1988 reveals the more significant role of the barter terms of trade (see table 28–3).

The period of economic stagnation was characterized by increased exogeneity of the growth process both directly (in terms of contribution to value added) and indirectly via import compression. This is particularly significant for monocultural economies where exports contribute a large proportion of GDP and import capacity. Typical cases are Zambia, Nigeria, Gabon, Congo, Mauritius, and Ethiopia.

Dependence on foreign resources ties growth performance to the economic performance and conditions of donor countries. As Rattso (1990) shows conceptually and as trends in resource inflows show, the pattern of flows has tended to be procyclical. World economic recessions weaken demand for Sub-Saharan Africa exports and reduce resource trans-

Table 28-3. *Sources of variation in the purchasing power of exports: Sub-Saharan Africa, 1980 to 1988*

Period	Growth rate of purchasing power (perentage)	Quantum growth rate (percentage)	Barter terms of trade growth (percentage)	Variation of purchasing power explained by quantum (percentage)	Variation of purchasing power explained by terms of trade (percentage)
1980-87	-5.8	-1.6	-4.2	27.6	72.4
1980-85	-5.6	-3.3	-2.3	58.9	41.1
1985-88*	-7.3	0.7	-7.7	-9.0	109.0

*Data for 1988 are projections.
Sources: Data from World Bank 1989a *World Development Report*; World Bank 1989b *African Economic and Financial Data*

fers. This occurred during the late 1970s and the first half of the 1980s. In addition, this period was characterized by increased real interest on external debt, which further reduced net inflows to finance growth.

Public Sector and Growth

The public sector's significant role in the growth process in Sub-Saharan Africa stems from its participation in investment and overall resource allocation. In addition to providing economic and social infrastructure, governments in Africa have participated in directly productive activities through statal and parastatal entities. Through direct controls of allocation of financial resources and foreign exchange, they have sought to control the pattern of resource allocation beyond budgetary operations.

Two issues arise with respect to the impact of high government participation in resource allocation for growth. The first relates to the impact of public sector investment programs on private investment. On the expenditure side, public sector investment, especially in infrastructure, raises the profitability of private investment. On the financing side, however, if public sector borrowing exceeds new deposits with the banking system, the public sector investment programs crowd out private investment. The net impact of public sector investment programs will thus depend empirically on the relative magnitudes of the two effects. Evidence from several studies on this has shown a net crowding in effect of public sector expenditure on investment.[3] The variation in the size of the crowding in effect of public sector expenditure on investment is most likely explained by the extent of infrastructural expenditure in total. Blejer and Khan (1984) show for developing countries as a group that longer term infrastructural expenditures, rather than short term public investment, positively induce private investment. The issue is not that of comparative direct effects on growth measured by relative productivity of private and public investment as discussed and empirically tested by Khan and Reinhard (1990). More important are the qualifications to the results offered by these two authors regarding the indirect effects of public investment on growth via raising profitability of private investment and absorption capacity of the economy more broadly.

The second issue relates to the link between fiscal deficit financing and the monetary process and hence inflation. Where public sector savings fall short of financing requirements for public sector investment, the gap will have to be financed either by noninflationary borrowing, at least in the short run (net new foreign resources to government and non-bank domestic borrowing), or through monetization. The gap between public investment and public sector savings determines the public sector financing requirement and the proportion of it monetized will depend on the availability of the other two sources.

Depending on how the fiscal deficit is financed, a trade-off between growth and inflation will ensue. If other sources are not available, an inflation tax becomes a major instrument for financing growth via monetization of the fiscal deficit, but with the cost of price instability. Such a strategy is nonsustainable, as has been well documented in the Olivera-Tanzi effect, due to fiscal drags as a result of collection lags and declining revenue yields beyond a certain level of inflation rate [Olivera (1967), Tanzi (1977, 1988), Aghevli and Khan (1977), Roe (1989)]. Increases in net foreign resource inflows and reductions in foreign debt servicing largely borne by governments have important implications for reducing monetization of the fiscal deficit and hence inflation.

Public sector deficits are not all related to investment programs, since large proportions have recently been associated with current dissaving. Subsidies and transfers via absorption of losses of distribution and marketing parastatals have significantly widened the financing gap. Part of this is directly absorbed into the fiscal budget through subventions to parastatals and part of it shows up in a build up on non-serviced overdrafts in the official banking system. The latter part of absorption has recently threatened viability of the banking system in some countries (e. g. Tanzania). This is a result of a large mandatory accumulation of non-performing assets in the banks against continued honoring of its obligations on the liabilities side, albeit at increasing interest rates, as efforts towards achieving positive real rates are being implemented.

Growth and Resource Gaps

A prototype growth model which incorporates the peculiar conditions obtaining in Sub-Saharan Africa is presented in this section. The model is a variant of the one developed by Taylor (1988, 1989) and is based on a consistent macroeconomic accounting framework. It could also be considered an extension of the World Bank's Revised Minimum Standard Model (RMSM) and is even more closely similar in spirit to the integrated model developed by Khan and Haque (1990). Its distinctiveness from the latter two models lies in two areas. First, it explicitly incorporates capacity utilization in the determination of resource gaps for capacity growth. This enables consideration of trade-offs between capacity growth and capacity utilization, and therefore of the relationship between capacity growth and actual growth for economies starting with low capacity utilization rates. Second, by including the consolidated fiscal gap and its financing, it provides an explicit but approximate link of growth and financing to the inflationary process via monetization.

In order to allow consideration for alleviation of the foreign exchange constraint through improved export performance, the version of the model presented here includes the real exchange rate as a relative price (Solimano 1990, Ndulu 1990). In this sense it reduces the deterministic nature of the other models. Unlike in the models by Hemphill (1974) and Moran (1990), import capacity is not considered exogenous in entirety. Export earnings, a component of the capacity to import, is itself influenced by policy stance.

The discussion first relates the model to the peculiar features of the growth process in Sub-Saharan Africa. An empirical calibration of the model is done for a single country (Tanzania), to bring out the range of strategic choices for closure of growth gaps and their implications for sustainability. Two empirical tests are carried out to demonstrate the impact of import compression on growth and the relationship between financing gaps and inflation. These tests are performed using cross-sectional data for Sub-Saharan Africa, and are largely indicative in view of wide variation across countries. The response to shocks and closure of resource gaps are discussed in the context of the model for the 1980 to 1985 and post-1985 periods in Sub-Saharan Africa.

Growth Model

Here the model is used to demonstrate the major factors which enhance and constrain real output growth in Sub-Saharan Africa.

Growth and investment. In summary form, the growth-investment relationship is presented in equations (28–2) through (28–6):

(28–2) $$g_p = g_0 + ki$$

(28–3) $$i = i_p + i_g$$

(28–4) $$i_p = i_o + i_1 u + i_2 i_g$$

(28–5) $$g_p = g_0 + k(i_o + i_1 u) + k(1 + i_2) i_g$$

(28–6) $$i_g = [1/(1 + i_2)] [(g_p - g_0) k - (i_o + i_1 u)]$$

Capacity output growth g_p is assumed to be determined by the level of investment i and its productivity k. Behind the growth-investment relationship is an imperfect substitution production function with capital being the most constraining factor.

Total investment is the sum of private i_p and public investment i_g. Public investment is determined exogenously by government decision and constrained by financing sources. Private investment is determined by availability of profitable investment opportunities and constrained by availability of funds. We shall assume for now that public investment's crowding in effect on private investment exceeds its crowding out effect. This seems to be plausible currently in view of the deterioration of economic and social infrastructure over the last decade. The net public expenditure crowding in effect on private investment is represented by $i_2 i_g$. The pervasive financial repression in Sub-Saharan Africa makes investible funds quantity rather than price rationed. The impact of the interest rate on private investment is thus assumed to be negligible.

All variables are normalized by capacity output y_p and hence y/y_p is capacity utilization u. The simple accelerator is $i_1 u$ and equation (28–6) is the consistent decision rule for public investment.

Growth constraints. Three potential constraints to capacity growth related to investment finance are private savings, public sector savings, and overall foreign savings. Overall accumulation, which is total savings, is represented by *s* in the following:

(28–7) $$s_p = b_o + b_1(1 - w)u$$

$$s_g = z_g + t - j^*$$

(28–8) $$z_g = z_o + z_1 u$$

$$s_f = m_c + m_z + m_k + j^* - e - t$$

$$m_z = a_o + a_1 u, \; m_k = (1-c)i$$

c = Domestic Content Of Investment
with u being a scale factor,

(28–9) $$e = e_o + e_1 rer$$

(28–10) $$s = s_p + s_g + s_f$$

All variables are again expressed in normalized terms.

Private savings s_p constitute the private sector's changes in commodity stock hoarding, enterprise provision for depreciation, and net additions to financial assets held with the banking system. Aggregate private savings depend on disposable income flows $[(1\text{-}w)u]$.

Public sector saving s_g in gross terms is defined here to include domestic and foreign components. Their sum is the public sector surplus including grants. The domestic component z_g is dependent on income flows, given the tax rate and domestic recurrent expenditure. The foreign component includes net official foreign transfers *t* and interest on foreign debt j^*.

Overall foreign savings s_f is equal to the negative of the current account deficit plus unrequited net transfers *t*. For simplicity it is assumed that only government borrows abroad and is a recipient of transfers. We separate the current account balance into two components. The exogenous component is interest on foreign debt j^*. The behavioral component is based on trade account determinants. Foreign savings increase with various components of imports and debt servicing, given exports and exogenous net foreign transfers.

Imports *m* are assumed to be quantity rationed. Imports of capital goods m_k are demanded in fixed proportion to investment. Intermediate imports m_z are determined by a derived demand of the capacity utilization rate, since domestic inputs are imperfect substitutes for imported intermediate goods (Khn and Knight 1988). Competitive imports m_c are assumed to be exogenous and dependent on the quantity rationing behaviour of the authorities.

Exports *e* are assumed to be supply-constrained and responsive to the internal terms of trade and foreign competitiveness, which is represented by the real exchange rate *rer*. It is assumed here that exports respond positively to the real exchange rate measured in domestic currency units. The capacity utilization rate *u* is considered a scale factor of total exports.

Equilibrium growth conditions consistent with financing constraints. The equilibrium condition consistent with the overall financing (saving) constraint is given as:

(28–11) $$i = s_p + s_g + s_f$$

Substitution yields the growth rate consistent with the overall savings constraint g_s:

(28–12) $$g_s = \frac{k}{c\,[d_o + d_1 u - d_{2rer}]}$$

$$d_o = \frac{g_o}{k + b_o + z_o + m_c + a_o + j^* + t - e_o}$$

$$d_1 = b_1(1-w) + z_1 + a_1 > 0$$

$$d_2 = e_1 > 0$$

The stability condition for the growth equation is that the domestic component of investment *c* is less than one. Savings-constrained growth will increase with capacity utilization through increased domestic savings $s_p + s_g$, which increase with capacity utilization *u*. In economies with excess capacity, the increase in net exports will not crowd out investment since domestic saving increases with u. The effects of changes in offsetting exogenous net foreign transfers, domestic saving that rises with *u*, and real depreciation impacts via exports have been separated in equation (28–12).

Based on a foreign exchange-constrained investment equation with a fixed import content of investment, the growth rate consistent with the balance of resources and uses of foreign exchange g_f is

(28–13)
$$g_f = \frac{k}{(1-c)\,[\emptyset + T = e_o - m_c - j^* - a_o - a_1 u + e_2 rer] + g_o}.$$

The stability condition is that $1\text{-}c < 1$.

Capacity output growth will fall as the capacity utilization rate increases. This is because the increase in intermediate imports required for a given import capacity would imply a reduction in imports of capital goods for investment. This is the trade-off between capacity growth and capacity utilization in an import compressed economy. In the case where $a_1 < 1\text{-}c$, actual growth will increase. More specifically, starting with a situation of capacity underutilization, as long as $(1 - c)/a_1 u > k$, a unit of foreign exchange switched from capital goods imports to intermediate imports will raise actual growth. Since domestic savings depend on actual output, an increase in u will eventually lead to increased investment. This would further relax the overall savings constraint.

Real depreciation of the exchange rate would lead to higher growth through increased exports. Through increased foreign exchange earnings, real depreciation would allow increased importation of intermediates and capital goods. Increases in net official transfers would strengthen the real exchange rate effect; increases in foreign debt servicing would reduce it. Under foreign exchange-constrained growth, the impact of real exchange rate depreciation on capacity growth is unambiguously positive.

The fiscal constraint to growth is derived based on the public sector borrowing requirement *psbr*. The psbr is determined for a given exogenously determined public sector investment rate, public sector savings which is dependent on the capacity utilization rate, and exogenous recurrent expenditure levels. The normalized public sector borrowing requirement is then given as:

$$(28\text{–}14) \quad pu = i_g - z_g + j^* - t = i_g - z_1 u + j^* - t$$

$$p = psbr/y \text{ and hence } pu = (psbr/y)(y/yp)$$

The *psbr* increases with public investment and foreign debt servicing; it declines with increases in domestic public sector savings and net official foreign transfers to the government.

Substituting for i_g from (28–6) and solving for the growth rate consistent with the fiscal constraint yields:

$$(28\text{-}15) \quad g_p = k(1 + i_2)\,[(\pi + z_1) + i_1]\,u + k(1 + i_2)\,(z_o - j^* + t) + g_o + ki_o\,.$$

Fiscal constrained growth, given a target *psbr*, increases with capacity utilization through increases in tax revenue and net official foreign transfers; it decreases with increases in foreign debt servicing.

Growth and inflation. The fiscal constraint provides a link to the inflationary process which depends on the monetized proportion of the borrowing requirement. Using a simple monetarist approach, the implication for inflation consistent with a given target growth rate and financing gaps can be approximated (cost-push factors are assumed to be incorporated into the growth rate of output on the supply side). Demand for money is assumed to be stable and changes only with real income in the medium term. In the short run, velocity changes and portfolio substitution may induce instability. In economies with undeveloped financial infrastructure, the dominance of transactions demand makes this assumption plausible.

Assuming constant velocity in the medium term, the equation of exchange is expressed in growth terms by equation (28–16). The change in money supply has three sources, which are represented in equation (28–17). Credit expansion targets consistent with inflation and growth targets can be set using equation (28–18), assuming sterilization of NFA.

$$(28\text{–}16) \quad \dot{M} = \dot{P} + \dot{X}$$

$\dot{M}$ = broad money supply growth
$\dot{P}$ = price level rate of change (inflation)
$\dot{X}$ = real output growth

$$(28\text{–}17) \quad \Delta M = h\pi u P y_p + \Delta L + \Delta NFA$$

ΔL = change in new bank credit to private sector
ΔNFA = change in net foreign assets
h = proportion of *psbr* monetized
$h\pi u P y_p$ = new bank credit to public sector
$uPy_p = p_x$ = actual output

$$(28\text{–}18) \quad \dot{p} = v[\Delta L/p_x + h\pi] - (g_u + gy_p)$$

v = costant velocity

Given the public sector borrowing requirement, the proportion of it monetized will depend partly on the magnitude of net official transfers. A higher t implies a lower h. The public sector borrowing requirement also depends on the extent of foreign debt servicing. Hence, reduction in inflationary pressure can be achieved via increased t and reduced j^*. In the longer term sustainable inflationary control will depend on reducing the domestic sources of high psbr.

Increased overall domestic savings not only helps reduce psbr directly via z_g, but also creates an opportunity for public sector financing from domestic non-inflationary sources.

Figure 28–1 depicts the three key constraints—savings, fiscal, and foreign exchange constraints—in the growth-capacity utilization space. It also depicts the trade-off between capacity utilization and inflation via the fiscal constraint. Ex post these three schedules must intersect at a given level of capacity utilization. In calibration of the model for a given year, this is achieved via appropriate adjustments in the intercepts if historically-estimated values of slope coefficients are to be preserved.

Figure 28-1. Constraints to Capacity Growth

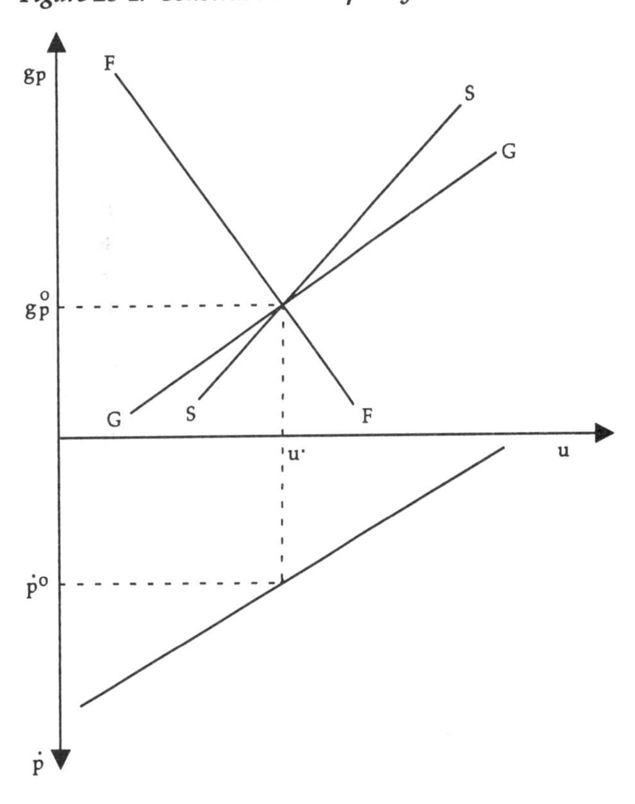

Implications for Growth and Adjustment Based on the Calibration of the Model and Simulations Using Tanzania as a Case

A variation of this model (holding the real exchange rate constant) has been calibrated for six Sub-Saharan African economies—Ghana, Nigeria, Uganda, Zambia, Zimbabwe, and Tanzania (Green 1989, Oyejide and Raheem 1989, Sepheri and Loxley 1989, Mkandawire 1989, Rattso and Davies 1989, and Ndulu 1989). A full version (including a variable real exchange rate) was also calibrated for Tanzania (Ndulu 1990). All six studies underscore the dominance of the foreign exchange constraint in the growth process. The trade-off between capacity growth and capacity utilization is emphasized under this constraint. Increased export performance to reduce import compression and increased capacity utilization to raise domestic savings are seen as the key measures for growth and stabilization. The magnitude of foreign debt service is also underscored by all six studies.

Behavioral parameters are obtained using historical data and least squares estimation. There are two approaches for achieving a consistent solution for the model. First, given a target growth rate, the saving and foreign exchange gaps are solved simultaneously for required changes in capacity utilization and the real exchange rate. The solution for capacity utilization is then used to obtain a fiscal balance consistent with the desired growth rate. This enables us to obtain a consistent public sector borrowing requirement. Using obtained values of the borrowing requirement, potential growth rate, growth in capacity utilization and given exogenous non-inflationary sources of finance, the extent of monetization and implications for inflation can then be drawn.

Alternatively, for a given real exchange rate, the solution for required transfers and increased capacity utilization can be obtained to achieve a target growth rate. This, however, can produce values of foreign resource inflows that are not sustainable and which could be symptomatic of growth without adjustment.

A comparison of the two strategies and the estimated parameters for the Tanzania economy are given in table 28–4. The target growth rate for potential output is set at 4 percent. The simulation entails the two strategies above compared to the actual base year (1986) performance and that of 1976, which historically reflected reasonable sustainability in terms of growth and macroeconomic balance.

Relative to the base year, the strategy that entails growth and adjustment (via the real exchange rate) comes out superior to the one that relies on foreign resource inflows to support growth. In the growth with adjustment strategy, the extent of self-financing for growth is much higher. The foreign savings requirement to achieve the target growth rate of potential output (nt) with adjustment is 13. 2 percent compared to 17. 5 percent with the other strategy. Since increased export performance entails greater capacity utilization (85 percent compared to 76 percent with the other strategy), domestic savings are significantly higher (13. 3 percent compared to 10. 7 percent). The external current account deficit also improves slightly from 6. 0 percent of potential output without exchange rate adjustment to 5. 4 percent with adjustment. Actual output growth is significantly higher with the adjustment scenario due to much higher growth of capacity utilization. In view of the lower public sector borrowing requirement in the growth with adjustment scenario, pressure for monetization and inflation decline. In brief, the strategy incorporating adjustment demonstrates much more sustainability.

Compared with actual performance in 1976 the above scenarios indicate a deterioration in the sustainability of growth. This significant drop in the yield of the adjustment process reflects a shift in relative prices. The foreign debt servicing ratio increased from 0. 8 percent in 1976 to 2. 1 percent of potential output in 1986. Higher levels are projected if debt relief is not forthcoming. For the same level of exports and imports, a higher current account deficit is implied. Since external debt is largely serviced by the government, its impact on the fiscal deficit is more pronounced. The domestic savings rate is significantly lower in the 1986 simulated scenarios than in 1976. This is a result of a lower savings effort, and particularly the public sector's reduced tax effort. Deterioration in supportive infrastructure for production and export performance are reflected in reduced public sector spending. This and deterioration in the terms of trade have reduced the yield of rela-

Table 28-4. *Comparative simulation for Tanzania with four percent target growth rate*

a) *Parameters (estimated using OLS)*

$g = -0.16 + 0.25i$

$i = 0.053 - 9.89u - 1.148i_g$

$s_p = -0.04 + 0.20u$

$z_g = -0.068 + 0.083u$

$s_f = 0.068 + 0.31u - 0.22\,rer + 0.33i$

$m_k = 0.33i$

$m_z = -0.156 + 0.31u$

$e = -0.234 + 0.22\,rer$

b) *Growth equations with intercepts adjusted for base year (1986) values*

(i) $g_s = -0.02 + 0.07u$ (without rer)

$g_s = -0.064 + 0.214u - 0.082\,rer$

(ii) $g_f = 0.165 - 0.193u$ (without rer)

$g_f = 0.077 - 0.234u + 0.166\,rer$

(iii) $g_p = -0.007 + 0.053u$

c) *Comparative simulations for 4% growth target (all variables normalized by y_p)*

Variable	*Base year (1986)*	*Projection with change in u and t*	*Projection with change in u and rer*	*Actual 1976 (good year)*
g	0.030	0.040	0.040	0.033
i	0.178	0.220	0.220	0.193
u	0.700	0.760	0.846	0.907
rer(1972=1)	0.703	0.703	0.970	—
t	0.068	0.096	0.068	0.014
	0.026	0.021	0.012	0.057
u	0.018	0.016	0.010	0.052
s_p	0.100	0.112	0.130	0.150
s_g	-0.010	-0.005	0.003	0.052
+t	0.126	0.155	0.122	0.034
+nt	0.146	0.175	0.132	0.045
i_g	0.055	0.086	0.080	0.110
i_p	0.123	0.134	0.140	0.083
e	0.061	0.061	0.120	0.200
m_z	0.038	0.054	0.083	0.098
j*	0.021	0.021	0.021	0.008

t = official transfers; nt = net transfers (including private)

tive price measures. The implications of the above changes are that, although a strategy of growth and adjustment is sine qua non for sustainability, it entails a long haul. External resource inflows are an important component of the adjustment process in the medium term. However, for these to be effective in promoting sustainability, continued adjustment is critical.

Empirical Testing of the Major Conclusions from the Model Using Sub-Saharan Africa Data

Two simple empirical tests are carried out using cross-sectional data for Sub-Saharan Africa for the period 1980 to 1985. The first demonstrates the impact of import compression on growth. The second demonstrates the relationship between financing gaps and inflation.

The impact of import compression on growth. Following the above model, real actual growth is postulated to depend on the rate of capacity creation and capacity utilization. The domestic investment rate determines the rate of capacity creation and hence capacity growth. Domestic investment makes the first claim to available import capacity. This contention is plausible at least for the period up to 1985, when the use of foreign savings was tied to investment projects.

The residual of the import capacity is allocated for current use, mainly for imported intermediates. The proportion of import capacity to support capacity utilization is thus more sensitive to the severity of import compression measured here as the growth rate of import volume. A third independent variable, growth in the terms of trade, is included to capture the effect of exogenous shocks. To the extent that this is already captured via import compression, this variable may not have a separate significant statistical impact on growth. Possible multi-collinearity between these two variables is dampened by the strong presence of exogenous components in the total import capacity.

(28–19) $$g = g_0 + g_1 i + g_2 g_m + g_3 g_{tot}$$

i= investment rate
g_m = growth rate of import volume
g_{tot}= growth rate of terms of trade

This equation was estimated cross-sectionally for 20 countries[4] using average values for 1980 to 1985, which is a period long enough to accommodate lags and cyclical changes.[5] Although complete data was available for 28 countries, the Sahelian economies, Sudan, and Ethiopia were excluded due to large fluctuations in import volumes related to food imports.[6] The estimated equation for the impact of import compression on growth is the following:

(28–20) $$g = -2.378 + 0.309 + 0.210 g_m + 0.104 g_{tot}$$
(-1.803)(4. 872) (2. 918) (0. 397)

$R^2 = 0.647$ $F = 13.85$ $DW = 1.99$
t – statistics are in brackets

These results show that during 1980 to 1985, capacity continued to be an important source of growth across the countries included. Import compression impacted on real growth via capacity utilization. Since g_m was negative in most of the countries, decline in capacity utilization rate must have partially offset capacity growth g_p, generated via continued high investment rate. The terms of trade had an insignificant impact on growth apart from import compression.

The instrumental variable method was also used to estimate equation (28–20) to further check for possible interdependence between i and g. The results of the two-stage least squares method support our postulation of exogeneity.

The relationship between financing gaps and inflation. The relationship between the inflation rate and growth of domestic credit and growth of real output is estimated in equation (28–21). Again, the analysis uses the average data for the 1980 to 1985 period. This hypothesis is tested to provide some insight on the impact of monetization of the fiscal deficit on inflation in an attempt to close resource gaps for growth. The period 1980 to 1985 involved a large expansion of the fiscal deficit as a proportion of GDP which was monetized as evidenced by high domestic credit growth (table 28–5). This occurred in spite of deceleration in actual real growth.

This test was done cross-sectionally for 21 countries for which complete data was available.[7] The growth rate of the consumer price index for 1980 to 1985 was used as a measure of the inflation rate. Inflationary pressure was postulated to increase with the rate of expansion of domestic credit and to decrease with increases in real growth. The following results were obtained.

(28–21) $$g_{cpi} = 0.157 + 0.942 g_{dcr} - 1.007 g$$
(0.048)(7.748) (-1.980)

$R^2 = 0.76$ $F = 32.72$ $DW = 1.823$
g_{dcr} = growth rate of domestic credit

These results support the simple version of the inflationary process linked to growth and its financing gaps as discussed in the model. Depending on the extent of monetization of borrowing requirements for growth, resultant growth in domestic credit will be partly transmitted into higher inflation. The counteracting effect of increased real growth is inversely related to inflation (see equation 28–18)). These two factors explain the largest proportion of variation in inflation rates across the countries. The near-unity of the sizes of the coefficients strongly support the simple specification with its corresponding assumptions.

However, the results obtained above are only indicative of broad explanatory relationships across the continent. These may not be directly transferable to country specific analysis using time series data (see London, 1989). Inflation expectations, for example, are deemed important and can best be included using time series data. Diversity of cost structures, structural bottlenecks, and pricing systems across countries and their variations over time are important considerations for analysis of inflation (Chhibber et al, 1989).

Response to Shocks and Closure of Resource Gaps in the Context of the Model: Sub-Saharan Africa

The closure of resource gaps for growth can be analyzed in the context of the model. We focus here on the case of a foreign exchange-constrained scenario. As a result of involuntary import compression, the economy will settle at a new equilibrium with lower capacity utilization, lower capacity growth, and high inflation. Due to the diversity of the countries involved, the discussion below is an approximation of the dominant features of the responses in the region.

The 1980 to 1985 period. During 1980 to 1985, responses to resource gaps for growth were predominantly passive and recessionary. Pressure from these gaps was accommodated through import compression effected by quantitative restrictions and hence, deceleration in real growth. Lower export growth (table 28—3) and reduced net inflows of new foreign resources resulted in reductions in real imports. Capacity utilization fell drastically as imports of intermediates were reduced relatively more than capital goods.

Gross domestic investment rates were maintained at approximately 20 percent of GDP for Sub-Saharan Africa as a whole during this period. The gross domestic savings rate declined sharply from 21. 6 percent in 1980 to 11. 5 percent in 1982 (The World Bank, 1989b). Although the continued high rate of investment partially cushioned the deceleration in the rate of growth, the decline in capacity utilization rate more than offset capacity growth. Average annual actual growth during 1980 to 1983 was -0. 7 percent. During this period, policy actions to reverse the decline in export growth were not implemented and decline in the terms of trade reinforced the trend.

Table 28–5. *Indicators for closure of resource gaps: Sub-Saharan Africa excluding Nigeria, 1980-1987*

	Percentage of GDP							
Period	Current account deficit	Current account deficit less interest payments on foreign debt	Fiscal deficit	Real imports	Gross domestic investment	Real exchange rate index[2] (1978=100)	Net official transfers to the government as a percentage of the overall fiscal deficit[3]	Domestic credit growth (percentage)
1980	9.5	7.4	-9.9	26.8	20.0	105.0	28.9	21.1
1981	11.6	8.9	-10.8	24.7	20.0	112.0	26.8	21.9
1982	11.5	8.8	-11.6	23.3	19.2	114.0	23.5	20.5
1983	9.1	6.5	-10.4	20.9	16.5	118.0	31.1	17.0
1984	6.8	3.8	-10.2	19.8	16.0	132.0	33.3	12.0
1985	6.6	3.7	-10.9	19.6	15.7	124.0	31.0	16.2
1986	9.0	6.2	-11.4	19.8	15.9	90.0	32.3	15.5
1987	—	—	-11.4	19.3	15.9	70.0	33.9	—

Note::
1 Includes unrequited net transfers.
2 Foreign currency/domestic curency.
3 Overall fiscal deficit excluding grants.
Sources: IBRD 1989b *African Economic and Financial Data;* IBRD 1989c *World Devlopment Report*

As public revenue declined during 1980 to 1982 in the face of sustained high expenditure, the fiscal deficit widened. In a number of cases pressure from fiscal populism resulted in increased monetization of the fiscal deficit. Domestic credit growth averaged 22 percent annually during 1980 to 1982. After 1982 inflows of new foreign resources declined. The domestic investment rate declined from 18. 2 percent in 1982 to 12. 1 percent in 1985 (World Bank, 1989b). Domestic credit growth decreased to an average of 15 percent as cuts in investment were effected and the fiscal deficit declined from 11. 5 percent of GDP in 1982 to 10 percent in 1984 (see table 28–5 for magnitudes excluding Nigeria). Correspondingly, the inflation rate (GDP deflator) increased to an annual average of 18. 5 percent during 1980 to 1985 compared to 6. 8 percent during 1973 to 1980. In addition to increased monetization, inflationary pressure was exacerbated by negative growth in real output.

The resultant appreciation of the exchange rate (see table 28–5) had two additional impacts apart from discouraging exports. First, the real domestic value of foreign savings available in foreign currency units declined. Second, the increased relative attractiveness of holding foreign assets induced capital flight (World Bank, 1989b). In some countries, the decreased tax effort during this period was a result of the relative growth of the informal sector.

The 1980 to 1985 period can thus be characterized as one of passive accommodation to resource gaps for growth. The attempt to defend growth via increased monetization of government deficit, external borrowing, and, in many cases, build up of payment arrears, could not be sustained beyond 1983. Import compression took its course and adjustment in the savings-investment process followed suit with the end result being reduced real growth. The absence of active policy measures, especially on the side of relative prices, was particularly conspicuous. Continued decline in per capita incomes and living standards was politically unsustainable. Decline in real public expenditures led to intensified deterioration of economic and social infrastructure, which threatened to reverse earlier achievements.

The post-1985 period. In the post-1985 period a much more active policy stance was adopted, albeit with some pressure from international financial institutions and bilateral donors. Attempts were made to reduce import compression, raise domestic savings and investment, achieve a sustainable balance of payments, and reduce inflation.

Attempts to tackle the import compression problems were two-pronged. First, there was an earnest drive to increase exports through real exchange rate depreciation and increase in the share of foreign prices received by export producers. The annual growth rate for export volume increased to 0. 7 percent during 1986 to 1988. Of the 33 countries for which data is available, 21 registered significant export quantum increases during 1987 and 1988. This is in large contrast to the average annual quantum growth rate of -3. 3 percent during 1980 to 1985 and 0. 2 percent during 1973 to 1980. The real purchasing power of exports during 1986 to 1988 registered an annual decline of 6. 3 percent due to a steep deterioration of the barter terms of trade. This exogenous shock of the steep fall in world prices for primary commodities significantly reduced the yield of policy effort.

Second, as a result of compliance to conditionality, a resurgence of net new foreign resources was recorded (see table 28–5). The bulk of this was in net official transfers, although net private transfers also increased substantially. However, during the same period, foreign debt service payment reduced the amount of foreign savings available to support current expenditure (table 28–5). The foreign savings leakage via debt service payments in 1986 was 31. 1 percent compared to 23. 3 percent in 1981.

Import compression was quite diverse across countries depending on when adjustment programs were adopted and the extent of efforts to increase exports. There seems to have been a shift in the allocation of import capacity between capacity expansion and capacity utilization. There was more concern for raising capacity utilization, often presented as 'consolidation of previous investment'. Foreign resources have become more fungible in terms of use since most of it now comes in the form of import support.

Investment for Sub-Saharan Africa as a whole increased modestly from 11. 5 percent of GDP in 1984 to 16. 4 percent in 1987 (excluding Nigeria which almost remained stagnant). This increase was predominantly supported by an increase in foreign savings. The reversal in the downward trend in domestic savings has not matched the increase in investment.

The overall fiscal deficit (excluding grants) increased from 10. 6 percent of GDP in 1985 to 11. 8 percent in 1987. The gap between increased public spending and stagnant public saving was filled predominantly via an increase in net official transfers. Government overall deficit, including grants, decreased slightly from -7. 3 percent in 1985 to -7. 2 percent in 1987 in spite of an increase in spending which exceeded the increase in domestic revenue. Of the 40 countries for which data is available, 24 reduced their rates of domestic credit growth during 1986 to 1987 primarily by substituting net foreign transfers to the government for borrowing from the banking system. A combination of reduced monetization and an increase in the real growth rate (averaging 1. 4 percent during 1986 to 1987 compared to -0. 3 percent during 1980 to 1985 for Sub-Saharan Africa as a whole) led to a reduction in the rate of inflation (CPI) in 26 of 32 countries for which data is available.

Concluding Remarks

The process of growth in Sub-Saharan Africa entails peculiar features that have to be taken into account for revival of sustained growth in the future. There is need to emphasize not only the trade-off between current consumption and capacity expansion, but also between capacty growth and capacity utilization in investment decisions. This is particularly important given the current low rates of capacity utilization and continuing import compression. Rapid growth of capacity utilization will require credit expansion beyond that consistent with projected growth. At the same time, investment to rehabilitate supportive infrastructure is critical.

The continued need for additional net foreign resource inflows stems from two considerations. First, rapidly deteriorating terms of trade have significantly reduced the yield from adjustment measures. Second, the debilitating foreign debt service burden constitutes a very large leakage from current resources to support growth.

Central to sustained and stable growth in the longer run, however, is the need to raise the extent of own-financing for growth. This entails not only measures for raising savings, but particularly more measures for improved and diversified export performance. While reduction in import dependence can be achieved through structural transformation, this indeed will require a very long time frame. There is still scope for raising efficiency in the use of the scarce import capacity. This requires not only realistic pricing of this scarce resource, but also fungibility of use especially for external resources.

More often than not, the revival of growth during the last five years has occurred without stabilization. The apparent reductions in the external account and fiscal deficits are predicated on an increase in net foreign transfers for balance of payments and fiscal support. Moreover, the current reduction in inflation rate (relative to 1980 to 1985) owes its realization in many countries partly to a decrease in monetization of fiscal deficits as a result of foreign financing.

The apparent fragility of the recent adjustment with growth is thus a major concern in the context of sustainability. Similar growth rates in the 1960s and first half of the 1970s had a much more sustainable financing structure. While noting the current hostile growth environment due to exogenous constraints, efforts to encourage domestic savings and a diversified export performance have to be sustained. Increased net inflows of foreign resources should be seen as a means for buying time to realize results from the adjustment measures.

Notes

1. Actual output y is obtained by multiplying capacity utilization u, with capacity output, y_p:

$y = uy_p \, ; dy = u \, dy_p + y_p du$.

Substitution and simplification yields:

$dy/y = g_y = dy_p/y_p + du/u = g_p + g_u$.

2. Imports of machinery and transport equipment were deflated by the import unit price index (World Bank 1989b). Investment was deflated by a weighted average of the import price index and the GDP deflator with assumed weights of 30 percent and 70 percent, respectively. Based on the six studies covering Nigeria, Tanzania, Zimbabwe, uganda, Zambia, and Ghana (op cit) the direct import content of investment averaged around 35 percent during 1985-86.

3. See Ndulu 1990 for Tanzania, Rattso and Davies 1990 for Zimbabwe, Oyejide and Raheem 1990 for Nigeria, Green 1990 for Ghana, Sepheri and Loxley 1990 for Uganda, Mkandawire 1990 for Zambia, and Boye 1990 for Senegal.

4. Countries in the sample are Burundi, Central African Republic, Ghana, Kenya, Liberia, Madagascar, Senegal, Sierra Leone, Tanzania, Togo, Zaire, Zambia, Botswana, Cote d'Ivoire, Mauritius, Zimbabwe, Cameroon, Congo, Gabon, and Nigeria.

5. Lavy and Pedroni (1990) have an insightful discussion on the various ways to deal with this problem, including the use of an explicitly dynamic framework to deal with cycles.

6. Data is from World Bank 1989b.

7. Countries in the sample are Burundi, Ethiopia, Gambia, Ghana, Kenya, Madagascar, Malawi, Niger, Senegal, Sierra Leone, Sudan, Togo, Zaire, Zambia, Cote d'Ivoire, Mauritius, Zimbabwe, Cameroon, Congo, Gabon, and Nigeria.

References

Aghevli, B. and M. Khan (1977): "Inflationary Finance and the Dynamics of Inflation: Indonesia 1951-72", *American Economic Review*, June 1977.

Balassa, B., E. Voloudakis, P. Fylaktos and S. T. Suh (1986): "Export Incentives and Economic Growth in Developing Countries: An Econometric Investigation", *World Bank DRD Discussion Paper* no. DRD 159, Washington, D.C.

Blejer, M. I. and M. S. Khan (1984): "Government Policy and Private Investment in Developing Countries", *International Monetary Fund Staff Papers*, vol. 31., no 2.

Chhibber, A., J. Cottani, R. Firuzabadi and M. Walton (1989): "Inflation, Price Controls and Fiscal Adjustment in Zimbabwe", *IBRD Working Papers*, WPS 192, Washington, D.C.

Chenery, H. and A. Strout (1966): "Foreign Assistance and Economic Development", *American Economic Review*, Vol. 56, No. 4.

Fischer, S. (1987): "Economic Growth and Economic Policy" in V. Corbo, M. Goldstein and M. Khan (eds.) *Growth Oriented Adjustment Programs* (Washington, D. C. : IMF and World Bank).

Green, R. H. (1989): "Medium Term Constraints to Growth: Ghana", a paper for *WIDER Project on Medium Constraints to Growth*, Helsinki.

Gulhati, R. and G. Datta (1983): "Capital Accumulation in Eastern and Southern Africa: A Decade of Setbacks", *World Bank Staff Working Papers*, No. 562, Washington, D.C.

Helleiner, G. (1986): "Outward Orientation, Import Stability and African Economic Growth: An Empirical Investigation" in Sanjaya Lall and F. Stewart (eds) *Theory and Reality in Development* (MacMillan).

Helleiner, G. K. (1990): "Structural Adjustment and Long-term Development in Sub-Saharan Africa". Paper presented to the workshop on Alternative Development Strategies in Africa. Queen Elizabeth House, Oxford. December 11-13.

Hemphill, W. (1974): "The Effects of Foreign Exchange Receipts on Imports of Less Developed Countries", *IMF Staff Papers*, 21:637-77.

World Bank (1989a): *Sub-Saharan Africa: From Crisis to Sustainable Growth*, (Washington, D. C.).

World Bank (1989b): *African Economic and Financial Data* (Washington, D. C.).

World Bank (1989c): *World Development Report*.

Khan, M. (1974): "Import and Export Demand in Developing Countries", IMF Staff Papers, 678-693.

Khan, M. and M. D. Knight (1988): "Import Compression and Export Performance in Developing Countries", *The Review of Economics and Statistics*, vol. LXX, No. 2.

Khan, M. and C. M. Reinhard (1990): "Private Investment and Economic Growth in Developing Countries", *World Development*, vol. 18, no. 1.

Khan, M. and N. Ul Haque (1990):"Adjustment with Growth:Relating the Analytical Approaches of the IMF and the World Bank," *Journal of Development Economics*.

Killick, T. (1990): "Structure, Development and Adaptation", *African Economic Research Consortium Special Paper*, no. 2, Nairobi.

Lavy, V. and P. Pedroni (1990): "Are Returns to Investment Really Lower in Sub-Saharan Africa?", Mimeo, World Bank, Washington, D.C.

London, A. (1989): "Money, Inflation and Adjustment Policy in Africa: Some Further Evidence", *African Development Review*, vol 1., no. 1.

Lipumba, N., B. Ndulu, S. Horton and A. Plourde (1988): "A Supply Constrained Macroeconometric Model of Tanzania", *Economic Modelling*, vol. 5, no. 4.

Mkandawire, T. (1990): "Growth Exercises on Zambia", a paper for *WIDER Project on Medium Term Adjustment Strategy*, Helsinki.

Moran, C. (1990): "Imports under a Foreign Exchange Constraint", *The World Bank Economic Review*, vol. 3, no. 2.

Ndulu, B. (1986): "Investment, Output Growth and Capacity Utilization in an African Economy: The Case of Manufacturing Sector in Tanzania", *East African Economic Review*, vol. 2.

Ndulu, B. (1990): "Macroeconomic Constraints to Growth: An Empirical Model for Tanzania", a paper for the *WIDER Project on Medium Term Adjustment Strategy*, Helsinki.

Ndulu, B. and N. Lipumba (1988): "International Trade and Economic Development in Tanzania" in R. Kanbur, J. F. Ansah and P. Svedberg (eds.) *International Trade and Economic Development in Sub-Saharan Africa* (forthcoming).

Olivera, J. (1967): "Money, Prices and Fiscal Lags: A Note on the Dynamics of Inflation", *Banca Nazionale del Lavoro Quarterly Review*.

Otani, I. and D. Villanueva (1989): "Determinants of Long-Term Growth Performance in Developing Countries", *IMF Staff Working Paper* (WP/88/97).

Oyejide, A. and M. Raheem (1990): "Macroeconomic Constraints and Growth Programming:Empirical Evidence from Nigeria", a paper for *WIDER Project on Medium Term Adjustment Strategy*, Helsinki.

Ram, R. (1985): "Exports and Economic Growth: Some Additional Evidence", *Economic Development and Cultural Change*, vol. 33, no. 1.

Rattso, J. (1988): "Import Compression Macro-dynamics: Macroeconomic Analysis for Sub-Saharan Africa", (University of Trondheim).

Rattso, J. (1989): "The Asymmetric Relation Between Sub-Saharan Africa and the Rest of the World: A Theoretical Analysis of the Role of Import Compression", (University of Trondheim).

Rattso, J. and R. Davies (1990): "Growth Programming for Zimbabwe", a paper for *WIDER Project on Medium Term Adjustment Strategy*, Helsinki.

Robinson, S. (1971): "Sources of Growth in Less Developed Countries: A Cross Section Study", *Quarterly Journal of Economics*, vol. 85, no. 3.

Roe, A. (1990): "Internal Debt Management in Africa", African Economic Research Consortium Special Papers no. 4. 77, Nairobi.

Sepheri, A. and J. Loxley (1990): "Medium Term Constraints to Growth: Uganda", a paper for *WIDER Project on Medium Term Adjustment Strategy*, Helsinki.

Solimano, A. (1990): "Macroeconomic Constraints for Medium Term Growth: A Model for Chile", *World Bank Working Papers*, WPS 400, Washington, D.C.

Solow, R. (1956): "A Contribution to the Theory of Economic Growth", *Quarterly Journal of Economics*, vol. 70, no. 1.

Svedberg, P. (1990): "The Export Performance of Sub-Saharan Africa", *Economic Development and Cultural Change*, vol. 38.

Tanzi, V. (1977): "Inflation, Lags in Collection and the Real Value of Revenues", *IMF Staff Paper*, vol. 24, 154-67.

Tanzi, V. (1988): "The Impact of Macroeconomic Policies on the Level of Taxation (and on the Fiscal Balance) in Developing Countries", *IMF Working Paper* WP/88/95.

Taylor, L. (1988): "Medium Term Development Strategy". Mimeo, Massachusetts Institte of Technology, Boston..

Taylor, L. (1989): "Gap Disequilibria: Inflation, Investment, Saving and Foreign Exchange", Mimeo, Massachusetts Institute of Technology, Boston..

Tutu, K. A. , N. K. Sowa and C. D. Jebuni (1989): "Real Exchange Rate and Macroeconomic Performance in Ghana", paper presented at African Economic Research Consortium Research Workshop (December, Harare).

Wangwe, S. M. (1983): "Industrialization and Resource Allocation in a Developing Country: The Case of Recent Experiences in Tanzania", *World Development*, vol. 11, no. 6.

Wheeler, D. (1984): "Sources of Stagnation in Sub-Saharan Africa", *World Development*, vol. 12, no. 1.

29

The Liberalization of Price Controls: Theory and an Application to Tanzania

Paul Collier and Jan Willem Gunning

A subsidy raises the price to producers at the same time as it lowers it to consumers. Thus it stimulates extra production and extra consumption and maintains market clearing. By contrast, price control works by lowering the price to both producers and consumers. It therefore depresses production at the same time as it induces extra consumption. Hence, if price controls succeed in altering prices, they give rise to excess demand. Although desired consumption rises, achieved consumption falls. Many African governments have been tempted into using price controls because they are free, whereas subsidies are a claim on the budget. Usually price controls are only applied to a few items. The frustrated excess demand for the price controlled items can then spill over into (satisfied) demand for other goods. However, sometimes governments have applied the controls over such a wide range of goods that the excess demand becomes generalized.

An example of this occurred in Tanzania from the late 1970s until 1984. Consumer prices for nearly all goods were held at a level well below that at which markets would have cleared. There was intense excess demand at these official prices. For the goods that people actually wanted a black market developed. In 1984 the price controls were abandoned. Helped by extra imports of consumer goods, the economy adjusted over the next two years back to market clearing.

This paper focuses on the transition back to market clearing. In the second section we set out a theory of how the economy behaves under shortages and the consequent difficulties liable to be encountered in the transition. The shortage economy turns out to have alarming properties. It implodes into subsistence. Response to producer price changes is perverse. The demand for money first rises and then falls. The longer transition is delayed, the larger the price changes will need to be to achieve market clearing. In the third section we show how this theory applies to the collapse of the Tanzanian economy between 1978 and 1984 and, more especially, to its rapid recovery between 1984 and 1989. The fourth section provides a brief conclusion.

The Theory of Rationing and the Transition to Market Clearing

Supply Response Under Rationing

In order to understand the transition to market clearing, it is necessary to investigate how the economy behaves under conditions of shortage.

The volume of crop sales. Under conditions of market clearing, the household is characterized as maximizing utility subject to relative prices, endowments, and technology. If the price of output (crops) changes relative to the price of consumer goods, the optimal volume of crop sales will change. Although the direction of change is a priori ambiguous, empirical studies generally find supply response to be positively

Figure 29-1. Influences on the Volume of Crop Sales

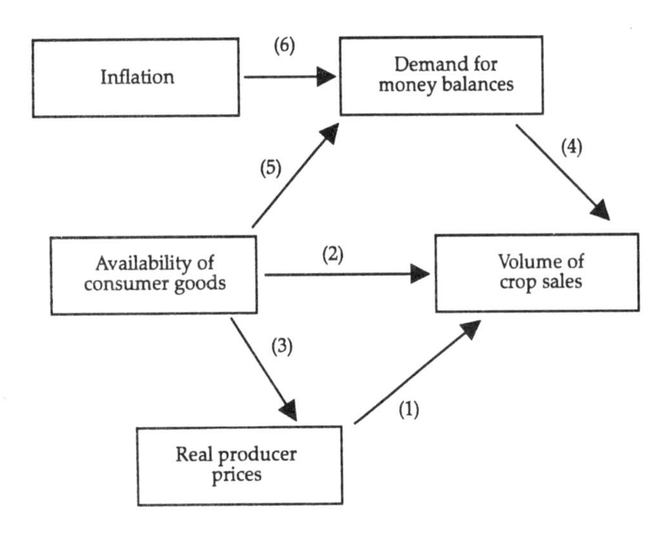

related to crop prices. In figure 29–1, which sets out a taxonomy of influences on crop sales, this is shown as relationship (1).

If the household faces generalized shortages of consumer goods, expenditure may not be a choice variable. This depends on how goods are rationed and how the black market operates. For the present we will abstract from the black market. If rationing is deterministic, then the household cannot significantly influence the amount of goods it is permitted to purchase, which, by assumption, is less than it would wish to buy. Knowing its entitlement, the household therefore equates its income to this constrained level of expenditure. The household's sale of crops will be proportional to the availability of consumer goods. In figure 29–1 this is shown as relationship (2). Thus, in an economy characterised by shortages of consumer goods, a sure way of increasing crop sales is by increasing the availability of consumer goods.

A surprising consequence of shortages of consumer goods is that, for a given entitlement, peasant supply response to crop prices becomes negative unit elastic. To maintain constant income, crop sales must be reduced proportionately to the increase in crop prices. Thus by increasing crop prices under conditions of shortages of consumer goods, a government will inadvertently intensify the problem. This change in the supply response to price is shown as relationship (3) in figure 29–1.

Assuming that the predominant reason for African peasants to hold money balances is to finance expenditure, we consider only the transactions demand for money and abstract from any asset demand. If the household's transactions demand for money increases, it must sell more crops to augment its cash balances. This is shown as relationship (4). In conditions of market clearing, the usual reason for a household to increase its cash balances is inflation. This is shown as relationship (5).

In conditions of shortage, the demand for money may be radically altered depending upon the nature of rationing. In most African contexts the more relevant form is stochastic rationing. Here the household faces only a probability distribution of possible availabilities of consumer goods which it might encounter. Within limits the household can, depending on the money balances which it chooses to hold, alter the expected value of its expenditure. Generally, as the household increases its money balances expected consumption increases and is an incentive for the household to increase its money balances. This behavior is a form of rent seeking.

Under market clearing conditions, peak money balances equal planned expenditure; under stochastic rationing, peak money balances exceed expected consumption for all nonzero values. The offsetting effect reduces the demand for money and the net effect is a priori ambiguous. If goods are sufficiently scarce, then the demand for money will be lower than under market clearing. However, for moderate degrees of shortage it is possible for money demand to be higher, even though expected expenditure is lower. Monte Carlo experiments reported in Bevan et al (1990) suggest that this is indeed likely to be the case.

The change in the demand for money consequent upon the regime switch from market clearing to stochastic rationing is referred to as the 'honeymoon effect' because of its implications for crop sales (Bevan et al 1987 and 1987a). If there is a once and for all increase in the demand for money, then peasants will choose to sell crops in order to build up their money balances. This is shown as relationship (6) in figure 29–1. Superficially this effect is like the inflation tax; however, there is an important difference. Inflation can be regarded as taxation because the government does not increase its liabilities in real terms. Under stochastic rationing with price controls, households increase their real and nominal money balances. This constitutes a liability for the government, which turns out to be central in the transition back to market clearing.

The market for goods and crops. So far we have focused only on the peasant household. To complete the framework for the analysis of the transition back to market clearing, we must specify the other agents in the economy. We assume that households trade only with a government trading agency, which buys crops from households in exchange for domestic currency. It sells these crops abroad for consumer imports which it sells to households for domestic currency. Consumer goods are not produced domestically. We assume that the trading agency is a price taker on world markets and that it is unable to borrow either abroad or domestically. The government sets the prices of goods and crops. It requires the trading agency to buy crops whenever it has the cash to do so and to sell goods whenever it has the foreign exchange to import them.

The markets for goods and crops are depicted in figure 29–2, which is a variant of the Malinvaud model (1977). The space is defined on the two nominal prices, P_c, for the cash crop, and P_g, for goods. The schedules *CC* and *GG* denote the loci of notional equilibrium in these two markets for a single period. Notional equilibria are defined on the assumption of equilibrium in all markets other than that of the locus being derived. Notional equilibrium in the goods market is achieved by variations in demand, supply being treated as exogenous within the period. Hence, an increase in the goods price leads to excess supply of goods, unless it is offset by the higher incomes generated by an increase in the price of cash crops. Proportionate increases in both prices are not neutral because of a real cash balance effect, which causes the *GG* locus to be steeper than a ray through the origin. Notional equilibrium in the cash crops market is defined by the trading agency holding just enough domestic currency to buy the cash crops supplied by households. Above the *CC* locus the trading agency holds insufficient currency to buy all of the offered crops. Below the locus there is quasi excess demand in the sense that the trading agency is willing to purchase more supplies than are forthcoming.

An increase in the price of crops leads to excess supply unless it is offset by an increase in the price of goods. Again proportionate increases in prices are not neutral because as prices rise the real cash balance effect induces an increase in crop supply. Hence, the *CC* locus is flatter than a ray through the origin. The two loci divide the space into four zones of disequilibrium, in which not all plans can be fulfilled, so that agents change their behavior. These changes in behavior alter the loci of market equilibria from notional to effective. The change implied by our previous theory is that, in response to an excess demand for goods, peasants revise downwards their marketing of crops. This expands the zone of excess

Figure 29-2a. National Equilibrium Loci in the Goods and Crops Markets

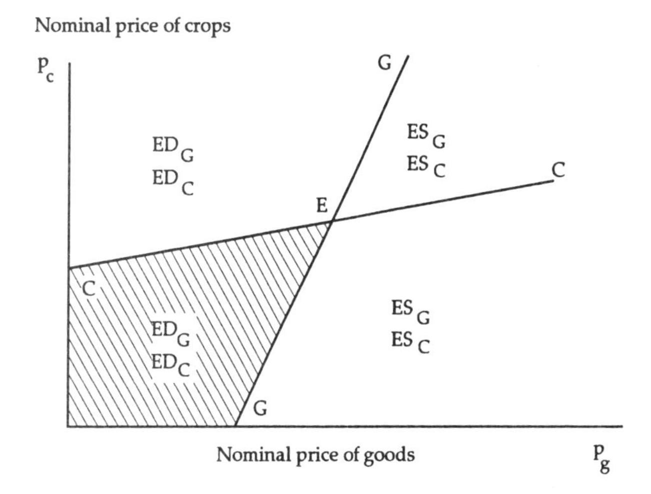

ES: Excess supply
ED: Excess demand

Figure 29-2b. Effective Equilibrium Loci in the Goods and Crops Markets

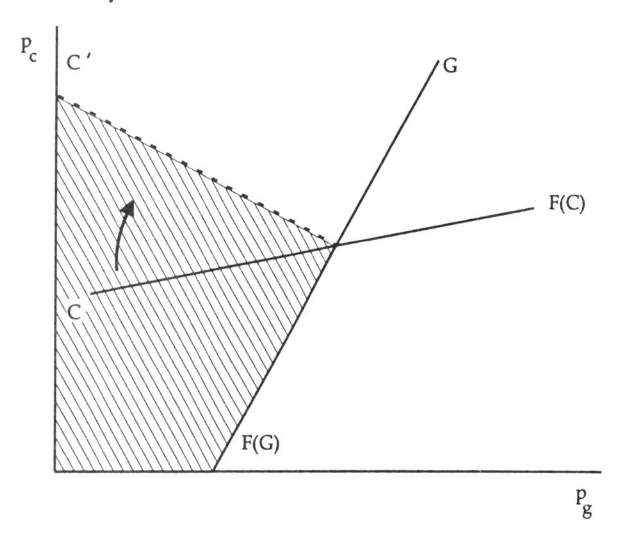

demand for crops, shifting the crops equilibrium locus to *C′C′* in figure 29–2.

This analysis can be used to distinguish different ways in which crop and goods prices may be set wrongly. For example, in an economy characterized by repressed inflation, the combination of excess demand for goods and crops is shown in the shaded zone in figure 29–3. Policy advice would generally infer that the excess demand for crops implies that their price is set too low. Prices P_c* and P_g* secure full market clearing. From figure 29–3 it is evident that repressed inflation does not imply that crop prices are below P_c^*. Three regions of repressed inflation can be distinguished. In the region bordered by C',E,P_c^* the nominal price of crops is too high even though there is an excess demand for them. In the region $P_c^*,E,0$ the nominal price of crops is too low, but the relative price of crops to goods is too high. Only in the third region $0,E,P'_g$ are both the nominal and the relative prices of crops too low. Hence, it is not possible to infer how crop prices differ from their equilibrium levels merely from the symptom of excess demand for crops once the goods market is also in disequilibrium. The comparison between the actual price of crops and their optimum price, P_c^*, is relevant only if the goods price is simultaneously raised to P_g^*. If instead the goods price is left unaltered at a level below P_g^*, then quite different considerations determine the appropriate change in the price of crops.

If the economy is characterized by repressed inflation, there is invariably a need to raise the price of goods. However, even if this is done, the nominal price of crops should be lowered if the economy starts in region C',E,P_c^*, and the relative price should be lowered if it is in region $P_c^*,E,0$. If the goods price is not raised, then a sustainable position can only be reached if the economy is in region P'_g,k,E. The sustainable position is reached by reducing the nominal price of crops until the goods market equilibrium locus is reached.

Dynamic analysis. If the economy enters the repressed inflation zone and prices are not altered so as to restore equilibrium, then marketed supply will contract as peasants reduce planned income to expected expenditure. For example, assume that the economy is in equilibrium at E (in figure 29–3). The government purchases from the trading agency a quantity of goods for its own consumption, which it finances by printing extra domestic currency. It freezes nominal prices at their initial levels and makes no further interventions. Although in subsequent periods goods supply to peasants reverts to normal levels, the peasant sector is in disequilibrium.

In aggregate peasants can reduce their excess money balances only by selling less, which will cause a reduction in the supply of goods in the next period. If the government has set P_c and P_g at world prices so that there is no taxation, the reduction in goods supply has the same value as the reduction in crop sales. The absolute magnitude of excess demand for goods is unaltered. Peasants continue attempting to adjust by reducing their income below expenditure.

Figure 29-3. Regions of Crop Mispricing

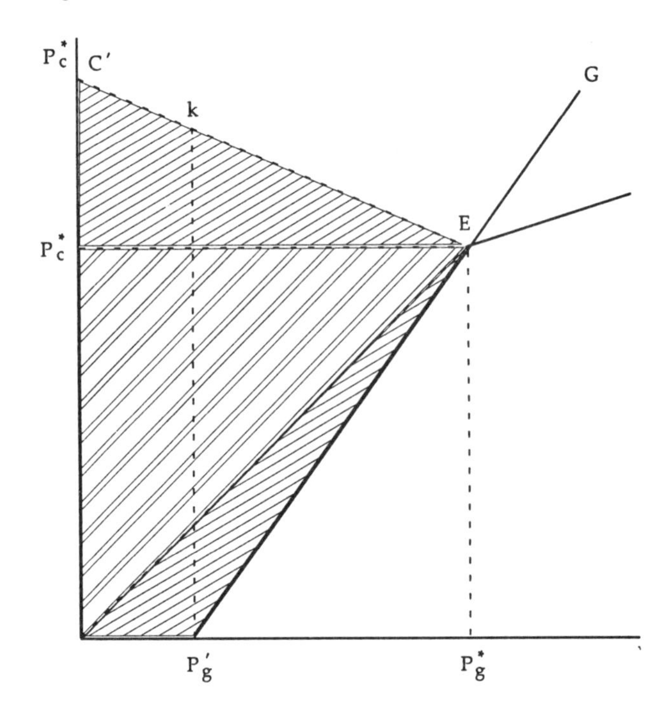

Figure 29-4. The Dynamics of Shortages

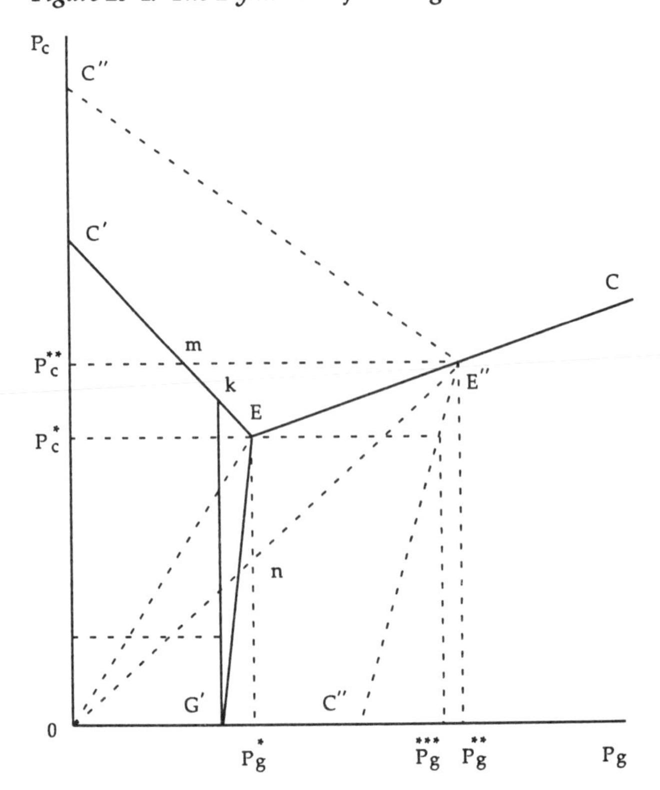

This response in aggregate causes an implosion into subsistence.

If, as is more likely, the government imposes some crop taxation so that $P_c/P_g < P_{cw}/P_{gw}$ (where w denotes world prices), then the pace of the implosion is increased.

The implosion to subsistence can occur without any government consumption. All that is required is government misregulation of the price level. Again starting from equilibrium, suppose that, instead of purchasing goods, the government had lowered prices such that $p_c < P_c^*$ and $p_g < P_g^*$, leaving relative prices unaffected. Peasants would experience a wealth effect and again have excess money balances. The consequences of such excess balances are as in the previous analysis.

When excess demand takes the form of stochastic shortages, this process of implosion can be divided into two phases. In the first phase, as availability deteriorates, desired money holdings may rise because peasants temporarily sell more crops than when their money stocks are at their chosen levels. Hence, the decline in output will initially be dampened. However, once availability deteriorates to the point at which peasants choose to reduce their money stocks, the long run tendency to implosive output reductions is accelerated by short run monetary dishoarding.

The Transition to Market Clearing

If the economy is in the repressed inflation zone, and hence imploding, it is imperative to make the transition back to market clearing. However, the design of liberalization policies has to take into account that the regime boundaries shift over time. This is due to three dynamic effects: the effect of crop sales via the balance of payments on future goods availability, the adjustment of expectations in response to rationing experiences, and the adjustment of money balances over time. Policies which aim at restoring equilibrium can, in principle, be grouped under two headings: policies which rely on changes in prices controlled by the government (i.e. P_c and P_g), and policies which aim at improving consumer goods supplies directly. Here we consider these in turn, and look at the effects of rational expectations.

Changes in prices controlled by the government. The honeymoon effect is likely to have increased holdings of real money balances relative to those which would be desired under market clearing conditions. Since the switch back to market clearing will reduce the real demand for money, it will lead to a permanent increase in the price level. The honeymoon effect is analogous to an special inflation tax arrangement in which the government acquires the resources now at the cost of inflation which is deferred until the regime switch. If the government wishes to avoid this, it must be in a position to supply consumer goods in exchange for the money which peasant households no longer wish to hold.

Because of the implosion of the economy, the transition begins from a position in which goods supplies are temporarily lower than they will be once market clearing is restored. This is why, if the transition is achieved only by price changes, goods prices must overshoot their long run equilibrium level. To see this we abstract from the honeymoon effect and imagine that, once goods supply is fully restored, prices can indeed revert to P_g^* and P_c^* (figure 29–4). At the onset of shortages, the repressed inflation zone is C, E_0, G_0. However, by the time of the proposed transition back to market clearing, the implosive contraction of the economy has reduced goods supply. The goods locus has shifted to G_1 and the repressed inflation zone has expanded to C, E_1, G_1. Reforming prices to P_g^* and P_c^* will therefore no longer achieve market clearing.

One sustainable sequence of reform is to increase prices to P_g^{**} and P_c^{**} in the present period, and lower them to P_c^* and P_g^* once expectations of goods market clearing have been reestablished. This would involve a reduction in the nominal price of crops, which would be offset by a greater proportionate reduction in the price of goods. Political considerations may make this suboptimal.

An alternative which avoids a subsequent reduction in crop prices involves setting prices in the present period at P_c^* and P_g^{***}. Goods prices would be lowered to P_g^* subsequently. In addition to eliminating the need to reduce crop prices, in this sequence goods prices do not have to overshoot as much because P_g^{***} is unambiguously lower than P_g^{**}. The lower price of goods does not cause peasants to make a larger claim on them, because the effect of the lower nominal price is offset by a higher relative price of goods to crops. (P_g^{***}/P_c^* is greater than P_g^{**}/P_c^{**}.)

Pricing policies in general have two drawbacks. First, they must temporarily overshoot their long run equilibrium levels. Once supply has contracted in response to shortages of goods, merely restoring prices to P_c^* and P_g^* is not sufficient. Second, pricing policies place a severe information burden on the government, which has no way of knowing price levels such as P_c^{**} and P_g^{***}. The government must therefore remove the controls and let prices find their own levels.

The increase in prices, once decontrolled, has three components. The first is the reversion to counterfactual prices. In our simple characterization in figure 29–4, controls depress prices below the original market clearing price, P_g*. The second is that, to the extent that the government has benefited from an increase in the demand for real money balances arising from stochastic rationing, it now reneges on the implied liabilities through an inflation tax. The new market clearing price level is thus permanently higher than the old one. The third is that, until peasants come to expect market clearing (to which they will respond with extra crop sales), goods supply is temporarily low and so goods prices will overshoot their long run level. The cumulative consequence of these three effects perhaps helps to explain why governments are so wary of price decontrol. However, since the controls cause the economy to implode, delay merely makes the problem more acute.

Improvements in consumer good supplies. The most common application of the fixed price macroeconomic theory we have deployed is the analysis of the Keynesian zone, in which the goods market clearing locus is shifted by the stimulation of demand. In its application to repressed inflation, the analogous policy is again the shifting of the goods market clearing locus. This is brought about by means of foreign aid (or an own funded imports scheme) to augment the supply of goods. The increase in goods supply reverses the endogenous implosion of the economy, and reduces the size of the repressed inflation zone.

Aid alone cannot achieve a sustainable equilibrium unless it is permanent. However, in conjunction with price changes, temporary aid can ease the path to equilibrium. The case for temporary aid is to avoid the setting of prices above the long run equilibrium level, which is otherwise likely to be necessary. The ideal deployment of aid would be to enable the economy to go straight to the long run equilibrium prices, P_c* and P_g*. For this to be sustainable, aid must shift the goods market locus back to its original position (i.e. that predicated upon expectations of market clearing). An aid policy of this magnitude is thus compatible with setting prices at P_c* and P_g* (figure 29–4).

Effects of rational expectations. So far we have considered only adaptive expectations as a characterization of how agents interpret the unfolding evidence of the implosion phase. However, were agents to have rational expectations of this process, the implosion would be more severe. Rational expectations are more likely to characterize a situation of large increases in foreign aid or the introduction of an own funded imports scheme. Both can be presented as major public events which constitute a regime change of which agents are quite likely to be aware. Since governments typically control information channels, it may be an option for the government to preannounce such events. We therefore consider the consequences of three distinct informational states.

In the first case, agents are unaware throughout that foreign aid is increasing goods supply, and they form their expectations adaptively on the basis of their individual experiences of goods availability. There may initially be no effect on crop sales. Sooner or later, however, individuals will recognize that there has been an improvement in the availability of goods. This has a positive effect on sales which is, if individuals believe the change to be permanent, mitigated by the downward adjustment of money balances.

In the second case, agents are informed of the aid only at the start of the period in which the goods supply is increased by aid. Agents will respond by increasing sales (provided that they correctly estimate the effect of the aid program on rural availability of consumer goods). This will cause the *GG* locus to shift inward. Compared to the first case full liberalization can be achieved in a shorter period, or with less aid (or a combination of the two).

In the third case, agents are informed one period in advance. Our analysis of this third case is somewhat analogous to the case of an anticipated change in monetary policy, which is analyzed by Neary and Stiglitz (1983).[1] This presupposes some form of rational expectations; otherwise the announcement of the availability of aid would not need to be true to be effective. Under rational expectations, the agent realizes that an individual increase in effort makes no sense unless extra aid is indeed available.

Black Markets Introduced

When governments attempt to control prices at below market clearing levels, illegal trade will generally spring up alongside official transactions. The extent and consequences of such trade depend on how rigorously the controls are enforced, and the origin of the goods being sold on the black market. There are three reasons why the black market can, in effect, be ignored.

First, if trading is illegal, its character is radically different from that of legal trade. Bevan et al (1989, 1990a) show how the rational maximizing black marketeer will set a price below that at which the market would clear. By setting a low price the trader achieves a quick sale and thus reduces the risk of detection. The aggregate consequences of this behavior are that many consumers will be rationed on the black market as well as on the official market. In the next section we will show that this was indeed the case in Tanzania. The black market then operates like a second (higher) tier of controlled prices.

Second, because black market prices are commonly far higher than official prices, many households will not make black market purchases. Such households are rationed on official markets and priced out of black markets. For such households the black market is irrelevant.

Third, if smuggling is negligible, the supplies of black market goods are likely to come from a diversion of supplies intended for the official market. It might therefore appear that the black market does not affect aggregate supply. However, the black marketeers' consumption out of their income must be considered. The typical black marketeer is someone with privileged access to goods at official prices, whose income per sale is the difference between the official and the black market price. Although black marketeers may trade among themselves, in aggregate they are able to buy for their own consumption at official prices. Hence, consumption by black marketeers can become very sizeable. Unless smuggling offsets this effect, the existence of a black market actually reduces the supply of goods to peasants. Hence, crop sales are not altered by the size or price premium of a black market as long as its source of supply is diversion from official supplies.

Figure 29-5. The Crop Supply Path, 1977-88

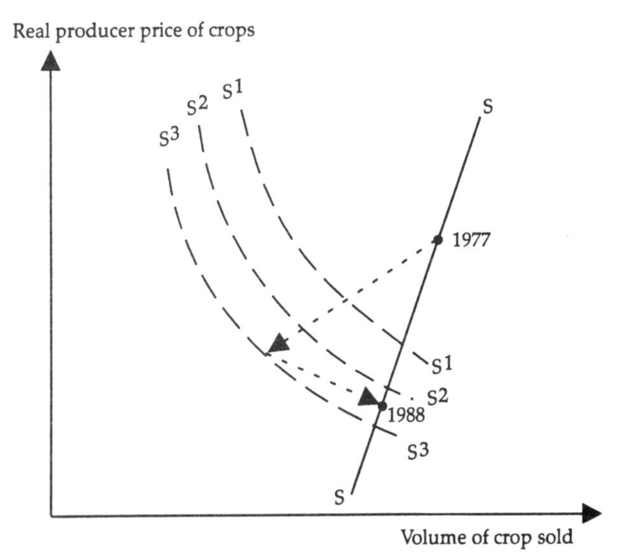

Application to Tanzania

Between the late 1970s and 1984, the Tanzanian economy suffered a remarkable implosion. Peasant living standards fell sharply (Bevan et al 1988) and crop sales declined (Bevan et al 1989). In 1984 the government introduced two radical reforms: consumer prices were decontrolled, and own funded imports were permitted (i.e. those with illegally acquired foreign currency were allowed to buy imports without punishment). In this section the theory of rationing and the transition to market clearing is applied to reform in Tanzania. We look at changes in crop sales related to peasants' need to finance expenditure and the demand for money. The impact of liberalization on consumer choice in Tanzania is then analyzed.

Changes in Crop Sales

Figure 29–5 shows the supply curve for marketed crops for the period from 1977 to 1988. Compared to 1977, in 1988 the crop price was lower in real terms and the rural population's crop supply per capita was lower. Hence, in the first instance, we can think of the economy as having moved down its supply curve. However, the supply changes during the period from 1977 to 1988 have not been movements down the supply curve, but rather movements off it and then back onto it as depicted by the dotted line.[2]

The pre-reform decline. Bevan et al (1987, 1987a, and 1989) argue that during the period from 1979 to 1984 there was a movement off the supply curve because of consumer goods shortages in rural markets. These shortages came about for three reasons. First, the national supply of consumer goods declined. The allocation of foreign exchange for consumer imports and the domestic production of manufactured consumer goods both fell severely. Second, Tanzania's cost-plus price control system did not permit prices to increase in response to the shortages. Third, these shortages were particularly acute in rural areas. Official allocations of consumer goods were increasingly skewed towards the cities. For example, in 1977 the Board of Internal Trade allocated only 22 percent of Matsushita radios to Dar es Salaam. By 1982, when total supply of the radios was less than half of its 1977 level, the allocation to Dar es Salaam was raised to around 40 percent (see Bevan et al 1989, table 3!6). Of the goods which left Dar es Salaam, a substantial proportion leaked onto the parallel market, however, there is evidence that these supplies were also skewed towards urban centres. For example, of the supply of soap to towns other than Dar es Salaam around 40% was probably via the parallel market, compared with only around 30% in villages. Such a bias in the parallel market is consistent with evidence that it was easier to operate as a black marketeer in urban than in rural areas (see Bevan et al 1989a).

The Board of Internal Trade data suggest that the official supplies diverted to the black market[3] remained largely in urban areas. Bevan et al (1990) found that in 1983 some three quarters of rural households in Tanzania did buy in black markets. However, the majority of these responded that they were unable to buy in the black market whenever they wanted to do so (at the black market price). Thus, these households were rationed in the official market and in the black market.

Prior to the shortages in the consumer goods market, peasant sales of crops were based upon the expectation that the resulting income could all be spent on desired purchases. As a result of shortages of consumer goods the supply curve has become negatively sloped (S_1) and, as shortages have intensified, it has shifted to the left (S_2, S_3).

Between 1977 and 1984 the supply of export crops per capita fell by 30 percent and the real producer price of export crops fell by 26 percent. According to our theory, without the fall in crop prices, the decline in crop sales would have been even larger. Econometric tests of supply response in 17 regions during the period from 1978 to 1984 have found the volume of crop sales to be positively and significantly related to the supply of consumer goods to rural areas and negatively (though insignificantly) related to real crop prices. This supports the above account (see Bevan et al 1989, Chapter 10).

The 1984 to 1988 period. As shown in table 29–1, since the 1984 reforms there appears to have been a substantial increase in the volume of crops sold. Between 1977 and 1985, the volume of official sales of export crops per capita of the rural population declined by an annual average of five percent; between 1985 and 1988 it expanded by an annual average of three percent. The recent growth in the volume of food sales has been far more dramatic, particularly through nonofficial channels. Total official and nonofficial food sales per capita of the rural population increased by 79 percent between 1983/84 and 1987/88. Overall, the volume of crop sales per capita of the rural population expanded at an annual rate of 10 percent during this four year period.

Table 29–1 shows that nominal crop prices increased dramatically during 1982/83 to 1987/88. To derive an estimate of real changes in crop prices requires the construction of a price index of goods bought by peasants from urban areas. However, in Tanzania there is no cost of living index specifically for peasants. There is only the national Consumer Price Index (CPI), which for our purposes is inappropriate for two reasons. First, although the typical peasant household has a consumption pattern similar to the national average, it has a radically different expenditure pattern because of subsistence consumption. Further, the peasant economy in aggregate has an even more different pattern of net purchases because most of the food which the typical peasant household buys is purchased directly or indirectly from other peasant households. Since food has a weight of 64 percent in the CPI, these differences are important. The CPI is dominated by changes in food prices, yet the peasant economy in aggregate does not purchase food.

This is particularly important when measuring the peasant terms of trade with the rest of the economy. The price of goods which peasants sell to the rest of the economy should be deflated by the price of the goods which they buy from it, rather than by the CPI. We have therefore constructed a peasant specific CPI based on the prices of those goods which peasants purchase from the rest of the economy.[4] The resulting series is shown in the second column of table 29–2. We use the series to deflate the nominal price series, assuming that crop income earned in 1982/83 is spent in calendar year 1983, etc. The result is that crop prices did not increase in real terms during 1982/83 to 1987/88 (see table 29–1). In fact they declined substantially: the index fell from 125 in 1984 to 87 in 1988. Hence, the increase in the volume of crop sales is not a conventional supply response. Prices declined by 30 percent in real terms in the period of the reforms. Based on our analysis of the pre-reform decline, we argue that the decline was caused by a deterioration in the availability of consumer goods in rural areas.

Changes in the Demand for Money

In addition to financing expenditure, peasants generate income to add to their cash balances. Cash balances are both the medium of exchange and the principal financial asset available to peasants. Should desired cash balances fall, the peasant can reduce income (i.e. crop sales) and maintain expenditure while money balances are being run down. Here we distinguish between real and nominal changes in money balances.

Table 29–1. *Crop sales: prices in quantities, 1982/83 to 1987/88*
(indices, 1982/1983 = 100)

Quantities	*1982/83*	*1983/84*	*1984/85*	*1985/86*	*1986/87*	*1987/88*
Grains	100	88	107	160	160	183
Export crops	100	101	102	90	119	121
All crops	100	94	105	15	139	151
Prices						
Grains (nominal)	100	193	314	249	252	326
Export crops (nominal)	100	132	174	254	325	416
All crops (nominal)	100	163	208	738	283	376
All crops (real)	100	125	110	92	83	87

Source: Bevan et al (1990a)

Real cash balances. The demand for real money balances is normally related to real expenditure and hence to real income. Between 1977 and 1983 peasant real cash income per capita fell substantially. Using a comparison of budget surveys, Bevan et al (1988, table 2–14) find a fall of 39 percent. The expectation is therefore that real cash balances per capita would also have fallen. This would have enabled peasants temporarily to reduce their cash income by more than the fall in their expenditure. The magnitude of such an effect depends upon the size of peasant cash holdings relative to income. Peasant cash holding was around 75 percent of cash income.[5] Hence, had peasants reduced their money holdings so as to restore this ratio of money to income, the 38.7 percent fall in income would have permitted the equivalent of a one-year reduction in cash income of 29 percent. Had this reduction in money holdings been completed by 1983, it would imply that over the six year period from 1977 to 1983, cash income would have been reduced by 5 percent. Hence, crop sales, as an important part of cash income, would also probably have been reduced by around this amount.

In fact, however, despite the large fall in real cash income, peasants appear to have increased their real holdings of money balances. Although there is no direct information on peasant holdings of cash balances, the mean cash holding of the Tanzanian household is a reasonable proxy. The latter is known from the total cash in circulation (from the Central Bank) and the number of households (from the Census). Since peasants make up 80 percent of the population, it is likely that their holdings are closely related to the Tanzanian average. Although peasants have lower cash incomes than the average household (and so would tend to hold less money), they receive income much less frequently and less predictably than wage earners, the other major group in the economy. This implies that, per shilling of income, they would tend to hold considerably more cash than wage earners. These two differences from the average Tanzanian household are therefore qualitatively offsetting. The series on real cash balances per capita is shown in table 29–3.

Between 1977 and 1981 there was a 21.2 percent increase in mean real cash balances. If cash income declined by around 39 percent during this period, as implied by the survey evidence, then the velocity of circulation halved in the peasant economy.[6] Our theory explains why, with the onset of rationing, it is possible for money balances to rise even though expenditure falls. We have also seen that as shortages intensify money holdings must eventually decline. The evidence presented in table 29–3 is consistent with this theory. Starting from market clearing in 1977, real balances increased as expenditure declined until 1981. Thereafter, there was a substantial decline. By 1983 real cash balances were slightly below their 1977 level. Hence, until 1981 crop supply was boosted by the change in real money balances. After 1981 real cash balances were reduced; this helps to account for the severe decline in crop sales during the early 1980s. The analysis implies that this effect

Table 29–2. *Peasant real crop income*
(indices, 1983 = 100)

Capita	*Nominal crop sales*	*Rural purchases from urban areas*	*Real income from crop sales per*
1983	100	100	100
1984	152	131	114
1985	257	189	130
1986	314	159	114
1987	398	340	107
1988	547	432	114

Source: World Bank data

would have continued as goods availability deteriorated.

We have noted that the effect of shortages on real money demand poses a potential problem for the transition back to market clearing. By 1986 goods were much more available in rural areas so that peasants could reduce money holdings per shilling of expenditure. As table 29–3 shows, real balances fell by around a third between 1983 and 1986, and have since been stable. The adjustment back onto the 1977 ratio of money to expenditure has been achieved partly by increased expenditure, but mainly by reducing real money holdings. This rapid reduction in real money holdings might have implied that cash income, and hence crop supplies, would temporarily be reduced. This was, however, avoided by offsetting changes in the price level.

The price level. The demand for money can rise for two reasons: either the desired holdings of real balances can increase (as it did during 1977 to 1981), or the price level can rise. The latter would require households to add to their money balances in order to keep them constant in real terms. In Tanzania the inflation tax was initially modest (see table 29–3). However, during 1980 to 1982 it became very powerful. This was because the rate of inflation increased and interacted with the honeymoon effect. Since peasants had increased their real money balances, a given rate of inflation was eroding a larger stock of money. Thereafter, as peasants started to reduce real balances, the yield from the inflation tax fell despite a generally rising inflation rate.

The net effect of the inflation tax and the change in the demand for real balances is shown in table 29–3. It is expressed as the addition of the two effects and as a percentage of cash income in each year. The latter is of interest because the income peasants need to devote to adding to money balances instead of to expenditure is analogous to a tax on cash income. Therefore it is an implicit income tax, which the government was levying on peasants and achieved by the changes in money holdings.

This analysis reveals substantial swings in the rate of implicit taxation. At the start of the period the rate was modest; it was 5 percent in 1977 and 8 percent in 1978. It then rose sharply to an average of 16 percent during 1979 to 1982. During this phase there was a change in the composition of the tax from the 'honeymoon effect.' In 1983 and 1984 there was a large fall to an average of 7 percent. During 1986 to 1988 the rate increased to an average of 12 percent. Despite the low level of real balances, the inflation rate was sufficiently high to restore the implicit tax rate to close to its previous peak. These large swings must have contributed to the changes in crop supply. The sharp fall in the tax rate during 1983 to 1984 helps to explain the severe economic collapse of those years; the restoration of tax rates by means of inflation explains why the economy was able to overcome the otherwise major transition problem posed by the decrease in real balances.

During the phase of decline in crop sales, 1977 to 1984, the volume of crop sales per capita fell by around 30 percent. We have argued that it is incorrect to interpret this decline as being caused by the fall in the real producer price of crops. A better explanation is provided by the special circumstances of shortages of consumer goods which prevailed during this period. During the four year period from 1983/84 to 1987/88, there was a spectacular recovery of 47 percent in the volume of crop sales per capita of the rural population. Yet the average crop price fell in real terms by 30 percent during the same period. Clearly, the remarkable growth in crop sales which reversed a long period of fairly continuous decline was not a movement up a normal supply curve. Between 1983/84 and 1987/88, the rural economy moved back to market clearing conditions. Hence, it reverted to something approaching the original supply curve (S-S in figure 29–1). As shown in table 29–4, this was achieved partly by larger imports of consumer goods, financed mainly by own-funded imports. The abandonment of price controls mainly raised the price level. The rise in the price level eroded real money balances, and lowered real crop prices (being only partly offset by nominal increases in producer prices).

Table 29–3. ***Improvements in consumer choice, 1983 to 1988***

	Availability of resources (1)	*Availablity of goods (2)*
1983	0.563	4.8
1984	—	4.7
1985	—	6.7
1986	2.544	6.1
1987	—	7.3
1988	3.825	7.4

Note: 1. Refers to how many varities of soap were available on a typical shopping trip. Data for soap varities is available only for 1983, 1986, and 1988. 2. Refers to eight goods: cooking oil, margerine, toilet soap, laundry soap, matches, khanga, kerosene, and cigarettes. Figure refers to how many of these eight goods were found to be available on the average 'shopping trip' by an enumerator of the Bureau of Statistics. The number is the unweighted average of 80 shopping trips per year; one per quarter for each of twenty regions.

Source: Derived from unpublished data gathered by the Bureau of Statistics for the construction of the CPI.

Hence our explanation for the paradoxical increase in output is that it reflects the return to market clearing. The abolition of price control made part of the money balances held by peasants unnecessary. This monetary overhang would have led to a reduction in output, as peasants would run down their excess money balances by producing less. However, their incentive to do this was offset by the rise in the inflation tax. The combination of the return to market clearing and price increases explains the paradox that output increased while producer prices fell in real terms.

Liberalization and Consumer Choice

We now turn to a final aspect of this liberalization: its effect on the range of choice of available goods.

The number of varities of a good. Price deflators are likely to mis-state the true change in the cost of living because they do not take into account changes in the range of choice available for the typical differentiated product. For example, even soap, a relatively standardized product, is differentiated by scent, quality and function (laundry or toilet). Between 1978 and 1984 the range of choice in Tanzania narrowed as part of the general contraction in supplies of consumer goods; since 1984 it has widened again. The range of choice affects the cost of living in two ways. First, since consumers have different requirements, the wider the range of choice the closer are the types of good available to what different people ideally want to buy. If people who want laundry soap have to make do with toilet soap, the true cost of washing clothes is increased. Second, for many differentiated products, the typical consumer wishes to buy more than one variety. People will wish to buy both laundry soap and toilet soap.

We now attempt to quantify the effects on the cost of living of the post-1984 widening of product choice. We use the theoretical framework developed by Dixit and Stiglitz (1977). They show how for a given total expenditure a consumer will get more welfare as variety is increased. In order to achieve a given level of welfare, a consumer will need to spend less as variety is increased. Thus, increased choice lowers the cost of living. The magnitude of this effect depends upon how closely one variety of a product can substitute for another, which is measured by the elasticity of substitution. When the consumer is completely indifferent between varieties, the elasticity of substitution is infinite and greater variety would not lower the cost of living.

For example, if the number of varieties doubles and the elasticity of substitution is 2.5 (the sort of magnitude that might be expected), then the implied increase in welfare is 32 percent.[7] This is equivalent to a reduction in the cost of living of 24 percent if the number of varieties is constant (1/(1-0.24) =1.32).

To apply this theory to Tanzania we need data on the range of product choice available to the typical consumer. For this we have adapted the underlying data gathered by the Bureau of Statistics for the Consumer Price Index. Our analysis is confined to the range of choice among different varieties of soap, but this is the single most important differentiated product in the expenditure pattern of peasants according to the 1977 Household Budget Survey. The Bureau samples 16 varieties of soap in twenty urban centres each quarter. Hence, each year there are 80 shopping visits. When a particular variety is not available, this is recorded by the statistical officer. Using this information, we have calculated for 1983, 1986 and 1988 the average number of varieties of soap found during these 80 visits. The results are shown in table 29–4. This example suggests that wider choice has substantially raised living standards. If the soap data are representative (and while for other goods the increase in the number of varieties may not have been quite as dramatic, for many goods other than soap one would expect the elasticity of substitution between varieties to be lower) the change between 1983 and 1988 would have raised welfare by more than 20 percent even if one adopted as high a value for the substitution elasticity as 10. Were the substitution elasticity 5 (still a rather high value) the improvement would be 47 percent. This would be equivalent to a 32 percent fall in the cost of living.

The number of different goods available. Choice can also be increased by increasing the number of goods. In many ways the number of different goods available is more important than the number of varieties of a single good. This is because the scope for substitution, say, between soap and matches is much lower than that between different varieties of soap, and so nonavailability has more deleterious effects upon living standards.

We attempt to quantify the widening in the choice of goods using the same Bureau of Statistics data on shopping visits. We consider eight important consumer goods (each usually represented by several varieties). A good is available if at least one of its varieties is available. The results in table 29–4 reveal a marked improvement in the availability of the range of goods. Between 1983 and 1988 the number of different goods available increased by 54 percent. The quarterly data reveal that availability deteriorated sharply between the first and third quarters of 1983 (from 6.1 to 3.8) and continued at this low level until the second quarter of 1984. From then on, i.e. cotemporaneous with the reforms which could be expected to have reduced shortages, there was a rapid and fairly continuous improvement. In the case of the increase in the number of varieties we were able to indicate the order of magnitude of the change in living standards and in the cost of living which it implied. To do the same thing for goods rather than varieties is more problematic since there are more likely to be large differences between goods in their substitutability. Ideally, we would need to estimate a demand system. Since this is not feasible on the data we can only follow the same procedure as that used for varieties. Again the magnitude of the effect is dependent upon the substitution elasticity chosen. Here we choose a value of 2.5. The improvement in availability between 1983 and 1988 (54 percent) would then imply a welfare improvement for given expenditure of 19 percent, or equivalently stated, a 16 percent reduction in the cost of living.

The effects which we have discussed above are cumulative. The wider choice of varieties may (using an elasticity of 5) have lowered the cost of living by 32 percent. The wider choice of goods may have further lowered the cost of living by 16 percent. Hence, together, they imply a fall in the cost of living of 43 percent (1 - (1-.32)(1-.16)). For a given level of cash expenditure, this implies an increase in the welfare derived from it of 75 percent. This is in addition to the more conventional changes in living standards which we will be analysing below, namely increases in cash income and increases in subsistence consumption. The figure of 75 percent is little more than a numerical illustration. However, it does suggest that there have been substantial gains from wider choice, which are ignored in the usual statistical standard of living analysis. Once we take into account improved availability producer prices would have fallen less than the 30 percent we estimated earlier. Indeed, if the 43 percent estimate in our example is correct, then producer prices would have remained constant rather than fallen in real terms (1.43 x 0.7 = 1.001).

Conclusion

Our theoretical analysis of the case of generalized price controls has disturbing policy implications. Conventional static analysis ceases to apply. Producer prices may be higher (in nominal or in real terms) than in a market clearing equilibrium, in spite of excess demand. Supply response is likely to be perverse as long as agents expect shortages to persist. Relying on changes in controlled prices in the transition to market clearing is problematic not only because of the informational requirements of such a policy, but also because the overshooting which will be involved may well confuse agents. In addition the transition can be frustrated by a monetary overhang. We have suggested that in these circumstances there is a case for aid.

Our analysis of Tanzania suggests that the substantial increase in agricultural output since 1984 was not due to producer price increases, but to the return to market clearing. Indeed, without taking into account that peasants were rationed prior to 1984, it is difficult to explain why output increased in the period from 1984 to 1988 since crop prices fell dramatically in real terms.

Notes

1. The policy presupposes that the government is credible, a condition which may not be satisfied in all countries for which the analysis is relevant.

2. There is a proviso to this. It is not safe to conclude that the economy can move back up the original supply curve simply by restoring prices. During the intervening years agricultural infrastructure has deteriorated so that it would probably be unable to handle sales at their 1977 level. This has, in effect, steepened the supply curve.

3. In Tanzania people usually speak of parallel markets. That term however is often used in other countries to denote legal transactions. In Tanzania parallel market activites were illegal and price control was enforced with considerable sanctions.

4. From the 1976/77 Household Budget Survey the Bureau of Statistics provided unpublished data on the consumption pattern of rural households. From this we identified goods supplied from outside the peasant economy (such as soap and bicycles). Virtually all of these goods were included in the price gathering process for the construction of the CPI. Hence, by going back to the underlying price data it was possible to construct a price index for rural purchases from the rest of the economy on 1977 weights. This was undertaken for the period 1983-88.

5. Mean cash income in 1980 per peasant household is known from a large sample rural survey (Collier, Radwan and Wangwe (1986)). Mean cash holdings from Central Bank and Census data.

6. Velocity of circulation (0.613/1.212) = 0.506).

7. In the Dixit-Stiglitz approach different varieties are modelled as constant elasticity of substitution substitutes

in demand. If different varieties are treated symmetrically, the consumer will choose to buy the same quantity of all available varieties. If availability improves in the sense that n, the number of available varieties, increases to n′ but that aggregate availability is unchanged (the consumer buys less of each variety, but his total expenditure remains constant when is the substitution elasticity), then the welfare change is given by

$$u'/u = (n'/n)^{1-\delta}$$

Hence welfare is increasing in the number of varieties.

References

Bank of Tanzania (1988), Economic Bulletin, Vol. XVIII no. 3.

Berthélemy, J.C. and Morrisson, C. (1987), "Manufactured Goods Supply and Cash Crops in Sub-Saharan Africa", World Development, Vol. 15, no. 10/11, pp. 353-1367.

---(1987a), East African Lessons on Economic Liberalization, Thames Essay, No. 48, Aldershot: Gower, for the Trade Policy Research Centre.

----(1988), "Incomes in the United Republic of Tanzania during the 'Nyerere Experiment'", in W. van Ginneken (ed) Trends in Employment and Labour Incomes, I.L.O. Geneva.

----Bevan, D.L., Collier, P. and Gunning, J.W., (1989(, Peasants and Governments: An Economic Analysis, Oxford: Oxford University Press.

----(1989a), "Black Markets: Information, Illegality and Rents", World Development, Vol. 17, no. 12, pp. 1955-63.

----(1990), Controlled Open Economies: a Neoclassical Approach to Structuralism, Oxford: Oxford University Press.

----(1990a), Price Controls and the Transition to Market Clearing: Theory and an Application to Tanzania, I.E.S. Applied Economics Discussion Paper No. 94

Collier, P., Radwan, S. and Wangwe, S. (1986); Labour and Poverty in Rural Tanzania; Oxford: Oxford University Press.

Dixit, A. and Stiglitz, J.E. (1977); "Monopolistic Competition and Optimum Product Diversity"; American Economic Review, Vol. 67, pp. 297-308.

Malinvaud, E. (1977); The Theory of Unemployment Reconsidered; Blackwell; Oxford.

Neary, J.P. and Stiglitz, J.E. (1983); "Towards a Reconstruction of Keynesian Economics: Expectations and Constrained Equilibria", Quarterly Journal of Economics, Vol. 98; Supplement; pp. 199-228.

30

The Prospects for an Outward Looking Industrialization Strategy Under Adjustment in Sub-Saharan Africa

S. Olofin

Sub-Saharan Africa continues to be in need of a viable and feasible long term industrialization strategy.[1] The inward looking industrialization strategy which many African countries have hitherto pursued has been a dismal failure. Its inherent inefficiencies have been well documented in Little et al (1970), Bhagwati and Krueger (1973), Krueger (1978), and Balassa (1978), among others. This paper considers Sub-Saharan African countries' prospects in looking forward to an Asian type of export oriented industrialization strategy.

Export oriented industrialization strategy is sometimes referred to as the G-4 model; its foremost practitioners are the so-called East Asian 'gang of four'—Hong Kong, South Korea, Singapore, and Taiwan. Most countries in Sub-Saharan Africa have begun to adopt some type of export oriented industrialization strategy in the context of World Bank and International Monetary Fund assisted structural adjustment programs. As of 1988 at least 30 countries in Sub-Saharan Africa had embarked upon structural adjustment programs (ECA 1989).

Few studies have been undertaken to analyze the implications of structural adjustment programs for industrialization efforts in Sub-Saharan Africa. The Asian newly industrialized countries, particularly the G-4 countries, have had great success with the export oriented industrialization strategy. In the structural adjustment programs being proposed to and adopted by several African countries, no explicit reference is made to export oriented industrialization strategy. However, since the 1970s the World Bank and the International Monetary Fund have favored this approached over inward looking industrialization strategy and have urged its emulation by other less developed countries (LDCs) (World Bank 1979, Hughes 1980, Benerji and Reidel 1980, Fei et al 1979). Perhaps what has come to be known as the Berg report (World Bank 1981) represents the World Bank's definitive statement regarding desirable economic policies for Sub-Saharan Africa.

The question as to whether the G-4 model can be replicated in other LDCs is not a new one. Numerous studies have been undertaken to assess the feasibility of its adoption by other LDCs. The fallacy of composition argument by Cline (1982) states that if a large number of LDCs were to simultaneously pursue an export oriented industrialization strategy, it would lead to an outpouring of manufactured exports. This in turn would generate a protectionist response from the industrialized countries. Hughes and Waelbroeck (1981) believe that the degree of accessibility to OECD markets is still such that generalization of the G-4 model by LDCs is feasible, even if progress would be slow. Studies by Schmitz (1984) and Evans and Alizadeh (1984) represent the dependency theory perspective. They argue that generalization of the G-4 model under neoclassical prescriptions may not be feasible, especially given the nonreproducible nature of the external circumstances which favored the G-4 countries during the 1970s.

The first major issue addressed in this paper is the extent to which the export oriented industrialization strategy—the G-4 model—may be expected to work in individual African countries. The fallacy of composition factor is discounted; focus is placed on the applicability of the model within the context of individual countries rather than all African countries as a group. It is assumed that some of the internal deficiencies and constraints facing individual African countries can be overcome. If several countries would embark on the strategy simultaneously, as was the case in the erstwhile import substitution strategy, the countries would be unlikely to achieve the same level of intensity of exports as is assumed in a typical G-4 model. However, the rate of progress would be likely to vary across countries. This would rule out the fallacy of composition factor, which relies strongly on the assumption of a uniform G-4 regime across countries (Cline 1982). It is assumed that no individual country would be able to influence the supply of manufactured goods so as to instigate protective barriers against its exports. For individual African countries, the constraints on international market demand would not be a major limiting factor.

The viability of an export oriented industrialization strategy in individual African countries is analyzed. It is assumed that international demand constraints and absorptive capacity in industrialized countries do not constitute major barriers.[2] Supply and demand potentials for successful export-led industrialization strategy are distinguished. Greater emphasis in this paper is placed on the former. It is assumed that, for the Sub-Saharan Africa region, ceteris paribus, intraindustry, intraregional, and South-South trade in manufactured exports could substitute for interindustry North-South trade. In an intraregional and South-South trade context, market access would not be likely to constitute a major problem. It is hypothesized that Sub-Saharan African countries' collective terms of trade would improve as intraregional and South-South trade grows.

The second major issue the paper seeks to address is the serious nature of the limiting internal constraints which have contributed to the ineffectiveness of inward looking industrialization strategy in Sub-Saharan Africa. The doctrinaire emphasis of a modified version of export oriented industrialization strategy would appear to be neglecting these internal constraints. This is partly due to ideological biases and partly due to preoccupation with pressing short term problems. The third major issue addressed is an alternative strategy that would focus on minimizing and possibly eliminating the effects of these internal constraints.

The paper is divided into six sections. Following this introduction, some theoretical issues are examined in the second section. There is a brief review of past inward looking industrialization strategies in the third section. In the fourth section, a quick review and evaluation of export oriented industrialization strategy is undertaken and related to the African context. This is followed in the fifth section by a broad outline of a rural oriented smallholder industrialization strategy. Possibly an interim alternative industrialization strategy, it would aim at dealing with some of the existing internal constraints. And it would possibly pave the way for adopting a sustainable long term industrialization strategy. A short conclusion in the final section rounds out the paper.

Some Theoretical Considerations

It would seem that the theoretical arguments of the orthodox general equilibrium school and the unorthodox disequilibrium approach are not directly relevant to the issues addressed in this paper. The approach being adopted by the World Bank and the International Monetary Fund in designing ongoing adjustment programs seems to have little to do with extreme doctrinaire considerations. It would appear that the underlying considerations are more pragmatic. However, the pragmatic approach is linked to some underlying theory. The neoclassical arguments surrounding free markets, trade liberalization, price incentives, and other issues provide the springboard for the Asian newly industrialized countries' type of export oriented industrialization strategy. These and the dependency theorists' counter arguments should be used in the search for an appropriate development strategy for Sub-Saharan Africa.

One of the central issues which needs to be addressed from a theoretical point of view relates to the role of the state vis-a-vis the invisible hand in the pursuit of standard neoclassical results. Some of the existing empirical studies, for example Chow (1987), tend to tilt the balance overwhelmingly in favor of standard neoclassical prescriptions. They attribute positive results to assigning a greater role to the invisible hand. The arguments and counterarguments regarding the role of the state are ideological. They tend to make little or no room for examining the realities in each country in relation to the minimum necessary and sufficient conditions for obtaining highly desirable standard neoclassical results.

However, such necessary and sufficient conditions do not exist in the majority of Sub-Saharan African countries. Too much focus on ideological biases clouds the real issues. The choice between market forces and state intervention may be purely secondary in light of the empirical experience of the region. Côte d'Ivoire, Malawi, and Senegal have followed market oriented strategies as opposed to the state interventionist strategies followed in Guinea, Ghana, Sudan, Zambia, and Ethiopia during the 1960s and 1970s. Both groups of countries experienced similar results—little or no industrialization, heavy indebtedness, and other characteristics of the underdevelopment trap. The evidence tends to suggest that across ideological lines the problems in the 1960s and 1970s had more to do with the wrong strategy of inward looking industrialization (Acharya 1981).

This paper focuses on two primary theoretical and empirical questions in relation to the challenges of underdevelopment in Sub-Saharan Africa. The first is how the necessary and sufficient internal conditions for deriving standard neoclassical results (as evidenced in the Asian newly industrialized countries) may be created in Sub-Saharan Africa. This may require some measure of strong state intervention. The second is how such considerations would go beyond the static efficiency concerns of neoclassical prescriptions, to dynamic efficiency considerations. To achieve this, social costs and profitability criteria would sometimes have to override private costs and benefit criteria.

Past Inward Looking Industrialization Strategy

From the late 1950s to the early 1980s, the industrialization programs that were adopted in Sub-Saharan Africa can all be characterized as inward looking. As it was in other LDCs, this approach to development was inspired by the mainstream theories of the 1950s and 1960s—such as in UN (1951), Singer (1950), Nurkse (1953, 1959), Scitovsky (1954), Leibenstein (1957), and Rostow (1960), among others. These theories advocated strategies which included the notion of the need for a big push, a productive role for the state, and rapid industrialization through inward looking import substitution.

Most of the Sub-Saharan Africa countries, regardless of their ideological leanings, have continued their preindependence pattern of relative neglect of the supply oriented productive sectors of the economy. They favor a weak export policy oriented towards the exports of a few agricultural cash crops. Finished consumer manufactures and raw materials for import substituting light manufactures are imported. Those countries with a mineral resource base —such as Zaire, Zambia, Liberia, and Nigeria—have tended to neglect agriculture, to take advantage of readily accessible rents and foreign exchange earnings, and to pursue similar inward looking, import substituting strategy. With the onset of declining commodity prices and terms of trade in general, this inward looking strategy has run into serious difficulties in the agricultural primary product exporting countries and in the rent earning countries. It has resulted in the unprecedented debt burden that is threatening to wipe out whatever modest growth gains may have been realized in the two or more decades of pursuing inward looking industrialization strategy.

In essence, inward looking industrialization development strategy used trade restrictions in the form of tariffs, quotas, and multiple exchange rates to promote industrialization and correct balance of payments difficulties. The model and the results derived from its implementation have been subject to much scrutiny. The overwhelming conclusion has been one of disenchantment, even by some of its early advocates (Prebisch 1964, Hirschman 1968, Briton 1970, Baer 1972, Sutcliffe 1971, and Diaz-Alejandro 1975).

The main arguments against inward looking industrialization strategy focus on the damaging effects of too much protectionism. This has invariably led to inefficient allocation of resources due to distortions in factor and product markets. Criticism is also made of this strategy's bias against exports and agriculture and its import intensity, which has often exacerbated the balance of payments pressure it was intended to relieve. There has been widespread disillusionment with inward looking industrialization policy across ideological and analytical divides, especially in its static efficiency aspects.

Most of the Sub-Saharan African countries, along with other debt ridden LDCs, have had no choice but to adopt World Bank and International Monetary Fund assisted structural adjustment programs. Many of these countries have balance of payments problems partly caused by the negative effects of inward looking industrialization strategy, and partly by domestic mismanagement of resources and unfavorable exogenous factors, such as declining terms of trade.

While the empirical findings on inward looking industrialization strategy are generally accepted, their orthodox interpretations, policy conclusions, and recommendations are not necessarily widely shared. Proponents of the dependency theory school argue for increased state intervention (Merhav 1969,

Sunkel 1973, Vaitos 1974, and Thomas 1974). However, despite the validity of some of these counterarguments and the political unpopularity of some of the conditionalities, structural adjustment programs have come to stay in most of these countries for a number of reasons.

First, the programs have inevitably become part of the policy packages or remedies most indebted LDCs must accept to obtain relief from a crushing debt burden. Second, and perhaps more importantly, the strong support from the advocates of the new policy, especially the International Monetary Fund and the World Bank, partly stems from the empirically observable success stories of those countries that have adopted the policy. In particular, the Asian newly industrialized countries, of which the G-4 is most prominent, have been successful. Inward looking industrialization strategy is shown to be ineffective and is discredited, while the record of those countries which switched to an outward looking, free market oriented policy is there for everyone to see. Although this success may be difficult to replicate in other developing countries, Cline (1982), the International Monetary Fund, and the World Bank continue to spread this "good news" through official and nonofficial channels—including Zulu and Nsouli (1989), World Bank (1989), Landell-Mills et al (1989), McCleary (1989), Thomas and Chhibber (1989), and Humphreys and Jaeger (1989).

There are at least two bothersome aspects about this new gospel which call for careful examination. The first has to do with the political economy of adjustment. The World Bank and the International Monetary Fund's role appears to be gradually moving away from that of a disinterested lender of the last resort to debt ridden nations to that of deep involvement in country policy. The level of involvement in day-to-day policy formulation and implementation is now such that the role and status of the resident representative is beginning to approximate that of a preindependence colonial governor in some countries. The second bothersome aspect relates to doubts as to which variant of neoclassical comparative cost advantage theory inspires the policy prescriptions under structural adjustment programs. Some of the fears expressed for example in Mkandawire (1988) imply that the seemingly deliberate deindustrialization stance of structural adjustment programs in many instances may be coming straight from Hla Myint (1958). This in effect would require that these countries go back to preindependence production and trade structures. It would not enable them to look forward to the type of outward looking industrialization achieved by the Asian newly industrialized countries. An attempt is made in this paper to address the second issue. For reasons of space, the over-involvement issue and related concerns pertaining to the political economy of adjustment are not addressed. According to Gulhati (1988), these should be of great concern not only to economists and policymakers in these countries, but also to the World Bank and the International Monetary Fund.

The inward looking industrialization strategy has produced modest growth in manufacturing output in some countries—such as Côte d'Ivoire and Kenya. However, over an extended period this growth has created very limited employment opportunities in relation to the magnitude of the investment involved. Emphasis has been on light manufacturing with little scope for growth in complementary exports to pay for raw material inputs, and little explored scope for sourcing output in local raw materials. The prospects for reviving growth based on relocation of production by industrialized countries is not bright, given the rising costs of imported energy and skills which have rendered such ventures uneconomical as relative prices have changed or foreign exchange squeeze has rendered them inoperable.

As table 30-1 shows, the manufacturing sector now contributes over 10 percent of GDP in countries such as Côte d'Ivoire, Senegal, and Kenya. However, its contribution to employment remains between 2 and 3 percent of the labor force. Invariably, 80 percent or more of industrial value added is in import substituting activities, principally in early stage lines such as food and beverage processing, textiles, garment manufacture, wood products, paper and printing, and a few exceptionally capital intensive projects such as cement manufacture, fertilizers, metal processing, and petroleum refining. There is also a heavy bias against exports (Acharya 1981).

Thus inward looking industrialization strategy as an industrialization and overall development strategy appears to have reached its limits. One doubts if an outward looking industrialization strategy or any variant of it would fair better, as long as the necessary basic preconditions for successful industrialization are not present.

Export Oriented Industrialization Strategy and the African Context

The characterization of the success in the Asian newly industrialized countries varies between the supportive neoclassical interpretations and the critical alternative dependency theory viewpoints. The literature varies on which countries are included in the newly industrialized countries. However, there

Table 30–1. ***Agriculture amd manufactuirng as a percentage of Gross Domestic Product in African countries***

	1960-1975[a]		*1980-1985*[a]	
Country	*Agriculture*	*Manufacturing*	*Agriculture*	*Manufacturing*
Algeria	10	14	6	12
Benin	39	5	36	7*
Botswana	34	12*	7	8*
Burkina Faso	40	13*	43	14*
Burundi	63	6	52	8
Cameroon	30	11	22	11
Central African Rep.	31	13*	na	na
Chad	41	11*	41	16*
Congo	13	8	7	4
Côte d'Ivoire	27	13	28	12
Djibouti	3	6	4	3
Egypt	28	20	18	1
Equatorial Guinea	64	1	59	1
Ethiopia	60	7	45	10
Gabon	26	10	5	7
Gambia	38	5	35	5
Ghana	47	11	41	11
Kenya	33	10	27	11
Lesotho	62	1	21	6
Liberia	24	2	17	7
Libya	5	2	2	3
Madagascar	37	.	3	15*
Malawi	61	6	52	9
Mali	53	7*	53	7*
Mauritania	27	2*	19	9*
Mauritius	20	12	13	17
Morocco	20	16*	18	17*
Niger	61	6	43	4
Nigeria	59	4	22	8
Reunion	11	8*	1	1*
Rwanda	52	13	43	16
Senega;	25	15*	18	21*
Seychelles	10	5	7	9
Sierra Leone	30	6	39	7
Somalia	48	.	43	5
South Africa	12	21	5	20
Sudan	52	4	34	73
Swaziland	24	12*	20	16*
Tanzania	42	7	53	5
Togo	46	5	27	6
Tunisia	25	8	15	12
Uganda	52	9	49	8
Zaire	17	8	na	na
Zambia	14	7	14	22*
Zimbabwe	18	14	12	26

Note: [a]The year varies from one country to another with the year for which data is available being selected within the range for each country. na indicates that data is not available. (.) indicates a value of less than one half of one percent. *Manufacturing combined with electricity, gas and water or mining and quarrying.

Source: *U.N. Yearbook of National Accounts* (various years).

has been particular focus on the so-called Asian Tigers or G-4 countries—South Korea, Taiwan, Hong Kong, and Singapore. While the G-4 model specifically refers to these four countries, to some extent their development strategy has been followed in a number of other countries, including Brazil and Mexico. Reference to the G-4 model in this paper refers to the broad experience of this group of countries, which accounted for 62 percent of LDC manufactured exports in 1975 (Lall 1980). During 1965 to 1978, it achieved annual real GDP growth rates of 11 percent, yearly increases in manufactured exports of between 20 and 40 percent, and annual increases in manufacturing employment of between four and eight percent (Schmitz 1984).

The dominant explanation for this success is hinged on the Heckscher-Ohlin-Samuelson factor proportions theory of comparative cost advantage (Balassa 1981, Bhagwati 1978, Krueger 1978, Westphal 1978, Little 1981, and Ranis 1981). The argument is that these countries adopted the right policies by liberalizing imports, adopting realistic exchange rates, and providing incentives for exports. Above all they managed to get factor prices right so that their economies, especially the manufacturing sector, could expand in line with their comparative cost advantage. It is further argued that their reliance on market forces and integration into the world economy yielded results superior to those under policies of protection and dissociation from the world economy (Schmitz 1984).

Others do not support this embodiment of the neoclassical parable which ascribes their successes primarily to export oriented policy changes. Instead they attribute the rise of the newly industrialized countries to cyclical and historical national and international factors. These have produced favorable access to advanced countries' markets, increased access to international finance, and increased relocation of production to the periphery by transnational corporations (Bienfeld 1982). A number of countries were able to take advantage of these favorable conditions, it is argued, due to their location and geopolitical significance, the existence of strong (repressive) internationally reliable regimes (Evans and Alizadeh 1984), and the existence of technological infrastructure resulting from earlier inward looking industrialization policy and state control over industrial development (Schmitz 1984).

The mid-1960s witnessed a tremendous surge in locational decisions of multinational corporations dealing in labor intensive manufactures. The multinationals took advantage of the cheap labor force and generous incentives in the form of tax exemptions and subsidised infrastructure in the newly industrialized countries. This was at a time when declining profits, increasing labor costs, and taxes made competition in the industrialized markets of the West more difficult. As special value added taxes were introduced it became more profitable to farm out part of the production processes to low wage countries. The result was an increase in the multinational corporations' share in newly industrialized countries' exports—31 percent for South Korea in 1974, 84 percent for Singapore in 1975, 51 percent for Brazil in 1973, and 34 percent for Mexico in 1974 (Lall 1980).

The degree and manner in which the newly industrialized countries became locations for multinational production varied from one country to another. But in general their becoming such locations formed part of the context for the expansion of their exports. This relocational policy was enhanced by the practice of permitting manufactured imports to enter national markets partially free of duties whenever raw material originated from the country of importation. It is estimated that United States imports from LDCs under these tariffs increased by 295 percent annually between 1966 and 1979, with the largest shares coming from Mexico, Taiwan, Singapore, and Hong Kong (Schmitz 1984).

Coinciding with this was the emergence of the Eurodollar market, whose supply of funds was fueled mainly by balance of payments deficits in the United States. These deficits resulted from massive military expenditure in Vietnam and, in the late 1970s, the recycling of surplus petrodollars from the OPEC countries. It is estimated that credit from private multinational banks, which became the main vehicle for recycling these funds, expanded by more than 50 times between 1966 and 1978. Over 50 percent of the loans went to LDCs. Brazil, Mexico, and South Korea accounted for about 50 percent of the total accumulated debt to multinational banks by 1980 (Griffith-Jones 1980). It is argued that access to this private capital market allowed these countries to avoid the International Monetary Fund's conditionality on economic policy. This is not the current experience of Sub-Saharan Africa countries (Schmitz 1984). The newly industrialized countries were able to maintain levels of imports which were not sustainable from exports in the early periods of their growth. This enabled them to maintain high growth rates in investment and output, despite their increasing debts, because they were able to finance their balance of payments deficits from external private capital flows. Studies have shown that liberal trade policies were accompanied by significant capital market controls, with governments controlling as much as 60 percent or more of investable funds as a means of state direction of economic activity (Wade 1982, Datta-Chaudri 1981).

There is also the viewpoint that inward looking industrialization strategy and outward looking export oriented industrialization strategy may not necessarily be substitutes. They could be complements, with the former achieving the accumulation of industrial experience, technical skills, and entrepreneurship, as was the case in Korea. The two strategies could be implemented together (Krueger 1981).

A study by Chow (1987), based on eight newly industrialized countries, concluded that there was overwhelming empirical evidence supportive of the a proiri argument in favor of export expansion as a development strategy for LDCs. However, there are some basic characteristics that are common to Sub-Saharan African economies that would be inimical to an export oriented industrialization strategy.[3] Some of these characteristics predate or have been exacerbated by colonial experience and would need to be tackled if any industrialization program or overall development strategy is to be sustainable over the long run. The most limiting characteristics include:

- absence of a viable entrepreneurial class,
- poor work ethic and the resulting low productivity of labor,
- acrimonious ethnic cultural plurality within artificial geographical boundaries that have made it difficult for cohesive nation states to evolve, and
- technological backwardness.

These four fundamental constraints have given rise to several others.

A Case for Rural Oriented Smallholder Industrialization Strategy

The four internal constraints on successful long term industrialization listed above explain why the erstwhile inward looking industrialization development strategy has not been sustainable in the Sub-Saharan Africa economies. The same would be true for an export oriented industrialization strategy or any urban oriented strategy based on capital intensive production. Given these constraints, a neoclassical-inspired liberalization policy would not cure market distortions; it would compound them in the absence of the prerequisites for taking advantage of market efficiency in a modern economy. Similarly, an urban oriented modernization effort aimed at realizing a Lewis-Fei-Ranis neoclassical type of rural-urban migration to enhance productivity may not be sustainable in the long run. The four internal constraints would constitute major obstacles to effective absorption of surplus rural labor into a modernized urban sector. In such circumstances, a reverse urban-rural migration would relieve the pressure in urban poverty enclaves.

A double pronged interim industrialization strategy could be an alternative to inward looking and export oriented industrialization strategies. It would first emphasize increasing labor productivity through improvements in skills, technology, and entrepreneurial ability in the rural setting, where 80 percent or more of the labor force resides in Sub-Saharan Africa countries. Second, it would take advantage of the relative abundance of labor and land to pursue labor intensive production techniques in smallholder units in rural areas. The rural oriented smallholder industrialization strategy advocated in this paper is referred to as the ROSH strategy. The ROSH strategy may be viewed as an interim measure that would pave the way for a longer term strategy. It could possibly make the adoption of an export oriented industrialization strategy feasible.

ROSH Development Strategy and State Intervention

The idea of a rural oriented smallholder industrialization strategy considered here is by no means entirely new.[4] However, it remains unpopular in African LDC policy circles for a number of reasons. First, there is a fascination with big projects and large-scale industries. There is misplaced confidence in their ability to enhance rapid economic growth. This attitude is related to the big push theories of the 1950s. Second, colonial and post-colonial patterns of investment have always favored large-scale investment in capital intensive enterprises, infrastructure, mining, industry, and agriculture. Third, the growing incidence of heavy foreign debt burden has left policy makers with little or no choice but to adopt strategies favored by their creditor countries and supporting multilateral agencies, particularly the World Bank and the International Monetary Fund.

The rationale for and content of ROSH strategy would in several respects be similar to its characterization in Acharya (1981); Daniel, Green, and Lipton (1985); and Johnston and Kilby (1975). The advocates of an export oriented industrialization paradigm have drawn inspiration from the successful experience of the Asian newly industrialized countries. The ideas behind ROSH strategy are drawn heavily from the experience of the Peoples Republic of China, India, and to some extent Pakistan. In these countries the smallholder approach to industrial development has recorded some measure of success, although not on the unprecedented scale associated with the G-4 countries.

Given that the basic elements of the small holder approach to industrialization are sufficiently familiar, they will not be repeated here.[5] What may be novel in this paper's characterization of ROSH strategy has more to do with the institutional framework for the implementation of a smallholder approach to development. Earlier advocates of a ROSH type of strategy (Oshima 1962 and World Bank 1989) have tended to favor an implementation strategy which assigns a major role to the private sector. However, strong government intervention may be crucial to the successful implementation of a ROSH strategy, especially within the context of African LDCs.

Although private sector initiative should be encouraged as much as possible, the standard neoclassical prescription of reliance on the profit motive is not likely to succeed. First, the shortage of managerial and technical skills and the absence of an entrepreneurial class would require that these cadres be created before one could expect market forces to operate efficiently. The responsibility for creating the conditions for technological innovation and acquisition of skills cannot be entrusted to the private sector. The provision of these minimum requirements by the state is akin to its providing necessary physical infrastructure and incentives to private investment capital. A strong case could be made for initial direct state participation. The state would transfer its holdings to private investment capital as the original deficiencies would be overcome. This would make such ventures sufficiently attractive to foreign and domestic private sector capital.

Second, the smallholder artisan units would likely initially produce low quality goods with lack of standardization. Government intervention would be required to ensure protection from higher quality, better standardized products from large- and medium-scale plants (Little and Mazumdar 1977).

Third, the observed failure of inefficient parastatals in these economies over the last two or three decades may not be entirely due to the inferiority of state intervention in private sector initiative. It may have had more to do with the deficiencies of inward looking industrialization as a development strategy, and its emphasis on large-scale capital intensive enterprise, which the overburdened bureaucracy was ill-equipped to operate efficiently. Carrying out necessary parastatal reforms and scaling down their activities to smallholder enterprises as envisaged under a ROSH strategy may enhance the efficiency of state intervention, especially at the early stages when such intervention may be crucial to successful implementation.

This argument for direct state intervention differs from the justification for such intervention by the dependency theorists. They argue for increased state intervention on purely ideological grounds. Instead in this paper initial state intervention is seen as inevitable to promote the necessary initial conditions for effectiveness of private sector initiative.

ROSH Strategy and Small-scale Manufacturing

Industrialization efforts in Sub-Saharan Africa have been faced with two major constraints in the past. First, there has been little or no attempt to link manufacturing to domestic sources of raw material. Second, the inward looking industrialization strategy form of manufacturing has not promoted development of domestic skills, technological know-how, and the effective adaptation of imported technology. This has followed from the import dependent nature of manufacturing.

For any long term industrialization strategy to be sustainable, the parallel development of agriculture and manufacturing must be broken. The two sectors should be made to complement each other by way of resource inputs and domestic acquisition of simple technical skills and domestic technological innovations. These are prerequisites to the effective mastering and adaptation of more sophisticated imported technology. Consequently the focus of policy would be on encouraging the growth of local small-scale manufacturing in rural areas. This would make use of direct inputs from agriculture and supply direct inputs into improving the technology in agriculture. The small-scale manufacturing of a wide range of hand tools and animal power farming implements—such as ploughs, cultivators, seed fertilizer drills, and planters—could provide the beginnings for an industrialization process that would have adeqaute inputs and foreign exchange. It would also increase the potential for rural nonfarm employment generation. The acquisition of technical skills could serve as a basis for a wider form of industrialization at a later stage. The historical experience of India and the Peoples' Republic of China shows that this form of rural industrialization could provide fertile ground for the development of efficient and increasingly sophisticated small-scale manufacturing enterprises (Johnston 1978).

The scope for this form of small-scale industry in Africa is largely uncharted. It could be significant, especially if small businesses can be induced to take advantage of the mutual supporting links between agriculture and rural industry to promote rural income growth (Acharya 1981). Exploiting this linkage would foster quick transformation of agriculture to more scientific land intensive methods of cultivation. Manufacturing could quickly progress from the manufacture of relatively simple implements such as ploughs and carts to that of seed drills, electric motors, diesel engines, stationary threshers, and power fillers. In addition, the manufacture of simple farm equipment has often led to the manufacture of consumer goods such as electric fans, bicycles, and sewing machines, and producer goods such as oil seed expellers, lathes, and hand drill presses (Johnston 1978). Because of the pervasive importance of metal-working skills, the technical and managerial capabilities bred in rural workshops could greatly enhance the development of a light engineering industry, which would be crucial for the assimilation of more complex and appropriate technologies.

Some Advantages of ROSH Strategy Over Earlier Strategies

While primarily making for acquisition of skills, technological know-how, and entrepreneurial skills that have been terribly deficient in these countries, a ROSH strategy would have a number of added advantages over inward looking and export oriented industrialization strategies. First, it would minimize the need for a massive infusion of net foreign capital inflows and free the development process from the familiar savings and foreign exchange constraints in two-gap models of development. Second, it would conform with the self-reliance philosophy which is dominant in these countries. Nationalist philosophy is inevitable in the face of increasing North-North trade links, with greater emphasis on high tech manufactures and reduced dependence on raw materials from LDCs.[6] Third, the reduction in manufacturing dependence on imported raw material and the corresponding self-sufficiency in food and basic manufactured goods could reduce balance of payments pressures. Fourth, successful implementation of a ROSH strategy would prepare these economies for fuller involvement and integration into the global economy on a more equitable basis.

Some Envisaged Implementation Problems

As the industrialized world moves into the age of the information revolution, it is questionable how feasible a deliberate isolationist strategy could be. It is doubtful that the rest of the world would allow Sub-Saharan Africa to replicate the experience of China and India with little or no external interference. China and India have pursued inward looking, self-reliant smallholder industrialization strategies as a prelude to opening up to the rest of the world. This preparatory phase may be far from complete in the cases of China and India, which remain essentially closed economies.

Another problem is the elite ruling class, which is a colonial legacy in Africa and has a track record of inept resource management. It remains doubtful that the commission agents who pass for entrepreneurs in these countries would support the emergence of true entrepreneurs who have to earn their profits. There would also be the problem of weaning the population of their taste for high quality and sophisticated imported manufactures. Low quality manufactures would be produced by small-scale units in the initial stages of their development. Chronic political and institutional instability are additional problems in these countries.

Perhaps the greatest obstacle that may militate against the successful adoption of ROSH strategy as a development strategy is the debt crisis in Sub-Saharan Africa. This obstacle will remain as long as the current debt burden relief measures continue to emphasize debt rescheduling and other short term palliatives rather than debt forgiveness, repudiation, or cancellation. As long as there is a debt problem there may be no room for any new industrialization or overall development strategy initiative. It may be that an approach like the ROSH strategy would not be seen as consistent with the short term interests of creditor countries. However, it is hoped that it would be obvious that whatever can be done to pull these fragile economies out of their chronic dependency and equip them to be able to respond more meaningfully to the demands of a changing global economy would in the longer run be in everybody's interest.

Conclusion

The Sub-Saharan Africa region continues to search for a viable, long term industrialization strategy. There may be very little to borrow from Asian newly industrialized countries—especially the G-4 countries—which would be of direct relevance to the region's peculiar characteristics. More appropriate lessons may come from India and the People's Republic of China, which have had a less spectacular, more gradual approach to industrial growth and development compared with the G-4 countries. It is important that the region try to tackle first its self-limiting internal constraints. Otherwise it may continue to find it difficult to take advantage of any favorable external conditions. A well implemented ROSH strategy would enhance the prospects for a sustainable long term industrialization strategy regardless of external conditions.

Notes

1. This presupposes that rapid industrialization is still regarded as a desirable primary component in the drive towards achieving self-sustaining economic growth and overall economic development. It would be safe to assume that this still holds in most economic development ministries in Africa, and probably all developing countries.

2. This assumption would have to be relaxed in an *n*-country case, where $n > 1$.

3. A fairly detailed examination of these characteristics has been undertaken by the author elsewhere; see Olofin (1989).

4. See for example Oshima (1962), UNDP (1974), and Kilby (1975). Although this idea is not new, it has not been incorporated into a rigorous and formal paradigm. From a policy point of view, it needs to be made more appealing than the current emphasis on large-scale projects.

5. For a lucid discussion of the main features of a smallholder approach, see Acharya (1981) and Daniel et al (1985).

6. There are ominous prospects for further decline in North-South trade, due to the planned economic integration of Europe in 1992 and the greater attention being focused on Eastern Europe.

References

Acharya, S.N. (1981): 'Perspectives and Problems of Development in Sub-Saharan Africa', *World Development*, Vol. 9, pp. 109-147.

Baer, W. (1972): *Industrialisation and Economic Development in Brazil*, (Homewood, Illinois).

Balassa, B. (1981): *The Newly Industrialising Countries in the World Economy*, (Oxford: Pergamon Press).

Banerji, Randadev and James Riedel (1980): 'Industrial employment expansion under alternative trade strategies: case of India and Taiwan: 1950-70', *Journal of Development Economics*, vol. 7, No. 4, pp. 567-577.

Bhagwati, J.N. and Anne Q. Krueger (1973): 'Exchange control, liberalization and economic development', *American Economic Review*, May.

Bhagwati, J.N. (1978): *Foreign Trade Regimes and Economic Development: Anatomy and Consequences of Exchange Control Regimes*, (Cambridge, MA.: Ballinger Press).

Bienfeld, M.A. (1982): 'The International Context for National Development Strategies: Constraints and Opportunities in a Changing World', in M. Bienfeld and M.Godrey (eds.) *The Struggle for Development: National Strategies in an International Context*. (Chichester: John Wiley).

Briton, H.J. (1970): 'The Import Substitution Strategy of Economic Development: A Survey", *Pakistan Development Review*, Vol. 10, No. 2.

Chow, P.C. (1987): 'Causality between export growth and industrial development: Empirical evidence from the NICs', *Journal of Development Economics*, Vol. 26, No. 1.

Cline, W.R. (1982): "Can the East Asian Model of Development be Generalized?", *World Development*, Vol. 10, No. 2, pp. 81-90.

Daniel, P., R.H. Green, and M. Lipton (1985): 'A Strategy for the Rural Poor', *Journal of Development Planning*, No. 15 (United Nations), pp. 113-136.

Datta-Chaudri, M.K. (1981): 'Industrialisation and Foreign Trade: The Development Experiences of South Korea and the Philippines' in E. Lee (ed.) *Export-led Industrialisation and Development*, Geneva: Asian Employment Programme, International Labour Office.

Diaz-Alejandro, C. (1975): 'Trade Policies and Economic Development' in Peter Kenen (ed.), *International Trade and Finance: Frontiers for Research*, (Cambridge: Cambridge University Press).

Evans, D. and P. Alizadeh (1984): 'Trade, Industrialisation and the Visible Hand', *Journal of Development Studies*, Special Issue on Industrialisation.

Fei, J.C., G. Ranis, and W.Y. Kuo (1979): *Growth with Equity: The Taiwan case*, (New York: Oxford University Press).

Griffith-Jones, S. (1980): 'The growth of Multinational Banking, the Euro-currency Market and the Effects on Developing Countries', *Journal of Development Studies*, Vol. 16, No. 2.

Gulhati, Ravi (1988): *The Political Economy of Reform in Sub-Saharan Africa*, EDI Policy Seminar Report No. 8, (Washington, D.C.: World Bank).

Hirschman, A.O. (1968): 'The Political Economy of Import Substituting Industrialisation in Latin America', *Quarterly Journal of Economics*, Vol. 82, No. 1, February.

Hughes, H. (1980): 'Achievements and objectives of industrialisation' in J. Cody, H. Hughes and D. Wall (eds.) *Policies for Industrial Progress in Developing Countries*, (New York: Oxford University Press), pp. 11-37.

Hughes, H. and J. Waelbroeck (1981): 'Can Developing Country Exports Keep Growing in the 1980s?', *The World Economy*, Vol. 4, No. 2. June. pp. 127-148.

Humphreys, C. and W. Jaeger (1989): 'Africa's Adjustment and Growth', *Finance & Development*, June.

Johnston, B.F. (1978): 'Agricultural production potentials and small farmer strategies in Sub-Saharan Africa', in S.N. Acharya and B.F. Johnston *Two Studies of Development in Sub-Saharan Africa*, World Bank Staff Working Paper No. 300 (October).

Kilby, P. (1975): 'Manufacturing in Colonial Africa' in P. Duiguan and L.H. Gann (eds.) *Colonialism in Africa: 1978-1960*, Vol. 4 (London: Cambridge University Press).

Krueger, A.O. (1978): *Foreign Trade Regimes and Economic Development: Liberalization Attempts* and Consequences, (Cambridge, Mass: Ballinger Press).

Krueger, A.O. (1981): 'Export-led Industrial Growth Reconsidered' in W. Hong and L.B. Krause (eds.) *Trade and Growth of the Advanced Developed Countries in the Pacific Basin*, (Seoul, Korea: KDI).

Lall, S., (1980): 'Exports of Manufactures by New Industrialising Countries: A Survey of Recent Trends', *Economic and Political Weekly*, December 6 and 13.

Landell-Mills, P., R. Agarwala, and Stanley Please (1989): 'From Crisis to Sustainable Growth in Sub-Saharan Africa'. *Finance & Development*, December.

Leibenstein, H. (1957): *Economic Backwardness and Economic Growth*, (New York: John Wiley).

Little, I., I. Scitovsky, and M. Scott (1970): *Industry and Trade in Some Developing Countries: A Comparative Study*, (Paris: OECD).

Little, I.M.D. and D. Mazumdar (1977): Research Proposal on Small scale enterprise development. mimeo, (The World Bank).

Little, I.M.D. (1981): 'The Experience and Causes of Rapid Labour-Intensive Development in Korea, Taiwan Province, Hong Kong and the Possibilities of Emulation' in E. Lee (ed.) *Export-led Industrialisation and Development*, Geneva, Asian Employment Programme, International Labour Office.

McCleary, William A. (1989): 'Policy Implementation under Adjustment Lending', *Finance & Development*. December.

Mkandawire, T. (1988): 'The Road to Crisis, Adjustment and De-Industrialisation: The African Case', *African Development*, Vol. XIII, No. 1.

Merhav, M. (1969): *Technological Dependence, Monopoly and Growth*, (Oxford: Pergamon Press).

Myint, Hla (1958): 'The Classical Theory of International Trade and the Underdeveloped Countries', *Economic Journal*, June. pp. 317-337.

Nurkse, R. (1953): *Problems of Capital Formation in Underdeveloped Countries*, (Oxford Blackwell).

Nurkse, R. (1959): *Patterns of Trade and Development*, (Wicksell Lectures).

Olofin, S. (1989): 'The Asian NICs Growth Model as an Alternative Development Strategy in Africa'. Paper presented at Project LINK conference held in Paris. August.

Oshima, H.T. (1962): *'A Strategy for Asian Development', Economic Development and Cultural Change*, 10, no. 3 .

Prebisch, R. (1964): 'Towards a New Trade Policy of Development', in Proceedings of the United Nations Conference on Trade and Development, vol. II (New York: United Nations).

Ranis, G. (1981): 'Challenges and Opportunities Posed by Asia's Super Exporters: Implications for Manufactured Exports from Latin America', in W. Baer and M. Gillis *Export Diversification and the New Protectionism - The Experience of Latin America*, National Bureau of Economic Research and the Bureau of Economic Research, University of Illinois.

Rostow, W.W. (1960): *The Stages of Economic Growth* (London: Cambridge).

Schmitz, H. (1984): 'Industrialisation Strategies in Less Developed Countries: Some Lessons of Historical Experience', *Journal of Development Studies*. Special issue on Industrialisation.

Scitovsky, T. (1954): 'Two concepts of external economies', *Journal of Political Economy*. April.

Singer, H.W. (1950): 'The Distribution of Gains Between Investing and Borrowing Countries', *American Economic Review*, papers and Proceedings, May.

Sunkel, O. (1973): 'Transnational Capitalism and National Disintegration in Latin America', *Social and Economic Studies*, Vol. 22, No. 1. March.

Sutcliffe, R.B. (1971): *Industry and Underdevelopment* (London: Addison-Wesley).

Thomas, C.Y. (1974): *Dependence and Transformation* (New York: Monthly Review Press).

Thomas, Vinod and Ajay Chhibber (1989): 'Experience with Policy Reforms Under Adjustment Lending', *Finance & Development*, December.

UNDP (1974): 'Sharing in Development: A Programme of Employment, Equity and Growth for the Philippines', Report of an intra-agency team financed by UNDP and organized by the ILO. (Geneva: ILO).

United Nations (1951): Dept. of Economic Affairs, Measures of Development of Underdeveloped Countries.

United Nations (Various years) *Yearbook of National Accounts*.

Vaitos, C.V. (1974): *Intercountry Income Distribution and Transnational Enterprises* (Oxford University Press).

Wade, R. (1982): *Irrigation and Agricultural Politics in South Korea* (Boulder, Colorado: Westview Press).

Westphal, L. (1978): 'The Republic of Korea's Experience with Export-led Industrial Development', *World Development*, Vol. 6, No. 3. March.

World Bank (1979) *World Development Report* 1979. (Washington, D.C.).

World Bank (1981): Accelerated Development in Sub-Saharan Africa: An Agenda for Action (Washington, D.C.).

World Bank and UNDP (1989): *Africa's Adjustment and Growth in the 1980s* (IBRD: Washington, D.C.).

World Bank (1989): *Sub-Saharan Africa: From Crisis to Sustainable Growth, A Long-Term Perspective Study* (Washington, D.C.).

Zulu, J. and S.M. Nsouli (1989): 'Adjustment Programs in Africa', *Finance & Development*, March.

31

Comments on Growth-Oriented Adjustments

Hafez Ghanem and C. Obidegwu

Hafez Ghanem

The Ndulu Paper

The Ndulu paper argues that development economists should pay more attention to capacity utilization rather than focusing exclusively on the rate of investment. It emphasizes the importance of the foreign exchange constraint and the necessity of maintaining a realistic real exchange rate.

The first comment I have on this paper concerns the difference between changing the level of output and changing the growth rate. We normally think of changing capacity utilization as a once-and-for-all change in the level of output and not necessarily as changing the growth rate of the economy. In most models, we talk about investment as affecting the growth rate of the economy. I think this is an issue that is not sufficiently discussed in the paper.

The second comment concerns the specification of the private investment equation. I have two sub-questions concerning that. The first is on the issue of crowding-in and crowding-out or the relationship between government and private investment. In the model specification, the author makes private investment a function of capacity utilization and a positive function of government investment. Shouldn't this really be an empirical issue? I assume, as is mentioned in the paper, that the crowding-in or crowding-out effect would differ from one country to another and would also differ according to the allocation of public investment.

The second sub-question concerns the private investment equation in the model and has to do with the lack of a profit variable. I understand the reduced form equation would include the real exchange rate since the structural equation includes capacity utilization. But shouldn't we explicitly include in the behavioral model some profit variable as affecting investment behavior, maybe the real exchange rate or the interest rate?

My second set of comments relates to the relationship between investment and imports where the author postulates that the relationship can be explained in terms of fixed coefficients. This could be a very realistic assumption in the short term but shouldn't we expect that over the long run the relationship between imports and investments, or the import content of investment, would be a function of relative prices?

A third comment concerns the behavior of savings. I was quite puzzled by the author's result where he says that savings constrained growth declines with a real depreciation, since the improved export performance will reduce foreign savings, i.e. the current account deficit will be smaller. The author does try to nuance his conclusion with talk of higher foreign transfers as a result of a real depreciation and by talking about an increase in the domestic value of foreign savings. I'm puzzled because most models I've seen have an accounting identity which says that domestic savings are the difference between exports of goods and non-factor services and imports of goods and non-factor services. So I don't see how any increase in exports can lead to a decrease in savings. I would have thought that foreign savings would fall as the deficit becomes smaller but this fall is fully compensated for by higher domestic savings.

Concerning the structure of the model, looking at the equations and diagram, I wondered about the conditions needed to ensure that the equilibrium reached in this model which is dynamic is a stable equilibrium. I feel that a discussion of the stability or instability of the equilibrium reached and what conditions are needed to ensure stability would be warranted.

Concerning the estimation techniques, my first question is why did the author take the average values of the different variables for a number of years for each country so that he has one observation per country rather than pool time-series and cross-section data. I would have thought that by pooling he would have had more efficient estimates. After going through the model and the different equations, I was struck that the author's estimating equations are not derived from the model. This raises issues of simultaneous equations bias in some instances and missing variables bias in others. The author mentions that in the paper, and also states that he used two stage least squares rather than OLS and that the results were not different. But the question in the reader's mind is, since we have gone through this model, why don't we try to estimate the structure of the equations using some maximum likelihood estimator, or at least use reduced form equations? I found the inflation equation especially problematic since it's derived from a money demand function where there's no interest rates and no expectation variables for future inflation.

The Collier and Gunning Paper

Concerning the Collier and Gunning paper, the reader's first reaction is to try to explain the preverse result they get from the theoretical model concerning the relationship between producer prices and output and the reaction of supply to changes in producer prices. In many countries in Africa, when there are price controls there are shortages, but there is also a black market. You can usually buy the goods, but at prices much higher than the administratively determined fixed prices. My feeling is that if we include this phenomena in the theoretical model we will not get the preverse supply response that the authors are getting.

I thought the most important point made by the paper concerned price deregulation. The paper shows that the welfare gains from price liberalization due to wider consumer choice and wider product availability can be substantial. We should not ignore them, however, it is rather disappointing that the authors could not actually estimate those gains for the Tanzania case because of data constraints.

There is an important link that needs to be stressed between price deregulation and the trade regime. In the case of Tanzania (and I know nothing about Tanzania except from reading this paper) price deregulation was accompanied by relaxing import controls through allowing for own imports. I think this relationship needs to be stressed since price deregulation, while maintaining quantitative restrictions on imports, would probably have a much smaller welfare enhancing effect than price deregulation with import liberalization. In fact, many governments would argue that they can not liberalize prices because of import controls which imply that certain producers have monopoly power on the local market. Once we link price deregulation and the trade regime, we immediately link price deregulation with overall consistent macro policies. I think this is an issue that needs to be stressed. If you have quantitative restrictions, exchange controls and other restrictions to imports, then you need to get rid of those prior to or simultaneous with price liberalization. In order to do that, you need "correct macro policies" and in particular, the real exchange rate has to be in equilibrium or the trade liberalization will not be sustainable. To get maximum benefits from price deregulation, you must have import liberalization which will only be sustainable in the long run if macro polices lead to a real exchange rate which is an equilibrium exchange rate.

The Olofin Paper

The Olofin paper states that an outward looking industrialization strategy would not be very successful in Africa and would not do much better than an inward looking industrialization strategy because of several constraints. Therefore the author argues for a ROSH strateg—a rural oriented smallholder industrialization strategy—with initial direct state participation. The author argues that industrialization in Africa, whether inward or outward oriented, faces four major problems which are:

- absence of an entrepreneurial class;
- poor work ethic;
- political unrest; and
- technological backwardness.

Once he posed the problem that way, my response was to wonder why the ROSH strategy would work.

Given those problems, it would seem that the appropriate response would be to tackle them directly, maybe through better education, better incentives for people to work, political reform and other incentives for technological transfer.

I don't fully understand why government as an entrepreneur would be more successful under the proposed ROSH strategy than under the failed inward looking industrialization strategy. It is true, as the author states, that maybe the types of projects that would be financed by the public sector under the ROSH strategy would be smaller and better adapted to the African environment, etc. But wouldn't the problems that plague public enterprises in Africa continue to exist under the new strategy? Namely, wouldn't the new projects be managed with the goal of obtaining some political and social objectives at the expense of economic efficiency? Hence, wouldn't it be more useful to try pushing the private sector from the beginning? Given the various problems African governments face when trying to privatize public enterprises, is it realistic to go ahead with a strategy based on public ownership with the hope that the projects would be eventually privatized? Wouldn't it be more realistic to go ahead from the beginning with a strategy based on private ownership and private participation while the government only plays a supportive role?

C. Obidegwu

It is my pleasure to comment on these three interesting and excellent papers. Two of them—the Ndulu and the Collier and Gunning papers—have much in common. Both use a disequilibrium framework in their analysis, and conclude that external financial assistance is an important element in the transition to growth and equilibrium.

The Collier and Gunning paper is a micro-economic analysis of peasant behavior in an economy faced with shortages of consumption goods as a result of the lack of foreign exchange for imports of production inputs and consumer goods. The paper analyzes the impact of this scarcity on peasant production decisions and examines the difficulties of stimulating production in such circumstances, as well as the transition to market clearing. The Ndulu paper is a macroeconomic analysis of a prototypical Sub-Saharan Africa economy which faces constrained imports and import-substitution possibilities. External assistance plays a direct role in stimulating production activity by raising the availability of imported inputs and consequently, domestic capacity utilization.

The two papers highlight the critical role that external finance can play in reviving an economy with fundamentally incorrect relative prices and few substitution possibilities for imported goods. In that sense, these papers lend support to the practice of multilateral agencies and bilateral donors in packaging policy advice and financial assistance. In the microeconomic framework of Collier and Gunning, the financial support increases the availability of consumer goods, which provides the incentive to peasants to produce. This improves supply response and the pace of adjustment.

The third paper, by Olofin, is concerned with strategies for development and industrialization for Sub-Saharan African countries. It discusses the failure of the post-independence import substitution industrialization strategies and the prospects for, and impact of, adopting the export oriented industrialization strategy by Sub-Saharan African countries. Olofin rejects both the import substitution and the export oriented industrialization strategies for Sub-Saharan Africa. According to him, import substitution industrialization strategy has already reached its limits in providing growth in industrial output and employment and export oriented industrialization strategy would not be appropriate since these countries do not possess the basic pre-conditions for industrialization. Olofin therefore advocates an alternative strategy -- the Rural Oriented Smallholder (ROSH) industrialization strategy. I shall now discuss each paper in turn.

The Collier and Gunning Paper

As the authors readily admit, this paper is really two papers put together. One part is a theory of peasant behavior under sustained disequilibrium (excess demand for goods) and repressed inflation. The other is an application of this theory to the Tanzanian economy during 1977 to 1988.

The theory makes some very restrictive assumptions about the supply and prices of market goods. There are two goods—a cash crop, produced by peasants and purchased by a government agent, and a consumption good, which is supplied by the state. The government fixes the price of both goods; there is a shortage of the consumption good at the prevailing relative price, controls are effective and therefore parallel markets do not develop. In effect, in this economy, the government or its agent can be regarded as the producer of crops and the peasants as the suppliers of labor, receiving a wage rate *Pc* (the price of crops). The real wage would therefore be *Pc/Pg* (where *Pg* is the price of goods). However, at this wage rate and the initial money holding, there is excess demand for both crops (labor) and goods, thus, the economy is operating in the zone of repressed inflation (see Cuddington et al 1984, page 26).

With these assumptions, and using a variant of the disequilibrium model of Malinvaud, the authors derive a number of policy-relevant conclusions for this economy.

In the presence of excess demand for goods and labor, the supply of labor is lower than it would be under market clearing conditions. Households withhold their labor because they are unable to get the goods they want.

2. Given the excess demand for goods, the supply response of crops to price changes is perverse. Thus, lowering crop prices would raise the supply of crops and reduce the national demand for goods. Therefore, raising the price of crops is counterproductive unless coordinated with the improved availability of consumption goods. The authors regard this as of very practical interest "since countries with predominantly agricultural exports are often advised to raise the producer prices of export crops."

3. In an economy characterized by repressed inflation it is not possible to infer how crop prices differ from their equilibrium merely from the symptom of excess demand for crops once the goods market is also in disequilibrium. It is very difficult to achieve equilibrium by official manipulation of the these prices, thus, sequenced price reform is heavily constrained.

4. The existence of monetary overhang undermines the adjustment to the abolition of price controls as it delays production response.

5. External aid facilitates the transition to market clearing prices by providing for an injection of goods into the market economy.

These results are consistent, except in one respect, with the analysis of the repressed inflation zone (excess demand for labor and goods) elaborated by Barro and Grossman (1976) and Cuddington et al (1984). The exception is that in the Collier and Gunning model, a decline in crop prices (equivalent to a fall in real wages) leads to a rise in labor supply; this perverse result is due the use of a one-period model. In this formulation, the only reason to work is to buy whatever goods are expected to be available; the peasants adjust their labor supply in order to acquire the goods. A decline of crops prices has no impact on supply of goods or its probability distribution. In the multi-period formulation such as that elaborated by Barro and Grossman, households expect that excess demand for goods will prevail for a finite period and therefore have an incentive to save. Thus, the effective labor supply curve remains positively sloping and is steeper than the notional labor supply curve.

In the second part of the paper, the authors apply their model to analyze the behavior of the Tanzanian economy during 1977 to 1988. This exercise makes heavy demands on data, and the authors had to make heroic assumptions to produce the required data. For instance, the computation of urban demand for grain assumes constant urban per capita consumption of grain during the period. An elaborate technique is used to calculate the "import" CPI for peasants; this CPI is used to deflate nominal income which is used for expenditures for goods including those produced in the peasant economy, for example, purchases of grain by coffee farmers who receive their income mostly from outside the peasant economy. Nevertheless, the authors should be congratulated for their patience in assembling all the required data.

The authors find that the decline in production of export crops between 1977 and 1984 was due to a shortage of consumer goods. They conclude that in times of shortage, crops sales are determined by the availability of consumer goods instead of the normal relationship to prices. However, it seems to me that the normal relationship to prices should not be expected when the system is out of equilibrium and with 'prices' that are really fictitious. The goods' prices do not reflect the transactions costs of acquiring these goods. For instance, consumers endure the frustrations of not being able to obtain desired goods and incur search costs which raise the price of the goods beyond the store price. As shortages become more acute, these costs rise, the implicit real price of cash crops (or the real wage) falls, and, *ceteris paribus*, output would be reduced. The authors' estimate that the real producer price of the export crops declined by 26 percent between 1977 and 1984 therefore considerably understates the real decline when the frustration factor is considered. Unfortunately this factor is not easily quantified.

The authors found that during 1984 to 1988 (the reform period), there was a large increase in crop sales despite a decline of real prices. They assert that a combination of the return to market clearing and nominal price increases explain the paradox that output rose while producer prices fell in real terms. In my view, there is not necessarily a paradox. Rising output could have been due to good weather, improved availability of inputs and transport, and lower costs of doing business due to some liberalization of prices and distribution; these factors would tend to shift the supply curve to the right. The agricultural supply elasticity with respect to non-price factors has been found to be higher than the supply elasticity with respect to prices, especially in poor countries with poor facilities and markets (see Chhibber 1988). While realized prices are generally used in time series econometric tests, the appropriate price variable in production decisions is expected prices; farmers decisions are likely to be based on long-term price trends. There is no evidence to suggest that the decline in real prices constituted a long-term trend. In any case, the decline in real prices of 20 percent over four years is probably exaggerated by the use of the peasants "import" price index, which excludes food prices.

The discussion in the paper of the evolution of real monetary balances seems quite complex. The authors attribute the holding of money to the precautionary motive -- the need for peasants, faced with random shortages of goods, to be prepared to avail themselves of those occasions when availability was atypically good. However, involuntary holding of money, as a result of the excess demand for goods and the fluctuations in the supply, is also a plausible explanation for the evolution of money balances.

This paper has a number of important lessons for development policy. One lesson is that widespread use of price and distribution controls could easily lead to severe policy distortions, undermining the use of prices as policy instruments. It is clear from this analysis that quantities rather than prices become more potent instruments, but in general, the manipulation of quantities is a far more difficult undertaking than prices. Another lesson is that controls could easily get the economy into a quagmire from which it would be very difficult to extricate itself. According to the authors, external aid can play a key role in the transition to market clearing conditions. In the Tanzanian situation, the abandonment of widespread price controls and confinement, and the legalization of own-funded imports, as well as increased flows of official external assistance to finance general imports, have been important elements in the economic recovery. The authors' finding that sequenced price reform is fraught with mistakes, has implications for the pace of price reform. It would indicate that rapid decontrol is preferable to an alternative pace in which the government attempts, over a period of time, to provide price incentives while decontrolling prices and distribution at a slow pace. This has implications for exchange rate adjustments where the emphasis has been on upward adjustments of the nominal exchange rate by a government agency, leaving the process of determination of the rate unchanged.

The Ndulu Paper

The Ndulu paper is concerned with adjustment and growth of Sub-Saharan African economies which face constrained imports, and low capacity utilization and growth. The paper notes that import compression affects not only current production but also the growth of investment, making the allocation of imports between capacity expansion and utilization a key policy issue. The author suggests that the centrality of improvement in export performance and net foreign resource inflows in the growth recovery process cannot be overemphasized. He further notes that growth in Sub-Saharan Africa is substantially influenced by a number of exogenous factors -- the changes in the barter terms of trade, external assistance, and the weather. For instance, in agriculture, growth fluctuations are mainly due to the vagaries of the weather. In another section of the paper, the author discusses the role of government in the economy, particularly in public investments and direct controls of financial resources, especially foreign exchange. He discusses evidence indicating that public investments, particularly in infrastructure, have a net crowding-in effect in several countries.

The central adjustment and growth issues raised by the paper relate to the growth implications of import compression and the related problem of rationing of imports, export growth and diversification, the role of government in the economy, and the vulnerability of the economies to exogenous shocks. The author specifies a macroeconomic accounting framework, presumably to analyze these issues. While I find the model interesting, the links between the policy issues and the accounting framework do not seem to be particularly strong. The author should make clear at the onset the kinds of policy issues the model is designed to elaborate.

I have some problems with the specification of some of the relationships in the model. For instance, since an import constrained economy is in disequilibrium, the effective imports should not be determined by desired investment and capacity utilization.

Rather, imports should be determined by the financial constraints; investments and capacity utilization would depend on imports, using some explicit import allocation criteria. In the approach used in the model, the author sidesteps the issue of the allocation of imports which he deemed important. The model has a simple supply function for exports which does not provide much guidance on the issues of export growth and diversification. One key adjustment policy issue relates to the factors that influence the response parameter. In an import constrained economy, import capacity would also tend to constrain export production and diversification. For instance, the lack of imported intermediate inputs to work the mines reduces the exports of mineral dependent economies and fertilizer shortages undermine agricultural production and diversification which directly or indirectly reduce agricultural exports. The paper implies that there is a behavioral specification for foreign saving, when indeed the equation is simply an identity. The impact of depreciation of the currency on foreign savings is not unambiguous as the author claims.

According to the author, the full model has been calibrated for Tanzania and partial models have been calibrated for a number of other countries. It would have been useful to present some simulation results from the Tanzanian model. The rest of the paper is devoted to econometric estimates of two of the relationships postulated in the model, and an evaluation of the progress of Sub-Saharan African countries in structural adjustment. The policy implications of the regression results are not clear, as they do not seem to throw any light on the major policy issues raised at the beginning of the paper. Finally, I feel that the paper is rather ambitious in trying to tackle a wide variety of issues simultaneously. It could benefit from a sharp focus on a few issues for which it could provide in-depth analysis.

The Olofin Paper

The Olofin paper argues that the export oriented industrialization strategy, which implicitly underpins the current adjustment programs in Sub-Saharan Africa, is inappropriate for these countries. The author therefore recommends instead, the Rural Oriented Smallholder (ROSH) industrialization strategy.

The author has two main lines of argument for rejecting export oriented industricalization strategy. The first relates to his interpretation of the success of Taiwan, Korea, Singapore and Hong-Kong (Asian G-4 countries) in industrialization. The second relates to some characteristics common to Sub-Saharan African economies, which the author regards as being inimical to an export oriented industrialization.

The paper outlines two main interpretations of the G-4 industrialization miracle. One is that the countries adopted the right policies to get factor prices right, relied on market forces and were integrated into the global economy through liberal trade policies. The other strand of argument is that the success of the G-4 countries was due to cyclical and historical national and international factors which produced favorable access to the markets of the advanced countries, increased their access to international finance, and encouraged the relocation of production within their boarders by transnational corporations.

The interpretations of the success of the G-4 countries are numerous. One participant in the debate (Amsden 1989) has even suggested that Korea succeeded by violating the cannons of economic wisdom—by getting prices wrong. Some others, such as (Nam 1988) and Rhee et al (1984) have attributed the success to, among other factors, the Confucian heritage, a belief system which places high value on hard work, thrift, loyalty, discipline and respect for authority. But Rhee et al have pointed out that the same Confucian heritage could be blamed for inhibiting economic development in Korea before the 1960s. However, under compatible policies, these human traits became a positive force for development. On the role of market forces Bradford (1986) points out that what "seems to distinguish the East Asian development experience is not the dominance of market forces, free enterprise and internal liberalization, but effective, highly interactive relationships between the public and private sectors characterized by shared goals and commitments embodied in the development strategy and economic policy of the government."

In spite of the plurality of interpretations, the factors to which the success is usually attributed are, in general, not mutually exclusive. A favorable external environment helped the G-4 countries as well as Brazil, Mexico, Turkey, etc. to achieve rapid industrialization. But when the global economy soured in the 1970s, the G-4 countries were able to sustain rapid growth while the others faltered. The one distinguishing factor between the G-4 countries and the others is the relentless pursuit of export growth promoted by credible export incentives, which were provided by a mixture of proven, effective measures. Export promotion is, of course, the central feature of outward orientation. If the success of the G-4 countries is regarded as exceptional, how about the recent successes of Mauritius, and the three ASEAN countries—Thailand, Malaysia and Indonesia? These are industrializing rapidly using export oriented industrialization strategy -- combining export promotion, liberal trade policies with reliance on the private sector, and markets. Their success is perhaps more instructive for African countries than that of the Asian G-4 since these ASEAN countries share many more characteristics with Sub-Saharan African countries.

The author outlines four characteristics which, in his view, limit the potential for industrialization in Sub-Saharan Africa—absence of an entrepreneurial class, poor work ethic and low productivity of labor, acrimonious ethnic rivalry and lack of national cohesion and identity, and technological backwardness. It is doubtful that these factors are so endemic that they cannot respond to policy change, investment and economic, institutional and political reform. It is not clear also that these constraints will be less debilitating to development under ROSH, or that ROSH will be more successful in ameliorating them. It is not obvious to me that reliance on the profit motive is "not likely to succeed" in spurring industrialization. Irrespective of the development strategy, incentives are needed for individuals and firms to be productive and these agents will only undertake economic activities if there is an expectation of profit.

There are many attractive features of ROSH. The proposals to emphasize growth in labor productivity, technology, entrepreneurial ability, labor intensive production techniques in smallholder units, and government investments in social and physical infrastructure are all excellent. These proposals, however, are not incompatible with structural adjustment programs or export oriented development strategies. Many structural adjustment and export development programs emphasize the development of small-scale industries, the use of labor intensive production and capacity building. In fact, the development of small-scale industries underpinned the rapid industrialization of Japan and the other successful Asian countries. In contrast, India, which is supposed to have had an explicit smallholder strategy, was less successful because the policies and regulatory environment under the ruling import substitution strategy inhibited the growth of small and large firms alike.

I agree with the author about the need to end the fascination with big projects, large-scale industry and capital intensive industrialization. I am in support of the emphasis on small-scale industry, although I do not see why it should be limited to rural areas. It is reasonable to expect that small-scale firms will make less demands on scarce managerial capacity as long as they are not part of a superstructure such as the government. There are many other details of the ROSH strategy that I have problems with, however. The paper advocates participation of the state in the direct development of the small-scale industries, at least at the initial stages. It is unlikely that the state, which has proved incapable of managing a relatively limited number of firms, will be successful at coordinating the activities and managing its investment in a multitude of diverse small firms. Besides, a transition from state ownership to the private sector may prove very difficult to effect as the current experience with privatization indicates. For successful industrialization in Sub-Saharan Africa, production, whether in small or large firms, will have to be in the domain of the private sector, with the state providing social and physical infrastructure. The development and operation of small firms in particular entail risks that demand flexibility and quick response; these are not the hallmarks of the bureaucratic environment in government. The paper advocates protection of small firms from competition from large- and medium-scale plants and, I assume, from imports. However, such differential protection, taxation and regulation are difficult to manage efficiently, and protection will remove an incentive for the firms to be efficient. It is not clear that the ROSH strategy will necessarily minimize foreign exchange constraints on growth. Small firms also make demands for imported capital and intermediate inputs; indeed some use of imported capital and inputs will be necessary for the technological advances that stimulate rapid industrialization. Experience shows that the only way to minimize foreign exchange crisis is sustained export growth, not avoidance of imports.

The paper alludes to the need to develop agriculture and industry in such a way that they complement each other. But it does not, in my view, vigorously emphasize the point that sustained rapid growth of agriculture is a prerequisite to industrialization, particularly a rural-based one. Structural adjustment measures programs in Sub-Saharan Africa have focussed on reviving agricultural growth by providing incentives for production and improving the access of farmers to the inputs that enhance their productivity. A rapid growing and profitable agricultural sector will lay a strong foundation for industrialization by meeting the basic food needs of the population, providing rural dwellers the resources they need to invest in rural industry, and generating the labor to work in such activities.

It seems to me that the isolationist ROSH strategy advocated by the author is a version of the import substitution strategy, but with a bias against large-scale urban production. The author realizes that there are severe obstacles to instituting and implementing this isolationist strategy. Even if it is the case that Africa is being increasingly marginalized in the world as some writers suggest, it still will be very difficult for any African country to isolate itself from the rest of the world. There are strong vested interests, both inside and outside Africa, to maintain and strengthen the links that exist. The best strategy for African countries is one that brings them the maximum benefits from being part of the global family.

References

Alice H. Amsden (1989). *Asia's Next Giant: South Korea and Late Industrialization*: Oxford University Press.

Robert Barro and Herschel Grossman (1976). *Money, Employment and Inflation,* Cambridge University Press.

Colin I. Bradford, Jr. (1986). "East Asian Models: Myths and Lessons", *Development Strategies Considered,* John P. Lewis, and Valeriana Kallab (eds.). Transaction Books.

Chhibber, Ajay (1988). "Aggregate Supply Response in Agriculture: A Survey", *Structural Adjustment in Agriculture: Theory and Practice,* S. Commander (ed.). James Curry Publishers.

Cuddington, John T., Per-Olov Johansson, Karl-Gustaf L?fgren (1984). *Disequilibrium Macroeconomics in Open Economics,* Basil Blackwell.

Sang-Woo Nam (1988). "Alternative Growth and Adjustment Strategies in Newly Industrializing Countries in Southeast Asia" *Beyond Adjustment: The Asian Experience,* Paul Streeten (ed.). International Monetary Fund.

Yung Whee Rhee, Bruce Ross-Larson, and Gary Pursell (1984). *Korea's Competitive Edge: Managing the Entry into World Markets,* John Hopkins University Press.